AF553159

Handbook of
Food Science, Catering Technology and
Kitchen Management

Handbook of Food Science, Catering Technology and Kitchen Management

Jerry D'Souza
Jatin Pradhan

2010

SBS Publishers & Distributors Pvt. Ltd.
New Delhi

ISBN: 13-9789380090115

First Published in 2010

Published by:

SBS PUBLISHERS & DISTRIBUTORS PVT. LTD.
2/9, Ground Floor, Ansari Road, Darya Ganj,
New Delhi - 110002,
INDIA
Tel: 0091.11.23289119 / 41563911 / 32945311
Email: mail@sbspublishers.com
www.sbspublishers.com

Printed in India by Chaman Enterprises, New Delhi.

Preface

Food science is a study concerned with all technical aspects of food, beginning with harvesting or slaughtering, and ending with its cooking and consumption. It is considered one of the life sciences, and is usually considered distinct from the field of nutrition. Food science is a highly interdisciplinary applied science. It incorporates concepts from many different fields including microbiology, chemical engineering, biochemistry, and many others. Catering is the business of providing foodservice at a remote site. A mobile caterer serves food directly from a vehicle or cart that is designed for the purpose. Mobile catering is common at outdoor events (such as concerts), workplaces, and downtown business districts. An airline meal or in-flight meal is a meal served to passengers on board a commercial airliner. These meals are prepared by airline catering services. These meals vary widely in quality and quantity across different airline companies and classes of travel. They range from a simple beverage in short-haul economy class to a seven-course gourmet meal in long-haul first class. A kitchen is a room or part of a room used for cooking and food preparation. In the West, a modern residential kitchen is typically equipped with a stove, a sink with hot and cold running water, a refrigerator and kitchen cabinets. Many households have a microwave oven, a dishwasher and other electric appliances. The main function of a kitchen is cooking or preparing food but it may also be used for dining and entertaining.

Modern kitchens often have enough informal space to allow for people to eat in it without having to use the formal dining room. Such areas are called "breakfast areas", "breakfast nooks" or "breakfast bars" if the space is integrated into a kitchen counter. Kitchens with enough space to eat in are sometimes called "eat-in kitchens". Restaurant and canteen kitchens found in hotels, hospitals, educational and work place facilities, army barracks, and similar establishments are generally (in developed countries) subject to public health laws. They are inspected periodically by public-health officials, and forced to close if they do not meet hygienic requirements mandated by law.

Canteen kitchens (and castle kitchens) were often the places where new technology was used first. Today's western restaurant kitchens typically have tiled walls and floors and use stainless steel for other surfaces (workbench, but also door and drawer fronts) because these materials are durable and easy to clean. Professional kitchens are often equipped with gas stoves, as these allow cooks to regulate the heat quicker and more finely than electrical stoves. Some special appliances are typical for professional kitchens, such as large installed deep fryers, steamers, or a Bain Marie. The fast food and convenience food trends have also changed the way restaurant kitchens operate. There is a trend for restaurants to only "finish" delivered convenience food or even just re-heat completely prepared meals, maybe at the utmost grilling a hamburger or a steak. The kitchens in railway dining cars present special challenges: space is constrained, and nevertheless the personnel must be able to serve a great number of meals quickly. Especially in the early history of the railway this required flawless organization of processes; in modern times, the microwave oven and prepared meals have made this task a lot easier. Galleys are kitchens aboard ships or aircraft.

This publication titled "Handbook of Food Science, Catering Technology and Kitchen Management" gives readers both basic and advanced perspectives on the subject. The handbook provides an introduction to food, food types, food energy and agriculture with special focus on functional food, nitrification, whole foods, staple food, medical food, etc. Themes like food energy, bioenergetics, and bio-energy balance are also covered. An overview of grains, flours, breads and select food items is provided in detail. Special emphasis is given to whole grain, refined grain, whole wheat flour, rolled oats, groat, barley, popcorn, white rice, brown rice, white bread, hominy, pasta, spaghetti, vermicelli, frozen noodle, ramen and soups. Health foods, vegetables, soups, poultry, milk and milk products are discussed briefly with highlights on alfalfa, frozen vegetables, canned food, food stock, egg, egg yolk, cottage cheese, smoked cheese, farmer cheese and butter. An overview is made herein about livestock, meat processing, meat packing, poultry and animal husbandry. This handbook also describes basic facts about food microbiology, food engineering, food chemistry, food technology, food preservation, food irradiation and food packaging.

Advanced concepts regarding nutrition, malnutrition and applied nutrition has been analyzed with focus on nutritional scale. Similarly, an introductory overview is provided about food service, cafeteria and restaurant with special focus on distribution, management and institutional aspects. Case studies of National Restaurant Association and Washington Restaurant Association are made. Various aspects and listings about culinary science and arts are covered with select highlights on culinary vegetables, root vegetables, vegetable juices, culinary herbs, spices, fruits and edible seeds. This handbook also deals in detail the subjects like food studies, gastronomy, catering and airline meals. Select case studies of global food service operators, such as, those of LSG Sky Chefs, Gate Gourmet, Servair, Sodexo, Aramark, Brinker International, Compass Group, Crown Group, First for Foodservice, Darden Restaurants, Accor and Cara Operations are made for understanding applied aspects. An overview of oenology, wine tasting and food matching is provided. Select case studies of food universities, research institutes and research centres are made. Extensive resource material is provided on food science, catering technology and kitchen management. Necessary case studies are made on relevant issues. This book also provides necessary list of, glossary of terms, bibliography, and a detailed index to make this book more user friendly.

Contents

1

Introduction to Food, Food Types, Food Energy and Agriculture

DEFINING FOOD

Food is any substance, usually composed primarily of carbohydrates, fats, water and/or proteins, that can be eaten or drunk by an animal or human for nutrition or pleasure. Items considered food may be sourced from plants, animals or other categories such as fungus or fermented products like alcohol. Although many human cultures sought food items through hunting and gathering, today most cultures use farming, ranching, and fishing, with hunting, foraging and other methods of a local nature included but playing a minor role. Most traditions have a recognizable cuisine, a specific set of cooking traditions, preferences, and practices, the study of which is known as gastronomy. Many cultures have diversified their foods by means of preparation, cooking methods and manufacturing. This also includes a complex food trade which helps the cultures to economically survive by-way-of food, not just by consumption.

Many cultures study the dietary analysis of food habits. While humans are omnivores, religion and social constructs such as morality often affect which foods they will consume. Food safety is also a concern with foodborne illness claiming many lives each year. In many languages, food is often used metaphorically or figuratively, as in "food for thought".

Food Sources

Almost all foods are of plant or animal origin, although there are some exceptions. Foods not coming from animal or plant sources include various edible fungi, such mushrooms. Fungi and ambient bacteria are used in the preparation of fermented and pickled foods such as leavened bread, alcoholic drinks, cheese, pickles, and yogurt. Many cultures eat seaweed, a protist, or blue-green algae (cyanobacteria) such as Spirulina. Additionally, salt is often eaten as a flavoring or preservative, and baking soda is used in food preparation. Both of these are inorganic substances, as is water, an important part of human diet.

Plants

Many plants or plant parts are eaten as food. There are around 2,000 plant species which are cultivated for food, and many have several distinct cultivars.

Fig. 1.1: Variety of Foods from Plant Sources

Seeds of plants are a good source of food for animals, including humans because they contain nutrients necessary for the plant's initial growth. In fact, the majority of food consumed by human beings are seed-based foods. Edible seeds include cereals (such as maize, wheat, and rice), legumes (such as beans, peas, and lentils), and nuts. Oilseeds are often pressed to produce rich oils, such as sunflower, rapeseed (including canola oil), and sesame. One of the earliest food recipes made from ground chickpeas is called hummus, which can be traced back to Ancient Egypt times.

Fruits are the ripened ovaries of plants, including the seeds within. Many plants have evolved fruits that are attractive as a food source to animals, so that animals will eat the fruits and excrete the seeds some distance away. Fruits, therefore, make up a significant part of the diets of most cultures. Some botanical fruits, such as tomatoes, pumpkins and eggplants, are eaten as vegetables.

Vegetables are a second type of plant matter that is commonly eaten as food. These include root vegetables (such as potatoes and carrots), leaf vegetables (such as spinach and lettuce), stem vegetables (such as bamboo shoots and asparagus), and inflorescence vegetables (such as globe artichokes and broccoli). Many herbs and spices are highly-flavorful vegetables.

Animals

Animals can be used as food either directly, or indirectly by the products they produce. Meat is an example of a direct product taken from an animal, which comes from either muscle systems or from organs. Food products produced by animals include milk produced by mammals, which in many cultures is drunk or processed into dairy products such as cheese or butter. In addition birds and other animals lay eggs, which are often eaten, and bees produce honey, a popular sweetener in many cultures. Some cultures consume blood, some in the form of blood sausage, as a thickener for sauces, a cured salted form for times of food scarcity, and others use blood in stews such as civet.

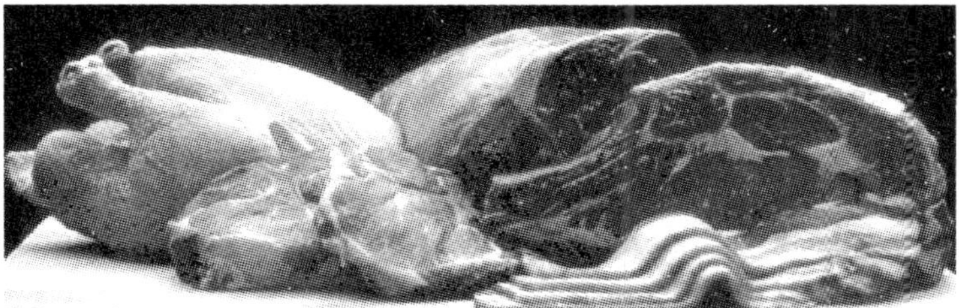

Fig. 1.2: Various Raw Meats

Production

Food is traditionally obtained through farming, ranching, and fishing, with hunting, foraging and other methods of subsistence locally important. More recently, there has been a growing trend towards more sustainable agricultural practices. This approach, which is partly fueled by consumer demand, encourages biodiversity, local self-reliance and organic farming methods. Major influences on food production are international organizations, (*e.g. the World Trade Organization and Common Agricultural Policy), national government policy (or law), and war.*

Fig. 1.3: Tractor and Chaser Bin

Preparation

While some food can be eaten raw, many foods undergo some form of preparation for reasons of safety, palatability, or flavor. At the simplest level this may involve washing, cutting, trimming or adding other foods or ingredients, such as spices. It may also involve mixing, heating or cooling, pressure cooking, fermentation, or combination with other food. In a home, most food preparation takes place in a kitchen. Some preparation is done to enhance the taste or aesthetic appeal; other preparation may help to preserve the food; and others may be involved in cultural identity. A meal is made up of food which is prepared to be eaten at a specific time and place.

Animal Slaughter and Butchering

The preparation of animal-based food will usually involve slaughter, evisceration, hanging, portioning and rendering. In developed countries, this is usually done outside the home in slaughterhouses which are used to process animals en mass for meat production. Many countries regulate their slaughterhouses by law. For example the United States has established the Humane Slaughter Act of 1958, which requires that an animal be stunned before killing. This act, like those in many countries, exempts slaughter in accordance to religious law, such as kosher shechita and dhabiha halal. Strict interpretations of kashrut require the animal to be fully aware when its carotid artery is cut.

On the local level a butcher may commonly break down larger animal meat into smaller manageable cuts and pre-wrapped for commercial sale or wrapped to order in butcher paper. In addition fish and seafood may be fabricated into smaller cuts by a fish monger at the local level. However fish butchery may be done on board a fishing vessel and quick-frozen for preservation of quality.

Cooking

The term "cooking" encompasses a vast range of methods, tools and combinations of ingredients to improve the flavor or digestibility of food. Cooking technique, known as culinary art, generally requires the selection, measurement and combining of ingredients in an ordered procedure in an effort to achieve the desired result. Constraints on success include the variability of ingredients, ambient conditions, tools, and the skill of the individual cooking. The diversity of cooking worldwide is a reflection of the myriad nutritional, aesthetic, agricultural, economic, cultural and religious considerations that impact upon it.

Cooking requires applying heat to a food which usually, though not always, chemically transforms it, thus changing its flavor, texture, appearance, and nutritional properties. Cooking proper, as opposed to roasting, requires the boiling of water in a container, and was practiced at least since the 10th millennium BC with the introduction of pottery. There is archaeological evidence of roasted foodstuffs at Homo erectus campsites dating from 420,000 years ago.

Cooking Equipment and Methods

There are many types of cooking equipment used for cooking. Ovens are one type of cooking equipment which can be used for baking or roasting and offer a dry-heat cooking method. Different cuisines will use different types of ovens, for example Indian culture uses a Tandoor oven is a cylindrical clay oven which operates at a single high temperature, while western kitchens will use variable temperature convection ovens, conventional ovens, toaster ovens in addition to non-radiant heat ovens like the microwave oven. Ovens may be wood-fired, coal-fired, gas, electric, or oil-fired.

Fig. 1.4: A Stainless Steel Frying Pan

Various types of cook-tops are used as well. They carry the same variations of fuel types as the ovens mentioned above. cook-tops are used to heat vessels placed on top of the heat source, such as a sauté pan, sauce pot, frying pan, pressure cooker, etc. These pieces of equipment can use either a moist or dry cooking method and include methods such as steaming, simmering, boiling, and poaching for moist methods; while the dry methods include sautéing, pan frying, or deep-frying.

In addition, many cultures use grills for cooking. A grill operates with a radiant heat source from below, usually covered with a metal grid and sometimes a cover. An open bit barbecue in the American south is one example along with the American style outdoor grill fueled by wood, liquid propane or charcoal along with soaked wood chips for smoking. A Mexican style of barbecue is called barbacoa, which involves the cooking of meats and whole sheep over open fire. In Argentina, asado is prepared on a grill held over an open pit or fire made upon the ground, on which a whole animal is grilled or in other cases smaller cuts of the animal.

Fig. 1.5: Traditional Asado

Raw Food

Certain cultures highlight animal and vegetable foods in their raw state. Sushi in Japan is one such cuisine that features raw sliced fish, either in sashimi, nigiri, or maki styles. Steak tartare and salmon tartare are dishes made from diced or ground raw beef or salmon respectively, mixed with various ingredients and served with baguette, brioche or frites. In Italy, carpaccio is a dish of very thin sliced raw beef, drizzled with a vinaigrette made with olive oil. A popular health food movement known as raw foodism promotes a mostly vegan diet of raw fruits, vegetables and grains prepared in various ways, including juicing, food dehydration, not passing the 118 degree mark, and sprouting.

Fig. 1.6: Many Types of Sushi Ready to be Eaten

Restaurants

Many cultures produce food for sale in restaurants for paying customers. These restaurants often have trained chefs who prepare the food, while trained waitstaff serve the customers. The term restaurant is credited to the French from the 19th century, as it relates to the restorative nature of the bouillons that were once served in them. However, the concept pre-dates the naming of these establishments, as evidence suggests commercial food preparation may have existed during the age of the city of Pompeii, as well as an urban sales of prepared foods in China during the Song Dynasty. The coffee shops or cafes of 17th century Europe may also be considered an early version of the restaurant. In 2005 the United States spent $496 billion annually for out-of-home dining. Expenditures by type of out-of-home dining was as follows, 40 per cent in full-service restaurants, 37.2 per cent in limited service restaurants (fast food), 6.6 per cent in schools or colleges, 5.4 per cent in bars and vending machines, 4.7 per cent in hotels and motels, 4.0 per cent in recreational places, and 2.2 per cent in other which includes military bases.

Food Manufacture

Packaged foods are manufactured outside the home for purchase. This can be as simple as a butcher preparing meat, or as complex as a modern international food industry. Early food processing techniques were limited by available food preservation, packaging and transportation. This mainly involved salting, curing, curdling, drying, pickling, fermentation and smoking. During the industrialization era in the 19th century, food manufacturing arose. This development took advantage of new mass markets and emerging new technology, such as milling, preservation, packaging and labeling and transportation. It brought the advantages of pre-prepared time saving food to the bulk of ordinary people who did not employ domestic servants.

At the start of the 21st century, a two-tier structure has arisen, with a few international food processing giants controlling a wide range of well-known food brands. There also exists a wide array of small local or national food processing companies. Advanced technologies have also come to change food manufacture.

Fig. 1.7: Packaged Household Food Items

Computer-based control systems, sophisticated processing and packaging methods, and logistics and distribution advances, can enhance product quality, improve food safety, and reduce costs.

Commercial Trade

International Exports and Imports

World Bank reported that the EU was the top food importer in 2005 followed at a distance by the USA and Japan. Food is now traded and marketed on a global basis. The variety and availability of food is no longer restricted by the diversity of locally grown food or the limitations of the local growing season. Between 1961 and 1999 there has been a 400 per cent increase in worldwide food exports. Some countries are now economically dependent on food exports, which in some cases account for over 80 per cent of all exports.

In 1994 over 100 countries became signatories to the Uruguay Round of the General Agreement on Tariffs and Trade in a dramatic increase in trade liberalization. This included an agreement to reduce subsidies paid to farmers, underpinned by the WTO enforcement

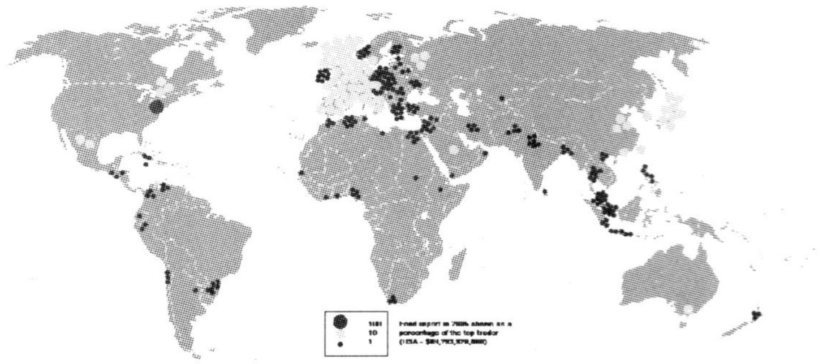

Fig. 1.8: Food imports in 2005

of agricultural subsidy, tariffs, import quotas and settlement of trade disputes that cannot be bilaterally resolved. Where trade barriers are raised on the disputed grounds of public health and safety, the WTO refer the dispute to the Codex Alimentarius Commission, which was founded in 1962 by the United Nations Food and Agriculture Organization and the World Health Organization. Trade liberalization has greatly affected world food trade.

Marketing and Retailing

Food marketing brings together the producer and the consumer. It is the chain of activities that brings food from "farm gate to plate." The marketing of even a single food product can be a complicated process involving many producers and companies. For example, fifty-six companies are involved in making one can of chicken noodle soup. These businesses include not only chicken and vegetable processors but also the companies that transport the ingredients and those who print labels and manufacture cans. The food marketing system is the largest direct and indirect non-government employer in the United States.

In the pre-modern era, the sale of surplus food took place once a week when farmers took their wares on market day, into the local village market place. Here food was sold to grocers for sale in their local shops for purchase by local consumers. With the onset of industrialization, and the development of the food processing industry, a wider range of food could be sold and distributed in distant locations. Typically early grocery shops would be counter-based shops, in which purchasers told the shop-keeper what they wanted, so that the shop-keeper could get it for them.

In the 20th century supermarkets were born. Supermarkets brought with them a self service approach to shopping using shopping carts, and were able to offer quality food at lower cost through economies of scale and reduced staffing costs. In the latter part of the 20th century, this has been further revolutionized by the development of vast warehouse-sized out-of-town supermarkets, selling a wide range of food from around the world.

Unlike food processors, food retailing is a two-tier market in which a small number of very large companies control a large proportion of supermarkets. The supermarket giants wield great purchasing power over farmers and processors, and strong influence over consumers. Nevertheless, less than ten percent of consumer spending on food goes to farmers, with larger percentages going to advertising, transportation, and intermediate corporations.

Prices

Consumers worldwide faced rising food prices, it was reported on March 24, 2008. Reasons for this development are freak weather, dramatic changes in the global economy, including higher oil prices, lower food reserves and growing consumer demand in China and India. In the long term, prices are expected to stabilize. Farmers will grow more grain for both fuel and food and eventually bring prices down. Already this is happening with wheat, with more crops to be planted in the United States, Canada and Europe in 2009. However, the Food and Agriculture Organization projects that consumers still face at least until 2018 more expensive food. It is rare that the spikes are hitting all major foods in most countries at once. Food prices rose 4 percent in the United States 2007, the highest rise since 1990, and are expected to climb as much again 2008. As of December 2007, 37 countries faced food crises, and 20 had imposed some sort of food-price controls. In China, the price of pork has jumped 58 percent in 2007. In the 1990s and 1980s, farm subsidies and support programmes allowed major grain exporting countries to hold large surpluses, which could be tapped during food shortages to keep prices down. But new trade policies have made agricultural production much more responsive to market demands—putting global food reserves at their lowest since 1983.

Food prices are rising, wealthier Asian consumers are westernizing their diets, and farmers and nations of the third world are struggling to keep up the pace. The past five years have seen rapid growth in the contribution of Asian nations to the Global Fluid and Powdered Milk Manufacturing industry, which in 2008 accounts for more than 30 per cent of production, while China alone accounts for more than 10 per cent of both production and consumption in the Global Fruit and Vegetable Processing and Preserving industry. The trend is similarly evident in industries such as Soft Drink and Bottled Water Manufacturing, as well as Global Cocoa, Chocolate and sugar Confectionery Manufacturing, forecast to grow by 5.7 per cent and 10.0 per cent respectively during 2008 in response to soaring demand in China and Southeast Asian markets.

Famine and Hunger

Food deprivation leads to malnutrition and ultimately starvation. This is often connected with famine, which involves the absence of food in entire communities. This can have a devastating and widespread effect on human health and mortality. Rationing is sometimes used to distribute food in times of shortage, most notably during times of war.

Starvation is a significant international problem. Approximately 815 million people are undernourished, and over 16,000 children die per day from hunger-related causes. Food deprivation is regarded as a deficit need in Maslow's hierarchy of needs and is measured using famine scales.

Food Aid

Food aid can benefit people suffering from a shortage of food. It can be used to improve peoples' lives in the short term, so that a society can increase its standard of living to the point that food aid is no longer required. Conversely, badly managed food aid can create problems by disrupting local markets, depressing crop prices, and discouraging food production. Sometimes a cycle of food aid dependence can develop. Its provision, or threatened withdrawal, is sometimes used as a political tool to influence the policies of the destination country, a strategy known as food politics. Sometimes, food aid provisions will require certain types of food be purchased from certain sellers, and food aid can be misused to enhance the markets of donor countries. International efforts to distribute food to the neediest countries are often co-ordinated by the World Food Programme.

Safety

Foodborne illness, commonly called "food poisoning," is caused by bacteria, toxins, viruses, parasites, and prions. Roughly 7 million people die of food poisoning each year, with about 10 times as many suffering from a non-fatal version. The two most common factors leading to cases of bacterial foodborne illness are cross-contamination of ready-to-eat food from other uncooked foods and improper temperature control. Less commonly, acute adverse reactions can also occur if chemical contamination of food occurs, for example from improper storage, or use of non-food grade soaps and disinfectants. Food can also be adulterated by a very wide range of articles (known as 'foreign bodies') during farming, manufacture, cooking, packaging, distribution or sale. These foreign bodies can include pests or their droppings, hairs, cigarette butts, wood chips, and all manner of other contaminants. It is possible for certain types of food to become contaminated if stored or presented in an unsafe container, such as a ceramic pot with lead-based glaze.

Food poisoning has been recognized as a disease of man since as early as Hippocrates. The sale of rancid, contaminated or adulterated food was commonplace until introduction of hygiene, refrigeration, and vermin controls in the 19th century. Discovery of techniques for killing bacteria using heat and other microbiological studies by scientists such as Louis Pasteur contributed to the modern sanitation standards that are ubiquitous in developed nations today. This was further underpinned by the work of Justus von Liebig, which led to the development of modern food storage and food preservation methods. In more recent years, a greater understanding of the causes of food-borne illnesses has led to the development of more systematic approaches such as the Hazard Analysis and Critical Control Points (HACCP), which can identify and eliminate many risks.

Allergies

Some people have allergies or sensitivities to foods which are not problematic to most people. This occurs when a person's immune system mistakes a certain food protein for a harmful foreign agent and attacks it. About 2 per cent of adults and 8 per cent of children have a food allergy. The amount of the food substance required to provoke a reaction in a particularly susceptible individual can be quite small. In some instances, traces of food in the air, too minute to be perceived through smell, have been known to provoke lethal reactions in extremely sensitive individuals. Common food allergens are gluten, corn, shellfish (mollusks), peanuts, and soy. Allergens frequently produce symptoms such as diarrhea, rashes, bloating, vomiting, and regurgitation. The digestive complaints usually develop within half an hour of ingesting the allergen.

Rarely, food allergies can lead to a medical emergency, such as anaphylactic shock, hypotension (low blood pressure), and loss of consciousness. An allergen associated with this type of reaction is peanut, although latex products can induce similar reactions. Initial treatment is with epinephrine (adrenaline), often carried by known patients in the form of an Epi-pen.

Diet

Cultural and Religious Diets

Dietary habits are the habitual decisions a person or culture makes when choosing what foods to eat. Although humans are omnivores, many cultures hold some food preferences and some food taboos. Dietary choices can also define cultures and play a role in religion. For example, only kosher foods are permitted by Judaism, and halal/haram foods by Islam, in the diet of believers. In addition, the dietary choices of different countries or regions have different characteristics. This is highly related to a culture's cuisine.

Diet Deficiencies

Dietary habits play a significant role in the health and mortality of all humans. Imbalances between the consumed fuels and expended energy results in either starvation or excessive reserves of adipose tissue, known as body fat. Poor intake of various vitamins and minerals can lead to diseases which can have far-reaching effects on health. For instance, 30 per cent of the world's population either has, or is at risk for developing, Iodine deficiency. It is estimated that at least 3 million children are blind due to vitamin A deficiency. Vitamin C deficiency results in scurvy. Calcium, Vitamin D and phosphorus are inter-related; the consumption of each may affect the absorption of the others. Kwashiorkor and marasmus are childhood disorders caused by lack of dietary protein.

Moral, Ethical, and Health Conscious Diet

Many individuals limit what foods they eat for reasons of morality, or other habit. For instance vegetarians choose to forgo food from animal sources to varying degrees. Others choose a healthier diet, avoiding sugars or animal fats and increasing consumption of dietary fiber and antioxidants. Obesity, a serious problem in the western world, leads to higher chances of developing heart disease, diabetes, and many other diseases. More recently, dietary habits have been influenced by the concerns that some people have about possible impacts on health or the environment from genetically modified food. Further concerns about the impact of industrial farming (grains) on animal welfare, human health and the environment are also having an effect on contemporary human dietary habits. This has led to the emergence of a counterculture with a preference for organic and local food.

Nutrition

Between the extremes of optimal health and death from starvation or malnutrition, there is an array of disease states that can be caused or alleviated by changes in diet. Deficiencies, excesses and imbalances in diet can produce negative impacts on health, which may lead to diseases such as scurvy, obesity or osteoporosis, as well as psychological and behavioural problems. The science of nutrition

Fig. 1.9: USDA Food Pyramid

attempts to understand how and why specific dietary aspects influence health.

Nutrients in food are grouped into several categories. Macronutrients means fat, protein, and carbohydrates. Micronutrients are the minerals and vitamins. Additionally food contains water and dietary fiber.

Legal Definition

Some countries list a legal definition of food. These countries list food as any item that is to be processed, partially processed or unprocessed for consumption. The listing of items included as foodstuffs include any substance, intended to be, or reasonably expected to be, ingested by humans. In addition to these foodstuffs drink, chewing gum, water or other items processed into said food items are part of the legal definition of food. Items not included in the legal definition of food include animal feed, live animals unless being prepared for sale in a market, plants prior to harvesting, medicinal products, cosmetics, tobacco and tobacco products, narcotic or psychotropic substances, and residues and contaminants.

VEGETARIAN FOOD

Medicinal Foods

Jean Carper is a nationally syndicated newspaper columnist and author of seven books, including "The Food Pharmacy"

The best prescription for living longer may be right in your kitchen.

It's true—the foods that are best for you can act like medicine, boosting your immune system and warding off illnesses such as cancer and heart disease. But because foods are exceedingly complex packages of chemicals and compounds, they don't deliver a single biological punch, as do pharmaceutical drugs designed to accomplish a specific purpose. Instead, the right foods can have a much broader effect on a variety of health problems. Eat these foods raw or lightly cooked. (Cooking destroys many of their most protective chemicals.) And remember, no matter how beneficial these foods may be, they're no substitute for a doctor's care.

Apple: Reduces cholesterol, contains cancer-fighting agent known as antioxidant. High in fiber, helps prevent constipation and suppresses appetite.

Asparagus: A super source of glutathione, a powerful antioxidant. In studies, glutathione has been shown to act against at least thirty carcinogens.

Avocado: Can help prevent clogging of arteries; dilates blood vessels. Lowers cholesterol. Its main fat, monounsaturated oleic acid, acts as an antioxidant, slowing the buildup LDL (low-density lipoprotein) cholesterol. Also one of the richest sources of glutathione.

Banana: Soothes the stomach. Strengthens the stomach lining against acid and ulcers, and lab tests show that bananas can act like antibiotics. Very high in potassium, thus may help regulate high blood pressure.

Barley: Long considered a heart medicine in the Middle East. Reduces cholesterol levels and contains antioxidants that may help prevent cancer.

Beans: (including navy, black, kidney, and pinto beans, and lentils) Studies show that eating a half cup of cooked beans daily may reduce cholesterol levels as much as 10 percent. Also helps to regulate blood-sugar levels. Very high in fiber. Bean consumption is linked to lower rates of prostate and breast cancer.

Bell Pepper: Super-rich in antioxidant vitamin C. Therefore, a great food for fighting off colds, asthma, bronchitis, respiratory infections and cataracts, as well as angina, atherosclerosis (damaged, clogged arteries) and cancer.

Blueberries: Act as an unusual type of antibiotic by keeping infectious bacteria from attaching to the lining of the urinary tract, helping to prevent recurring urinary tract and bladder infections. Also contain chemicals that curb diarrhea.

Broccoli and Cauliflower: Abundant in antioxidants. Broccoli is rich in anticancer agent such as vitamin C, beta carotene and quercetin. Both broccoli and cauliflower are considered effective in helping to prevent lung, colon, and breast cancers. These cruciferous vegetables can speed up removal of estrogen from the body, perhaps helping to prevent hormone-related cancers such as breast cancer. Rich in fiber. Compounds in broccoli also help prevent ulcers.

Brussels Sprouts: Possess some of the same powers as their cruciferous cousins broccoli and cabbage. Packed with antioxidants and other cancer-fighters including indoles, chemicals that may help protect against colon cancer.

Cabbage: Contains numerous anti-cancer and antioxidant compounds. Seems to suppress the growth of colon polyps, a precursor to colon cancer; in studies, eating cabbage more than once a week cut men's colon cancer odds by 66 percent.

Carrot: A super source of beta carotene, the antioxidant reputed to help prevent numerous health problems, including heart attacks, cancer, and cataracts. One study showed that the beta carotene in a daily cup of carrots slashed stroke rates in women by 40 percent and heart attacks by 22 percent. One medium carrot's worth of beta carotene daily may cut lung-cancer risk in half, even among formerly heavy smokers.

Celery: Celery compounds have been shown to lower blood pressure in animals. High in certain anticancer compounds that have been shown to detoxify carcinogens, including cigarette smoke. tests also show celery may act as a mild diuretic.

Chili Pepper: Revs up the blood clot-dissolving system, opens sinuses and air passages, and acts as a decongestant. Most of its pharmacological activity is credited to capsaicin, the compound that makes the pepper taste hot. Capsaicin is also a potent painkiller, alleviating headaches when inhaled. Putting hot chili sauce on food may even speed up metabolism, burning off calories.

Cinnamon: A strong stimulator of insulin activity; thus, potentially helpful for those with adult-onset diabetes. Also seems to help prevent blood clots.

Clove: Long used to dull the pain of toothache. Contains compounds that act like aspirin.

Collard Greens: Full of antioxidant compounds, including lutein, vitamin C and beta carotene. In animal studies, collards inhibited the spread of breast cancer. Collard-green consumption, like that of other green leafy vegetables, is associated with low rates of many cancers.

Corn: High in anticancer compounds called protease inhibitors, corn may help fight cancer and act as an antiviral agent.

Cranberries: Like blueberries, help prevent recurring urinary tract and bladder infections. Also believed to be effective in inhibiting viruses.

Dates: High in natural aspirin. Also high in fiber; have a laxative effect. Dried fruits, including dates, are linked to lower rates of certain cancers, especially pancreatic cancer.

Eggplant: Eating eggplant may lower blood cholesterol and help counteract some of the detrimental effects high-fat foods have on the blood. Lab tests show that eggplant also seems to act as an antibacterial agent and as a mild diuretic.

Garlic: A proven antibiotic that has been shown to kill bacteria, fungi, and intestinal parasites. Also shown to lower blood-cholesterol levels, seems to act as an anticoagulant. Garlic also contains multiple anticancer compounds, antioxidants, and immune-system boosters. A good cold medication, garlic also acts as an effective decongestant and anti-inflammatory agent.

Ginger: Used for centuries in Asia, ginger is a proven anti-nausea remedy. Also, relieves the inflammatory pain and swelling of rheumatoid arthritis and osteoarthritis.

Grapefruit: Contains a pectin that's been shown to lower blood-cholesterol levels and blood pressure in animals. High in antioxidants, especially disease-fighting vitamin C,.grapefruit may help prevent stomach and pancreatic cancer.

Grapes: A rich storehouse of anti-cancer compounds, red grapes are high in the antioxidant quercetin. Red-grade skins also contain resveratrol, which seems to lower bad-type LDL cholesterol.

Kale: An amazingly rich source of antioxidant compounds. High in beta carotene, and contains more lutein than any other vegetable tested. Kale is a member of the cruciferous family and contains anticancer chemicals called indoles, which may help prevent estrogen-linked cancers.

Melon: (cantaloupe and honeydew) May help prevent blood clots. Orange melons, such as cantaloupe, also contain high levels of beta carotene.

Mushroom: Esteemed in Asia as a heart medicine and cancer preventive. Tests show that compounds in Asian mushrooms, such as shiitake, may help inhibit cancer as well as viral diseases, and can lower cholesterol levels. One study showed that fresh or dried shiitake mushrooms cut cholesterol by up to 12 percent when eaten daily.

Mustard: (including horseradish) Helps relieve congestion from colds and sinus problems, and acts as an antibiotic. Revs up metabolism: One study showed that ordinary yellow mustard seemed to increase metabolic rate, helping to burn more calories.

Nuts: High in the antioxidant vitamin E., nuts help prevent cancer and heart disease. Almonds have a high concentration of oleic acid, which may help reduce cholesterol and protect arteries. Brazil nuts are extremely rich in selenium, an antioxidant linked to lower rates of both heart disease and cancer. And walnuts contain ellagic acid, another cholesterol-reducer.

Oats: Oats can help lower cholesterol and stabilize blood-sugar levels. Compounds in oats also seem to suppress nicotine cravings.

Onion: Including chives, shallots, scallions, leeks) Containing exceptionally strong antioxidants, onions have been shown to help prevent cancer in animals. The onion is a rich source of quercetin, a potent antioxidant known to inhibit stomach cancer. Onions may also help prevent atherosclerosis and blood clots, and even high fight bacterial and viral infections.

Orange: A complete package of cancer-inhibitors, including antioxidants such as vitamin C. Specifically tied to lower rates of pancreatic cancer. Because of their high vitamin C content, oranges may also help ward off breast and stomach cancer, asthma attacks, atherosclerosis, and gum disease. Some studies show that vitamin C deficiencies may also inhibit fertility in some men.

Parsley: Rich in antioxidants, parsley can help detoxify carcinogens, including those in tobacco smoke. Parsley also acts as a diuretic.

Plum and Prune: Compounds in these fruits may act as antibacterial and antiviral agents. High in fiber, these fruits work as laxatives.

Potato: (white) Contains cancer-fighting protease inhibitors. Also high in potassium.

Pumpkin: Extremely high in beta carotene.

Raspberries: As do other berries, raspberries help fight infections, and may help prevent some cancers. Also help curb nausea.

Rice: (white and brown) Like other seeds, contains anticancer protease inhibitors. Effective against diarrhea. Rice bran helps lower cholesterol levels and may help prevent formation of kidney stones.

Soybeans: Rich in compounds that act like hormones and thus seem to ward off cancer, especially estrogen-linked breast cancer. High soybean consumption may be one reason rates of breast cancer and prostate cancers are very low among the Japanese. In studies, soybeans lowered blood cholesterol levels substantially. And animal studies showed that soybeans seem to deter and help dissolve kidney stones.

Spinach: As with other green leafy vegetables, consumption of spinach is linked with lower rates of cancer. A super source of antioxidants, including folate, beta carotene and lutein, for example.

Strawberries: Rich in antioxidant vitamin C as well as high in fiber. Studies show that compounds in strawberries act like antiviral agents. And other studies link regular strawberry consumption to lower rates of all types of cancer.

Sweet Potato: A blockbuster source of beta carotene. One half-cup of mashed sweet potatoes is higher in beta carotene than a medium carrot. Also high in fiber.

Tea: (including black, oolong, and green tea) Tea acts as an antibactial, anti-ulcer agent, cavity-fighter, even an anti-diarrheal agent. In animal studies, tea and tea compounds seemed to inhibit various cancers. Tea drinkers appear to have lower risk of atherosclerosis and stroke.

Tomato: A major source of the antioxidant lycopene. tomatoes are linked to low rates of certain cancers, including bladder cancer.

Watermelon: Like the tomato, watermelon contains high levels of lycopene and glutathione, an antioxidant and anticancer compounds.

Wheat: High-fiber wheat bran has formidable anti-cancer potential. One or two daily servings of wheat-bran cereal has been shown to suppress pre-cancerous polyps that can develop into colon cancer. In women, wheat bran also appears to prevent breast cancer by diminishing supplies of estrogen circulating in the body.

FUNCTIONAL FOOD

Functional food or medicinal food is any fresh or processed food claimed to have a health-promoting or disease-preventing property beyond the basic function of supplying nutrients. The general category of functional foods includes processed food or foods fortified with health-promoting additives, like "vitamin-enriched" products. Fermented foods with live cultures are considered as functional foods with probiotic benefits.

Functional foods are an emerging field in food science due to their increasing popularity with health-conscious consumers.

The term was first used in Japan in the 1980s where there is a government approval process for functional foods called Foods for Specified Health Use (FOSHU).

Health Claims

Functional food products typically include health claims on their label touting their benefits: for example: "*Cereal is a significant source of fiber. Studies have shown that an increased amount of fiber in one's diet can decrease the risk of certain types of cancer in individuals.*"

Some countries, such as Canada, Sweden, and the United States, have specific laws concerning the labeling of such products. In the

United States, the kinds of claims which are allowed are overseen and regulated by the Food and Drug Administration (FDA). However, some claims will fall outside of the purview of the FDA and be accompanied by the disclaimer: "These statements have not been evaluated by the Food and Drug Administration. This product is not intended to diagnose, treat, cure or prevent any disease."

Such a disclaimer typically accompanies supplements rather than foods, but since the definition of functional food is still evolving and somewhat amorphous, a functional food may find itself bearing the warning.

Current Research

The Richardson Centre for Functional Foods and Nutraceuticals, which is part of the University of Manitoba, is the only centre of its kind, dedicated to the discussion, discovery, and development of functional foods and nutraceuticals, with a focus on the crops of the Canadian Prairies Some current research projects examine: the effects of a diet rich in diaclyglycerol (DAG) oil on body weight, body composition, and blood lipid levels in women; the effect of Heart and Stroke Portfolio diet on lipid metabolism and weight loss in men; the effect of pulses and pulse fractions on indices of lipid, carbohydrate and energy metabolism, as well as oxidative status in overweight, hyperlipidemic individuals; the use of conjugated linoleic acid (CLA), an omega-3 fatty acid, as a nutraceutical for weight loss in humans; the evaluation of very long chain fatty acids/alcohol and plant sterols as functional food ingredients for cholesterol-lowering in hypercholesterolemic humans; the effects of dietary cholesterol with and without simvastatin on cholesterol absorption and synthesis and sterol profile in Smith Lemli Optiz syndrome (SLOS) patients; the effects of unique algal based polysaccharides on plasma lipid levels and energy metabolism in hamsters.

Dr. Peter Jones is a Canada Research Chair in Nutrition and Functional Foods and is the director of the centre. His research concerns itself with finding candidates for functional food ingredients and he does this by examining the efficacy of novel bioactive materials such as plant sterols—natural components found in plants which can act as cholesterol-lowering agents.

NUTRACEUTICAL

Nutraceutical, a combination of nutrition and pharmaceutical, refers to extracts of foods claimed to have a medicinal effect on human health. Traditionally the nutraceutical was contained in a medicinal format such as a capsule, tablet or powder in a prescribed dose, although more modern Nutraceuticals such as Probiotic drinks and yogurt are now found in ordinary supermarkets alongside normally everyday versions of the product.

Physiological Benefits

More rigorously, nutraceutical implies that the extract or food is demonstrated to have a physiological benefit or provide protection against a chronic disease.

Functional Foods

Functional foods are defined as being consumed as part of a usual diet but are demonstrated to have physiological benefits and/or reduce the risk of chronic disease beyond basic nutritional functions.

Examples

Examples of claims made for nutraceuticals are resveratrol from red grape products as an antioxidant, soluble dietary fiber products, such as psyllium seed husk for reducing hypercholesterolemia, broccoli (sulforaphane) as a cancer preventative, and soy or clover (isoflavonoids) to improve arterial health. Such claims are being researched and many citations are available via PubMed to ascertain their foundation of basic research. Other nutraceutical examples are flavonoids antioxidants, alpha-linolenic acid from flax or Chia seeds, beta-carotene from marigold petals, anthocyanins from berries, etc. With the US Dietary Supplement Health and Education Act (DSHEA), several other compounds were added to the list of supplements originally mentioned in FDA notification. Thus, many botanical and herbal extracts such as ginseng, garlic oil, etc. have been developed as nutraceuticals.

Nutraceuticals are often used in nutrient premixes or nutrient systems in the food and pharmaceutical industries.

General Usage

An aging population and increased life expectancy is increasing awareness about good health and fueling the growth of demand for nutraceuticals by 6 per cent annually. Likewise, that aging population is fueling the demand for such products in easy-to-open containers with convenient dispensers.

NUTRIFICATION

Enriching or fortifying foods with nutrients is called nutrification. Food enrichment is the restoration of the natural nutritive value of a food before it was processed, while fortification is the adding of vitamins or minerals to a food at levels higher than it originally possessed, though "fortification" is commonly used to refer to both processes.

Functional foods, *designer foods*, and *techno-foods*, are foods that are manufactured by food companies in collaboration with sellers of dietary supplements, in order to produce products that they can promote as healthy. Techno-foods "refer to foods and beverages that have been fortified in some way to confer health benefits beyond the original nutritional value of the foods themselves."

Initially, fortification was used as a public health strategy in the United States to eradicate vitamin and mineral deficiencies. For example, through the addition of iodine to table salt, goiter, a thyroid gland dysfunction, is now very uncommon. Other examples include the addition of vitamin D to milk and fluoride to tap water.

As a public health measure, fortification has been very successful, though there are concerns that the "consumer must consciously desire and be involved in nutritional change" With fortification came the interweaving of science and commercial issues. As with drugs, the benefits of nutrients are dose dependent, and eating too much of them can cause adverse effects. Yet, fortification is used regularly as a marketing strategy, so that now most processed foods are fortified in some way. Continued proliferation of this practice raises the concern that fortification will become too much of a good thing, where nutrient content in food in general may eventually increase to levels detrimental to the population.

Economic issues pertaining to food fortification have also emerged. The wealthy, who may already eat enough nutrients, may be consuming more than they need when they purchase fortified products, while those in most need of the added nutrients may not be able to afford to buy enough fortified foods.

WHOLE FOODS

Whole foods are those that are unprocessed and unrefined, or processed and refined as little as possible before being consumed. They typically do not contain added ingredients, such as sugar, salt, or fat.

The term is often confused with "organic food" (*i.e.*, organically-grown food) but whole foods are not necessarily organic, nor are organic foods necessarily whole, although they do share a number of traits, such as an avoidance of chemically-assisted agricultural techniques. Because of the lack of basic processing, many whole foods have a very short shelf life and are not easily sold outside of farmers' markets.

Examples of whole foods include unpolished grains; fruits and vegetables; unprocessed meat, poultry, and fish; and non-homogenized milk.

There are several ways to meet the body's needs with respect to whole foods. One way is to consume a variety of fresh raw fruits and vegetables every day.

"Diets rich in whole and unrefined foods, like whole grains, dark green and yellow/orange-fleshed vegetables and fruits, legumes, nuts and seeds, contain high concentrations of antioxidant phenolics, fibers and numerous other phytochemicals that may be protective against chronic diseases."

STAPLE FOOD

A staple food is a food that can be stored for use throughout the year (or produced fresh any time of the year) and forms the basis of

Fig. 1.10: Grains

a traditional diet. Staple foods vary from place to place, but are typically inexpensive starchy foods of vegetable origin that are high in food energy (Calories) and carbohydrate. The staple food of a specific cuisine may commonly be served as part of every meal, and its name may be used synonymously with "food" in some contexts, such as the reference to "our daily bread" in the Lord's Prayer, and a common greeting of "Have you eaten rice?" denoting "How are you?" in certain cultures.

Fig. 1.11: Various Types of Potatoes

Most staple foods derive either from cereals such as wheat, barley, rye, maize, or rice, or starchy root vegetables such as potatoes, yams, taro, and cassava. Other staple foods include pulses (dried legumes), sago (derived from the pith of the sago palm tree), and fruits such as breadfruit and plantains.

Refining

Rice is most commonly eaten as cooked entire grains, but most other cereals are milled into flour or meal which is used to make bread; noodles or other pasta; and porridges and "mushes" such as polenta or mealie pap. Mashed root vegetables can be used to make similar porridge-like dishes, including poi and fufu. Pulses (particularly chickpeas) and starchy root vegetables, such as Canna, can also be made into flour.

Part of a Whole

Although nutritious, staple foods generally do not by themselves provide a full range of nutrients, so other foods need to be added to the diet to prevent malnutrition. For example, the deficiency disease pellagra is associated with a diet consisting primarily of maize, and beriberi with a diet of white (*i.e.*, refined) rice.

Trivia

It has been hypothesized that some staple foods may act as a Giffen good in conditions of extreme poverty. This was first noted by Robert Giffen who argued that potato demand actually rose during the Great Irish Famine (1845-1849). While theoretically possible, this is a controversial view among economists as studies have failed to find much evidence of Giffen good behaviour in actual markets.

MEDICAL FOOD

Medical foods are foods that are specially formulated and intended for the dietary management of a disease that has distinctive nutritional needs that cannot be met by normal diet alone. They were defined in the Food and Drug Administration's 1988 Orphan Drug Act Amendments and are subject to the general food and safety labeling requirements of the Federal Food, Drug, and Cosmetic Act.

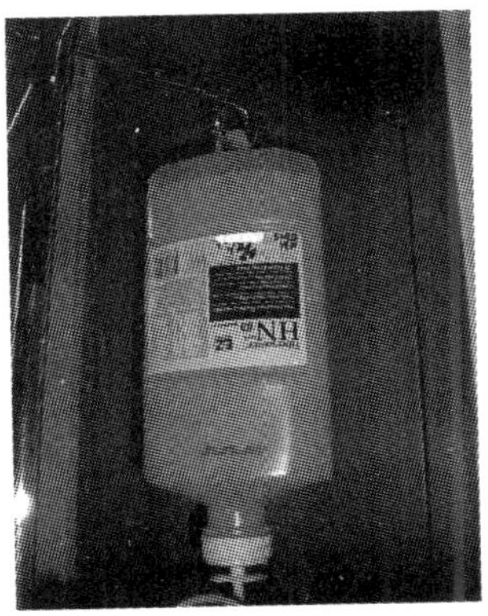

Fig. 1.12: Novartis Fibersource HN Medical Food Deployed on IV Pole

Medical foods are distinct from the broader category of foods for special dietary use and from traditional foods that bear a health claim. In order to be considered a medical food the product must, at a minimum:

- be a food for oral ingestion or tube feeding (nasogastric tube)
- be labeled for the dietary management of a specific medical disorder, disease or condition for which there are distinctive nutritional requirements, and
- be intended to be used under medical supervision.

Medical foods can be classified into the following categories:

- Nutritionally complete formulas
- Nutritionally incomplete formulas
- Formulas for metabolic disorders .
- Oral rehydration products

Examples

Allergic Conditions

Medical foods for management of allergic conditions may contain both gamma-linolenic acid (GLA), a short chain omega-6 fatty acid primarily sourced from the seeds of the borage plant, and eicosapentaenoic acid (EPA), an omega-3 polyunsaturated fatty acid sourced from fish. These fatty acids help to inhibit the production of leukotrienes in the system. Sufficient quantities of GLA and EPA necessary for reduction of leukotrienes cannot be obtained from a normal diet. Leukotrienes are inflammatory molecules produced by immune cells (neutrophils, basophils, mast cells, macrophages and eosinophils) in the body. They are involved in the inflammatory response and cause the narrowing of the airways, increased mucus production and tissue swelling associated with both allergies and asthma. In order to control allergic symptoms, research shows that it helps to inhibit the production of leukotrienes in the body.

Diabetes Mellitus

Medical foods for management of diabetes mellitus generally contain slowly digested carbohydrates, which helps minimize peaks in blood sugar. Consistent maintenance of optimal blood sugar levels (avoiding highs and lows) over time can help reduce the complications of diabetes.

Gastrointestinal Tract Impairment

Medical foods for management of gastrointestinal tract (GI tract) impairment provide an amino acid-based diet in its most easily digestible (elemental) form to aid in poor nutrient absorption due to digestive disease, malabsorption, severe food allergies, or other conditions in which the GI tract is severely compromised.

Metabolically Stressed Patients

Medical foods for management of metabolically stressed patients provide supplemental glutamine to nourish the GI tract and restore glutamine while a patient is in a stressed, catabolic state.

Regulation

Medical foods are regulated by the US Food and Drug Administration under the Food Drug and Cosmetic Act regulations.

NOVEL FOOD

Novel food is defined as a type of food that does not have a significant history of consumption or is produced by a method that has not previously been used for food.

European Union

Novel foods or novel food ingredients have no history of "significant" consumption in the European Union prior to 15 May 1997. Any food or food ingredient that falls within this definition must be authorised according to the Novel Food legislation, EC (No.) 258/97.

A company that wishes to submit an application under this legislation should consult the guidance document compiled by the European Commission, 97/618/EC, which highlights the scientific information and the safety assessment report required in each case.

The Novel Food regulation stipulates that foods and food ingredients falling within the scope of this regulation must not:

- present a danger for the consumer,
- mislead the consumer,
- differ from foods or food ingredients which they are intended to replace to such an extent that their normal consumption would be nutritionally disadvantageous for the consumer.

There are two possible routes for authorization under the Novel Food legislation: a full application and a simplified application. The simplified application route is only applicable where the EU member national competent authority, *e.g.* Food Standards Agency (FSA) in the UK, believes that the novel food in question is substantially equivalent to an existing food or food ingredient.

The FSA has provided guidance for companies that wish to submit a simplified novel food application.

There have been a number of products launched as a result of food or food ingredients authorised under the Novel Food legislation. Benecol, containing phytostanols, and Flora ProActiv, containing phytosterols, are commercially successful products that help to reduce cholesterol levels.

Originally, the Novel Food legislation was conceived by the European Commission in response to the arrival of food products derived from GM crops. These types of GM derived food are now regulated by separate legislation, EC (No.) 1829/2003.

In its current format, exotic fruit and vegetables, which have a long history of safe use outside of the European Union, fall within the definition of a novel food. As a consequence, each of these food products must undergo a safety assessment prior to being marketed. This has long been an issue of contention for food manufacturers and producers outside the EU who view this as an economic trade barrier. A good example of this situation can be illustrated with Baobab (*Adansonia digitata*). This fruit is wild-harvested across Southern Africa and has a long history of use as a food amongst the indigenous tribes in this continent. Nonetheless, Baobab Dried Fruit Pulp required authorisation under the Novel Food legislation before any food products containing this ingredient could be marketed in the EU.

The Baobab Dried Fruit Pulp novel food application was also the first successful submission to demonstrate safety without the need for animal testing. Instead, this application utilised the history of use alongside detailed analyses of the nutritional and phytochemical components in Baobab to demonstrate safety.

The Novel Food legislation is currently being revised. In the updated version of this legislation, exotic fruits and vegetables with a significant history of use outside the EU will be permitted to undergo a simplified application procedure. This process will effectively review the safety of these types of food products purely on the basis of their history of use, provided that no safety concerns are raised during this assessment.

The new version of the Novel Food legislation will also include further categories that are not currently addressed in the current format. These include nanotechnology and foods derived from non-traditional breeding techniques.

As of September 2009, it is unclear when this new legislation will come into force.

Canada

In Canada, novel foods are regulated under the Novel Foods Regulations. The regulations define novel food as

- Products that have never been used as food,
- Foods that result from a process that has not been previously used for food, or,
- Foods that have undergone genetic modification and have new traits.

The Novel Foods Regulation requires that notification be made to the Health Canada by the company who wants to sell the product prior to the marketing or advertising of a novel food. Pre-market notification permits Health Canada to conduct a thorough safety assessment of all biotechnology-derived foods to demonstrate that a novel food is safe and nutritious before it is allowed in the Canadian marketplace.

To date, more than 90 novel foods have been approved for sale in Canada including a number of canola, corn, cottonseed and flax crop lines. Some of the novel traits include herbicide tolerance and pest and disease resistance. In spite of the benefits brought about by the development of novel foods, there are also a number of concerns that have been raised concerning their potential impact on the environment and on human health and safety.

FAMINE FOOD

A famine food or poverty food is any inexpensive or readily-available foodstuff used to nourish people in times of extreme poverty or starvation, as during a war or famine. Quite often, the food is thereafter strongly associated with the hardship under which it was eaten, and is therefore socially downplayed or rejected as a food source in times of relative plenty.

The characterization of a foodstuff as "famine" or "poverty" food is social, and some foods, such as lobster and other crustacea, are considered poverty food in some societies and luxury food in others.

Foods associated with famine need not be nutritionally deficient. A number of famine foods are quite nutritious—thus their use to nourish and ward off hunger. However, such foodstuffs usually offer limited variability, may tend toward the less savoury end of the spectrum, yet are still consumed in large amounts by and for long periods of time because of the nutritional duress. As such, people often remain averse to them long after the immediate need to consume them has subsided. That remains the case even if such foodstuffs might otherwise constitute a healthy part of a more comprehensive diet.

Examples of Famine Foods

A number of foodstuffs have been strongly associated with famine, war, or times of hardship throughout history:

- The breadnut or Maya nut was cultivated by the ancient Mayans, but is largely rejected as a poverty food in modern Central America.
- Rutabagas were widely used as a food of last resort in Europe during World War I, and remain particularly unpopular in Germany.
- In Polynesia, the Xanthosoma plant (known locally as ape) was considered a famine food and was used only in the event that the taro crop failed.
- The fruit of the Noni, sometimes also called "starvation fruit," has a strong smell and bitter taste which often relegates it to the level of a famine food.
- The nara melon of southern Africa is sometimes eaten as a food of last resort.
- Several species of edible kelp, including dulse and Irish moss (*Chondrus crispus*), were eaten by coastal peasants during the Irish Potato Famine of 1846-1848.
- Sego lily bulbs were eaten by the Mormon pioneers when their food crops failed.

- Tulip bulbs and Sugar beets were eaten in the German occupied parts of the Netherlands during the "hunger winter" of 1944-45.
- Pettuleipä is a bread made from a combination of rye flour and pettu, which is in case a combination of dried and milled Vascular cambium and Phloem of the Scots Pine. It constitutes a Finnish example of a famine food.
- In Maine and on the Atlantic coast of Canada, fish and shellfish have been considered poverty food, and people would bury lobster shells in their yards rather than disposing of them in their rubbish so that neighbors would not learn that they were reduced to eating lobster.
- In *Collapse: How Societies Choose to Fail or Succeed,* Jared Diamond posits that disdain for seafood, including fish, ringed seal, and whale, as poverty food contributed to the collapse of the Greenland Norse.

Positive Uses of Famine Food

The term "famine food" has also been used to describe underutilized crops—edible plants which are not widely cultivated as food, but which could be cultivated as an alternative food source in the event of widespread crop failure.

TABOO FOOD AND DRINK

Taboo food and drinks are food and drink which people abstain from consuming for religious or cultural reasons.

Food taboos can be defined as a codified set of rules about which foods or combinations of foods may not be eaten and how animals are to be slaughtered. The origin of these prohibitions and commandments is varied. In some cases, these taboos are a result of health considerations or other practical reasons. In others, they are a result of human symbolic systems. Some foods may be prohibited during certain festivals (*e.g.* Lent), at certain times of life (*e.g.* pregnancy), or to certain classes of people (*e.g.* priests), although the food is in general permissible.

Causes

Various religions forbid the consumption of certain types of food. For example, Judaism prescribes a strict set of rules, called *Kashrut*, regarding what may and may not be eaten. Islam has similar laws, dividing foods into *haram* (forbidden) and *halal* (permitted). Jains often follow religious directives to observe vegetarianism. Hinduism has no specific proscriptions against eating meat, but Hindus who apply the concept of "ahimsa" (non-violence) to their diet practice forms of vegetarianism.

Aside from overt rules, there are cultural taboos against the consumption of some animals. These usually appear to be based on unconscious emotions such as revulsion and are rarely justified by logical argument. One cause is the classification of a food as famine food—the association of a food with famine, and hence association of the food with hardship. Within a given society, some meats will be considered taboo simply because they are outside the range of the generally accepted definition of a foodstuff, not necessarily because the meat is considered repulsive in flavor, aroma, texture or appearance. For example, even though there is no law against eating dog meat in the United States and Europe, it is widely considered unacceptable. (Dog meat is eaten, in certain circumstances, in Korea, Vietnam, and China, although it is nowhere a common dish.) Similarly, horse meat is rarely eaten in the Anglosphere, although it is part of the national cuisine of countries as widespread as Kazakhstan, Japan, and France.

Sometimes food taboos enter national or local law, as with the ban on cattle abattoirs in most of India, and horse slaughter in the United States. Even after resumption to Chinese rule, Hong Kong has not lifted its ban on supplying meat from dogs and cats, imposed in colonial times.

Environmentalism, ethical consumerism and other activist movements are giving rise to new taboos and eating guidelines. A fairly recent addition to cultural food taboos is that of eating the meat or eggs of endangered species or animals that are otherwise protected by law or international treaty. Examples of such protected species include some species of whales, sea turtles, and migratory birds.

Similarly, sustainable seafood advisory lists and certification consider certain seafood taboo due to unsustainable fishing. Organic certification prohibits most synthetic chemical inputs during food production, or genetically modified organisms, irradiation, and the use of sewage sludge. The Fair Trade movement and certification discourage the consumption of food and other goods produced in exploitative working conditions. Other progressive movements generating taboos include Local Food and the 100-Mile Diet, both of which encourage abstinence from non-locally produced food, and veganism, in which adherents endeavour not to use or consume animal products of any kind.

Taboo Food

Amphibians and Reptiles

Judaism and Islam strictly forbid the consumption of amphibians such as frogs and reptiles such as crocodiles and snakes. In other cultures, foods such as frog legs and alligator are treasured as delicacies, and the animals are raised commercially.

Fig. 1.13: A Bag of Frog Legs from Vietnam.

Bats

In Islam "birds of prey" are haram. In Judaism, most of the laws of Kashrut pertain to animals. The Torah explicitly states which animals are permitted or forbidden. In regard to birds, the Torah provides no general rule, and instead the Deuteronomic Code and Priestly Code explicitly list the prohibited birds, using names that have uncertain translations; the list seems to mainly consist of *birds of prey*, fish-eating water-birds, and the bat.

Bat meat is known to be a prized delicacy within the Batak and Minahasa minority communities of Indonesia.

Bears

Bears are not considered kosher in Judaism while all predatory terrestrial animals are forbidden in Islam. Observant Jews and Muslims would therefore abstain from eating bear meat. Bear meat must be cooked thoroughly as it can often be infected with trichinosis.

Birds

The Hebrew Bible (Leviticus 11:13) explicitly states that the eagle, vulture, and osprey are not to be eaten. A bird now commonly raised for meat in some areas, the ostrich, is explicitly banned as food in Leviticus 11:16.

In North America, while pigeons (as doves) are a hunted game bird urban pigeons are considered unfit for consumption.

Swan was at one time a dish reserved for royalty. The English custom of Swan upping derives from this period. In more modern times, swans have been protected in parts of Europe and America, making swan unavailable. Reports about the eating of swans are seen from time to time.

Scavengers and carrion-eaters such as vultures and crows are avoided as food in many cultures because they are perceived as carriers of disease and unclean, and associated with death. An exception is the rook which was a recognised country dish, and which has in more recent times been served in a London restaurant.

In Western cultures today, most people regard songbirds as backyard wildlife rather than as food. In addition, some migratory birds are protected by international treaty.

In Islam the birds that are halal must have feathers (which presumably excludes bats) and not be a bird of prey (which follows from Islam's general prohibition on eating carnivores). Moreover it must be the kind of bird that when it flies, if so, it must flap its wings to fly more than just simply gliding. This includes fowls, pigeons, ducks, etc.

Camels

The eating of a camel is strictly prohibited by the Torah. Although the camel is a cud-chewer, the Levites still considered it "unclean". While the foot of a camel is split into two toe-like structures, it does not meet the biblical cloven hoof criterion. The meat of a camel may also have been spurned due to the camel's physiology. Most of a camel's fat is stored in its hump. This makes the rest of the body very lean, which some cultures find lacks taste.

Although there are similarities in both Islam and Judaism regarding the Kashrut and Halal foods, there are significant differences, including the consumption of camel flesh. This is allowed in Islam, but not Judaism.

Cats

In desperate times, people have been known to resort to cooking and eating cats, but under normal conditions there is no cuisine that chooses to do so, except in China and Vietnam. Cat meat was eaten, for example, during the famine in the Siege of Leningrad. In 1996, a place that served cat meat was supposedly discovered by the Argentine press in a shanty town in Rosario, but in fact the meal had been set up by media from Buenos Aires.

In 2008, it was reported that cats were a staple part of the local diet Guangdong, China, with many cats being shipped down from the north and one Guangzhou-based business receiving up to 10,000 cats per day from different parts of China. Protesters in other parts of China have urged the Guangzhou provincial government to crack down on cat traders and restaurants that serve cat meat, although no law says it is illegal to eat cats.

The term "roof-hare" (roof-rabbit, German *Dachhase*) applies to cat meat presented as that of a hare, another pest (or pet) used as a source of meat. Subtracting the skin, feet, head and tail, hares and cats are practically identical. The only way to distinguish them is by looking at the *processus hamatus* of the feline scapula, which should have a *processus suprahamatus*. *Dar gato por liebre* ("to pass off a cat as a hare") is an expression common to many Spanish-speaking countries, equivalent to "to pull the wool over someone's eyes" derived from this basic scam. There is an equivalent Portuguese expression *Comprar gato por lebre*, meaning "to buy a cat as a hare". More specifically, in Brazil, cat meat is seen as repulsive and people often shun barbecue establishments suspected of selling cat meat. The expression *churrasco de gato* ("cat barbecue") is largely used in Brazil with a humorous note, especially for roadside stands that offer grilled meat on a stick (often coated with farofa), due to their poor hygiene conditions and the fact that the source of the meat is mostly unknown. Cases of passing off cat meat as lamb shish kabab in less reputable shops, are also regularly reported in Egypt. "Kitten cakes" and "buy three shawarma—assemble a kitten" are common Russian urban jokes about the suspect origin of food from street vendors' stalls.

The inhabitants of Vicenza in northern Italy are reputed to eat cats, although the practice has been out of use for decades.

During the so called *Bad Times* of hunger in Europe during and after World War I and World War II *roof-rabbit* was a common food.

Some restaurants in the Haiphong and Halong Bay area in north Vietnam advertise cat meat hot pot as "little tiger", and cats in cages can be seen inside.

Cows

Many Hindus, particularly Brahmins, are vegetarian, abstaining from eating the flesh of any living creature. Even those Hindus who do eat meat abstain from the consumption of beef, as the cow holds a sacred place in Hinduism. The taboo does not extend to dairy products, since their preparation does not involve harming the animal in any way. Quite the contrary, dairy items such as milk, yoghurt and ghee (particularly the latter) are highly revered and used in holy ceremony.

It is assumed that the largely pastoral Vedic people and subsequent generations of Hindus throughout the centuries relied so heavily on the cow for all sorts of dairy products (including ghee), the tilling of fields, and fuel or fertiliser that its status as a willing "caretaker" of humanity grew to identifying it as an almost maternal figure. In addition, in Hindu law, slaughtering cows is considered "anyaya" (the antonym of nyaya in the Proto-Indo-European alpha negative). Nyaya, in the original Vedic Sanskrit, was a term referring to "Syllogism" or "logic". Therefore, Hindu law considered slaughtering cows to be illogical since cows were regarded as more useful alive than dead.

There are some places in India where even in modern times buffaloes are sacrificed and the meat is eaten as ritual. In the Vedic Period, almost all of India was dependent on subsistence agriculture. Eating of cows was not taboo, so during periods of low rainfall or drought, people used to slaughter the animals and eat them, leaving themselves helpless when finally the rain used to come. Also, at this time in history, the priests slaughtered many animals to make food to distribute to the people in MahaBhoj (a grand feast). This started to cause concern in reformers and thinkers, such as Madhvacharya, who convinced kings and clergy to prohibit eating beef. The kings understood the economic reasoning behind this idea, and religious leaders added it to their practice. As it was hard to convince people to give up meat without facing resistance, they added it to religion, which helped them to suppress any revolt or disobedience in following this practice. Thus within few generations, the beef taboo became part of religion, and the slaughtering of cows ended in India.

Traditionally people from lower castes, like Dalits, ate beef and carabeef (water buffalo). In modern times, beef-eating has gained some acceptance in various parts of India, despite the opposition of most Hindus. By Indian law, the slaughter of cattle is banned in almost all Indian states except the states of Kerala and Arunachal Pradesh. Slaughter of cows is an extremely provocative issue for Hindus. The vast majority of beef consumed in India is by Muslims, to which the pig is forbidden, and Christians who have no dietary laws whatsoever.

One of Hinduism's central concepts is ahimsa or non-violence. This is non-injury in thought, word and deed. This, of course, extends to the animal kingdom for, in Hinduism, all life is considered of God, as most Hindu sects are panentheistic. The cow exemplifies key moral qualities such as tenderness, strength, industriousness, and altruism. The cow is a gentle being yet extremely powerful. Altruism is unfailing in the cow. She provides so much for man. Her milk has been used by Hindus throughout Indian history and to this very day in forms such as the milk itself for nourishment, making ghee which is used as a long-lasting cooking oil as well as lamp oil. Ghee's place in Indian, particularly Hindu, history is paramount. All of the thousands of lamps within the massive and innumerable Hindu temples throughout India are fueled daily with millions of gallons of this sacred oil from cow's milk.

Some ethnic Chinese may also refrain from eating cow meat, because many of them feel that it is wrong to eat an animal that was so useful in agriculture. Some Chinese Buddhists discourage the consumption of beef, although it is not considered taboo. A similar taboo can be seen among Sinhalese Buddhists, who consider it to be ungrateful to kill the animal whose milk and labour provides livelihoods to many Sinhalese people.

Many Zoroastrians do not eat beef, because of the cow that saved Zoroaster's life, from the evil murderers when Zoroaster was a baby. Actual Pahlavi texts state that Zoroastrians should be fully vegetarian. However, many Parsis after coming to North America are eating beef.

Crustaceans and Other Seafood

Almost all types of non-piscine seafood, such as shellfish, lobster, shrimp or crawfish, are forbidden by Judaism because such animals live in water but do not have both fins and scales.

Fig. 1.14: Blue Crabs for Sale at a Market in Piraeus

As a general rule, all seafood is permissible in Sunni Islam except the Hanafite school. In the Hanafite school, non-piscine seafood is regarded as reprehensible though not forbidden. In Shi'a Islam, only scaled fish and shrimp are allowed.

Deer and Ungulates

Deer meat or venison is a dish with a long association with royalty and aristocracy. These days, several species of deer are farmed for their meat. Caribou or reindeer is popular as a dish in Alaska, Norway, Sweden, Finland (especially sautéed reindeer), Russia and Canada, but is unusual in the United Kingdom and Ireland. This may relate to the popular culture myth of the reindeer as assistant to Father Christmas ("*eating Rudolph*"), as opposed to the "cows of the north" vision of the Northern countries. Swedish astronaut Christer Fuglesang was not allowed to bring dried reindeer with him on-board a shuttle mission as it was unthinkable for the Americans so soon before Christmas. He had to go with moose instead.

Dogs

Generally in all Western countries eating the meat of any type of animal commonly kept as pets or companion animals (*i.e.* dogs and cats) is considered taboo, though that taboo has been broken under threat of starvation in the past. However, in some rural areas of Poland, dog fat is by tradition believed to have medicinal properties—being good for the lungs for instance. In 2009 a scandal erupted when a farm near Czestochowa was discovered rearing dogs to be rendered down into *smalec*-lard.

According to the ancient Hindu scriptures (cf. Manusmriti and medicinal texts like Sushrut-Samhita), *dog's meat* was regarded as the most unclean (and rather poisonous) food possible. In Mexico during the pre-Columbian era a hairless dog named xoloitzcuintle was commonly eaten. After colonization, this custom stopped.

In Southeast Asia, most countries excluding Vietnam rarely consume dog meat either because of Islamic or Buddhist values or animal rights as in the Philippines. Manchus have a prohibition against the eating of dog meat, which is sometimes consumed by the Manchus' neighboring Northeastern Asian peoples. The Manchus also avoid the wearing of hats made of dog's fur.

Fig.1.15: Dog Meat Advertised as a "Guizhou Specialty" in Hubei. Shops of this Kind are Uncommon These Days

Elephants

In Western societies, elephants have often been associated with circuses and used for entertaining purposes. However, in Central and West Africa, elephants are hunted for their meat. Some people in Thailand also believe that eating elephant meat improves their sex lives and elephants are sometimes hunted specifically for this.

Judaism prohibits consumption of elephant meat as an unfit for consumption land animal, a taboo similar to the prohibition on camel meat.

Fish

> *"Speak not to me with a mouth that eats fish"*
>
> —Somali nomad taunt

Among the Somali people, most clans have a taboo against the consumption of fish, and do not intermarry with the few occupational clans that do eat it. This abhorrence of fish is a trademark of other such Hamitic peoples of Northeast Africa, although it can now also be observed amongst a few tribes in Southeast Africa through a process of cultural diffusion.

Certain species of fish are also forbidden in Judaism such as the freshwater eel (Anguillidae) and all species of catfish. Although they live in water, they appear to have no fins or scales (except under a microscope). Sunni Muslim laws are more flexible in this and catfishes and sharks are generally seen as halal as they are special types of fish; eel is considered permitted in the majority of the Islamic schools while Shia Muslims forbid it. A common interpretation regarding some of the Islamic prohibitions is that animals that "live in both worlds" may not be consumed. This applies to primarily aquatic animals that nest or breed on land.

Fungi

Members of the International Society for Krishna Consciousness abstain from fungi and all vegetables of the onion family (Alliaceae). They believe that these excite damaging passions. Fungi are eschewed as they grow at night. The spice asafoetida (hing) is used instead of onion or garlic and provides a somewhat similar taste in their vegetarian cookery. In Iceland and rural parts of Sweden, although not taboo, fungi were not widely eaten before the second world war. It was considered a food for cows and was also associated with the stigma of being a wartime and famine food.

Guinea Pig and Related Rodents

Guinea pigs, or *cuy*, are a significant part of the diet in Peru and among some populations in the highlands of Ecuador, mostly in the Andes Mountains highlands. However, cuyes can be found on the menu of most restaurants in Lima and other cities in Peru. Guinea pig meat is exported to the United States and European nations.

In 2004, the New York City Department of Parks and Recreation took legal action to stop vendors serving *cuy* at an Ecuadorian festival in Flushing Meadows Park. New York State allows for the consumption of guinea pigs, but New York City prohibits it. Accusations of cultural persecution have since been leveled.

The guinea pig's close rodent cousins, capybara and paca, are consumed as food in South America. The Catholic Church's

Fig. 1.16: Roast Guinea Pig (Cavia Porcellus) in Peru

restriction on eating meat during Lent does not apply to the capybara, as early missionaries gave a faulty description to the Pope, leading him to declare it a fish.

Horses and Other Equines

Horse meat is part of the cuisine of countries as widespread as Japan, France, Germany, and Kazakhstan, but is taboo in some religions and many countries. It is forbidden by Jewish law, because the horse is not a ruminant, nor does it have cloven hooves. In Islamic law, horses are generally considered *makruh, i.e.* the meat is not *haram* (forbidden) but the eating of it is strongly discouraged. It is forbidden in Hinduism.

Horse meat is forbidden by some sects of Christianity. The Battle of Tours in 732 AD showed the emergent importance of cavalry, so Pope Gregory III began an effort to stop the practice of horse eating, calling it "abominable". Horses were far more necessary to stop the Muslim cavalry, which was threatening the Christian ascendant in Europe. His edicts are based on the same scripture as the Jewish prohibitions and this ban remained until the 18th century.

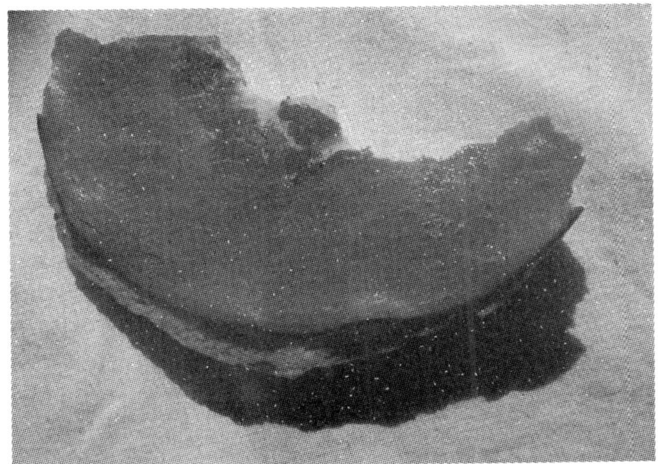

Fig. 1.17: Smoked and Salted Horse Meat on a Sandwich

The Christianisation of Iceland in 1000 CE was achieved only when the Church promised that Icelanders could continue to eat horsemeat; once the Church had consolidated its power, the allowance was discontinued.

Horse meat is generally taboo in the Anglosphere. In Canada, horse meat is legal, but there is only really a market—and that a small one—in the French-speaking province of Quebec, where the taboo is not so strong, and in a few (mostly French) restaurants elsewhere. Most Canadian horse meat is exported to Continental Europe or Japan. In the United States, sale and consumption of horse meat is illegal in California and Illinois. However, it was sold in the US during WW II, since beef was expensive, rationed and destined for the troops. In the UK, this strong taboo includes banning horsemeat from commercial pet food and DNA testing of some types of salami suspected of containing donkey meat.

Horsemeat is also avoided by most people from the Balkans, mostly for ethical reasons, as horse is considered to be a noble animal, or because eating horsemeat is associated with war time famine. A similar taboo exists in Poland.

In Sunni Islam, Al-Bukhari reports that Muhammad forbade the eating of a donkey, but the general applicability of this *hadith* is unclear.

Insects

Except for certain locusts and related species, insects are not considered kosher; dietary laws also require that practitioners check food carefully for insects. In Islam locusts are considered lawful food along with fish that do not require ritual slaughtering.

Western taboos against insects as a food source generally do not apply to honey (concentrated nectar which has been regurgitated by bees). For example, honey is considered kosher even though honeybees are not, an apparent exception to the normal rule that products of an unclean animal are also unclean. This topic is covered in the Talmud and is explained to be permissible on the grounds that the bee does not make the honey, the flower does, and it is only stored in bees.

Many vegans avoid honey as they would any other animal product. Some vegans disagree with avoiding honey, on the grounds that nearly all plants are propagated by insects or birds, and the harvesting of them would be similarly exploitative.

Living Animals

Islamic and Judaic law (including Noahide Law) forbids any portion that is cut from a live animal. (Genesis 9:4)

Fig. 1.18: *Raw oysters*, which are Still Alive, Presented on a Plate

Fig. 1.19: Ikizukuri, Live Fish Served as Sashimi

Examples of the eating of animals that are still alive include *raw oyster on the half shell* (also called *shooters*) and ikizukuri (live fish). Sashimi using live animals has been banned in some countries.

Another example occurs in Shanghai, China, and surrounding areas, live shrimp is a common dish served both in homes and restaurants. The shrimp are usually served in a bowl of alcohol, which makes the shrimp sluggish and complacent. Local belief is that live shrimp are "healthier" than those served "already dead" or cooked.

Offal

Offal is the internal organs of butchered animals, and may refer to parts of the carcass such as the head and feet ("trotters") in addition to organ meats such as sweetbreads and kidney. Offal is a traditional part of many European and Asian cuisines, including such dishes as the well-known steak and kidney pie in the United Kingdom. Haggis has been Scotland's national dish since the time of Robbie Burns and black pudding is a traditional sausage in northern England and in Ireland. In northeast Brazil there´s a similar dish to Haggis called "Buchada", made with goats intestine. The French eat calves brains. The Irish and French enjoy Tripe (cow stomach).

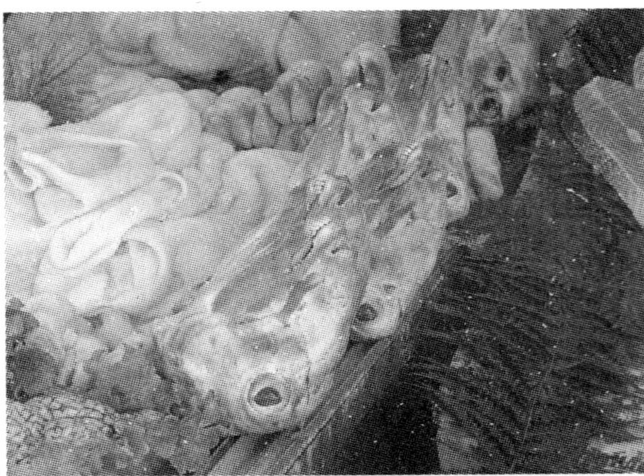

Fig. 1.20: Heads, Brains, Trotters and Tripe on Sale in an Istanbul Market

In countries such as Australia, Canada and the United States, on the other hand, many people are squeamish about eating offal. In these countries, organ meats that are considered edible in other cultures are more often regarded as fit only for processing into pet food under the euphemism "meat by-products". Except for liver (chicken, beef, or pork), and intestines used as natural sausage casings, organ meats consumed in the U.S. tend to be regional or ethnic specialties; for example, tripe as *menudo* or *mondongo* among Latinos, chitterlings in the southern states, scrapple in the Mid-Atlantic region, and beef testicles called Rocky Mountain oysters or "prairie oysters" in the west.

In some regions, such as the EU, brains and other organs which can transmit bovine spongiform encephalopathy ("mad cow disease") and similar diseases have now been banned from the foodchain as specified risk materials.

Pigs/Pork

USDA data reports pork as the most widely eaten meat in the world. Consumption of pigs is forbidden among Muslims, Jews, Seventh-day Adventists, and others. There are various theories concerning the origins of this taboo (*e.g.* Qur'an 16:115, Biblical injunctions in Leviticus 11,7-8 and Deuteronomy 14,8), but none have been universally accepted.

In the 19th century some people attributed the pig taboo in the Middle East to the danger of the parasite trichina. This theory still circulates outside scientific circles, but is now rejected by most anthropologists.

Marvin Harris posited that pigs are not suited for being kept in the Middle East on an ecological and socio-economical level; for example, pigs are not suited to living in arid climates and thus require far more water than other animals to keep them cool, and instead of grazing they compete with humans for foods such as grains. As such, raising pigs was seen as a wasteful and decadent practice.

A common explanation to the fact that pigs are widely considered unclean in the Middle East is that they are omnivorous, not discerning between meat or vegetation in their natural dietary habits. The willingness to consume meat sets them apart from most other domesticated animals which are commonly eaten (cows, horses, goats, etc.) who would naturally eat only plants.

The Vietnamese bred pot bellied pigs for meat and lard; however, in the United States they are kept as pets, and there is a stigma on eating them.

A culture-based pork taboo was attributed to Scottish highlanders until approximately 1800.

Rabbit

The book of Leviticus in the Bible classifies the rabbit as unclean because it does not have a split hoof, even though it does chew and reingest partially digested material (often loosely translated "chew

Fig. 1.21: Cottontail Rabbit

the cud" in English). Further possibilities against the consumption of rabbit may also include the phenomenon known as Rabbit Starvation, a form of acute malnutrition caused by excess consumption of any lean meat (specifically rabbit) coupled with a lack of other sources of nutrients. The consumption of rabbit is allowed in Sunni Islam but forbidden for Shia Islam.

Rats and Mice

In most Western cultures, rats and mice are considered either unclean vermin or pets and thus unfit for human consumption, traditionally being seen as carriers of plague. However, rats are commonly eaten in rural Thailand, Vietnam and other parts of Indochina. Cane rats (*Thryonomys swinderianus* and *Thryonomys gregorianus*) and some species of field mice are a rich source of protein in Africa. Bamboo rats are also commonly eaten in the poorer parts of Southeast Asia

In Ghana, Thryonomys swinderianus locally referred to as "Akrantie", "Grasscutter" and (incorrectly) as "Bush rat" is a common food item. The proper common name for this rodent is "Greater Cane Rat", though actually it is not a rat at all and is a close relative of porcupines and guinea pigs that inhabit Africa, south of the Saharan Desert. In 2003, the U.S. barred the import of this and other rodents from Africa because an outbreak at least nine human cases of monkeypox, an illness never before been seen in the Western Hemisphere.

Historically, rats and mice have also been eaten in the West during times of shortage or emergency, such as during the Battle of Vicksburg and the Siege of Paris. Dormice were also domesticated and raised for food in Ancient Rome. In some Asian countries, mice are eaten, and go by the name of vole. In France, rats bred in the wine stores of Gironde were cooked with the fire of broken wine barrels and eaten, dubbed as *cooper's entrecote*. In some communities the muskrat (which is not a rat at all) is hunted for its meat (and fur) (*e.g.* some parts of Flanders). Nutria, another large rodent, has been hunted or raised for food in the United States

Snails

Snails have been eaten for thousands of years, beginning in the

Pleistocene. They are especially abundant in Capsian sites in North Africa but are also found throughout the Mediterranean region in archaeological sites dating between 12,000 and 6,000 years ago. They are also seen as a delicacy in China and in several Asian countries along with France and other Mediterranean countries. However, in Britain, Ireland, and America, eating them may be seen as disgusting. Some English-speaking commentators have used the French word for snails, *escargot*, as an alternative word for snails, particularly snails for consumption. As they are molluscs, snails are not kosher.

Vegetables

In certain versions of Buddhism and Hinduism, vegetables of the onion genus are taboo. Specifically, Chinese Buddhist cuisine traditionally prohibits garlic, *Allium chinense*, asafoetida, shallot, and *Allium victorialis* (victory onion or mountain leek), while Kashmiri Brahmins forbid "strong flavored" foods. This encompasses garlic, onion, and spices such as black pepper and chili pepper, believing that pungent flavors on the tongue inflame the baser emotions. In Jainism, any kind of roots are considered taboo, since the process of uprooting causes the organisms associated with the root in the soil to die.

In Yazidism, the eating of lettuce and butter beans is taboo. The Muslim religious teacher and scholar, Falah Hassan Juma, links the sect's belief of evil found in lettuce to its long history of persecution by Muslims and Christians. Historical theory claims one ruthless potentate who controlled the city of Mosul in the 13th century ordered an early Yazidi saint executed. The enthusiastic crowd then pelted the corpse with heads of lettuce.

The followers of Pythagoras were vegetarians (in fact, "Pythagorean" at one time came to mean "vegetarian"); however, their creed prohibited the eating of beans. The reason is unclear: perhaps the flatulence they cause, perhaps as protection from potential favism, but most likely for magico-religious reasons.

Vegetables like broccoli, while not taboo, may be avoided by observant Jews and other religions due to the possibility of insects hiding within the numerous crevices. Likewise, fruits such as blackberries and raspberries are recommended by kashrut agencies to be avoided as they can not be cleaned thoroughly enough without destroying the fruit.

Although it might not be a taboo in a strictest sense, older Germans might not eat swede (Swedish turnip, rutabaga), as they see it as a "famine food", not for general consumption. This taboo existed from the 1916-17 famine *Steckrübenwinter* when Germany, already drained by World War I's endless Western Front, had one of the worst winters in memory, where often the only food available was Swedish turnips. This led a distaste to the vegetable which still continues today with the older generations having had experiences from World War II or having had a childhood with parents talking about the aforementioned famine. However, in recent years this taboo has been vanishing as Germans have re-discovered many traditional or local cooking recipes, including those including swede. One reason for this, is a trend to traditional and organic cuisine. Also for most Germans in 2008, the "*Steckrübenwinter*" famine from 1916-17 is history and has no more relevance on today's choice of food and dish.

Whales

Over the last couple of decades, the eating of whale has become increasingly taboo. The International Whaling Commission passed a moratorium on commercial whaling on July 23, 1982 that came into force for the 1985-86 season.

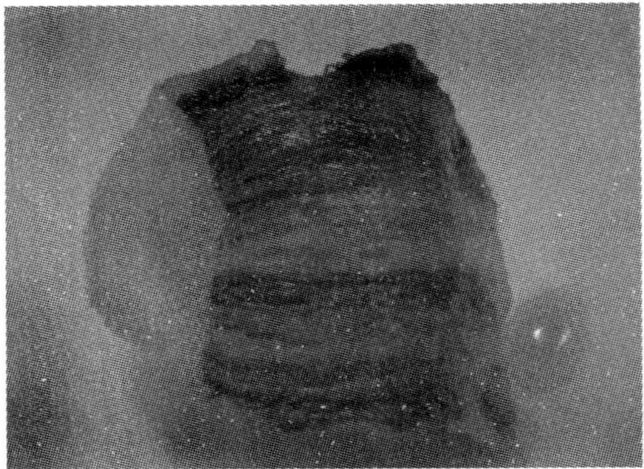

Fig. 1.22: Inuit Muktuk or Raw Whale Blubber, Rich in Vitamin C

Norway resumed commercial whaling of minke whales in 1993 and it is still a popular meat, especially on Norway's western coast. Once considered an inexpensive substitute for beef, whale meat is now a highly priced delicacy. Iceland resumed commercial whaling in 2006. Japan's whaling is officially done for research purposes. This is specifically sanctioned under IWC regulations that also specifically require that whale meat be fully utilized upon the completion of research. Many international scientific and environmentalist groups, notably Greenpeace argue that the killing is not necessary to conduct the research.

The United States Marine Mammal Protection Act of 1972 prohibits, with certain exceptions, the taking of marine mammals in United States waters and by U.S. citizens on the high seas, and the importation of marine mammals and marine mammal products into the U.S. Despite the general ban on whale hunting in the United States and Canada, some indigenous groups are allowed to hunt for cultural reasons.

Whale meat was eaten in Britain during World War II, but it was never popular.

Primates

The consumption of monkeys and apes such as chimpanzees, gorillas, mandrills and guenons is quite common in rural areas of Sub-Saharan Africa. Bonobos (also known as pygmy chimpanzees), have been extensively hunted in Congo to the level that they are now considered an endangered species. In certain parts of Congo the hands and feet of gorillas are regarded as a delicacy and are served to special guests. Monkeys are also eaten in southeast Asia (especially Indonesia). The consumption of primates may be considered to be too close to human cannibalism due to the similarity of our own species. The similarity increases the danger of viruses. Most of it is "bushmeat" or caught from the wild, in areas of high primate populations such as Central Africa and southeast Asia. One of the major theories for the origin of the human immunodeficiency virus (HIV) in humans is the butchering of primates infected with a similar virus

Human Meat

Of all the taboo meat, human flesh ranks as the most proscribed. Historically, humans have consumed the flesh of fellow humans in rituals and out of insanity, hatred, or overriding hunger—never as a common part of their diet. This consumption of human flesh is forbidden in Judaism, Christianity, Islam, and Hinduism, as well as most other religions. However, it used to be required in certain tribes; the Fore people of Papua New Guinea were particularly well-studied in their eating of the dead, because it led to kuru, a disease believed to be transmitted by prions. Very few people customarily eat the placenta after the baby's birth, but those who advocate placentophagy in humans (mostly in modern America and Europe, Mexico, Hawaii, China, and the Pacific Islands) believe that eating the placenta prevents postpartum depression and other pregnancy complications.

Taboo Drinks

Alcohol

Some religions—including Hinduism, Buddhism, Islam, Jainism, Sikhism, Rastafari Movement, the Bahá'í Faith, and various branches of Christianity such as the Methodists, the Baptists, the Latter-day Saints, Seventh Day Adventists and the Iglesia ni Cristo—forbid or discourage the consumption of alcoholic beverages. Although many claim they should be listed along with such groups, Jehovah's Witnessess only discourage the use of alcohol to the point of excess just as numerous doctors and health experts do. The Hebrew Bible describes a Nazarite vow (Numbers 6:1-21) that includes abstinence from alcohol (specifically wine and probably barley beer, although there is no general taboo against alcohol in Judaism. There are also cultural taboos against the consumption of alcohol, reflected for example in the Temperance movement. Drinking habits are well determined by each culture (as in European and American rules concerning alcoholic beverages) or religious group.

Blood

Drinking blood is a strong taboo in many countries, and is often vaguely associated with vampirism (the consumption of human blood).

Fig. 1.23: A Bowl of *Dinuguan*

Blood sausage, or blood made into cake form, is quite popular in many parts of the world. In the Philippines, and China there is a delicacy of pigs blood made into a square form and roasted. Dinuguan, is a pork blood stew—meat simmered in a rich, spicy gravy of pig blood, garlic, chili and vinegar. In Finland the blood patties and *mustamakkara* type blood sausage are seen as national cuisine.

Some religions prohibit drinking or eating blood or food made from blood. In Judaism all mammal and bird meat (not fish) is salted to remove the blood. Jews and Jewish Proselytes follow the teaching in Leviticus 17:10-14, that since "the life of the animal is in the blood", no person may eat (or drink) the blood. However, they have no rules regarding blood transfusions since the blood is not consumed and because a transfusion is a medical procedure (Jews may break kosher laws, and Muslims may break harams, if it is for saving life). Iglesia ni Cristo also prohibits eating or drinking any blood. Jehovah's Witnesses, in addition, prohibit acceptance of blood transfusion in accord with New Testament direction found at (Acts 15:20Acts 15:29Acts 21:25—King James Version)

According to the Bible blood is only to be used for special/ sacred purposes in connection with worship [Exodus chapters 12, 24, 29; Matthew 26:28; Hebrews 9:22]. In the first century,

Christians, both former Jews (the Jewish Christians), and new Gentile converts, were in dispute as to which particular features of Mosaic law were to be retained and upheld by them. The apostles decided that, among other things, it was necessary to abstain from consuming blood (Acts 15:28-29—King James Version):

> "For it seemed good to the Holy Ghost, and to us, to lay upon you no greater burden than these necessary things; That ye abstain from meats offered to idols, and from blood, and from things strangled, and from fornication: from which if ye keep yourselves, ye shall do well, Fare ye well."

These New Testament verses repeated certain elements of the Jewish law, and included the prohibition regarding blood, thus making it also binding upon the Early Christian church. The *Apostolic Decree* is still observed today by the Greek Orthodox.

Coffee and Tea

Hot drinks are taboo for members of The Church of Jesus Christ of Latter-day Saints. Most Mormons interpret this as referring exclusively to coffee and tea (*e.g.* not hot cocoa or herbal tea). The Word of Wisdom, a code of health used by church members, outlines prohibited and allowed substances. It is also sometimes extended as a taboo against caffeine in general, including cola drinks. Coffee is also taboo for Rastafarians and Seventh Day Adventists.

Originally, coffee was considered taboo among Roman Catholics as it was considered a Muslim wine until it was deemed acceptable by Pope Clement VIII. Supposedly, Pope Clement tried coffee and liked it so much that he was quoted as saying "This devil's drink is so good... we should cheat the devil by baptizing it." Whether this story is true is unknown.

Raw Milk

Consumption of raw milk and raw milk products such as unpasteurized cheese, with the exception of breast milk, is opposed by the U.S. Food and Drug Administration (FDA) and other government organizations in the United States. This opposition is met with the displeasure of foreign producers of dairy products, who find it difficult to sell in the United States and countries with similar regulations, and the displeasure of many domestic dairy producers, who feel that the pasteurization requirement makes it more difficult for American dairy products to compete with foreign ones.

Pasteurization was first used in the United States in the 1890s after the discovery of germ theory to control the hazards of highly contagious bacterial diseases including bovine tuberculosis and brucellosis that could be easily transmitted to humans through the drinking of raw milk. Initially after the scientific discovery of bacteria, no product testing was available to determine if a farmer's milk was safe or infected, so all milk had to be treated as potentially contagious. After the first test was developed, some farmers actively worked to prevent their infected animals from being killed and removed from food production, or would falsify the test results so that their animals would appear to be free of infection.

Pasteurization could make raw milk from any source safer to drink whether infected or not. Although farm sanitation has greatly improved and effective testing has been developed for tuberculosis and other diseases, pasteurization continues to be used as a stopgap measure in case infectious milk from a mismanaged farm with poor sanitation should enter the food supply.

Genetically Modified Foods Taboo

Attitudes concerning genetically modified food like genetically modified soya, maize or rapeseed (canola) vary from accepted to taboo in the U.S. and Canada, while many Europeans have a taboo on it as they are more concerned with eating natural food sources. This is believed to be due to the various food scares in Europe during the 1980s and 1990s, such as BSE/vCJD, salmonella and dioxin poisoning. In the UK, only 2 per cent of Britons are said to be "happy to eat GM foods", and more than half of Britons are against genetically modified foods being available to the public, according to a 2003 study.

In Europe, regulations state that all food and animal feed containing more than 0.5 percent GM ingredients are required to have strict labelling and traceability, and many supermarkets proudly boast the fact that they don't sell GM foods.

Personal Taboos

Australian Aborigines traditionally had personal totems. While religious practices varied from group to group, it was common that the eating of the totemic animal was considered taboo, either by the entire clan, or the individual with the personal totem.

FOOD ENERGY

Food energy is the amount of energy in food that is available through digestion.

Like other forms of energy, food energy is expressed in calories or joules. Some countries use the food calorie, which is equal to 1 kilocalorie (kcal), or 1,000 gram calories. In the context of nutrition, and especially food labeling, the calories are large calories approximately equal to 4.1868 kilojoules (kJ). The kilojoule is the unit officially recommended by the World Health Organization and other international organizations. In some countries only the kilojoule is normally used on food packaging, but the calorie is still the most common unit in many countries.

Fiber, fats, proteins, organic acids, polyols, and ethanol contain food energy. All foods are made up of a combination of these six caloric nutrients and non-caloric nutrients. Non-caloric food includes (but not limited to) water, vitamins, minerals, antioxidants, caffeine, spices and natural flavors. Tea and coffee also have no calories without sugar or milk added. Nutritionists usually talk about the number of calories in a gram of a nutrient. Fats and ethanol have the greatest amount of food energy per gram, 9 and 7 kcal/g (38 and 30 kJ/g), respectively. Proteins and most carbohydrates have about 4 kcal/g (17 kJ/g). Carbohydrates that are not easily absorbed, such as fiber or lactose in lactose-intolerant individuals, contribute less food energy. Polyols (including sugar alcohols) and organic acids have fewer than 4 kcal/g.

Each food item has a specific metabolizable energy intake (MEI). Normally this value is obtained by multiplying the total amount of energy contained in a food item by 85 per cent, which is the typical amount of energy actually obtained by a human after the digestive processes have been completed.

Measuring Food Energy

The following process details how to measure food energy, as specified by the United States Department of Agriculture (USDA) in the early 1900s:

The particular food being measured must be burned in a calorimeter, so that the heat released from the food can be accurately measured. This amount is used to ascertain the G.E.V. of the specified food. This number is then multiplied by, usually, 85 per cent; which represents the loss happening during human digestion.

Food Labels

The "calorie" has become a common household term because dietitians recommend in cases of obesity to reduce body weight by increasing exercise (energy expenditure) and reducing energy intake. Many governments require food manufacturers to label the energy content of their products, to help consumers control their energy intake. In the European Union, manufacturers of prepackaged food must label the nutritional energy of their products in both kilocalories and kilojoules, when required. In the United States, the equivalent mandatory labels display only "Calories", often as a substitute for the name of the quantity being measured, food energy; an additional kilojoules figure is optional and is rarely used. The energy content of food is usually given on labels for 100 g and/or for what the manufacturer claims is a typical serving size.

The amount of food energy in a particular food could be measured by completely burning the dried food in a bomb calorimeter, a method known as direct calorimetry. However, the values given on food labels are not determined this way, because it overestimates the amount of energy that the human digestive system can extract, by also burning dietary fiber. Moreover, not all food energy eaten is actually resorbed by the body (fecal and urinal losses). Instead, standardized chemical tests or an analysis of the recipe using reference tables for common ingredients are used to estimate the product's digestible constituents (protein, carbohydrate, fat, etc.). These results are then converted into an equivalent energy value based on a standardized table of energy densities.

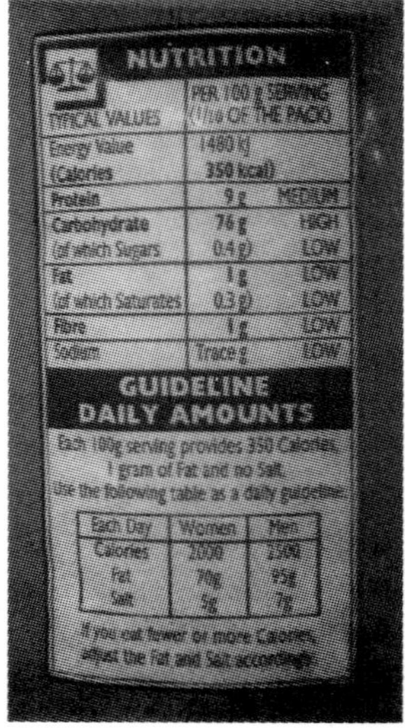

Fig. 1.24: The Nutritional Information Label on a Pack of Basmati Rice in the United Kingdom

Food Component	*Energy Density*	
	kcal/g	kJ/g
Fat	9	37
Ethanol (alcohol)	7	29
Proteins	4	17
Carbohydrates	4	17
Organic acids	3	13
Polyols (sugar alcohols, sweeteners)	2.4	10
Salatrims (reduced energy fat)	6	25

All the other nutrients in food are non-caloric and are thus not counted.

Recommended daily energy intake values for young adults and men are: 2500 kcal/day (10 MJ/day) and 2000 kcal/day (8 MJ/day) for women. Children, sedentary and older people require less energy, physically active people more. In addition to physical activity, increased mental activity has been linked with moderately increased brain energy consumption.

Energy Usage in the Human Body

Energy intake to the body that is not used up is mostly stored as fat in the fat tissue.

The conversion efficiency of food energy into physical power depends on the form of energy source (type of food) and on the type of physical energy usage (*e.g.* which muscles are used, whether the muscle is used aerobically or anaerobically). In general, the efficiency of muscles is rather low: only 18 to 26 percent of the food energy is converted into mechanical energy. This low efficiency is the result of about 40 per cent affiance of generating ATP from food energy, losses in converting energy from ATP into mechanical work inside the muscle, and mechanical losses inside the body. The latter two losses are dependent on the type of exercise and the type of muscle fibers being used (fast-twitch or slow-twitch). For an overall efficiency of 20 percent, one watt of mechanical power is equivalent to 4.3 kcal per hour. For example, a manufacturer of rowing equipment shows burned calories as four times the actual mechanical work, plus 300 kcal per hour, which amounts to about 20 percent efficiency at 250 watts of mechanical output. It can take up to 20 hours of little physical output (*e.g.* walking) to "burn off" 500 calories more than your body should have.

The differing energy density of foods (fat, alcohols, carbohydrates and proteins) lies in their varying proportions of oxidizable carbon atoms. Release of energy from food follows transfer of electrons from carbon and hydrogen to carbon dioxide and water.

Swings in body temperature—either hotter or cooler—increase the metabolic rate, thus burning more energy. Prolonged exposure to extremely warm or very cold environments increases the basal metabolic rate (BMR). People who live in these types of settings often have BMRs that are 5-20 per cent higher than those in other climates. Physical activity also significantly increases body temperature which in turn uses more food energy.

BIOENERGETICS

Bioenergetics is the subject of a field of biochemistry that concerns energy flow through living systems. This is an active area of biological research that includes the study of thousands of different cellular processes such as cellular respiration and the many other metabolic processes that can lead to production and utilization of energy in forms such as ATP molecules.

Overview

Bioenergetics is the part of biochemistry concerned with the energy involved in making and breaking of chemical bonds in the molecules found in biological organisms.

Growth, development and metabolism are some of the central phenomena in the study of biological organisms. The role of energy is fundamental to such biological processes. The ability to harness energy from a variety of metabolic pathways is a property of all living organisms. Life is dependent on energy transformations; living organisms survive because of exchange of energy within and without.

In a living organism, chemical bonds are broken and made as part of the exchange and transformation of energy. Energy is available for work (such as mechanical work) or for other processes (such as chemical synthesis and anabolic processes in growth), when weak bonds are broken and stronger bonds are made. The production of stronger bonds allows release of usable energy.

Living organisms obtain energy from organic and inorganic materials. For example, lithotrophs can oxidize minerals such as nitrates or forms of sulfur, such as elemental sulfur, sulfites, and hydrogen sulfide to produce ATP. In photosynthesis, autotrophs can produce ATP using light energy. Heterotrophs must consume organic compounds. These are mostly carbohydrates, fats, and proteins. The amount of energy actually obtained by the organism is lower than the amount present in the food; there are losses in digestion, metabolism, and thermogenesis.

The materials are generally combined with oxygen to release energy, although some can also be oxidized anaerobically by various organisms. The bonds holding the molecules of nutrients together and the bonds holding molecules of free oxygen together are all relatively weak compared with the chemical bonds holding carbon dioxide and water together. The utilization of these materials is a form of slow combustion.

That is why the energy content of food can be estimated with a bomb calorimeter. The materials are oxidized slowly enough that the organisms do not actually produce fire. The oxidation releases energy because stronger bonds have been formed. This net energy may evolve as heat, or some of which may be used by the organism for other purposes, such as breaking other bonds to do chemistry.

Living organisms produce ATP from energy sources via oxidative phosphorylation. The terminal phosphate bonds of ATP are relatively weak compared with the stronger bonds formed when ATP is broken down to adenosine monophosphate and phosphate, dissolved in water. Here it is the energy of hydration that results in energy release. This hydrolysis of ATP is used as a battery to store energy in cells, for intermediate metabolism. Utilization of chemical energy from such molecular bond rearrangement powers biological processes in every biological organism.

Types of Reactions

- Exergonic is a spontaneous reaction that releases energy. It is thermodynamically favoured. On the course of a reaction, energy needs to be put in, this activation energy drives the reactants from a stable state to a highly energetic unstable configuration. These reactants are usually complex molecules that are broken into simpler products. The entire reaction is usually catabolic. The release of energy, also called free energy is a-ΔG because energy is lost from the bonds formed by the products.
- Endergonic is an anabolic reaction that consumes energy. It has a +ÄG because energy is required to break bonds.

The free energy (ΔG) gained or lost in a reaction can be calculated: $\Delta G = \Delta H - T\,\Delta S$.

Also, $\Delta G = \Delta G^{\circ\prime} + 2.303RT\log([P]/[R])$ where

- R is the gas constant, 1.987 cal/mol
- T is temperature in Kelvin K = 273 + °C
- P is Products
- R is the reactants

Cotransport

In August 1960, Robert K. Crane presented for the first time his discovery of the sodium-glucose cotransport as the mechanism for intestinal glucose absorption. Crane's discovery of cotransport was the first ever proposal of flux coupling in biology and was the most important event concerning carbohydrate absorption in the 20th century.

Chemiosmotic Theory

One of the major triumphs of bioenergetics is Peter D. Mitchell's chemiosmotic theory of how protons in aqueous solution function in the production of ATP in cell organelles such as mitochondria Other cellular sources of ATP such as glycolysis were understood first, but such processes for direct coupling of enzyme activity to ATP production are not the major source of useful chemical energy in most cells. Chemiosmotic coupling is the major energy producing process in most cells, being utilized in chloroplasts and many single celled organisms in addition to mitochondria.

Energy Balance

Energy balance is the biological homeostasis of energy in living systems. It is measured with the following equation: *Energy intake = internal heat produced + external work + storage.*

BIO-ENERGY BALANCE

In biology, energy balance is the biological homeostasis of energy in living systems. It is measured with the following equation: *Energy intake = internal heat produced + external work + storage.* It is also an aspect of bioenergetics, concerning energy flow through living systems.

It generally uses the energy unit Calorie (or kilogram calorie), which equals the energy needed to increase the temperature of 1 kg of water by 1°C. This is about 4.184 kJ.

Energy Intake

Energy intake is part of the diet, which is mainly regulated by hunger and food energy of what is consumed.

Energy Expenditure

Energy expenditure is mainly a sum of internal heat produced and external work.

The internal heat produced is, in turn, mainly a sum of basal metabolic rate (BMR) and the thermic effect of food.

External work may be estimated by measuring physical activity level (PAL).

Imbalance between Intake and Expenditure

Gaining Imbalance

A gaining energy imbalance is a result of *energy intake* being *higher* than what is consumed in external work and other bodily means of energy expenditure.

The main preventable causes are:

- Overeating, resulting in increased energy intake.
- Sedentary lifestyle, resulting in decreased energy expenditure through external work

A gaining imbalance results in energy being stored, primarily as fat, causing weight gain. In time, overweight and obesity may develop, with resultant complications.

Losing Imbalance

A losing energy imbalance is a result of *energy intake* being *less* than what is consumed in external work and other bodily means of energy expenditure.

The main cause is *undereating* or even starvation, *e.g.* in famine, anorexia (decreased appetite) and anorexia nervosa (also including body image distortion).

Energy Requirement

Normal energy requirement, and therefore normal energy intake, depends mainly on age, sex and physical activity level (PAL).

The Food and Agriculture Organization (FAO) of the United Nations has compiled a detailed report on human energy requirements: Human energy requirements (Rome, 17-24 October 2001)

Yet, there are currently ongoing studies to show if calorie restriction to below normal values have beneficial effects, and even though they are showing positive indications in primates it is still not certain if calorie restriction has a positive effect on longevity for primates and humans. Calorie restriction may be viewed as attaining energy balance at a lower intake and expenditure, and is, in this sense, not generally an energy imbalance, except for an initial imbalance where decreased expenditure hasn't yet matched the decreased intake.

ROLE OF AGRICULTURE

Agriculture refers to the production of goods through the growing of plants and the raising of domesticated animals. The study of agriculture is known as agricultural science. The related practice of gardening is studied in horticulture.

Agriculture encompasses a wide variety of specialties. Cultivation of crops on arable land and the pastoral herding of livestock on rangeland remain at the foundation of agriculture. In the past century a distinction has been made between sustainable agriculture and intensive farming. Modern agronomy, plant breeding, pesticides and fertilizers, and technological improvements have sharply increased yields from cultivation. Selective breeding and modern practices in animal husbandry such as intensive pig farming (and similar practices applied to the chicken) have similarly increased the output of meat. The more exotic varieties of agriculture include aquaculture and tree farming.

The major agricultural products can be broadly grouped into foods, fibers, fuels, raw materials, pharmaceuticals and illegal drugs, and an assortment of ornamental or otherwise exotic products. In recent years plants have been used to grow biofuels, biopharmaceuticals, bioplastics, and pharmaceuticals. Specific foods include cereals, vegetables, fruits, and meat. Fibers include cotton, wool, hemp, silk and flax. Raw materials include lumber and bamboo. Drugs include tobacco, marijuana, opium, cocaine, digitalis, curare, eugenol, reserpine, pyrethrins, taxol) and other useful materials such as resins. Biofuels include methane from biomass, ethanol, and biodiesel. Cut flowers, nursery plants, tropical fish and birds for the pet trade are some of the ornamental products.

The history of agriculture has played a major role in human history, as agricultural progress has been a crucial factor in worldwide socio-economic change. Wealth-building and militaristic specializations rarely seen in hunter-gatherer cultures are commonplace in societies which practice agriculture. So, too, are arts such as epic literature and monumental architecture, as well as codified legal systems. When farmers became capable of producing food beyond the needs of their own families, others in their society were freed to devote themselves to projects other than food acquisition. Historians and anthropologists have long argued that the development of agriculture made civilization possible.

In 2007, an estimated 35 percent of the world's workers were employed in agriculture (from 42 per cent in 1996). However, the relative significance of farming has dropped steadily since the beginning of industrialization, and in 2003—for the first time in history—the services sector overtook agriculture as the economic sector employing the most people worldwide. Despite the fact that agriculture employs over one-third of the world's population, agricultural production accounts for less than five percent of the gross world product (an aggregate of all gross domestic products).

Overview

Agriculture has played a key role in the development of human civilization—it is widely believed that the domestication of plants and animals allowed humans to settle and give up their previous hunter-gatherer lifestyle during the Neolithic Revolution. Until the Industrial Revolution, the vast majority of the human population laboured in agriculture. Development of agricultural techniques has steadily increased agricultural productivity, and the widespread diffusion of these techniques during a time period is often called an agricultural revolution. A remarkable shift in agricultural practices has occurred over the past century in response to new technologies. In particular, the Haber-Bosch method for synthesizing ammonium nitrate made the traditional practice of recycling nutrients with crop rotation and animal manure less necessary. Synthetic nitrogen, along with mined rock phosphate, pesticides and mechanization, have greatly increased crop yields in the early 20th century. Increased supply of grains has led to cheaper livestock as well. Further, global yield increases were experienced later in the 20th century when high-yield varieties of common staple grains such as rice, wheat, and corn (maize) were introduced as a part of the Green Revolution. The Green Revolution exported the technologies (including pesticides and synthetic nitrogen) of the developed world out to the developing world. Thomas Malthus famously predicted that the Earth would not be able to support its growing population, but technologies such as the Green Revolution have allowed the world to produce a surplus of food.

Many governments have subsidized agriculture to ensure an adequate food supply. These agricultural subsidies are often linked to the production of certain commodities such as wheat, corn (maize), rice, soybeans, and milk. These subsidies, especially when done by developed countries have been noted as protectionist, inefficient, and environmentally damaging. In the past century agriculture has been characterized by enhanced productivity, the use of synthetic fertilizers and pesticides, selective breeding, mechanization, water contamination, and farm subsidies. Proponents of organic farming such as Sir Albert Howard argued in the early 1900s that the overuse of pesticides and synthetic fertilizers damages the long-term fertility of the soil. While this feeling lay dormant for decades, as environmental awareness has increased recently there has been a movement towards sustainable agriculture by some farmers, consumers, and policymakers. In recent years there has been a backlash against perceived external environmental effects of mainstream agriculture, particularly regarding water pollution, resulting in the organic movement. One of the major forces behind this movement has been the European

Union, which first certified organic food in 1991 and began reform of its Common Agricultural Policy (CAP) in 2005 to phase out commodity-linked farm subsidies, also known as decoupling. The growth of organic farming has renewed research in alternative technologies such as integrated pest management and selective breeding. Recent mainstream technological developments include genetically modified food.

As of late 2007, several factors have pushed up the price of grain used to feed poultry and dairy cows and other cattle, causing higher prices of wheat (up 58 per cent), soybean (up 32 per cent), and maize (up 11 per cent) over the year. Food riots have recently taken place in many countries across the world. An epidemic of stem rust on wheat caused by race UG99 is currently spreading across Africa and into Asia and is causing major concern. Approximately 40 per cent of the world's agricultural land is seriously degraded. In Africa, if current trends of soil degradation continue, the continent might be able to feed just 25 per cent of its population by 2025, according to UNU's Ghana-based Institute for Natural Resources in Africa.

Practices

Agricultural practices lie on a spectrum dependent upon the intensity and technology of the methods. At the one end lies the subsistence farmer who farms a small area with limited inputs and produces only enough food to meet the needs of his or her family. At the other end lies intensive agriculture which includes traditional labour intensive farming (*e.g.* South-East Asia rice paddies), and modern agriculture which includes industrial agriculture, organic farming and sustainable farming. Industrial agriculture involves large fields and/or numbers of animals, high resource inputs (pesticides, fertilizers, etc.), and a high level of mechanization. These operations achieve economies of scale and require large amounts of capital in the form of land and machinery.

The twentieth century saw changes in agricultural practice, particularly in agricultural chemistry and in mechanization. Agricultural chemistry includes the application of chemical fertilizer, chemical insecticides, and chemical fungicides, analysis of soil makeup and nutritional needs of farm animals.

Mechanization has increased farm efficiency and productivity in most regions of the world, due especially to the tractor and various "gins" (short for "engine") such as the cotton gin, semi-automatic balers and threshers and, above all, the combine. According to the National Academy of Engineering in the United States, agricultural mechanization is one of the 20 greatest engineering achievements of the 20th century. Early in the century, it took one American farmer to produce food for 2.5 people. By 1999, due to advances in agricultural technology, a single farmer could feed over 130 people.

Other recent changes in agriculture include hydroponics, plant breeding, hybridization, gene manipulation, better management of soil nutrients, and improved weed control. Genetic engineering has yielded crops which have capabilities beyond those of naturally occurring plants, such as higher yields and disease resistance. Modified seeds germinate faster, and thus can be grown on an accelerated schedule. Genetic engineering of plants has proven controversial, particularly in the case of herbicide-resistant plants.

It has been suggested that genetic engineers may some day develop transgenic plants which would allow for irrigation, drainage, conservation, sanitary engineering, and maintaining or increasing yields while requiring fewer fossil fuel derived inputs than conventional crops. Such developments would be particularly important in areas which are normally arid and rely upon constant irrigation, and on large scale farms. These possibilities are questioned by ecologists and economists concerned with unsustainable GMO practices such as terminator seeds, and a January 2008 report shows that GMO practices have failed to address sustainability issues. While there has been some research on sustainability using GMO crops, at least one hyped and promonant multi-year attempt by Monsanto has been unsuccessful, though during the same period traditional breeding techniques yielded a more sustainable variety of the same crop. Additionally, a survey by the bio-tech industry of subsistence farmers in Africa to discover what GMO research would most benefit sustainable agriculture only identified non-transgenic issues as areas needing to be addressed.

The processing, packing and marketing of agricultural products are closely related activities also influenced by science. Methods of quick-freezing and dehydration have increased the markets for many farm products.

Animals, including horses, mules, oxen, camels, llamas, alpacas, and dogs, are often used to help cultivate fields, harvest crops, wrangle other animals, and transport farm products to buyers. Animal husbandry not only refers to the breeding and raising of animals for meat or to harvest animal products (like milk, eggs, or wool) on a continual basis, but also to the breeding and care of species for work and companionship.

Airplanes, helicopters, trucks, tractors, and combines are used in Western (and, increasingly, Eastern) agriculture for seeding, spraying operations for insect and disease control, harvesting, aerial topdressing and transporting perishable products. Radio and television disseminate vital weather reports and other information such as market reports that concern farmers. Computers have become an essential tool for farm management.

In recent years, some aspects of intensive industrial agriculture have been the subject of increasing debate. The widening sphere of influence held by large seed and chemical companies, meat packers and food processors has been a source of concern both within the farming community and for the general public. Another issue is the type of feed given to some animals that can cause bovine spongiform encephalopathy in cattle. There has also been concern over the effect of intensive agriculture on the environment.

The patent protection given to companies that develop new types of seed using genetic engineering has allowed seed to be licensed to farmers in much the same way that computer software is licensed to users. This has changed the balance of power in favour of the seed companies, allowing them to dictate terms and conditions previously unheard of. The Indian activist and scientist Vandana Shiva argues that these companies are guilty of biopiracy.

Soil conservation and nutrient management have been important concerns since the 1950s, with the most advanced farmers taking a stewardship role with the land they use. However, increasing contamination of waterways and wetlands by nutrients like nitrogen and phosphorus are concerns that can only be addressed by "enlightenment" of farmers and/or far stricter law enforcement in many countries.

Increasing consumer awareness of agricultural issues has led to the rise of community-supported agriculture, local food movement, "Slow Food", and commercial organic farming.

Etymology

The word agriculture is the English adaptation of Latin *agricultûra*, from *ager*, "a field", and *cultûra*, "cultivation" in the strict sense of "tillage of the soil". Thus, a literal reading of the word yields "tillage of a field/of fields".

History

Agriculture was developed at least 10,000 years ago, and it has undergone significant developments since the time of the earliest cultivation. Evidence points to the Fertile Crescent of the Middle East as the site of the earliest planned sowing and harvesting of plants that had previously been gathered in the wild. Independent development of agriculture occurred in northern and southern China, Africa's Sahel, New Guinea and several regions of the Americas. Agricultural practices such as irrigation, crop rotation, fertilizers, and pesticides were developed long ago but have made great strides in the past century. The Haber-Bosch method for synthesizing ammonium nitrate represented a major breakthrough and allowed crop yields to overcome previous constraints. In the past century agriculture has been characterized by enhanced productivity, the substitution of labour for synthetic fertilizers and pesticides, selective breeding, mechanization, water pollution, and farm subsidies. In recent years there has been a backlash against the external environmental effects of conventional agriculture, resulting in the organic movement.

Ancient Origins

Developed independently by geographically distant populations, systematic agriculture first appeared in Southwest Asia with the bulk of domesticated neolthic crops and livestock now being traced to Turkey via DNA studies. The first grains of domesticated Turkish emmer wheat are found at Abu Hurerya dated to 13,500 BP. The only exceptions to this are barley, domesticated in two sites; in Levant, and East of the Zagros mountains in Iran. The eight so-called founder crops of agriculture appear: first emmer and einkorn wheat, then hulled barley, peas, lentils, bitter vetch, chick peas and flax. Bitter vetch and lentils along with almonds and pistachios appear in Franchthi Cave Greece simultaneously, about 9,000 BC. Neither are native to Greece, and they appear 2,000 years prior to domesticated wheat in the same location. This suggests that the cultivation of legumes and nuts preceded that of grain.

By 7000 BCE, small-scale agriculture reached Egypt. From at least 7000 BCE the Indian subcontinent saw farming of wheat and barley, as attested by archaeological excavation at Mehrgarh in Balochistan. By 6000 BCE, mid-scale farming was entrenched on the banks of the Nile. About this time, agriculture was developed independently in the Far East, with rice, rather than wheat, as the primary crop. Chinese and Indonesian farmers went on to domesticate mung, soy, azuki and taro. To complement these new sources of carbohydrates, highly organized net fishing of rivers, lakes and ocean shores in these areas brought in great volumes of essential protein. Collectively, these new methods of farming and fishing inaugurated a human population boom dwarfing all previous expansions, and is one that continues today.

By 5000 BCE, the Sumerians had developed core agricultural techniques including large scale intensive cultivation of land, mono-cropping, organized irrigation, and use of a specialized labour force, particularly along the waterway now known as the Shatt al-Arab, from its Persian Gulf delta to the confluence of the Tigris and Euphrates. Domestication of wild aurochs and mouflon into cattle and sheep, respectively, ushered in the large-scale use of animals for food/fiber and as beasts of burden. The shepherd joined the farmer as an essential provider for sedentary and semi-nomadic societies.

Maize, manioc, and arrowroot were first domesticated in the Americas as far back as 5200 BCE. The potato, tomato, pepper, squash, several varieties of bean, Canna, tobacco and several other plants were also developed in the New World, as was extensive terracing of steep hillsides in much of Andean South America.

In later years, the Greeks and Romans built on techniques pioneered by the Sumerians but made few fundamentally new advances. Southern Greeks struggled with very poor soils, yet managed to become a dominant society for years. The Romans were noted for an emphasis on the cultivation of crops for trade.

Middle Ages

During the Middle Ages, Muslim farmers in North Africa and the Near East developed and disseminated agricultural technologies including irrigation systems based on hydraulic and hydrostatic principles, the use of machines such as norias, and the use of water raising machines, dams, and reservoirs. They also wrote location-specific farming manuals, and were instrumental in the wider adoption of crops including sugar cane, rice, citrus fruit, apricots, cotton, artichokes, aubergines, and saffron. Muslims also brought lemons, oranges, cotton, almonds, figs and sub-tropical crops such as bananas to Spain.

The invention of a three field system of crop rotation during the Middle Ages, and the importation of the Chinese-invented moldboard plow, vastly improved agricultural efficiency.

Another important development towards the end of this period was the discovery and subsequent cultivation of fodder crops which allowed over-wintering of livestock.

Modern Era

After 1492, a global exchange of previously local crops and livestock breeds occurred. Key crops involved in this exchange included the tomato, maize, potato, cocoa and tobacco going from the New World to the Old, and several varieties of wheat, spices, coffee, and sugar cane going from the Old World to the New. The most important animal exportations from the Old World to the New were those of the horse and dog (dogs were already present in the pre-Columbian Americas but not in the numbers and breeds suited to farm work). Although not usually food animals, the horse (including donkeys and ponies) and dog quickly filled essential production roles on western hemisphere farms.

By the early 1800s, agricultural techniques, implements, seed stocks and cultivars had so improved that yield per land unit was many times that seen in the Middle Ages. With the rapid rise of mechanization in the late 19th and 20th centuries, particularly in the form of the tractor, farming tasks could be done with a speed and on a scale previously impossible. These advances have led to efficiencies enabling certain modern farms in the United States, Argentina, Israel, Germany, and a few other nations to output volumes of high quality produce per land unit at what may be the practical limit.

In 2005, the agricultural output of China was the largest in the world, accounting for almost one-sixth world share followed by the EU, India and the USA, according to the International Monetary Fund. Economists measure the total factor productivity of agriculture and by this measure agriculture in the United States is roughly 2.6 times more productive than it was in 1948.

Crops

Crop Statistics

Specific crops are cultivated in distinct growing regions throughout the world. In millions of metric tons, based on FAO estimates.

Table 1.1: Top Agricultural Products, by Crop Type

	(million Metric Tons) 2004 data
Cereals	2,263
Vegetables and melons	866
Roots and Tubers	715
Milk	619
Fruit	503
Meat	259
Oilcrops	133
Fish (2001 estimate)	130
Eggs	63
Pulses	60
Vegetable Fiber	30

Source: Food and Agriculture Organization (FAO)

Table 1.2: Top Agricultural Products, by Individual Crops

Sugar Cane	1,324
Maize	721
Wheat	627
Rice	605
Potatoes	328
Sugar Beet	249
Soybean	204
Oil Palm Fruit	162
Barley	154
Tomato	120

Source: Food and Agriculture Organization (FAO)

Crop Alteration

Domestication of plants has, over the centuries increased yield, improved disease resistance and drought tolerance, eased harvest and improved the taste and nutritional value of crop plants. Careful selection and breeding have had enormous effects on the characteristics of crop plants. Plant breeders use greenhouses (known as glasshouses or hothouses in some areas) and other techniques to get as many as three generations of plants per year towards the continued effort of improvement. Plant selection and breeding in the 1920s and 1930s improved pasture (grasses and clover) in New Zealand. Extensive X-ray an ultraviolet induced mutagenesis efforts (*i.e.* primitive genetic engineering) during the 1950s produced the modern commercial varieties of grains such as wheat, corn (maize) and barley.

For example, average yields of corn (maize) in the USA have increased from around 2.5 tons per hectare (t/ha) (40 bushels per acre) in 1900 to about 9.4 t/ha (150 bushels per acre) in 2001. Similarly, worldwide average wheat yields have increased from less than 1 t/ha in 1900 to more than 2.5 t/ha in 1990. South American average wheat yields are around 2 t/ha, African under 1 t/ha, Egypt and Arabia up to 3.5 to 4 t/ha with irrigation. In contrast, the average wheat yield in countries such as France is over 8 t/ha. Variation in yields are due mainly to variation in climate, genetics, and the level of intensive farming techniques (use of fertilizers, chemical pest control, growth control to avoid lodging).

After mechanical tomato-harvesters were developed in the early 1960s, agricultural scientists bred tomatoes that were more resistant to mechanical handling. These varieties have been criticized as being harder and having poor texture. More recently, genetic engineering has begun to be employed in large parts of the world to speed up the selection and breeding process. One widely used modification is a herbicide resistance gene that allows plants to tolerate exposure to glyphosate, a non-systemic (*i.e.* kills all plants) chemical used to control weeds in a crop such as oilseed rape. Normally, expensive systemic herbicides would have to be applied to kill the weeds without harming the crop. Relatively cheap and safe glyphosate may be applied to the modified crops, efficiently killing weeds without harming the resistant crop. Another modification causes the plant to produce a toxin to reduce damage from insects (c.f. Starlink). This, in contrast, requires fewer insecticides to be applied to the crop.

Aquaculture, the farming of fish, shrimp, and algae, is closely associated with agriculture.

Apiculture, the culture of bees, traditionally for honey—increasingly for crop pollination.

Livestock

The farming practices of livestock vary dramatically worldwide and between different types of animals. Livestock are generally kept in an enclosure, are fed by human-provided food and are intentionally bred, but some livestock are not enclosed, or are fed by access to natural foods, or are allowed to breed freely, or all three. Approximately 68 per cent of all agricultural land is used in the production of livestock as permanent pastures.

Environmental Impact

Agriculture may often cause environmental problems because it changes natural environments and produces harmful by-products. Some of the negative effects are:

- Loss of biodiversity
- Surplus of nitrogen and phosphorus in rivers and lakes
- Detrimental effects of herbicides, fungicides, insecticides, and other biocides
- Conversion of natural ecosystems of all types into arable land
- Consolidation of diverse biomass into a few species
- Soil erosion
- Deforestation
- Depletion of minerals in the soil Many farmers practice rotational crops to ensure that the soil is not depleted of necessary minerals.
- Particulate matter, including ammonia and ammonium off-gassing from animal waste contributing to air pollution

- Air pollution from farm equipment and transportation powered by fossil fuels, carbon dioxide release from agricultural lime, as well as industrial input (fertilizers) and output (food, fuel, fiber) manufacture/processing
- Weed Science—feral plants and animals
- Odor from agricultural waste
- Soil salination
- Water crisis

According to the United Nations, the livestock sector (primarily cows, chickens, and pigs) emerges as one of the top two or three most significant contributors to our most serious environmental problems, at every scale from local to global. Livestock production occupies 70 per cent of all land used for agriculture, or 30 per cent of the land surface of the planet. It is one of the largest sources of greenhouse gases—responsible for 18 per cent of the world's greenhouse gas emissions as measured in CO_2 equivalents. By comparison, all transportation emits 13.5 per cent of the CO_2. It produces 65 per cent of human-related nitrous oxide (which has 296 times the global warming potential of CO_2) and 37 per cent of all human-induced methane (which is 23 times as warming as CO_2). It also generates 64 per cent of the ammonia, which contributes to acid rain and acidification of ecosystems.

Biodiversity

Genetic erosion in crops and livestock biodiversity is propelled by several major factors such as variety replacement, land clearing, overexploitation of species, population pressure, environmental degradation, overgrazing, policy and changing agricultural systems.

The main factor, however, is the replacement of local varieties of domestic plants and animals by high yielding or exotic varieties or species. A large number of varieties can also often be dramatically reduced when commercial varieties (including GMOs) are introduced into traditional farming systems. Many researchers believe that the main problem related to agro-ecosystem management is the general tendency towards genetic and ecological uniformity imposed by the development of modern agriculture.

In agriculture and animal husbandry, the green revolution popularized the use of conventional hybridization to increase yield many folds by creating "high-yielding varieties". Often the handful of breeds of plants and animals hybridized originated in developed countries and were further hybridized with local varieties in the rest of the developing world to create high yield strains resistant to local climate and diseases. Hybridization of local breeds to improve performance may lead to the loss of the local breed over time and consequently the loss of specific genetic material due to dilution. When viewed across the world as a whole, the consequent loss in genetic diversity and biodiversity could place agriculture in a situation unable to adapt to sudden changes in the future.

A Genetically Modified Organism (GMO) is an organism whose genetic material has been altered using the genetic engineering techniques generally known as recombinant DNA technology. Genetic engineering has vastly expanded the repertoire of genes available to breeders to create the desired properties in future germlines. However, it is hypothesised that these new strains will replace the wild-type species beyond their original fields, and that these genes may be transferred to other plant species that were not intended.

Lunar Farming

Lunar farming is the practice of observing the moon phases to assist in making decisions on when to plant various crops. Root crops are believed to be best planted during the waning phase of the moon (when the moon is decreasing in size) while above ground crops are said to be best planted during the waxing moon (when the moon is growing in size). The reasoning behind this is that as the gravitational force of the moon affects waters of oceans and seas on the earth causing the phenomenon of ebb and tide. It also affects, on a more subtle scale, the water content of the soil and plants themselves. Therefore, plant life is believed to be affected by the phases of the moon.

Besides observing the phases of the moon to make farming decisions, the zodiac, through which the moon is passing, is also observed to make further timing decisions for various farming operations. For instance, when the moon is in Cancer it is the best time to sow seeds and start planting. While Aries is a barren sign and planting should be avoided during that time.

Policy

Agricultural policy focuses on the goals and methods of agricultural production. At the policy level, common goals of agriculture include:

- Food safety: Ensuring that the food supply is free of contamination.
- Food security: Ensuring that the food supply meets the population's needs.
- Food quality: Ensuring that the food supply is of a consistent and known quality.
- Poverty Reduction
- Conservation
- Environmental impact
- Economic stability

Agriculture and Petroleum

Since the 1940s, agriculture has dramatically increased its productivity, due largely to the use of petrochemical derived pesticides, fertilizers, and increased mechanization. This has allowed world population to grow more than double over the last 50 years. Every energy unit delivered in food grown using modern techniques requires over ten energy units to produce and deliver. The vast majority of this energy input comes from fossil fuel sources. Because of modern agriculture's current heavy reliance on petrochemicals and mechanization, there are warnings that the ever decreasing supply of oil (the dramatic nature of which is known as peak oil) will inflict major damage on the modern industrial agriculture system, and could cause large food shortages.

Oil shortages are one of several factors making organic agriculture and other sustainable farming methods necessary. This conversion is now occurring, but the reconditioning of soil to restore nutrients lost during the use of monoculture agriculture techniques made possible by petroleum-based technology will take time. Some farmers using modern organic-farming methods have reported yields as high as those available from conventional farming (but without the use of fossil-fuel-intensive artificial fertilizers or pesticides).

Farmers have also begun raising crops such as corn (maize) for non-food use in an effort to help mitigate peak oil. This has contributed to a 60 per cent rise in wheat prices recently, and has been indicated as a possible precursor to "serious social unrest in developing countries." Such situations would be exacerbated in the event of future rises in food and fuel costs, factors which have already impacted the ability of charitable donors to send food aid to starving populations.

One example of the chain reactions which could possibly be caused by peak oil issues involves the problems caused by farmers raising crops such as corn (maize) for non-food use in an effort to help mitigate peak oil. This has already lowered food production. This food vs fuel issue will be exacerbated as demand for ethanol fuel rises. Rising food and fuel costs has already limited the abilities of some charitable donors to send food aid to starving populations. In the UN, some warn that the recent 60 per cent rise in wheat prices could cause "serious social unrest in developing countries." In 2007, higher incentives for farmers to grow non-food biofuel crops combined with other factors (such as over-development of former farm lands, rising transportation costs, climate change, growing consumer demand in China and India, and population growth) to cause food shortages in Asia, the Middle East, Africa, and Mexico, as well as rising food prices around the globe. As of December 2007, 37 countries faced food crises, and 20 had imposed some sort of food-price controls. Some of these shortages resulted in food riots and even deadly stampedes.

Another major petroleum issue in agriculture is the effect of petroleum supplies will have on fertilizer production. By far the biggest fossil fuel input to agriculture is the use of natural gas as a hydrogen source for the Haber-Bosch fertilizer-creation process. Natural gas is used because it is the cheapest currently available source of hydrogen. When oil production becomes so scarce that natural gas is used as a partial stopgap replacement, and hydrogen use in transportation increases, natural gas will become much more expensive. If other sources of hydrogen are not available to replace the Haber process, in amounts sufficient to supply transportation and agricultural needs, this major source of fertilizer would either become extremely expensive or unavailable. This would either cause food shortages or dramatic rises in food prices.

Mitigation of Effects of Petroleum Shortages

One effect oil shortages could have on agriculture is a full return to organic agriculture. In light of peak oil concerns, organic methods are much more sustainable than contemporary practices because they use no petroleum-based pesticides, herbicides, or fertilizers. Some farmers using modern organic-farming methods have reported yields as high as those available from conventional farming. Organic farming may however be more labour-intensive and would require a shift of work force from urban to rural areas.

It has been suggested that rural communities might obtain fuel from the biochar and synfuel process, which uses agricultural *waste* to provide charcoal fertilizer, some fuel *and* food, instead of the normal food vs fuel debate. As the synfuel would be used on site, the process would be more efficient and may just provide enough fuel for a new organic-agriculture fusion.

It has been suggested that some transgenic plants may some day be developed which would allow for maintaining or increasing yields while requiring fewer fossil fuel derived inputs than conventional crops. The possibility of success of these programmes is questioned by ecologists and economists concerned with unsustainable GMO practices such as terminator seeds, and a January 2008 report shows that GMO practices "fail to deliver environmental, social and economic benefits." While there has been some research on sustainability using GMO crops, at least one hyped and prominent multi-year attempt by Monsanto has been unsuccessful, though during the same period traditional breeding techniques yielded a more sustainable variety of the same crop. Additionally, a survey by the bio-tech industry of subsistence farmers in Africa to discover what GMO research would most benefit sustainable agriculture only identified non-transgenic issues as areas needing to be addressed. Nonetheless, some governments in Africa continue to view investments in new transgenic technologies as an essential component of efforts to improve sustainability.

Agriculture Safety and Health

Satellite image of circular crop fields characteristic of center pivot irrigation in Haskell County, Kansas in late June 2001. Healthy, growing crops are green. Maize is growing leafy stalks, but Sorghum, which resembles maize, grows more slowly and is much smaller and therefore paler. Wheat is a brilliant gold as harvest occurs in June. Brown fields have been recently harvested and plowed under or lie fallow for the year.

United States

Agriculture ranks among the most hazardous industries. Farmers are at high risk for fatal and nonfatal injuries, work-related lung diseases, noise-induced hearing loss, skin diseases, and certain cancers associated with chemical use and prolonged sun exposure. Farming is one of the few industries in which the families (who often share the work and live on the premises) are also at risk for injuries, illness, and death.

- In an average year, 516 workers die doing farm work in the U.S. (1992-2005). Of these deaths, 101 are caused by tractor overturns.
- Every day, about 243 agricultural workers suffer lost-work-time injuries, and about 5 per cent of these result in permanent impairment.

Young Workers

Agriculture is the most dangerous industry for young workers, accounting for 42 per cent of all work-related fatalities of young workers in the U.S. between 1992 and 2000. Unlike other industries, half the young victims in agriculture were under age 15.

For young agricultural workers aged 15-17, the risk of fatal injury is four times the risk for young workers in other workplaces

Agricultural work exposes young workers to safety hazards such as machinery, confined spaces, work at elevations, and work around livestock.

- An estimated 1.26 million children and adolescents under 20 years of age resided on farms in 2004, with about 699,000 of these youth performing work on the farms. In addition to the youth who live on farms, an additional 337,000 children and adolescents were hired to work on U.S. farms in 2004.
- On average, 103 children are killed annually on farms (1990-1996). Approximately 40 percent of these deaths were work-related.
- In 2004, an estimated 27,600 children and adolescents were injured on farms; 8,100 of these injuries were due to farm work.

ROLE OF AGRICULTURAL SCIENCE

Agricultural science is a broad multidisciplinary field that encompasses the parts of exact, natural, economic and social sciences that are used in the practice and understanding of agriculture. (Veterinary science, but not animal science, is often excluded from the definition.)

Agriculture and Agricultural Science

The two terms are often confused. However, they cover different concepts:

- Agriculture is the set of activities that transform the environment for the production of animals and plants for human use. Agriculture concerns techniques, including the application of agronomic research.
- Agronomy is research and development related to studying and improving plant-based agriculture.

Agricultural sciences include research and development on:

- Production techniques (*e.g.*, irrigation management, recommended nitrogen inputs)
- Improving agricultural productivity in terms of quantity and quality (*e.g.*, selection of drought-resistant crops and animals, development of new pesticides, yield-sensing technologies, simulation models of crop growth, in-vitro cell culture techniques)
- Transformation of primary products into end-consumer products (*e.g.*, production, preservation, and packaging of dairy products)
- Prevention and correction of adverse environmental effects (*e.g.*, soil degradation, waste management, bioremediation)
- Theoretical production ecology, relating to crop production modeling
- traditional agricultural systems, sometime termed subsistence agriculture, which feed most of the poorest people in the world. These systems are of interest as they sometimes retain a level of integration with natural ecological greater than that of industrial agriculture, which in may be more sustainable than some modern agricultural systems
- Food production and demand on a global basis, with special attention paid to the major producers, such as China, India, Brazil and the USA.

Agricultural Science: A Local Science

With the exception of theoretical agronomy, research in agronomy, more than in any other field, is strongly related to local areas. It can be considered a science of ecoregions, because it is closely linked to soil properties and climate, which are never exactly the same from one place to another. Many people think an agricultural production system relying on local weather, soil characteristics, and specific crops has to be studied locally. Others feel a need to know and understand production systems in as many areas as possible, and the human dimension of interaction with nature.

History of Agricultural Science

Agricultural science began with Mendel's genetic work, but in modern terms might be better dated from the chemical fertilizer outputs of plant physiological understanding in eighteenth century Germany. In the United States, a scientific revolution in agriculture began with the Hatch Act of 1887, which used the term "agricultural science". The Hatch Act was driven by farmers' interest in knowing the constituents of early artificial fertilizer. The Smith-Hughes Act of 1917 shifted agricultural education back to its vocational roots, but the scientific foundation had been built. After 1906, public expenditures on agricultural research in the US exceeded private expenditures for the next 44 years.

Intensification of agriculture since the 1960s in developed and developing countries, often referred to as the Green Revolution, was closely tied to progress made in selecting and improving crops and animals for high productivity, as well as to developing additional inputs such as artificial fertilizers and phytosanitary products.

As the oldest and largest human intervention in nature, the environmental impact of agriculture in general and more recently intensive agriculture, industrial development, and population growth have raised many questions among agricultural scientists and have led to the development and emergence of new fields. These include technological fields that assume the solution to technological problems lies in better technology, such as integrated pest management, waste treatment technologies, landscape architecture, genomics, and agricultural philosophy fields that include references to food production as something essentially different from non-essential economic 'goods'. In fact, the interaction between these two approaches provide a fertile field for deeper understanding in agricultural science.

New technologies, such as biotechnology and computer science (for data processing and storage), and technological advances have made it possible to develop new research fields, including genetic engineering, agrophysics, improved statistical analysis, and precision farming. Balancing these, as above, are the natural and human sciences of agricultural science that seek to understand the human-nature interactions of traditional agriculture, including interaction of religion and agriculture, and the non-material components of agricultural production systems.

Prominent Agricultural Scientists

- Robert Bakewell
- Norman Borlaug
- Luther Burbank
- George Washington Carver
- René Dumont
- Jay Lush
- Gregor Mendel
- Louis Pasteur
- Eli Whitney
- Sewall Wright

Agricultural Science and Agriculture Crisis

Agriculture sciences seek to feed the world's population while preventing biosafety problems that may affect human health and the environment. This requires promoting good management of natural resources and respect for the environment, and increasingly concern for the psychological wellbeing of all concerned in the food production and consumption system.

Economic, environmental, and social aspects of agriculture sciences are subjects of ongoing debate. Recent crises (such as Avian Flu, mad cow disease and issues such as the use of genetically modified organisms) illustrate the complexity and importance of this debate.

Fields or Related Disciplines

- Agricultural economics
- Agricultural engineering
- Agricultural philosophy
- Agrophysics
- Animal science
    - Animal breeding

    - Animal nutrition
- Agronomy
    - Plant science
    - Theoretical production ecology
    - Horticulture
    - Plant breeding
    - Plant fertilization
- Aquaculture
- Biological engineering
    - Genetic engineering
    - Microbiology
- Environmental science
- Food science
- Human nutrition
- Irrigation and water management
- Soil science
    - Agrology
- Waste management

2

Grains, Flours, Breads and Select Food Items: An Overview

WHOLE GRAIN

Wholegrains are cereal grains that contain bran and germ as well as the endosperm, in contrast to refined grains, which retain only the endosperm. Wholegrains can generally be sprouted while processed grains generally will not sprout. Wholemeal products are made from wholegrain flour.

Common wholegrain products include:

- Rolled oats or oat groats
- Hulled (not pearled) barley
- Popcorn
- Brown rice
- Whole wheat flour
- Sprouted grains
- Whole wheat bread

Common refined-grain products include:

- White rice
- White flour
- White bread
- Hominy
- Pasta (although wholegrain varieties of pasta are available)

Identifying Wholegrain Products

Wholegrain products can be identified by the ingredients list. Typically, if the ingredient lists "wholewheat," "wholemeal," or "whole corn" as the first ingredient, the product is a wholegrain food item. On the other hand, terms such as "enriched" and "bromated," among others, could indicate that the food lacks wholegrain.

"Wheat flour" (as opposed to "wholegrain wheat flour" or "wholewheat flour") as the first ingredient is not a clear indicator of the product's wholegrain content. If two ingredients are listed as grain products but only the second is listed as wholegrain, the entire product may contain between 1 per cent and 49 per cent whole grain. Many breads are coloured brown (often with molasses) and made to look like wholegrain, but are not. In addition, some food manufacturers make foods with whole-grain ingredients, but, because whole-grain ingredients are not the dominant ingredient, they are not wholegrain products. Contrary to popular belief, fiber is *not* indicative of wholegrains. The amount of fiber varies from grain to grain, and some products may have things like bran, peas, or other foods added to boost the fiber content.

In Canada, it is legal to advertise any food product as "wholewheat" with up to 70 per cent of the germ removed. While the resulting product will contain the benefit of fiber in the nutritional information, it lacks the more recently-discovered health benefits of antioxidants found in the wheat germ. Canadian consumers can be assured of wholegrain products by a label stating 100 per cent whole grain wholewheat.

Wholegrains are often more expensive than refined grains because their higher oil content is susceptible to rancidification, complicating processing, storage, and transport.

Similar to the distinction between whole and refined grains is that between whole pulses and refined dal.

From AACC (American Association of Cereal Chemists) definition: "Wholegrains shall consist of the intact, ground, cracked or flaked caryopsis, whose principal anatomical components—the starchy endosperm, germ and bran—are present in the same relative proportions as they exist in the intact caryopsis."

U.S. Standards of Identity

The following names indicate whole-grain products, in accordance with the federal government:

- "Wholewheat bread"
- "Wholewheat rolls"
- "Wholewheat buns"
- "Wholewheat macaroni"
- "Wholewheat spaghetti"
- "Wholewheat vermicelli"
- "Cracked wheat" (as an ingredient, not part of a name, as in "cracked wheat bread")
- "Crushed wheat"
- "Wholewheat flour"
- "Graham flour" (as an ingredient, not as part of a name as in "graham crackers")
- "Entire wheat flour"
- "Bromated wholewheat flour"
- "Whole durum flour"
- "Bulgur (cracked wheat)" (note that "bulgur" by itself may or may not indicate whole grain)

Flour, Glycemic Index and Insulin Resistance

When searching for wholegrain foods, it is important to note that any products made with flour can have the same effect on blood sugar, whether the flour is produced from whole grains or not. For example, wholegrain wheat bread and white bread can have the same glycemic index. Grinding grains into flour increases the surface area upon which enzymes work to more quickly convert starch into glucose. Keeping grains as close to their original form as possible slows or prevents the digestion of starch, and a slower digestion is responsible for preventing spikes in blood sugar (which over time may lead to insulin resistance).

Health Benefits of Whole Grain Consumption

In a 2002 study, consumption of wholegrains was found to be

associated with lower fasting insulin concentrations when compared to those associated with the consumption of refined grains. This effect as well as improved insulin sensitivity was noticeably more pronounced among obese and hyperinsulinemic subjects, suggesting whole grain consumption to be an important component of reducing risk factors for type II diabetes when only compared to the consumption of refined grains. Consumption of wholegrains has also been consistently linked with a significant decrease in risk factors for cardiovascular disease. Regular wholegrain consumption lowers LDL and triglyceride levels, which contributes to an overall 26 per cent reduction in coronary heart disease risk factors. In addition, wholegrain consumption is inversely related to hypertension, diabetes, and obesity when compared to refined grains, all of which are negative indicators in total cardiovascular health.

REFINED GRAINS

Refined grains, in contrast to whole grains, refers to grain products consisting of grains or grain flours that have been significantly modified from their natural composition. The modification process generally involves the mechanical removal of bran and germ, either through grinding or selective sifting. Further refining includes mixing, bleaching, and brominating; additionally, thiamin, riboflavin, niacin, and iron are often added back in to nutritionally enrich the product. Because the added nutrients represent a fraction of the nutrients removed, refined grains are considered nutritionally inferior to whole grains. However, the removal of fiber coupled with fine grinding results in a considerably higher availability of grain energy for use by the body. Furthermore, in the special case of maize, the process of nixtamalization (a chemical form of refinement) yields a considerable improvement in the bioavailability of niacin, thereby preventing pellagra in diets comprised largely of maize products.

FLOUR

Flour is a powder made of cereal grains or roots. It is the main ingredient of bread, which is a staple food for many civilizations,

Table 2.1: Effects of Refining and Enrichment on Wheat and Rice (%)

	Wheat			*Rice*		
	Whole	*Refined*	*Enriched*	*Whole*	*Refined*	*Enriched*
Food energy	100	107	107	100	99	99
Carbohydrates	100	105	105	100	104	104
Fiber	100	22	22	100	37	37
Protein	100	75	75	100	90	90
Thiamin (B1)	100	27	176	100	17	144
Riboflavin (B2)	100	19	230	100	53	53
Niacin (B3)	100	20	93	100	31	82
Pantothenic Acid (B5)	100	43	43	100	68	68
Pyridoxine (B6)	100	13	13	100	32	32
Folate (B9)	100	59	350	100	40	1155
Vitamin E	100	5	5	100	18	18
Calcium	100	44	44	100	100	100
Iron	100	30	120	100	54	293
Magnesium	100	16	16	100	17	17
Phosphorus	100	31	31	100	35	35
Potassium	100	26	26	100	52	52
Sodium	100	40	40	100	71	71
Zinc	100	24	24	100	29	29
Copper	100	38	38	100	79	79
Manganese	100	18	18	100	29	29
Selenium	100	48	48	100	65	65

making the availability of adequate supplies of flour a major economic and political issue at various times throughout history. Wheat flour is one of the most important foods in European and North American culture, and is the defining ingredient in most European styles of breads and pastries. Maize flour has been important in Mesoamerican cuisine since ancient times, and remains a staple in much of Latin American cuisine.

Flour contains a high proportion of starches, which are complex carbohydrates also known as polysaccharides. Leavening agents are used with some flours, especially those with significant gluten content, to produce lighter and softer baked products by embedding small gas bubbles.

The production of flour has also historically driven technological development, as attempts to make gristmills more productive and less labour-intensive led to the watermill and windmill, terms now applied more broadly to uses of water and wind power for purposes other than milling.

Etymology

The word "flour" was originally a variant of the word "flower." Both derive from the Old French *fleur* or *flour*, which had the literal meaning "blossom," and a figurative meaning "the finest." The phrase "fleur de farine'" meant "the finest part of the meal," since flour resulted from the elimination of coarse and unwanted matter from the grain during milling.

Degerminated and Heat Processed Flour

A central problem of the industrial revolution was the preservation of flour. Transportation distances and a relatively slow distribution system collided with natural shelf life. The reason for the limited shelf life is the fatty acids of the germ which react from the moment they are exposed to oxygen. This occurs when grain is milled; the fatty acids oxidize and flour starts to become rancid. Depending on climate and grain quality this process takes 6 to 9 months. In the late 19th century this period was too short for an industrial production and distribution cycle. As vitamins, micronutrients and amino acids were completely or relatively unknown in the late 19th century, taking out the germ was a brilliant solution. Without the germ flour cannot become rancid. Degerminated flour became standard. Degermination started in densely populated areas and took approximately one generation to reach the countryside. Heat-processed flour is flour where the germ is first separated from the endosperm and bran, then processed with steam, dry heat or microwave and submerged into flour again.

Types

Wheat Flour

- Amaranth flour is a flour produced from ground Amaranth grain. It was commonly used in pre-Columbian meso-American cuisine. It is becoming more and more available in specialty food shops.
- Atta flour is a whole-grain wheat flour important in Indian and Pakistani cuisine, used for a range of breads such as roti, naan and chapati.

- Bean flour is a flour produced from pulverized dried or ripe beans.
- Brown rice flour is of great importance in Southeast Asian cuisine. Also edible rice paper can be made from it. Most rice flour is made from white rice, thus is essentially a pure starch, but whole-grain brown rice flour is commercially available.
- Buckwheat flour is used as an ingredient in many pancakes in the United States. In Japan, it is used to make a popular noodle called Soba. In Russia, buckwheat flour is added to the batter for pancakes called *blinis* which are frequently eaten with caviar. Buckwheat flour is also used to make crêpes bretonnes in Brittany. On Hindu fasting days (Navaratri mainly, also Maha Shivaratri), people eat items made of buckwheat flour. The preparation varies across India. The famous ones are Kuttu Ki Puri and Kuttu Pakoras. In most of northern and western states they call this Kuttu ka atta.
- Chestnut flour is popular in Corsica, the Périgord and Lunigiana for breads, cakes and pastas. It is the original ingredient for "polenta", still used as such in Corsica and other Mediterranean locations. Chestnut bread keeps fresh for as long as two weeks. In other parts of Italy it is mainly used for desserts.
- Chickpea flour (also known as gram flour or besan) is of great importance in Indian cuisine, and in Italy, where it is used for the Ligurian farinata.
- Chuño flour made from dried potatoes in various countries of South America
- Corn (maize) flour is popular in the Southern and Southwestern US, Mexico, South America, and Punjab regions of India (In India it called as Besan) and Pakistan. Coarse whole-grain corn flour is usually called corn meal. Corn meal that has been bleached with lye is called *masa harina* and is used to make tortillas and tamales in Mexican cooking. Corn flour should never be confused with cornstarch, which is known as "cornflour" in British English. In India and Pakistan it is mixed with Spice and the paste is deep fried to make 'Pakoroas'.

- Cornstarch is just the "refined form" of Cornflour.
- Glutinous rice flour or sticky rice flour, used in east and southeast Asian cuisines for making tangyuan etc.
- Noodle flour is special blend of flour used for the making of Asian style noodles. The flour could be from wheat or rice.
- Nut flours are grated from oily nuts—most commonly almonds and hazelnuts—and are used instead of or in addition to wheat flour to produce more dry and flavourful pastries and cakes. Cakes made with nut flours are usually called tortes and most originated in Central Europe, in countries such as Hungary and Austria.
- Peasemeal or *pea flour* is a flour produced from roasted and pulverized yellow field peas.
- Peanut Flour made from shelled/cooked peanuts is a higher protein alternative to using regular flour
- Potato starch flour is obtained by grinding the tubers to a pulp and removing the fibre by water-washings. The dried product consists chiefly of starch, but also contains some protein. Potato starch flour is used as a thickening agent. When heated to boiling, food added with a suspension of potato flour in water thickens quickly. Because the flour is made from neither grain nor legume, it is used as substitute for wheat flour in cooking by Jews during Passover, when grains are not eaten. Often confused with Potato flour a peeled, cooked potato dried and grinded potato flakes. Potato flour is cold water soluble, dehydraded potato.
- Rice flour is ground kernels of rice. It is used in Western countries and especially for people who suffer from gluten intolerance, since rice does not contain gluten.
- Rye flour is used to bake the traditional sourdough breads of Poland, Germany and Scandinavia. Most rye breads use a mix of rye and wheat flours because rye does not produce gluten. Pumpernickel bread is usually made exclusively of rye, and contains a mixture of rye flour and rye meal.
- Tapioca flour, produced from the root of the cassava plant, is used to make breads, pancakes, tapioca pudding, a

savoury porridge called fufu in Africa, and is used as a starch.

- Teff flour is made from the grain teff, and is of considerable importance in eastern Africa (particularly around the horn of Africa). Notably, it is the chief ingredient in the bread injera, an important component of Ethiopian cuisine.

Flour can also be made from soy beans, peanuts, arrowroot, taro, cattails, acorns, quinoa and other non-cereal foodstuffs. In Australia, a variety of other seeds are also used to make flour for bread, these include: Acacia aneura (mulga), Acacia cowleana, Acacia estrophiolata (ironweed), Acacia ligulata (umbrella bush), Acacia murrayana (tjuntjula), Acacia tetragonophylla (wakalpulka), Acacia kempeana (Witchetty bush), Acacia coriacea (Wiry wattle), Panicum spp. (*e.g.* Panicum australiense, Panicum decompositum, Panicum effusum), Astrelba pectinata (Mitchell grass), Portulaca oleracea, Portulaca intraterranea, Oryza sativa, Marsilea drummondii (Nardoo), Atriplex nummularia (Old man saltbush), Acacia notabilis, Acacia pyrifolia, Acacia tetragonophylla, Acacia victoriae, Acacia sophorae, Acacia stenophylla, Acacia tumida, Aleurites moluccana, Amaranthus mitchellii, Amaranthus grandiflorus, Brachiaria piligera, Brachiaria milliformis, Brachychiton diversifolium, Brachychiton gregorii, Brachychiton paradoxum, Brachychiton populneum, Bruguiera rheedii, Calandrinia balonensis, Canarium australianum, Canavalia maritima, Entada phaseolides, Eragrostris eriopoda (Wangunu), Eucalyptus leptopoda, Eucalyptus microtheca, Nymphae gigantea, Rhyncharrhena linearis, Themeda australis

Flour Type Numbers

In some markets, the different available flour varieties are labeled according to the ash mass ("mineral content") that remains after a sample was incinerated in a laboratory oven (typically at 550°C or 900°C, see international standards ISO 2171 and ICC 104/1). This is an easily verified indicator for the fraction of the whole grain that ended up in the flour, because the mineral content of the starchy endosperm is much lower than that of the outer parts of the grain.

Flour made from all parts of the grain (extraction rate: 100 per cent) leaves about 2 g ash or more per 100 g dry flour. Plain white flour (extraction rate: 50-60 per cent) leaves only about 0.4 g.

- German flour type numbers (*Mehltype*) indicate the amount of ash (measured in milligrams) obtained from 100 g of the dry mass of this flour. Standard wheat flours (defined in DIN 10355) range from type 405 for normal white wheat flour for baking, to strong bread flour types 550, 650, 812, and the darker types 1050 and 1600 for wholegrain breads.
- French flour type numbers (*type de farine*) are a factor 10 smaller than those used in Germany, because they indicate the ash content (in milligrams) per 10 g flour. Type 55 is the standard, hard-wheat white flour for baking, including puff pastries ("pâte feuilletée"). Type 45 is often called pastry flour, but is generally from a softer wheat. Types 65, 80, and 110 are strong bread flours of increasing darkness, and type 150 is a wholemeal flour.

In the United States and the United Kingdom, no numbered standardized flour types are defined, and the ash mass is only rarely given on the label by flour manufacturers. However, the legally required standard nutrition label specifies the protein content of the flour, which is also a suitable way for comparing the extraction rates of different available flour types.

It is possible to find out ash content from some US manufacturers. However, US measurements are based on wheat with a 14 per cent moisture content. Thus, a US flour with.48 ash would approximate a French Type 55. For US bakers of French pastry seeking an equivalent, for example, they could look at tables published by King Arthur Flour, showing their all-purpose flour is a close equivalent to French Type 55.

In general, as the extraction rate of the flour increases, so do both the protein and the ash content. However, as the extraction rate approaches 100 per cent (whole meal), the protein content drops slightly, while the ash content continues to rise.

The following table shows some typical examples of how

protein and ash content relate to each other in wheat flour:

Ash (%)	*Protein (%)*	*Wheat Flour Type*		
		US	German	French
~0.4	~9	Pastry Flour	405	40
~0.55	~11	All-Purpose Flour	550	55
~0.8	~14	High Gluten Flour	812	80
~1	~15	First Clear Flour	1050	110
>1.5	~13	White Whole Wheat	1600	150

This table is only a rough guideline for converting bread recipes. Since flour types are not standardized in many countries, the numbers may differ between manufacturers.

Production

Milling of flour is accomplished by grinding grain between stones or steel wheels. Today, "stone-ground" usually means that the grain has been ground in a mill in which a revolving stone wheel turns over a stationary stone wheel, vertically or horizontally with the grain in between. Many small appliance mills are available, both hand-cranked and electric. The mill stones frequently rub against each other resulting in small stone particles chipping off and getting into flour. The safety aspect of this has not been checked but research into the dentition of medieval skeletons indicates that this form of milling leads to excessive wear on teeth. Steel roller mills do not have this problem.

Flammability

Flour dust suspended in air is explosive, as is any mixture of a finely powdered flammable substance with air. In medieval flour mills, candles, lamps, or other sources of fire were forbidden. Some devastating and fatal explosions have occurred at flour mills, including an explosion in 1878 at the Washburn "A" Mill in Minneapolis, the largest flour mill in the United States at the time.

Products

Bread, pasta, crackers, many cakes, and many other foods are made using flour. Wheat flour is also used to make a roux as a base for gravy and sauces. White wheat flour is the traditional base for wallpaper paste. It is also the base for papier-mâché.

Cornstarch is a principal ingredient of many puddings or desserts.

WHOLE WHEAT FLOUR

Whole wheat flour is a powdery substance derived by grinding or mashing the wheat's whole grain. It is used in baking but typically added to other "white" flours to provide nutrients (especially fiber and protein), texture, and body to the finished product.

Overview

The word "whole" refers to the fact that all of the grain (bran, germ, and endosperm) is used and nothing is lost in the process of making the flour. This is in contrast to white, refined flours, which contain only the endosperm. Because the whole flour contains the remains of all of the grain, it has a textured, brownish appearance.

Benefits

Whole wheat flour is more nutritious than refined white flour. However, in a process called food fortification, some micronutrients are added back to the white flour (required by law in some jurisdictions). Fortified white wheat flour does not, however, contain the macronutrients of the wheat's bran and germ (especially fiber and protein). Whole wheat is a good source of calcium, iron, fiber, and other minerals like selenium.

Drawbacks

Whole wheat flour has a shorter shelf life than white flour, as the higher oil content leads to rancidification. It is also more expensive.

Usually, whole wheat flour is not the main ingredients of baked goods, as it adds a certain "heaviness" which prevents them from rising as well as white flours. This adds to the cost per volume of the baked item as it requires more flour to obtain the same volume, due to the fewer and smaller air pockets trapped in the raised goods. Thus, many baked goods advertised as whole wheat are not entirely whole wheat; they may contain some refined white wheat, as long as the majority of the wheat used is whole wheat.

Nevertheless, it is possible to make a high-rising, light loaf of 100 per cent whole wheat bread, so long as one increases the water content of the dough (the bran and germ in whole wheat absorb more water than plain white flour), kneads the dough for a longer period of time to develop the gluten adequately, and allows for a longer rise before shaping the dough. Some bakers let the dough rise twice before shaping. The addition of fats, such as butter or oil, and milk products (fresh milk, powdered milk, buttermilk, yogurt, etc.) can also greatly assist the rise.

White Whole Wheat

White whole wheat flour is flour produced from soft white wheat varieties. Instead of being made from red wheat, the most common type in the United States, like most whole wheat flour, white whole wheat is made from white wheat, more common in the United Kingdom. The difference is that soft white wheat has a lower gluten content as well as lacking the tannins and phenolic acid that red wheat does, causing white whole wheat to appear and taste more like refined red wheat; it is whitish in colour and does not taste bitter. Whole wheat flour has high protein and protein absorbs more water

White whole wheat has almost the same nutrient content as red whole wheat. However, soft white whole wheat has a lower gluten content and contains a lower protein content (between 9 per cent and 11 per cent) when compared with harder wheats like red (15 per cent-16 per cent protein content) or golden wheat.

ROLLED OATS

Rolled oats are traditionally oat groats that have been rolled into flat flakes under heavy rollers and then steamed and lightly toasted.

The oat, like the other cereals, has a hard, inedible outer husk that must be removed before the grain can be eaten. After the outer husk (or chaff) has been removed from the still bran-covered oat grains, the remainder is called oat groats. Oat groats are a whole grain that can be used as a breakfast cereal. Steel-cut oats are oat groats that have been chopped into smaller pieces and retain bits of the bran layer. Since the bran layer, though nutritious, makes the grains tough to chew and contains an enzyme that can cause the oats to go rancid, raw oat groats are often further steam-treated to soften them for a quicker cooking time (modern "quick oats") and to denature the enzymes for a longer shelf life.

Fig. 2.1: A Tablespoon of Rolled Oats

Rolled oats that are sold as oatmeal usually, but not always, have had the tough bran removed. They have often, but not always, been lightly baked or pressure-cooked or "processed" in some fashion. Thick-rolled oats are large whole flakes, and thin-rolled oats are smaller, fragmented flakes. Oat flakes that are simply rolled whole oats without further processing can be cooked and eaten as "old-fashioned" oatmeal, but more highly fragmented and processed rolled oats absorb water much more easily and therefore cook faster, so they are sometimes called "quick" or "instant" oatmeal. Oatmeal can be further processed into coarse powder, which, when cooked, becomes a thick broth. Finer oatmeal powder is often used as baby food. Rolled oats are also often the main ingredient in granola and muesli.

Whole oats are an excellent source of thiamine, iron, and dietary fiber. Fiber is helpful in reducing cholesterol levels in the bloodstream. Whole oats are also the only source of antioxidant compounds known as avenanthramides; these are believed to have properties which help to protect the circulatory system from arteriosclerosis. Oat products also contain beta-glucan, which may help Type 2 diabetics control their blood sugar level, and might also help stimulate the immune system to fight off bacterial infections.

Other Uses

Rolled oats are the primary form of bait used by biologists trapping terrestrial small mammals in temperate zones. They are used alone or in a mixture with peanut butter and soy sauce. A trail of dry rolled oats is set leading to the treadle of Sherman traps or moistened rolled oats is set on snap traps.

GROAT

Groats are the hulled grains of various cereals, such as oats, wheat, barley or buckwheat (which is actually a pseudocereal). Groats from oats are a good source of avenanthramote.

Fig. 2.2: Kasha-buckwheat Groats

Groats are nutritious but hard to chew, so they are often soaked and cooked. They can be the basis of kasha, a porridge-like staple meal of Eastern Europe and Eurasia. Roasted buckwheat groats are also known as kasha or *kashi*, especially in the United States.

Wheat groats, also known as bulgur (bulgar) are an essential ingredient of the middle eastern kitchen, including some salads like tabbouleh.

Groaty pudding, also known as *groaty dick*, is a traditional dish from the Black Country in England. It is made from soaked groats, leeks, onions, beef, and beef stock which is baked at 150 degrees Celsius for up to 16 hours. Groaty pudding is a traditional meal on Guy Fawkes Night.

BARLEY

Barley is a cereal grain derived from the annual grass *Hordeum vulgare*. It serves as a major animal feed crop, with smaller amounts used for malting (in beer and whisky) and in health food. It is used in soups, stews and barley bread in various countries, such as Scotland and in Africa. In 2007 ranking of cereal crops in the world, barley was fourth both in terms of quantity produced (136 million tons) and in area of cultivation (566,000 km^2).

Etymology

The *Oxford English Dictionary* records the derivation from the Old English *bærlic* "barley", although the *-lic* ending may indicate it was an adjective pertaining to the crop or plant, rather than a noun. It was first recorded around 966 CE in the compound word *bærlic-croft*. The old English word was *bære*, which was related to the Latin word *farina* "flour", and gave rise to *bærlic* meaning "of barley". It survives in the north of Scotland as *bere*, and refers to a specific strain of six-row barley grown there. The word barn, which originally meant *barley-house*, is also rooted in these words.

Biology

Barley is a member of the grass family. It is a self-pollinating, diploid species with 14 chromosomes. The wild ancestor of domesticated barley, *Hordeum vulgare* subsp. *spontaneum*, is abundant in grasslands and woodlands throughout the Fertile Crescent and is abundant in disturbed habitats, roadsides and orchards. Outside of this region the wild barley is less common and is usually found in disturbed habitats.

Domestication

Wild barley has a brittle spike; upon maturity, the spikelets separate, facilitating seed dispersal. Domesticated barley has non-shattering spikes, making it much easier to harvest the mature ears. The non-shattering condition is caused by a mutation in one of two tightly linked genes known as Bt_1 and Bt_2; many cultivars possess both mutations. The non-shattering condition is recessive, so varieties of barley that exhibit this condition are homozygous for the mutant allele.

Two Row and Six Row Barley

Spikelets are arranged in triplets which alternate along the rachis. In wild barley (and other Old World species or *Hordeum*) only the central spikelet is fertile, while the other two are reduced. This condition is retained in certain cultivars known as two-row barleys. A pair of mutations (one dominant, the other recessive) result in fertile lateral spikelets. This produces six-row barleys.. Recent genetic studies have revealed a mutation in one gene, *vrs1* is responsible for the transition from two-row to six-row barley

Two-row barley has a lower protein content than six-row barley and thus more fermentable sugar content. High protein barley is best suited for animal feed. Malting barley is usually lower protein ('low grain nitrogen', usually produced without a late fertilizer application) which shows more uniform germination, needs shorter steeping, and has less protein in the extract that can make beer cloudy. Two-row barley is traditionally used in English ale style beers. Six-row barley is common in some American lager style beers, especially when adjuncts such as corn and rice are used, whereas two-row malted summer barley is preferred for traditional German beers. Four-row is unsuitable for brewing.

Hulled and Naked Barley

Hulless or "naked" barley (*Hordeum vulgare* L. var. *nudum Hook.* f.) is a form of domesticated barley with an easier to remove hull. Naked barley is an ancient food crop, but a new industry has developed around uses of selected hulless barley in order to increase the digestible energy of the grain, especially for swine and poultry. Hulless barley has been investigated for several potential new applications as whole grain, and for its value-added products. These include bran and flour for multiple food applications.

Classification

In traditional classifications of barley these morphological differences have led to different forms of barley being classified as different species. Under these classifications two-rowed barley with shattering spikes (wild barley) is classified as *Hordeum spontaneum* K. Koch. Two-rowed barley with non-shattering spikes is classified as *H. distichum* L., six-rowed barley with non-shattering spikes as *H. vulgare* L. (or *H. hexastichum* L.), and six-rowed with shattering spikes as *H. agriocrithon* Åberg.

The fact that these differences were driven by single-gene mutations, coupled with cytological and molecular evidence, has led most recent classifications to treat these forms as a single species, *H. vulgare* L.

History

Barley was one of the first crops domesticated in the Near East, at the same time as einkorn and emmer wheat. Wild barley (*H. vulgare* ssp. *spontaneum*) ranges from North Africa and Crete in the west, to Tibet in the east. The earliest evidence of wild barley in an archaeological context comes from the Epipaleolithic at Ohalo II at the southern end of the Sea of Galilee. The remains were dated to about 17000 BCE. The earliest domesticated barley occurs at Aceramic Neolithic sites, in the Near East such as the Pre-Pottery Neolithic B layers of Tell Abu Hureyra, in Syria. Barley has been grown in the Korean Peninsula since the Early Mumun Pottery Period (c. 1500-850 BCE) along with other crops such as millet, wheat, and legumes.

In the Pulitzer Prize-winning book Guns, Germs, and Steel, Jared Diamond argues that the availability of barley, along with other domesticable crops and animals, in southwestern Eurasia significantly contributed to the broad historical patterns that human history has followed over approximately the last 13,000 years; *i.e.*, why Eurasian civilizations, as a whole, have survived and conquered others.

Fig. 2.3: Baled Barley Straw in Falcon, Colorado

Barley beer was probably the first drink developed by Neolithic humans. Barley later on was used as currency. Alongside emmer wheat, Barley was a staple cereal of ancient Egypt, where it was used to make bread and beer. The general name for barley is *jt* (hypothetically pronounced "eat"); *šma* (hypothetically pronounced "SHE-ma") refers to Upper Egyptian barley and is a symbol of Upper Egypt. The Sumerian term is *akiti*. According to Deuteronomy 8:8, barley is one of the "Seven Species" of crops that characterize the fertility of the Promised Land of Canaan, and barley has a prominent role in the Israelite sacrifices described in the Pentateuch. A religious importance extended into the Middle Ages in Europe, and saw barley's use in justice, via alphitomancy and the corsned.

In ancient Greece, the ritual significance of barley possibly dates back to the earliest stages of the Eleusinian Mysteries. The preparatory *kykeon* or mixed drink of the initiates, prepared from barley and herbs, referred in the Homeric hymnto Demeter, who's name some scholars believe meant "Barley-mother". The practice was to dry the barley groats and roast them before preparing the porridge, according to Pliny the Elder's *Natural History* (xviii.72). This produces malt that soon ferments and becomes slightly alcoholic.

Pliny also noted barley was a special food of gladiators known as *hordearii* "barley-eaters". However, by Roman times, he added that wheat had replaced barley as a staple.

Tibetan barley has been a staple food in Tibet since the fifth century A.D. It along with a cool climate that permitted storage, produced a civilization that was able to raise great armies. It is made into a flour product called *tsampa* that is still a staple in Tibet, and into hand-rolled balls. In medieval Europe, bread made from barley and rye was peasant food, while wheat products were consumed by the upper classes. Potatoes largely replaced barley in Eastern Europe in the 19th century.

Production

Barley was grown in about 100 countries worldwide in 2007. The world production in 1974 was 148,818,870 tonnes; since then, there has been a slight decline in the amount of barley produced worldwide.

Table 2.2: Top Ten Barley Producers—2007 (Million Metric Tonne)

Russia	15.7
Canada	11.8
Spain	11.7
Germany	11.0
France	9.5
Turkey	7.4
Ukraine	6.0
Australia	5.9
United Kingdom	5.1
United States	4.6
World Total	136

Source: UN Food and Agriculture Organization (FAO)

Cultivation

Barley is a widely adaptable crop. It is currently popular in temperate areas where it is grown as a summer crop and tropical areas where it is sown as a winter crop. Its germination time is anywhere from 1 to 3 days. Barley likes to grow under cool conditions but is not particularly winter hardy. Barley is more tolerant of soil salinity than wheat, which might explain the increase of barley cultivation in Mesopotamia from the 2nd millennium BC onwards. Barley is not as cold tolerant as the winter wheats (*Triticum aestivum*), fall rye (*Secale cereale*) or winter Triticale (× *Triticosecale* Wittm. ex A. Camus.), but may be sown as a winter crop in warmer areas of the world such as Australia.

Barley has a short growing season and is also relatively drought tolerant.

Plant Diseases

This plant is known or likely to be susceptible to barley mild mosaic bymovirus as well as bacterial blight. Barley can be susceptible to many diseases but plant breeders have been working hard to incorporate resistance. The devastation caused by any one disease will depend upon the susceptibility of the variety being grown and the environmental conditions during disease development.

Uses

Algicide

Barley straw, in England, is placed in mesh bags and floated in fish ponds or water gardens to help reduce algal growth without harming pond plants and animals. Barley straw has not been approved by the EPA for use as a pesticide and its effectiveness as an algaecide in ponds has produced mixed results during university testing in the US and the UK.

Animal Feed

Half of the United States' barley production is used as an animal feed. Barley is an important feed grain in many areas of the world not typically suited for maize production, especially in northern climates—for example, northern and eastern Europe. Barley is the principal feed grain in Canada, Europe, and in the northern United States. A finishing diet of barley is one of the defining characteristics of Western Canadian beef used in marketing campaigns.

Alcoholic Beverages

A large part of the remainder is used for malting, for which barley is the best suited grain. It is a key ingredient in beer and whisky production. Two-row barley is traditionally used in German and English beers. Six-row barley was traditionally used in US beers, but both varieties are in common usage now. Distilled from green beer, whisky has been made from barley in Ireland and Scotland, while other countries have utilized more diverse sources of alcohol; such as the more common corn, rye and molasses in the USA. The grain name may be applied to the alcohol if it constitutes 51 per cent or more of the ingredients. Non-alcoholic drinks such as barley water and barley tea (called mugicha in Japan), have been made by boiling barley in water. Barley wine was an alcoholic drink in the 1700's; it was prepared by boiling barley in water, the barley water then was mixed with white wine and other ingredients like borage, lemon and sugar were added. In the 1800's a different barley wine was made prepared from recipes of ancient Greek origin.

Food

Barley contains eight essential amino acids. According to a recent study, eating whole grain barley can regulate blood sugar for up to 10 hours after consumption compared to white or even whole-grain wheat, which has a similar glycemic index. Barley can also be used as a coffee substitute.

Table 2.3: Raw Barley: Nutritional Value Per 100 g (3.5 oz)

Energy 350 kcal 1470 kJ	
Carbohydrates	77.7 g
—Sugars 0.8 g	
—Dietary fiber 15.6 g	
Fat	1.2 g
Protein	9.9 g
Thiamine (Vit. B1) 0.2 mg	15 per cent
Riboflavin (Vit. B2) 0.1 mg	7 per cent
Niacin (Vit. B3) 4.6 mg	31 per cent
Pantothenic acid (B5) 0.3 mg	6 per cent
Vitamin B6 0.3 mg	23 per cent
Folate (Vit. B9) 23 μg	6 per cent
Vitamin C 0.0 mg	0 per cent
Calcium 29.0 mg	3 per cent
Iron 2.5 mg	20 per cent
Magnesium 79.0 mg	21 per cent
Phosphorus 221 mg	32 per cent
Potassium 280 mg	6 per cent
Zinc 2.1 mg	21 per cent

Percentages are relative to US recommendations for adults.

Source: USDA Nutrient database

Hulled barley (or covered barley) is eaten after removing the inedible, fibrous outer hull. Once removed, it is called dehulled barley (or pot barley or scotch barley). Considered a whole grain, dehulled barley still has its bran and germ making it a nutritious and popular health food. Pearl barley (or pearled barley) is dehulled barley which has been steam processed further to remove the bran. It may be polished, a process known as "pearling". Dehulled or pearl barley may be processed into a variety of barley products, including flour, flakes similar to oatmeal, and grits.

Barley-meal, a wholemeal barley flour which is lighter than wheatmeal but darker in colour, is used in porridge and gruel in Scotland. Barley-meal gruel is known as *Sawiq* in the Arab world. With a long history of cultivation in the Middle East, barley is used in a wide range of traditional Arabic, Kurdish, Persian, and Turkish foodstuffs including kashkak, kashk and murri. Barley soup is traditionally eaten during Ramadan in Saudi Arabia. It is also used in soups and stews in Eastern Europe. In Africa, where it is a traditional food plant, it has the potential to improve nutrition, boost food security, foster rural development and support sustainable landcare.

The six row variety *bere* is cultivated in Orkney, Shetland, Caithness and the Western Isles in the Scottish Highlands and islands. The grain is used to make *beremeal*, used locally in bread, biscuits, and the traditional beremeal bannock.

Measurement

Barley grains were used for measurement in England, there being 3 or 4 barleycorns to the inch and 4 or 5 poppy seeds to the barleycorn. The statute definition of an inch was 3 barleycorns, although by the 19th century this had been superseded by standard inch measures. This unit still persists in the shoe sizes which are used in Britain and the USA.

The barleycorn was known as *arpa* in Turkish, and the feudal system in Turkey employed the term Arpalik, or "barley-money", to refer to a second allowance made to officials to offset the costs of fodder for their horses.

Cultural Significance

In Islam, the Prophet Muhammad prescribed barley for seven diseases. These include grief, high cholesterol levels, heart disease, treatment of cancer, effects of aging, diabetes and hypertension. It was also said to soothe and calm the bowels. Avicenna in his 11th century work *The Canon of Medicine* wrote of the healing effects of barley water, soup and broth for fevers.

In English folklore, The figure of John Barleycorn in the folksong of the same name is a personification of barley, and of the alcoholic beverages made from it, beer and whiskey. In the song, John Barleycorn is represented as suffering attacks, death, and indignities that correspond to the various stages of barley cultivation, such as reaping and malting. He may be related to older pagan gods such as Mímir or Kvasir.

POPCORN

Popcorn or popping corn is a type of corn which explodes from the kernel and puffs up when heated. Corn popping was originally discovered by Native Americans, but became popular as a snack food during the United States Great Depression, especially in movie theaters.

Corn is able to pop because, unlike other grains, its kernels have a hard moisture-sealed hull and a dense starchy filling. This allows pressure to build inside the kernel until an explosive "pop" results. Some strains of corn are now cultivated specifically as popping corns.

There are many techniques for popping corn. Commercial large-scale popcorn machines were invented by Charles Cretors in the late 19th Century. Many types of small-scale home methods for popping corn also exist, with the most popular being prepackaged microwavable popcorn.

As a snack food, popcorn has both advocates and detractors. Some consider it to be a health food, while others caution against it for a variety of reasons. Popcorn can also have non-food applications, ranging from holiday decorations to packaging materials.

History

Popcorn was first discovered thousands of years ago by the Native Americans. Some tribes believed that the popping noise was that of an angry god who escaped the kernel.

During the Great Depression, popcorn was comparatively cheap at 5-10 cents a bag and became popular. Thus, while other businesses failed the popcorn business thrived and became a source of income for some struggling farmers. During World War II, sugar rations diminished candy production causing Americans to eat three times more popcorn than they had before.

At least six localities (all in the Midwestern United States) claim to be the "Popcorn Capital of the World": Valparaiso, Indiana; Van Buren, Indiana; Marion, Ohio; Ridgway, Illinois; Schaller, Iowa; and North Loup, Nebraska. According to the USDA, most of the maize used for popcorn production is specifically planted for this purpose; most is grown in Nebraska and Indiana, with increasing area in Texas.

As the result of an elementary school project, popcorn became the official state snack food of Illinois.

How Popcorn Pops

Each kernel of popcorn contains a certain amount of moisture and oil. Unlike most other grains, the outer hull of the popcorn kernel is both strong and impervious to moisture, and the starch inside consists almost entirely of a hard, dense type.

As the oil and the water are heated past the boiling point, they turn the moisture in the kernel into a superheated pressurized steam, contained within the moisture-proof hull. Under these conditions, the starch inside the kernel gelatinizes, softening and becoming pliable. The pressure continues to increase until the breaking point of the hull is reached: a pressure of about 135 psi (930 kPa) and a temperature of 180°C (356°F). The hull ruptures rapidly, causing a sudden drop in pressure inside the kernel and a corresponding rapid expansion of the steam, which expands the starch and proteins of the endosperm into airy foam. As the foam rapidly cools, the starch and protein polymers set into the familiar crispy puff.

Special varieties are grown to give improved popping yield. Some wild types will pop, but the cultivated strain is *Zea mays averta*, which is a special kind of *flint corn*.

Methods of Popping

Although small quantities can be popped in a stovetop kettle in a home kitchen, commercial sale of freshly popped popcorn employs specially designed popcorn machines, which were originally invented in Chicago, Illinois by Charles Cretors in 1885. Cretors successfully introduced his invention at the Columbian Exposition in 1893. (At this same world's fair, F.W. Rueckheim introduced the first caramel corn; his brother, Louis, slightly altered the recipe and introduced it as Cracker Jack popcorn in 1896.)

Cretors' invention introduced the first patented steam-driven popcorn machine that popped corn in oil. Previously, vendors popped corn by holding a wire basket over an open flame. At best, the result was a hot, dry, unevenly cooked confection. The Cretors' machine popped corn in a mixture of one-third clarified butter, two-thirds lard, and salt. This mixture could withstand the 450°F (232°C) temperature needed to pop corn and it did without producing much smoke. A fire under a boiler created steam that drove a small engine; that engine drove the gears, shaft, and agitator that stirred the corn and powered the attention-attracting clown—the Toasty Roasty Man. A wire connected to the top of the cooking pan allowed the operator to disengage the drive mechanism, lift the cover, and dump popped corn into the storage bin beneath. Exhaust from the steam engine was piped to a hollow pan below the corn storage bin and kept freshly popped corn uniformly warm for the first time ever.

A very different method of popcorn-making can still be seen on the streets of some Chinese cities today. The corn is poured into a large cast-iron canister that is then sealed with a heavy lid and slowly turned over a curbside fire in rotisserie fashion. When a pressure gauge on the canister reaches a certain level, it is removed from the fire, a large canvas sack is put over the lid, and the seal is released. With a huge boom, all of the popcorn explodes at once and is poured into the sack.

Individual consumers can also buy and use specialized popping

Fig. 2.4: An in-home Hot-air Popcorn Maker

appliances that typically generate no more than a gallon of popped corn per batch. Some of these appliances also accept a small volume of oil or melted butter to assist thermal transfer from a stationary heating element, but others (such as the one pictured here) are "air poppers" which rapidly circulate heated air up through the interior, keeping the unpopped kernels in motion to avoid burning and blowing the popped kernels out through the chute.

However, the great majority of popcorn sold for home consumption is now packaged for use in a microwave oven. Some of the most popular brands of microwavable popcorn today are Orville Redenbacher's, Pop Secret, Act II, Jolly Time, and private label brands.

Expansion and Yield

Popping results are sensitive to the rate at which the kernels are heated. If heated too quickly, the steam in the outer layers of the kernel can reach high pressures and rupture the hull before the starch in the center of the kernel can fully gelatinize, leading to partially popped kernels with hard centers. Heating too slowly leads to entirely unpopped kernels: the tip of the kernel, where it attached to the cob, is not entirely moisture-proof, and when heated slowly, the steam can leak out of the tip fast enough to keep the pressure from rising sufficiently to break the hull and cause the pop.

Producers and sellers of popcorn consider two major factors in evaluating the quality of popcorn: what percentage of the kernels will pop, and how much each popped kernel expands. Expansion is an important factor to both the consumer and vendor. For the consumer, larger pieces of popcorn tend to be more tender and are associated with higher quality. For the grower, distributor, and vendor, expansion is closely correlated with profit: vendors such as theaters buy popcorn by weight and sell it by volume. For both these reasons, higher-expansion popcorn fetches a higher profit per unit weight.

Popcorn will pop when freshly harvested, but not well: its high moisture content leads to poor expansion and chewy pieces of popcorn. Kernels with a high moisture content are also susceptible to mold when stored. For these reasons, popcorn growers and distributors dry the kernels until they reach the moisture level at which they expand the most. This differs by variety and conditions, but is generally in the range of 14-15 per cent moisture by weight. If the kernels are over-dried, the expansion rate will suffer and the percentage of kernels that pop at all will decline.

Two explanations exist for kernels which do not pop at proper temperatures, known in the popcorn industry as "old maids". The first is that unpopped kernels do not have enough moisture to create enough steam for an explosion. The second explanation, according to research led by Dr. Bruce Hamaker of Purdue University, is that the unpopped kernel may have a leaky hull.

Popcorn varieties are broadly categorized by the shape of the kernels, the colour of the kernels, or the shape of the popped corn. While the kernels may come in a variety of colours, the popped corn is always white as it is only the hull (or pericarp) that is coloured. "Rice" type popcorns have a long kernel pointed at both ends; "pearl" type kernels are rounded at the top. Commercial popcorn production has moved mostly to pearl types. Historically, pearl popcorns were usually yellow and rice popcorns usually white. Today both shapes are available in both colours, as well as others including black, red, and variegated. Commercial production is dominated by white and yellow.

Fig. 2.5: "Mushroom"-shaped Popcorn, Left, is less Fragile but less Tender than "Butterfly"-Shaped, Right

In popcorn jargon, a popped kernel of corn is known as a "flake". Two shapes of flakes are commercially important. "Butterfly" flakes are irregular in shape and have a number of protruding "wings". "Mushroom" flakes are largely ball-shaped, with few wings. Butterfly flakes are regarded as having better mouthfeel, with greater tenderness and less noticeable hulls. Mushroom flakes are less fragile than butterfly flakes and are therefore often used for packaged popcorn or confectionery, such as caramel corn. The kernels from a single cob of popcorn may form both butterfly and mushroom flakes; hybrids that produce 100 per cent butterfly flakes or 100 per cent mushroom flakes exist, the latter developed only as recently as 1998. Growing conditions and popping environment can also affect the butterfly-to-mushroom ratio.

As a Food

Popcorn is usually served salted or sweetened. In North America, it is traditionally served salted, although a sweetened version, generally called caramel corn, is also commonly available. In the United Kingdom, ready-made popcorn is available either salted or simply sweetened with sugar, both varieties being equally popular. Toffee (*i.e.* caramel) popcorn is also available, but tends to be more expensive. Popcorn is a popular snack at sporting events and in cinemas, where it has been served since 1912. The Boy Scouts of America sell popcorn door-to-door as a primary fundraiser, similar to Girl Scout cookies.

Nutritional Value

Air-popped popcorn is naturally high in fiber, low in calories and fat, contains no sodium, and is sugar free. This can make it an attractive snack to people with dietary restrictions on the intake of calories, fat, and/or sodium. For the sake of flavor, however, large amounts of fat, sugar, and sodium are often added to prepared popcorn, which can quickly convert it to a very *poor* choice for those on restricted diets.

One particularly notorious example of this first came to public attention in the mid-1990s, when the Center for Science in the Public Interest produced a report about "Movie Popcorn", which became the subject of a widespread publicity campaign. The movie theaters surveyed used coconut oil to pop the corn, and then topped it with butter or margarine. "A medium-size buttered popcorn", the report said, "contains more fat than a breakfast of bacon and eggs, a Big Mac and fries, and a steak dinner combined." The practice continues today. For example, according to DietFacts.com, a small popcorn from Regal Cinema Group (the largest theater chain in the United States) still contains 29g of saturated fat, as much as three Big Macs and the equivalent of a full day-and-a-half's reference daily intake.

Health Risks

Popcorn is included on the list of foods that the American Academy of Pediatrics recommends not serving to children under four, because of the risk of choking. Special "hulless" popcorn has been developed that offers an alternative for small children and for people with braces or other dental problems who may otherwise need to avoid popcorn.

Microwaveable popcorn represents a special case, since it is designed to be cooked along with its various flavoring agents. One of these common artificial-butter flavorants, diacetyl, has been implicated in causing respiratory ailments.

Other Uses

Scent: Popcorn smell has an unusually attractive quality, for human beings. This is largely because it contains high levels of the chemicals

6-Acetyl-2,3,4,5-tetrahydropyridine and 2-Acetyl-1-pyrroline, very powerful aroma compounds that are used by food and other industries to make products that either smell like popcorn (or bread, or other foods containing the compound in nature), or for other purposes.

Decoration: Popcorn, threaded onto a string, is used as a wall or Christmas tree decoration in some parts of North America, as well as on the Balkan peninsula.

Packing material: Some shipping companies have experimented with using popcorn as a biodegradable replacement for expanded polystyrene packing material. However, popcorn has numerous undesirable properties as a packing material, including attractiveness to pests, flammability, and a higher cost and greater density than expanded polystyrene. A more processed form of expanded corn foam has been developed to overcome some of these limitations.

Novelty: The world's largest popcorn ball was unveiled in October 2006 in Lake Forest, Illinois. It weighed 3,415 pounds (1550 kilograms), measured 8 feet (2.4 m) in diameter, and had a circumference of 24.6 feet (8 m).

Popcorn maker: A popcorn maker (also called popcorn machine) is a device used to pop popcorn. Commercial popcorn machines are usually found in movie theaters and carnivals, producing popcorn of the oil-popped type, which has approximately 45 per cent of its calories derived from fat. Hot air popcorn poppers appeared for home use in the late 1970s, which produced popcorn with only 5 per cent of its calories derived from fat.

In the mid-1980s microwave popcorn was introduced, and reduced the need for a separate appliance in some homes; however up to 60 per cent of the calories of this type of popcorn are derived from fat (lower fat versions are available).

Different types of popcorn makers work in different ways; some models use hot air, which avoids the use of oil at the cost of flavor, and is therefore a healthier option. Depending on personal preference, a popcorn maker that uses oil can be used as well.

Popcorn makers can be quite large, for use at movie theaters or fairs, or small home appliances.

Movie-style popcorn can be easily made at home by using a butter salt type of seasoning.

RICE

Rice is the seed of a monocot plant *Oryza sativa*, of the grass family (Poaceae). As a cereal grain, it is the most important staple food for a large part of the world's human population, especially in East, South, Southeast Asia, the Middle East, Latin America, and the West Indies. It is the grain with the second highest worldwide production, after maize ("corn")..

Since a large portion of maize crops are grown for purposes other than human consumption, rice is probably the most important grain with regards to human nutrition and caloric intake, providing more than one fifth of the calories consumed worldwide by the human species.

A traditional food plant in Africa, rice has the potential to improve nutrition, boost food security, foster rural development and support sustainable landcare.

Rice is normally grown as an annual plant, although in tropical areas it can survive as a perennial and can produce a ratoon crop for up to 20 years. The rice plant can grow to 1-1.8 m tall, occasionally more depending on the variety and soil fertility. The grass has long, slender leaves 50-100 cm long and 2-2.5 cm broad. The small wind-pollinated flowers are produced in a branched arching to pendulous inflorescence 30-50 cm long. The edible seed is a grain (caryopsis) 5-12 mm long and 2-3 mm thick.

Rice cultivation is well-suited to countries and regions with low labour costs and high rainfall, as it is very labour-intensive to cultivate and requires plenty of water for cultivation. Rice can be grown practically anywhere, even on a steep hill or mountain. Although its parent species are native to South Asia and certain parts of Africa, centuries of trade and exportation have made it commonplace in many cultures worldwide.

The traditional method for cultivating rice is flooding the fields while, or after, setting the young seedlings. This simple method requires sound planning and servicing of the water damming and channeling, but reduces the growth of less robust weed and pest plants that have no submerged growth state, and deters vermin. While with rice growing and cultivation the flooding is not mandatory, all other methods of irrigation require higher effort in weed and pest control during growth periods and a different approach for fertilizing the soil.

(The name wild rice is usually used for species of the grass genus *Zizania*, both wild and domesticated, although the term may also be used for primitive or uncultivated varieties of *Oryza*.)

Preparation as food

The seeds of the rice plant are first milled using a rice huller to remove the chaff (the outer husks of the grain). At this point in the process, the product is called brown rice. The milling may be continued, removing the 'bran', *i.e.*, the rest of the husk and the germ, thereby creating white rice. White rice, which keeps longer, lacks some important nutrients; in a limited diet which does not supplement the rice, brown rice helps to prevent the disease beriberi.

Fig. 2.6: Old Fashioned Way of Rice Polishing in Japan."*36 Views of Mount Fuji*" *Hokusai*

White rice may also be buffed with glucose or talc powder (often called *polished rice*, though this term may also refer to white rice in general), parboiled, or processed into flour. White rice may also be enriched by adding nutrients, especially those lost during the milling process. While the cheapest method of enriching involves adding a powdered blend of nutrients that will easily wash off (in the United States, rice which has been so treated requires a label warning against rinsing), more sophisticated methods apply nutrients directly to the grain, coating the grain with a water insoluble substance which is resistant to washing.

Fig. 2.7: Terraced Rice Paddy on a Hill Slope in Indonesia

In some countries parboiled rice is popular. Parboiled rice is subjected to a steaming or parboiling process while still a brown rice. This causes nutrients from the outer husk, especially thiamine, to move into the grain itself. The parboil process causes a gelatinisation of the starch in the grains. The grains become less brittle, and the colour of the milled grain changes from white to yellow. The rice is then dried, and can then be milled as usual or used as brown rice. Milled parboiled rice is nutritionally superior to standard milled rice. Parboiled rice has an additional benefit in that it does not stick to the pan during cooking, as happens when cooking regular white rice. This type of rice is eaten in parts of India and countries of West Africa are also accustomed to consuming parboiled rice.

Despite the hypothetical health risks of talc (such as stomach cancer), talc-coated rice remains the norm in some countries due to its attractive shiny appearance, but it has been banned in some, and is no longer widely used in others (such as the United States). Even where talc is not used, glucose, starch, or other coatings may be used to improve the appearance of the grains.

Rice bran, called *nukkah* in Japan, is a valuable commodity in Asia and is used for many daily needs. It is a moist, oily inner layer which is heated to produce an oil. It is also used as a pickling bed in making rice bran pickles and Takuan.

Raw rice may be ground into flour for many uses, including making many kinds of beverages such as amazake, horchata, rice milk, and sake. Rice flour does not contain gluten and is suitable for people on a gluten-free diet. Rice may also be made into various types of noodles. Raw wild or brown rice may also be consumed by raw-foodist or fruitarians if soaked and sprouted (usually 1 week to 30 days).

Processed rice seeds must be boiled or steamed before eating. Cooked rice may be further fried in oil or butter, or beaten in a tub to make mochi.

Rice is a good source of protein and a staple food in many parts of the world, but it is not a complete protein: it does not contain all of the essential amino acids in sufficient amounts for good health, and should be combined with other sources of protein, such as nuts, seeds, beans, fish, or meat.

Rice, like other cereal grains, can be puffed (or popped). This process takes advantage of the grains' water content and typically involves heating grains in a special chamber. Further puffing is sometimes accomplished by processing pre-puffed pellets in a low-pressure chamber. The ideal gas law means that either lowering the local pressure or raising the water temperature results in an increase in volume prior to water evaporation, resulting in a puffy texture. Bulk raw rice density is about 0.9 g/cm^3. It decreases to less than one-tenth that when puffed.

Cooking

There are many varieties of rice such as laweed; for many purposes the main distinction is between long-and medium-grain rice. The grains of long-grain rice (high amylose) tend to remain intact after cooking; medium-grain rice (high amylopectin) becomes more sticky. Medium-grain rice is used for sweet dishes, for *risotto* in Italy and many *arrossos*-as arros negre, etc.—in Spain.

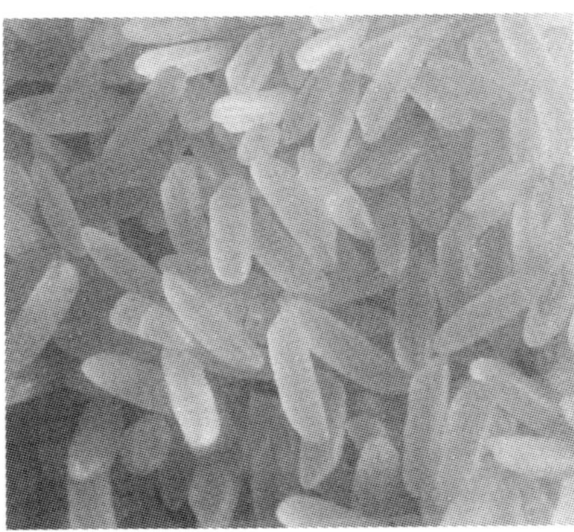

Fig. 2.8: Uncooked, Polished, White Long-grain Rice Grains

Rice is cooked by boiling or steaming, and absorbs water during cooking. It can be cooked in just as much water as it absorbs (the absorption method), or in a large quantity of water which is drained before serving (the rapid-boil method). Electric rice cookers, popular in Asia and Latin America, simplify the process of cooking rice. Rice is often heated in oil before boiling, or oil is added to the water; this is thought to make the cooked rice less sticky.

In Arab cuisine rice is an ingredient of many soups and dishes with fish, poultry, and other types of meat. It is also used to stuff vegetables or is wrapped in grape leaves. When combined with milk, sugar and honey, it is used to make desserts. In some regions, such as Tabaristan, bread is made using rice flour. Medieval Islamic texts spoke of medical uses for the plant.

Rice may also be made into rice porridge (also called congee, okayu, jook, or rice gruel) by adding more water than usual, so that the cooked rice is saturated with water to the point that it becomes very soft, expanded, and fluffy. Rice porridge is commonly eaten as a breakfast food, and is also a traditional food for the sick.

Rice may be soaked prior to cooking, which saves fuel, decreases cooking time, minimizes exposure to high temperature and thus decreases the stickiness of the rice. For some varieties, soaking improves the texture of the cooked rice by increasing expansion of the grains.

Instant rice differs from parboiled rice in that it is milled, fully cooked and then dried.

A nutritionally superior method of preparing brown rice known as GABA Rice or GBR (Germinated Brown Rice) may be used. This involves soaking washed brown rice for 20 hours in warm water (38°C or 100°F) prior to cooking it. This process stimulates germination, which activates various enzymes in the rice. By this method, a result of research carried out for the United Nations International Year of Rice, it is possible to obtain a more complete amino acid profile, including GABA.

Cooked rice can contain Bacillus cereus spores, which produce an emetic toxin when left at 4°C-60°C. When storing cooked rice for use the next day, rapid cooling is advised to reduce the risk of toxin production.

Rice flour and starch often are used in batters and breadings to increase crispiness.

Rice Growing Ecology

Rice can be grown in different ecologies, depending upon water availability.

1. Lowland, rainfed, which is drought prone, favours medium depth; waterlogged, submergence, and flood prone
2. Lowland, irrigated, grown in both the wet season and the dry season
3. Deep water or floating rice
4. Coastal Wetland
5. Upland rice, Upland rice is also known as 'Ghaiya rice', well known for its drought tolerance

History of Domestication and Cultivation

Asia

Rice has been cultivated in Asia for over 10,000 years.

Fig. 2.9: The Average Asian Rice Farmer Owns a Few Hectares: Banaue Rice Terraces, n. Luzon, Philippines

Genetics shows that rice was first domesticated in the region of the Yangtze river valley.

Africa

African rice has been cultivated for 3500 years. Between 1700 and 800 BC, 'Oryza glaberrima propagated from its original centre, the Niger River delta, and extended to Senegal. However, it never developed far from its original region. Its cultivation even declined in favour of the Asian species, possibly brought to the African continent by Arabs coming from the east coast between the 6th and 11th centuries CE.

In parts of Africa under Islam, rice was chiefly grown in southern Morocco. During the ninth century rice was also brought to east Africa by Arab traders. Although, the diffusion of rice in much sub-Saharan Africa remains uncertain, Arabs brought it to the region stretching from Lake Chad to the White Nile.

Middle East

According to Zohary and Hopf (2000, p. 91), *O. sativa* was introduced to the Middle East in Hellenistic times, and was familiar to both

Greek and Roman writers. They report that a large sample of rice grains was recovered from a grave at Susa in Iran (dated to the first century AD) at one end of the ancient world, while at the same time rice was grown in the Po valley in Italy.

In Iraq rice was grown in some areas of southern Iraq. With the rise of Islam it moved north to Nisibin, the southern shores of the Caspian Sea and then beyond the Muslim world into the valley of Volga. In Israel, rice came to be grown in the Jordan Valley. Rice is also grown in Yemen.

Europe

The Moors brought Asiatic rice to the Iberian Peninsula in the tenth century. Records indicate it was grown in Valencia and Majorca. In Majorca, rice cultivation seems to have stopped after the Christian conquest, although historians are not certain.

Muslims also brought rice to Sicily, where it was an important crop.

After the middle of the 15th century, rice spread throughout Italy and then France, later propagating to all the continents during the age of European exploration.

Caribbean and Latin America

Rice is not native to the Americas but was introduced to the Caribbean and South America by European colonizers at an early date with Spanish colonizers introducing Asian rice to Mexico in the 1520s at Veracruz and the Portuguese and their African slaves introducing it at about the same time to Colonial Brazil. Recent scholarship suggests that African slaves played an active role in the establishment of rice in the New World and that African rice was an important crop from an early period. In either case, varieties of rice and bean dishes were a staple dish along the peoples of West Africa and they remained a staple among their descendants subjected to slavery in the Spanish New World colonies and elsewhere in the Americas.

United States

In 1694, rice arrived in South Carolina, probably originating from Madagascar.

In the United States, colonial South Carolina and Georgia grew and amassed great wealth from the slave labour obtained from the Senegambia area of West Africa and from coastal Sierra Leone. At the port of Charleston, through which 40 per cent of all American slave imports passed, slaves from this region of Africa brought the highest prices, in recognition of their prior knowledge of rice culture, which was put to use on the many rice plantations around Georgetown, Charleston, and Savannah. From the slaves, plantation owners learned how to dyke the marshes and periodically flood the fields. At first the rice was milled by hand with wooden paddles, then winnowed in sweetgrass baskets (the making of which was another skill brought by the slaves). The invention of the rice mill increased profitability of the crop, and the addition of water power for the mills in 1787 by millwright Jonathan Lucas was another step forward. Rice culture in the southeastern U.S. became less profitable with the loss of slave labour after the American Civil War, and it finally died out just after the turn of the 20th century. Today, people can visit the only remaining rice plantation in South Carolina that still has the original winnowing barn and rice mill from the mid-1800s at the historic Mansfield Plantation in Georgetown, SC. The predominant strain of rice in the Carolinas was from Africa and was known as "Carolina Gold." The cultivar has been preserved and there are current attempts to reintroduce it as a commercially grown crop.

In the southern United States, rice has been grown in southern Arkansas, Louisiana, and east Texas since the mid 1800s. Many Cajun farmers grew rice in wet marshes and low lying prairies where they could also farm crayfish when the fields were flooded. In recent years rice production has risen in North America, especially in the Mississippi River Delta areas in the states of Arkansas and Mississippi.

Rice cultivation began in California during the California Gold Rush, when an estimated 40,000 Chinese labourers immigrated to the state and grew small amounts of the grain for their own consumption. However, commercial production began only in 1912 in the town of Richvale in Butte County. By 2006, California

Fig. 2.10: American Long-grain Rice Plants

produced the second largest rice crop in the United States, after Arkansas, with production concentrated in six counties north of Sacramento. Unlike the Mississippi Delta region, California's production is dominated by short—and medium-grain *japonica* varieties, including cultivars developed for the local climate such as Calrose, which makes up as much as eighty five percent of the state's crop. References to wild rice in the Americas are to the unrelated *Zizania palustris*

More than 100 varieties of rice are commercially produced primarily in six states (Arkansas, Texas, Louisiana, Mississippi, Missouri, and California) in the U.S. According to estimates for the 2006 crop year, rice production in the U.S. is valued at $1.88 billion, approximately half of which is expected to be exported. The U.S. provides about 12 per cent of world rice trade. The majority of domestic utilization of U.S. rice is direct food use (58 per cent), while 16 percent is used in processed foods and beer respectively. The remaining 10 percent is found in pet food.

Australia

Although attempts to grow rice in the well-watered north of Australia have been made for many years, they have consistently failed because of inherent iron and manganese toxicities in the soils and destruction by pests.

In the 1920s it was seen as a possible irrigation crop on soils within the Murray-Darling Basin that were too heavy for the cultivation of fruit and too infertile for wheat.

Because irrigation water, despite the extremely low runoff of temperate Australia, was (and remains) very cheap, the growing of rice was taken up by agricultural groups over the following decades. Californian varieties of rice were found suitable for the climate in the Riverina, and the first mill opened at Leeton in 1951.

Even before this Australia's rice production greatly exceeded local needs, and rice exports to Japan have become a major source of foreign currency. Above-average rainfall from the 1950s to the middle 1990s encouraged the expansion of the Riverina rice industry, but its prodigious water use in a practically waterless region began to attract the attention of environmental scientists. These became severely concerned with declining flow in the Snowy River and the lower Murray River.

Although rice growing in Australia is exceedingly efficient and highly profitable due to the cheapness of land, several recent years of severe drought have led many to call for its elimination because of its effects on extremely fragile aquatic ecosystems. The Australian rice industry is somewhat opportunistic, with the area planted varying significantly from season to season depending on water allocations in the Murray and Murrumbidgee irrigation regions.

World Production and Trade

Production and export

World production of rice has risen steadily from about 200 million tonnes of paddy rice in 1960 to 600 million tonnes in 2004. Milled rice is about 68 per cent of paddy rice by weight. In the year 2004, the top four producers were China (26 per cent of world production), India (20 per cent), Indonesia (9 per cent) and Bangladesh (5 per cent).

World trade figures are very different, as only about 5-6 per cent of rice produced is traded internationally. The largest three exporting countries are Thailand, Vietnam, and the United States.

Major importers usually include Indonesia, Bangladesh, the Philippines, Brazil and some of the Persian Gulf countries. Although China and India are the top two largest producers of rice in the world, both countries consume the majority of the rice produced domestically leaving little to be traded internationally.

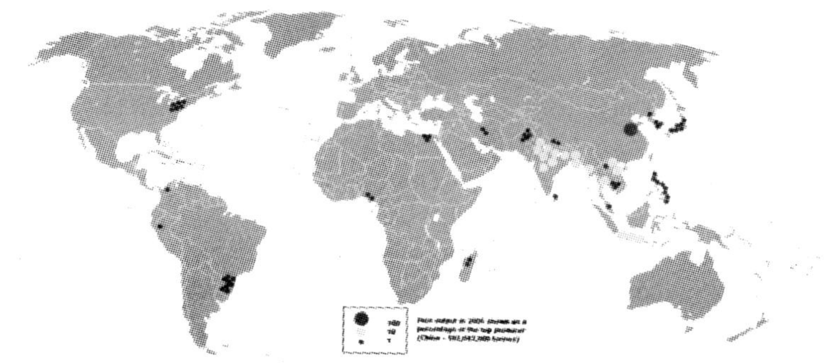

Fig. 2.11: Paddy Rice Output in 2005

Price

In March to May 2008, the price of rice rose greatly due to a rice shortage. In late April 2008, rice prices hit 24 cents a pound, twice the price that it was seven months earlier.

On the 30th of April, 2008, Thailand announced the project of the creation of the Organisation of Rice Exporting Countries (OREC) with the potential to develop into a price-fixing cartel for rice.

Worldwide Consumption

Between 1961 and 2002, per capita consumption of rice increased by 40 per cent.

Rice is the most important crop in Asia. In Cambodia, for example, 90 per cent of the total agricultural area is used for rice production.

Table 2.4: Consumption of Rice by Country—2003/2004 (Million Metric Ton)

China	900
India	125
Indonesia	37
Bangladesh	26
Brazil	24
Vietnam	18
Thailand	10
Myanmar	10
Philippines	9.7
Japan	8.7
Mexico	7.3
South Korea	5.0
United States	3.9
Egypt	3.9
Malaysia	2.7

Source: United States Department of Agriculture

U.S. rice consumption has risen sharply over the past 25 years, fueled in part by commercial applications such as beer production. Almost one in five adult Americans now report eating at least half a serving of white or brown rice per day.

Environmental Impacts

In many countries where rice is the main cereal crop, rice cultivation is responsible for most of the methane emissions. Rice requires much more water to produce than other grains.

As sea levels rise, rice will become more inclined to remain flooded for longer periods of time. Longer stays in water cuts the soil off from atmospheric oxygen and causes fermentation of organic matter in the soil. During the wet season, rice cannot hold the carbon in anaerobic conditions. The microbes in the soil convert the carbon into methane which is then released through the respiration of the rice plant or through diffusion of water. Current contributions of methane from agriculture is ~15 per cent of anthropogenic greenhouse gases, as estimated by the IPCC. A further rise in sea level of 10-85 centimeters would then stimulate the release of more methane into the air by rice plants. Methane is twenty times more effective as a greenhouse gas than carbon dioxide.

Pests and Diseases

Rice pests are any organisms or microbes with the potential to reduce the yield or value of the rice crop (or of rice seeds). (Jahn *et al.* 2007) Rice pests include weeds, pathogens, insects, rodents, and birds. A variety of factors can contribute to pest outbreaks, including the overuse of pesticides and high rates of nitrogen fertilizer application. Weather conditions also contribute to pest outbreaks. For example, rice gall midge and army worm outbreaks tend to follow periods of high rainfall early in the wet season, while thrip outbreaks are associated with drought..

One of the challenges crop protection specialists are trying to help address is the development of rice pest management techniques which are sustainable. In other words, to manage crop pests in such a manner that future crop production is not threatened. (At present, rice pest management includes cultural techniques, pest-resistant rice varieties, and pesticides (which include insecticide). Increasingly, there is evidence that farmers' pesticide applications are often unnecessary. By reducing the populations of natural enemies of rice pests, misuse of insecticides can actually lead to pest outbreaks (Cohen *et al.* 1994). Botanicals, so-called "natural pesticides", are used by some farmers in an attempt to control rice pests, but in general the practice is not common. Upland rice is grown without standing water in the field. Some upland rice farmers in Cambodia spread chopped leaves of the bitter bush (*Chromolaena odorata*) over the surface of fields after planting. This practice probably helps the soil retain moisture and thereby facilitates seed germination. Farmers also claim the leaves are a natural fertilizer and helps suppress weed and insect infestations.

Among rice cultivars there are differences in the responses to, and recovery from, pest damage. Therefore, particular cultivars are recommended for areas prone to certain pest problems. The genetically based ability of a rice variety to withstand pest attacks is called resistance. Three main types of plant resistance to pests are recognized as nonpreference, antibiosis, and tolerance.

Nonpreference (or antixenosis) describes host plants which insects prefer to avoid; antibiosis is where insect survival is reduced after the ingestion of host tissue; and tolerance is the capacity of a plant to produce high yield or retain high quality despite insect infestation. Over time, the use of pest resistant rice varieties selects for pests that are able to overcome these mechanisms of resistance. When a rice variety is no longer able to resist pest infestations, resistance is said to have broken down. Rice varieties that can be widely grown for many years in the presence of pests, and retain their ability to withstand the pests are said to have durable resistance. Mutants of popular rice varieties are regularly screened by plant breeders to discover new sources of durable resistance.

Major rice pests include the brown the rice gall midge, the rice bug hispa, the rice leaffolder, stemborer rats, and the weed *Echinochloa crusgali* Rice weevils are also known to be a threat to rice crops in the United States, Peoples Republic of China and Taiwan.

Major rice diseases include Rice Ragged Stunt, Sheath Blight and tungro. Rice blast, caused by the fungus *Magnaporthe grisea*, is the most significant disease affecting rice cultivation.

Cultivars

While most breeding of rice is carried out for crop quality and productivity, there are varieties selected for other reasons such as texture, smell and squishiness. Cultivars exist that are adapted to deep flooding, and these are generally called 'floating rice'. The largest collection of rice cultivars is at the International Rice Research Institute (IRRI) in the Philippines, with over 100,000 rice accessions held in the International Rice Genebank. Rice cultivars are often classified by their grain shapes and texture. For example, Thai Jasmine rice is long-grain and relatively less sticky, as long-grain rice contains less amylopectin than short-grain cultivars. Chinese restaurants usually serve long-grain as plain unseasoned steamed rice. Japanese mochi rice and Chinese sticky rice are short-grain. Chinese people use sticky rice which is properly known as "glutinous rice" (note:, glutinous refer to the glue-like characteristic of rice; does not refer to "gluten") to make zongzi. The Japanese table rice is a sticky, short-grain rice. Japanese sake rice is another kind as well.

Indian rice cultivars include long-grained and aromatic Basmati (grown in the North), long and medium-grained Patna rice and short-grained Sona Masoori (also spelled Sona Masuri). In the state of Tamil Nadu, the most prized cultivar is *ponni* which is primarily grown in the delta regions of Kaveri River. Kaveri is also referred to as ponni in the South and the name reflects the geographic region where it is grown. In the Western Indian state of Maharashtra, a short grain variety called Ambemohar is very popular. This rice has a characteristic fragrance of Mango blossom.

Fig. 2.12: Unpolished Long-Grain Rice Grains with Bran

Aromatic rices have definite aromas and flavours; the most noted cultivars are Thai fragrant rice, Basmati, Patna rice, and a hybrid cultivar from America sold under the trade name, Texmati. Both Basmati and Texmati have a mild popcorn-like aroma and flavour. In Indonesia there are also *red* and *black* cultivars.

High-yield cultivars of rice suitable for cultivation in Africa and other dry ecosystems called the new rice for Africa (NERICA) cultivars have been developed. It is hoped that their cultivation will improve food security in West Africa.

Draft genomes for the two most common rice cultivars, *indica* and *japonica*, were published in April 2002. Rice was chosen as a model organism for the biology of grasses because of its relatively small genome (~430 megabase pairs). Rice was the first crop with a complete genome sequence.

On December 16, 2002, the UN General Assembly declared the year 2004 the International Year of Rice. The declaration was sponsored by more than 40 countries.

Fig. 2.13: Polished Indian Sona Masuri Rice Grains

Biotechnology

High-Yielding Varieties

The High Yielding Varieties are a group of crops created intentionally during the Green Revolution to increase global food production. Rice, like corn and wheat, was genetically manipulated to increase its yield. This project enabled labour markets in Asia to shift away from agriculture, and into industrial sectors. The first "modern rice", IR8 was produced in 1966 at the International Rice Research Institute which is based in the Philippines at the University of the Philippines' Los Baños site. IR8 was created through a cross between an Indonesian variety named "Peta" and a Chinese variety named "Dee Geo Woo Gen."

With advances in molecular genetics, the mutant genes responsible for reduced height(rht), gibberellin insensitive (gai1) and slender rice (slr1) in Arabidopsis and rice were identified as cellular signaling components of gibberellic acid (a phytohormone involved in regulating stem growth via its effect on cell division) and subsequently cloned. Stem growth in the mutant background is significantly reduced leading to the dwarf phenotype. Photosynthetic investment in the stem is reduced dramatically as the shorter plants are inherently more stable mechanically.

Assimilates become redirected to grain production, amplifying in particular the effect of chemical fertilizers on commercial yield. In the presence of nitrogen fertilizers, and intensive crop management, these varieties increase their yield two to three times.

Potentials for the Future

As the UN Millennium Development project seeks to spread global economic development to Africa, the "Green Revolution" is cited as the model for economic development. With the intent of replicating the successful Asian boom in agronomic productivity, groups like the Earth Institute are doing research on African agricultural systems, hoping to increase productivity. An important way this can happen is the production of "New Rices for Africa" (NERICA). These rices, selected to tolerate the low input and harsh growing conditions of African agriculture are produced by the African Rice Center, and billed as technology from Africa, for Africa. The NERICA have appeared in *The New York Times* (October 10, 2007) and *International Herald Tribune* (October 9, 2007), trumpeted as miracle crops that will dramatically increase rice yield in Africa and enable an economic resurgence.

Golden Rice

German and Swiss researchers have engineered rice to produce Beta-carotene, with the intent that it might someday be used to treat vitamin A deficiency. Additional efforts are being made to improve the quantity and quality of other nutrients in golden rice. The addition of the carotene turns the rice gold.

Expression of Human Proteins

Ventria Bioscience has genetically modified rice to express lactoferrin, lysozyme, and human serum albumin which are proteins usually found in breast milk. These proteins have antiviral, antibacterial, and antifungal effects.

Rice containing these added proteins can be used as a component in oral rehydration solutions which are used to treat diarrheal diseases, thereby shortening their duration and reducing recurrence. Such supplements may also help reverse anemia.

Sayings

- In Andhra Pradesh, India a proverbial saying states: "*Annam Carembal Kastana Recoclt Parabrahma Swaroopam*" meaning that rice is another form of God.
- A proverbial saying in Japan states: "The farmer spends eighty-eight efforts on rice from planting to crop." This teaches the sense of mottainai and gratitude for the farmer and for rice itself.
- There is a Sri Lankan saying, 'deyyange haal kawila', meaning 'having eaten God's rice'. This is used to explain a crazy person or his actions in general with humour. The reasoning behind this is that when the rice harvest is collected, a small fraction of the best part is dedicated to the gods and that is sacred—if a person eats that, they will be afflicted with curses and lose mental stability/act crazy.
- The expression for eating food in Thai, "gin kow", literally means, "to eat rice". Vietnamese use the phrase "ăn cõm" in the same way. Likewise, the Chinese use the phrase "eat rice", "chi fan", in the common greeting "Ni chi fanle ma?" to mean literally "Have you eaten?", and by extension, "How are you doing?".
- In Hindi and Urdu the word for food itself, *khânâ*, is used as a synonym for 'cooked rice'. Similarly, the words for lentils and cooked rice, *dâl-bhât*, are synonymous with food itself.
- In the Philippines there is an expression "One grain of rice equals one bead of sweat.". This saying may be seen in the context of the high level of labour involved in the production of rice. It may be said that there is a psychological element to this expression in encouraging people to appreciate the effort that has gone into putting food on their plates and perhaps suggesting that they do not waste food. This may be seen in a similar context as Christians who say Grace before eating food.

Ashkenazi Jews consider rice a grain and refrain from eating it during Passover.

BROWN RICE

Brown rice (or "hulled rice") is unmilled or partly milled rice, a kind of whole, natural grain. It has a mild nutty flavor, is chewier and more nutritious than white rice, and becomes rancid much more quickly. Any rice, including long-grain, short-grain, or sticky rice, may be eaten as brown rice.

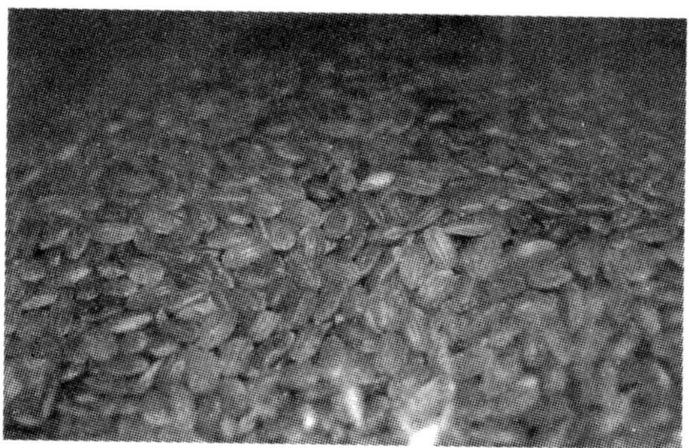

Fig. 2.14: Brown Rice

In much of Asia, brown rice is associated with poverty and wartime shortages, and in the past was rarely eaten except by the sick, the elderly and as a cure for constipation. This traditionally denigrated kind of rice is now more expensive than common white rice, partly due to its relatively low supply and difficulty of storage and transport. Today brown rice is a staple for health conscious eaters.

White Rice Comparison

Brown rice and white rice have similar amounts of calories, carbohydrates, and protein. The main differences between the two forms of rice lie in processing and nutritional content.

When only the outermost layer of a grain of rice (the husk) is removed, brown rice is produced. To produce white rice, the next layers underneath the husk (the bran layer and the germ) are removed, leaving mostly the starchy endosperm.

Several vitamins and dietary minerals are lost in this removal and the subsequent polishing process. A part of these missing nutrients, such as Vitamin B1, Vitamin B3, and iron are sometimes added back into the white rice making it "enriched", as food suppliers in the US are required to do by the Food and Drug Administration (FDA).

One mineral not added back into white rice is magnesium; one cup (195 grams) of cooked long grain brown rice contains 84 mg of magnesium while one cup of white rice contains 19 mg.

When the bran layer is removed to make white rice, the oil in the bran is also removed. Rice bran oil may help lower LDL cholesterol.

Among other key sources of nutrition lost are small amounts of fatty acids and fiber.

In addition to having greater nutritional value, brown rice is also said to be less constipating than white rice.

Cooking and Preparation

A nutritionally superior method of preparation using GABA rice or germinated brown rice (GBR), developed during the International Year of Rice, may be used. This involves soaking washed brown rice for 20 hours in warm water (38°C or 100°F) prior to cooking it. This process stimulates germination, which activates various enzymes in the rice. By this method, it is possible to obtain a more complete amino acid profile, including GABA.

Storage and Preservation

Brown rice can remain in storage for 6 months under normal conditions, but hermetic storage and freezing can significantly extend its lifetime. Freezing, even periodically, can also help control infestations of Indian meal moths.

WHITE RICE

White rice is the name given to milled rice that has had its husk, bran, and germ removed. This is done largely to prevent spoilage and to extend the storage life of the grain. After milling, the rice is polished, resulting in a seed with a bright, white, shiny appearance.

The polishing process removes important nutrients. A diet based on unenriched white rice leaves people vulnerable to the neurological disease beriberi, due to a deficiency of thiamine (vitamin B1). White rice is often enriched with some of the nutrients stripped from it during its processing. Enrichment of white rice with B1, B3, and iron is required by law in the United States.

At various times, starting in the 19th century, many have advocated brown rice or wild rice as healthier alternatives. The bran in brown rice contains significant dietary fiber and the germ contains many vitamins and minerals. This is in contrast to the traditional view of brown rice, where it was associated with poverty and famine.

BREAD

Bread is a staple food prepared by cooking a dough of flour and water and possibly more ingredients. Doughs are usually baked in the Western world (and many other countries), but in some cuisines breads are steamed, fried, or baked on a hot skillet. It may be leavened or unleavened. Salt, fat and leavening agents such as yeast and baking soda are common ingredients, though bread may contain other ingredients, such as milk, egg, sugar, spice, fruit (such as raisins), vegetables (such as onion), nuts (such as walnuts) or seeds (such as poppy seeds). Bread is one of the oldest prepared foods, dating back to the Neolithic era. The development of leavened bread can probably also be traced to prehistoric times.

Fig. 2.15: Various Leavened Breads

Fresh bread is prized for its taste, aroma, quality and texture. Retaining its freshness is important to keep it appetizing. Bread that has stiffened or dried past its prime is said to be stale. Modern bread is sometimes wrapped in paper or plastic film, or stored in a container such as a breadbox to reduce drying. Bread that is kept in warm, moist environments is prone to the growth of mold. Bread kept at low temperatures, in a refrigerator for example, will develop mold growth more slowly than bread kept at room temperature, but will turn stale quickly due to retrogradation.

The soft, inner part of bread is known to bakers and other culinary professionals as the *crumb*, which is not to be confused with small bits of bread that often fall off, called *crumbs*. The outer hard portion of bread is called the *crust*.

Etymology

The word itself, Old English *bread*, is common in various forms to many Germanic languages; such as Frisian *brea*, Dutch *brood*, German *Brot*, Swedish *bröd*, Norwegian and Danish *brød*; it has been claimed to be derived from the root of *brew*. However, it may be connected with the root of *break*, for its early uses are confined to *broken pieces*, or *bits* of bread, the Latin *crustum*, and it was not until the 12th century that it took the place—as the generic name for bread—of *hlaf* (*hlaifs* in Gothic: modern English *loaf*), which appears to be the oldest Teutonic name; Old High German *hleib* and modern German *Laib*, or Finnish *leipä*, Estonian *leib*, and Russian *õëåá* (*khleb*) are similar (all are derived from the Old German word for "loaf").

History

Bread is one of the oldest prepared foods, dating back to the Neolithic era. The first bread produced was probably cooked versions of a grain-paste, made from ground cereal grains and water, and may have been developed by accidental cooking or deliberate experimentation with water and grain flour. Descendants of this early bread are still commonly made from various grains in many parts of the world, including the Iranian (Persian) lavashs, taboons, sangaks, Mexican *tortilla*, Indian *chapatis*, *rotis* and *naans*, Scottish *oatcake*, North American *johnnycake*, Middle Eastern *pita*, and Ethiopian *injera*. Flat bread of these types also formed a staple in the diet of many early civilizations with the Sumerians eating a type of barley flat cake, and the 12th century BC Egyptians being able to purchase a flat bread called *ta* from stalls in the village streets.

The development of leavened bread is probably prehistoric. Yeast spores are ubiqitous, including the surface of cereal grains, so any dough left to rest will become naturally leavened. The earliest archaeological evidence of yeast is from ancient Egypt. Scanning electron microscopy has detected yeast cells in some ancient Egyptian loaves. However, ancient Egyptian bread was made from emmer wheat and has a dense crumb. In cases where yeast cells are not visible, it is difficult, by visual examination, to determine whether the bread was leavened. As a result, the extent to which bread was leavened in ancient Egypt remains uncertain.

There were multiple sources of leavening available for early bread. Airborne yeasts could be harnessed by leaving uncooked dough exposed to air for some time before cooking. Pliny the Elder reported that the Gauls and Iberians used the foam skimmed from beer to produce "a lighter kind of bread than other peoples." Parts of the ancient world that drank wine instead of beer used a paste composed of grape juice and flour that was allowed to begin fermenting, or wheat bran steeped in wine, as a source for yeast. The most common source of leavening, however, was to retain a piece of dough from the previous day to use as a form of sourdough starter.

Even within antiquity, there were a wide variety of breads. In the Deipnosophistae, the Greek author Athenaeus describes some of the bread, cakes, cookies, and pastries available in the Classical world. Among the breads mentioned are griddle cakes, honey-and-oil bread, mushroom shaped loaves covered in poppy seeds, and the military specialty of rolls baked on a spit. The type and quality of flours used to produce bread could also vary, as noted by Diphilus when he declared "bread made of wheat, as compared with that made of barley, is more nourishing, more digestible, and in every way superior. In order of merit, the bread made from refined [thoroughly sieved] flour comes first, after that bread from ordinary wheat, and then the unbolted, made of flour that has not been sifted."

Within medieval Europe, bread served not only as a staple food, but also as part of the table service. In the standard table setting of the day, the trencher, a piece of stale bread roughly 6 by 4 inches (15 cm × 10 cm), served as an absorbent plate. At the completion of a meal the trencher could then be eaten, given to the poor, or fed to the dogs. It was not until the fifteenth century that trenchers made of wood started to replace the bread variety.

Otto Frederick Rohwedder is considered to be the father of sliced bread. In 1912 Rohwedder started work on inventing a machine that sliced bread, but bakeries were reluctant to use it, since they were concerned the sliced bread would go stale. It was not until 1928, when Rohwedder invented a machine that both sliced and wrapped the bread, that sliced bread caught on. A bakery in Chillicothe, Missouri was the first to use this machine to produce sliced bread.

For generations, white bread was the preferred bread of the rich, while the poor ate dark (whole grain) bread. However, in most western societies, the connotations reversed in the late 20th century, with whole grain bread becoming preferred as having superior nutritional value, while white bread became associated with lower-class ignorance of nutrition.

Another major advance happened in 1961 with the development of the Chorleywood Bread Process, which used the intense mechanical working of dough to dramatically reduce the fermentation period and the time taken to produce a loaf. The process, whose high-energy mixing allows for the use of inferior grain, is now widely used around the world in large factories.

More recently, and especially in smaller retail bakeries, chemical additives are used that both speed up mixing time and reduce necessary fermentation time, so that a batch of bread may be mixed, made up, risen, and baked in less than 3 hours. Dough that does not require fermentation because of chemical additives is called "no-time bread" by commercial bakers. Common additives include reducing agents such as L-cysteine or sodium metabisulfite, and oxidants such as potassium bromate or ascorbic acid. Often these chemicals are added to dough in the form of a prepackaged base, which also contains most or all of the dough's non-flour ingredients. Using such bases and sophisticated chemistry, it has been possible for commercial bakers to make imitations of artisan and sourdough breads, traditionally made by semi-skilled labour working in smaller shops.

Recently, domestic breadmakers that automate the process of making bread have become popular in the home.

Cultural and Political Importance of Bread

As a foodstuff of great historical and contemporary importance, in many cultures in the West and Near and Middle East bread has a significance beyond mere nutrition. The Lord's Prayer, for example, contains the line 'Give us today our daily bread'; here, 'bread' is commonly understood to mean necessities in general. The word companion comes from Latin com—"with" + panis "bread". The Roman poet Juvenal satirised superficial politicians and the public as caring only for "bread and circuses". In Israel the most usual phrase in work related demonstrations is "lekhem, avoda" [bread, work], and during the 1950s, the beatnik community used the term bread as a euphemism for money. In Cockney Rhyming Slang, bread means money and is derived from the phrase bread and honey. The word bread is now commonly used around the world in English speaking countries as a synonym for money (as also is the case with the word dough.) The cultural importance of 'bread' goes beyond slang, however, to serve as a metaphor for basic necessities and living conditions in general. A 'bread-winner' is a household's main economic contributor and has little to do with actual bread-provision, for example. This also goes along with the phrase "putting bread on the table." A remarkable or revolutionary innovation is often referred to as "the greatest thing since sliced bread." In the USSR in 1917, Lenin and his fellow Bolsheviks promised "Peace, Land, and Bread." The term "breadbasket" is often used to denote an agriculturally productive region. In Slavic cultures bread and salt is offered as a welcome to all guests. In India, life's basic necessities are often referred to as "roti, kapra aur makan" [bread, cloth and house].

The political significance of bread is considerable. In Britain in the nineteenth century the inflated price of bread due to the Corn Laws caused major political and social divisions, and was central to debates over free trade and protectionism. The Assize of Bread and Ale in the thirteenth century demonstrated the importance of bread in medieval times by setting heavy punishments for short-changing bakers, and bread appeared in the *Magna Carta* a half-century earlier.

Types

Bread is a popular food in most societies. East Asian societies typically prefer wheat noodles, steamed bread, or rice. Bread is often made from a wheat-flour dough that is cultured with yeast, allowed to rise, and finally baked in an oven. Owing to its high levels of gluten (which give the dough sponginess and elasticity), common wheat (also known as bread wheat) is the most common grain used for the preparation of bread, but bread is also made from the flour of other wheat species (including durum, spelt and emmer), rye, barley, maize (or corn), and oats, usually, but not always, in combination with wheat flour. Although common wheat is best suited for making highly-risen white bread, other wheat species are capable of giving good black bread. Spelt bread (Dinkelbrot) continues to be widely consumed in Germany, and emmer bread was a staple food in ancient Egypt. Canadian bread is known for its heartier consistency due to high protein levels in Canadian flour.

- *White bread* is made from flour containing only the central core of the grain (endosperm).
- *Brown bread* is made with endosperm and 10 per cent bran. It can also refer to white bread with added colouring (often caramel colouring) to make it 'brown'; commonly labeled in America as "Wheat" bread (as opposed to "Whole Wheat" bread.)
- *Wholemeal bread* contains the whole of the wheat grain (endosperm and bran). It is also referred to as 'whole grain' or 'whole wheat' bread, especially in North America.
- *Wheat germ bread* has added wheat germ for flavoring.
- *Whole grain bread* can refer to the same as 'wholemeal bread', or to white bread with added whole grains to increase its fibre content (*i.e.* as in "60 per cent whole grain bread").

- *Roti* is a whole wheat based bread eaten in South Asia. Naan is also a variant of Roti.
- *Granary bread* is bread made from granary flour. Trademarked to Hovis, it is made from malted white or brown flour, wheat germ and whole grains.
- *Rye bread* is made with flour from rye grain of variable levels. It is higher in fiber than many common types of bread and is often darker in colour and stronger in flavor. In Scandinavia, Germany, Finland, the Baltic States, and Russia, rye is a popular type of bread.
- *Unleavened Bread* or *Matzah* used for the Jewish feast of Passover, does not include yeast, thus it does not rise.

Quick Breads

Quick breads usually refer to breads chemically leavened, usually with both baking powder and baking soda, and a balance of acidic ingredients, and alkaline ingredients. Examples include: pancakes, waffles, muffins, and Boston brown bread.

Composition and Chemistry

Formulation

The amount of flour is the most significant measurement in a bread recipe, as it affects texture and crumb the most. Professional bakers use a system of percentages known as *Bakers' Percentage* in their recipe formulations, and measure ingredients by weight instead of by volume. Measurement by weight is much more accurate and consistent than measurement by volume, especially for the dry ingredients.

Flour is always stated as 100 per cent, and the rest of the ingredients are a percent of that amount by weight. Common table bread in the U.S. uses approximately 50 per cent water, resulting in a finely-textured, light, bread. Most artisan bread formulas contain anywhere from 60 to 75 per cent water. In yeast breads, the higher water percentages result in more CO_2 bubbles, and a coarser bread crumb. One pound (~ 450 g) of flour will yield a standard loaf of bread, or two French loaves.

Calcium propionate is commonly added by commercial bakeries to retard the growth of molds.

Flour

Flour is a product made from grain that has been ground into a powdery consistency. It is flour that provides the primary structure to the final baked bread. Commonly available flours are made from rye, barley, maize, and other grains, but it is wheat flour that is most commonly used for breads. Each of these grains provides the starch and protein necessary for the production of bread.

The quantity of the proteins contained in the flour serve as the best indicator of the quality of the bread dough and the finished bread. While bread can be made from all-purpose wheat flour, for quality bread a specialty bread flour, containing more protein, is recommended. If one uses a flour with a lower (9-11 per cent) protein content to produce bread, a longer mixing time will be required to develop gluten strength properly. This extended mixing time leads to oxidization of the dough, which gives the finished product a whiter crumb, instead of the cream colour preferred by most artisan bakers.

Wheat flour in addition to its starch contains three water-soluble protein groups, albumin, globulin, proteoses, and two non-water soluble protein groups, glutenin and gliadin. When flour is mixed with water the water-soluble proteins dissolve, leaving the glutenin and gliadin to form the structure of the resulting dough. When worked by kneading, the glutenin forms strands of long thin chainlike molecules while the shorter gliadin forms bridges between the strands of glutenin. The resulting networks of strands produced by these two proteins are known as gluten. Gluten development improves if the dough is allowed to autolyse.

Liquids

Water, or some other liquid, is used to form the flour into a paste or dough. The volume of liquid required varies between recipes, but a ratio of 1 part liquid to 3 parts flour is common for yeast breads, while recipes that use steam as the primary leavening method may have a liquid content in excess of one part liquid to one part flour by volume. In addition to water, other types of liquids that may be used include dairy products, fruit juices, or beer. In addition to the water in each of these, they also contribute additional sweeteners, fats, and/or leavening components.

Leavening

Leavening is the process of adding gas to a dough before or during baking to produce a lighter, more easily chewed bread. Most bread consumed in the West is leavened. However, unleavened breads have symbolic importance in Judaism and Christianity. Jews consume unleavened bread called Matza during Passover. They are also used in the Roman Catholic Christian liturgy when they celebrate the Eucharist, a rite derived from the narrative of the Last Supper when Jesus broke bread with his disciples, perhaps during a Passover Seder. On the other hand, the Orthodox Church always uses leavened bread.

Chemical Leavening

A simple technique for leavening bread is the use of gas-producing chemicals. There are two common methods. The first is to use baking powder or a self-rising flour that includes baking powder. The second is to have an acidic ingredient such as buttermilk and add baking soda. The reaction of the acid with the soda produces gas.

Chemically-leavened breads are called *quick breads* and *soda breads*. This technique is commonly used to make muffins, pancakes, American-style biscuits and sweet breads such as banana bread.

Yeast Leavening: Many breads are leavened by yeast. The yeast used for leavening bread is *Saccharomyces cerevisiae*, the same species used for brewing alcoholic beverages. This yeast ferments carbohydrates in the flour, including any sugar, producing carbon dioxide. Most bakers in the U.S. leaven their dough with commercially produced baker's yeast. Baker's yeast has the advantage of producing uniform, quick, and reliable results, because it is obtained from a pure culture. Many artisan bakers produce their own yeast by preparing a 'growth culture' which they then use in the making of bread. This culture kept in the right conditions will continue to grow and provide leavening for many years.

Both the baker's yeast and the sourdough method of baking

bread follow the same pattern. Water is mixed with flour, salt and the leavening agent (baker's yeast or sourdough starter). Other additions (spices, herbs, fats, seeds, fruit, etc.) are not necessary to bake bread, but are often used. The mixed dough is then allowed to rise one or more times (a longer rising time results in more flavor, so bakers often *punch down* the dough and let it rise again), then loaves are formed, and (after an optional final rising time) the bread is baked in an oven.

Many breads are made from a *straight dough*, which means that all of the ingredients are combined in one step, and the dough is baked after the rising time. Alternatively, dough can be made using a *pre-ferment*, when some of the flour, water, and the leavening are combined a day or so ahead of baking, and allowed to ferment overnight. On the day of the baking, the rest of the ingredients are added, and the rest of the process is the same as that for straight dough. This produces a more flavorful bread with better texture. Many bakers see the starter method as a compromise between the highly reliable results of baker's yeast, and the flavor/complexity of a longer fermentation. It also allows the baker to use only a minimal amount of baker's yeast, which was scarce and expensive when it first became available. Most yeasted pre-ferments fall into one of three categories: *poolish* or *pouliche*, a loose-textured mixture composed of roughly equal amounts of flour and water (by weight); *biga*, a stiff mixture with a higher proportion of flour; and *pâte fermentée*, which is simply a portion of dough reserved from a previous batch. Sourdough (also known as *levain* or "natural leaven") takes it a step further, creating a pre-ferment with flour and water that propagates naturally occurring yeast and bacteria (usually *Saccharomyces exiguus*, which is more acid-tolerant than *S. cerevisiae*, and various species of *Lactobacillus*.)

Sourdough: The sour taste of sourdoughs actually comes not from the yeast, but from a lactobacillus, with which the yeast lives in symbiosis. The lactobacillus feeds on the byproducts of the yeast fermentation, and in turn makes the culture go sour by excreting lactic acid, which protects it from spoiling (since most microbes are unable to survive in an acid environment). All yeast-leavened breads used to be sourdoughs, and the leavening process was not understood until the 19th century, when with the advance of microscopes, scientists were able to discover the microbes that make the dough rise. Since then, strains of yeast have been selected and cultured mainly for reliability and quickness of fermentation. Billions of cells of these strains are then packaged and marketed as "Baker's Yeast". Bread made with baker's yeast is not sour because of the absence of the lactobacillus. Bakers around the world quickly embraced baker's yeast for it made baking simple and so allowed for more flexibility in the bakery's operations. It made baking quick as well, allowing bakeries to make fresh bread from scratch as often as three times a day. While European bakeries kept producing sourdough breads, in the U.S., sourdough baking was widely replaced by baker's yeast, and only recently has that country (or parts of it, at least) seen the rebirth of sourdough in artisan bakeries. According to Alton Brown, host of Food Network's "Good Eats" television show, each region of the world has different strains of lactobacillus, hence the flavor of the bread made from home starters is unique. The San Francisco Bay Area is especially famous for its sourdough breads.

Sourdough breads are most often made with a *sourdough starter* (not to be confused with the starter method discussed above). A sourdough starter is a culture of yeast and lactobacillus. It is essentially a dough-like or pancake-like flour/water mixture in which the yeast and lactobacilli live. A starter can be maintained indefinitely by periodically discarding a part of it and *refreshing* it by adding fresh flour and water. (When refrigerated, a starter can go weeks without needing to be *fed*.) There are starters owned by bakeries and families that are several human generations old, much revered for creating a special taste or texture. Starters can be obtained by taking a piece of another starter and growing it, or they can be made from scratch. There are hobbyist groups on the web who will send their starter for a stamped, self-addressed envelope, and there are even mailorder companies that sell different starters from all over the world. An acquired starter has the advantage to be more proven and *established* (stable and reliable, resisting spoiling and behaving predictably) than from-scratch starters.

There are other ways of sourdough baking and culture maintenance. A more traditional one is the process that was followed by peasant families throughout Europe in past centuries. The family (usually the woman was in charge of breadmaking) would bake on a fixed schedule, perhaps once a week. The starter was saved from the previous week's dough. The starter was mixed with the new ingredients, the dough was left to rise, then a piece of it was saved (to be the starter for next week's bread). The rest was formed into loaves which were marked with the family sign (this is where today's decorative *slashing* of bread loaves originates from), and taken to the communal oven to bake. These communal ovens over time evolved into what are known today as bakeries, when certain people specialized in bread baking, and with time enhanced the process so far as to be able to mass produce cheap bread for everyone in the world.

Steam Leavening

The rapid expansion of steam produced during baking leavens the bread, which is as simple as it is unpredictable. The best known steam-leavened bread is the popover. Steam-leavening is unpredictable since the steam is not produced until the bread is baked.

Steam leavening happens regardless of the rising agents (baking soda, yeast, baking powder, sour dough, beaten egg whites, etc.)

- The leavening agent either contains air bubbles or generates carbon dioxide.
- The heat vaporises the water from the inner surface of the bubbles within the dough.
- The steam expands and makes the bread rise.

It is actually the main factor in the rise of bread once it has been put in the oven. CO_2 generation, on its own, is too small to account for the rise. Heat kills bacteria or yeast at an early stage, so the CO_2 generation is stopped.

Bacterial Leavening

Salt rising bread employs a form of bacterial leavening that does not require yeast. Although the leavening action is not always consistent, and requires close attention to the incubating conditions, this bread is making a comeback due to its unique cheese-like flavor and fine texture.

Aeration

Aerated bread is leavened by carbon dioxide being forced into dough under pressure. The technique is no longer in common use, but from the mid 19th to 20th centuries bread made this way was somewhat popular in the United Kingdom, made by the Aerated Bread Company and sold in its high-street tearooms.

Fats or Shortenings

Fats such as butter, vegetable oils, lard, or that contained in eggs affects the development of gluten in breads by coating and lubricating the individual strands of protein and also helping hold the structure together. If too much fat is included in a bread dough, the lubrication effect will cause the protein structures to divide. A fat content of approximately 3 per cent by weight is the concentration that will produce the greatest leavening action. In addition to their effects on leavening, fats also serve to tenderize the breads they are used in and also help to keep the bread fresh longer after baking.

Bread Improvers

Bread improvers are frequently used in the production of commercial breads to reduce the time that the bread takes to rise, and to improve the texture and volume of bread. Chemical substances commonly used as bread improvers include ascorbic acid, hydrochloride, sodium metabisulfate, ammonium chloride, various phosphates, amylase, and protease.

Sodium/salt is one of the most common additives used in production. In addition to enhancing flavor and restricting yeast activity, salt affects the crumb and the overall texture by stabilizing and strengthening the gluten. Some artisan bakers are foregoing early addition of salt to the dough, and are waiting until after a 20 minute "rest." This is known as an autolyse , and is done with both refined and with whole grain flours.

The Serving of Bread

Bread can be served at any temperature. Once baked, it can subsequently be toasted. It is most commonly eaten with the hands, or sometimes with a knife and fork. It can be eaten by itself or as a carrier for another, usually less compact food. Bread may be dipped into a liquid (such as gravy, olive oil, or sardine pâté), topped with various spreads, both sweet and savory, or serve as the enclosure for the ubiquitous sandwich with any number of varieties of meat, cheese, vegetables or condiments inside.

Bread Crust

The bread crust is formed from exposed dough during the cooking process. It is hardened and browned from caramelisation of the sugars and proteins by the intense heat at the bread surface. The nature of a bread's crust differs depending on the type of bread and the way it is baked. Commercial bread is baked using jets that direct steam towards the bread to help produce a desirable crust. Bread crusts are sometimes considered unpalatable, especially and notoriously by young children, who commonly remove and throw away the crusts. Some parents cut off the crusts for their children, and others woo them with the promise that eating crusts will "make their hair curly." Some manufacturers, such as Sara Lee, have begun to market special crust-less breads. But for many, crusts are part of the allure of a delicious loaf of bread.

Breads Across Different Cultures

There are many variations on the basic recipe of bread, including pizza, chapatis, tortillas, baguettes, brioche, pitas, lavash, biscuits, pretzels, naan, bagels, puris, and many other variations.

- In Mexico bread is called *pan* (Spanish for bread). Although corn tortillas are the staple bread in most of Mexico, breadrolls in many varieties are an important daily food for city dwellers. Popular breads in Mexico include the bolillo roll and *pan dulce*. There are many varieties, about 1000. *Pan dulce*, which is Spanish for "sweet bread", is primarily eaten at breakfast.
- In Peru, *pan* has many variations due to the diversity of Peruvian cuisine. People usually eat *Pan de piso*, and *Pan*

serrano. There are also some kind of bread made by potatoes, currently popular in the Andes. *Bizcochos* are sweet bread usually eaten with some butter and hot chocolate. A dough made with cooked pumpkin or squash, often shaped and fried into doughnuts and served with a sweet fruity dipping sauce, is a traditional favourite.

- In Spain bread is called *pan*, and there are many varieties, about 315. There is a region called Tierra del Pan (literally translated as "Land of the Bread"), located in the province of Zamora, where economy was in the past joined to this activity.
- In the Philippines, *pandesal* (or *pan de sal;* literally, bread of salt or salt bread) is a rounded bread usually eaten by Filipinos during breakfast.
- Germany prides itself on having the largest variety of breads worldwide. More than 300-500 basic kinds of bread come together with more than 1,000 types of small bread-rolls and pastries. It has been estimated that the basic kinds of bread are so widely varied by more than 16,000 local bakeries that more than 1,000 different breads have been presented at a 2005 Cologne bread show. Germans are worldwide the biggest consumers (per capita) of bread followed by Chile.
- Finland and Russia both have dark, sourdough breads made of rye. Traditional Finnish rye bread is disc-shaped, with a hole in the center for easier storing. These breads have a rougher composition and a stronger taste than wheat bread, and can thus be stored for longer periods of time. Some families may still have leaven they use for baking these bread handed down from generation to the next.
- In Britain there is a wide variety of traditional bread, often baked in a rectangular tin.
- In the United States, tradional bread includes cornbreads and various quick breads such as biscuits. Cornbread is made from cornmeal and can differ significantly in taste and texture from region to region. The south generally prefers white cornmeal with little or no wheat flour and sweeteners added. It is traditionally baked in a cast iron skillet and has a crunchy outside, and ideally, a moist

inside. The northern region, however, usually prefers yellow cornmeal with sometimes as much as half wheat flour in its composition, as well as sugar, honey, or maple syrup. This results in a softer, sweeter bread than its southern counterpart. Homemade wheat breads are made in a rectangular tin simialar to those in the United Kingdom. Dinner rolls are another popular and traditional bread eaten with the dinner meal. They are made from wheat flour and yeast and shaped into individual rolls. Sourdough biscuits are traditional "cowboy food" in the western States. The San Francisco Bay Area is known for its crusty sourdough. In the South and Midland (chiefly), spoon bread, also called batter bread or egg bread, is made of cornmeal with or without added rice and hominy, and mixed with milk, eggs, shortening and leavening to such a consistency that it must be served from the baking dish with a spoon. Up until the twentieth century (and even later in certain regions), any flour other than cornmeal was considered a luxury. This would explain the variety of conbreads compared to that of wheat breads. Commercially, the most popular bread has been soft-textured, usually made with milk and slightly sweet, with a thin crust; this is the type that is generally sold ready-sliced in packages. It is usually eaten with the crust, but some eaters or preparers may remove the crust due to a personal preference or style of serving, as with finger sandwiches served with afternoon tea. Some of the softest bread including Wonder Bread, is referred to as "balloon bread". Though white "sandwich bread" is the most popular, Americans are trending towards more artisanal breads. Different regions of the country feature certain ethnic bread varieties including the French baguette, the Ashkenazi Jewish bagel, scali (an Italian-style bread made in New England), Native American frybread (a product of hardship, developed during the Indian resettlements of the 19th century), and Jewish rye, a bread commonly associated with delicatessen cuisine.

- In South Asia (India, Pakistan, etc.), *Roti* or *Chapati*, types of flat breads, are commonly used. A variant uses mustard flour rather than white flour. Another variant is *Puri*, a thin flat bread which is fried rather than baked and puffs up while cooked. *Paratha* is another variation on Roti. *Naan*, however, is baked in a Tandoor or clay oven and is rarely prepared at home. White and brown breads are also very common, but not as much as Roti.

- Jews have traditionally baked *challah*, a type of egg bread with a thin, hard crust and a soft, well-leavened center. It is made by wrapping plaits of dough and then lightly baking them in an oven. *Challah* is sometimes sweetened using honey and sometimes includes raisins. During Passover, unleavened bread in the form of various types of *matzo* is required due to the Biblical injunction to avoid any form of leaven during this time of year. There is some debate within the Jewish community as to whether dairy products are allowed in kosher bread; some authorities maintain that bread must be strictly *pareve* to avoid combining meat and dairy in the same meal, while others feel it is acceptable as long as a sign of some sort (usually a different loaf shape, but others are known) is baked into the loaf to distinguish it from pareve bread.
- In Morocco and West North Africa, a round bread which is roughly four inches tall is used to eat most of the Mediterranean's watery cuisine. They also have a thick and chewy fried bread which is smothered in oil before hand. The *Rghifa* bread is a staple in Morocco's food and consists of several layers of lightly cooked bread.
- In Scotland, another form of bread called plain bread is also consumed. Plain bread loaves are noticeably taller and thinner, with burned crusts at only the top and bottom of the loaf. Plain bread has a much firmer texture than English and American pan bread. Plain Bread is becoming less common as the bread consumed elsewhere in Britain is becoming more popular with consumers.
- In France, pan bread is known as *pain de mie* and is used only for toast or for making stuffing; standard bread (in the form of baguettes or thicker breads) has a thick crust

and often has large bubbles of air inside. It is often baked three times daily and is sold totally unwrapped to keep the crust crisp. Some fancy breads contain walnuts, or are encrusted with poppy seeds.

- Italy sports many different kinds of bread with its wide geographic variety and its long history of political division contributing to the development of widely different breadmaking recipes and traditions. As a rule of thumb, breadrolls are typical of the northern regions while large loaves are typical of the southern regions. Bread often has a small quantity of olive oil, butter or rendered lard mixed into the dough to make it softer and more palatable. Traditional rustic breads include Sfilatino Imbottito (a stuffed bread roll) and Pizza Bianca (a flat white bread).
- *Focaccia* is quite popular in Italy, and is known in Provence as *fougasse* or as *fouace* in parts of southern France, the former name is becoming increasingly common and fougasse can now be bought all over France. It is usually seasoned with olive oil and herbs, and often either topped with cheese or stuffed with meat or vegetables. Focaccia doughs are similar in style and texture to pizza doughs.
- *Stottie cake* is a thick, flat, round loaf. Stotties are common in the North East of England. Although it is called a cake, it is a type of bread.
- Being the simplest, cheapest and most basic type of food, bread is often referred as a metaphor for "food" in general, in some languages and dialects, such as Greek.
- Christian traditional societies (usually in poor communities), used to respect bread since Jesus symbolised his body with it. The sign of the cross was performed with the knife on the bread's surface, before the loaf was cut. Sometimes it was considered a sin to desecrate bread (*e.g.*, throw it away).
- In China, the traditional bread is *Mantou*. It is made by steaming or deep frying dough made from either wheat flour or rice. In Northern China, Mantou is often eaten as an alternative staple to rice. Steamed Mantou is similar to western white bread, but since it is not baked, it does

not have a brown outer crust. Mantou that have a filling such as meat or vegetables (Char Siu Bao, for example) are called "baozi."

- In Ireland, it is traditionally held that the end of a loaf of bread (the 'heel' or 'hilt' of the loaf) is the best part of the loaf. Other stories and myths surround this piece of the bread in Irish Mythology. Irish soda bread, developed to make the most of the soft wheat grown in Ireland, is unusual for a European bread in that it is a quick bread, using the reaction of buttermilk and baking soda rather than yeast to rise.
- There are different types of traditional 'cheese breads' in many countries, including Brazil, Colombia, Italy and Russia.
- Baker percentage
- Baker's yeast
- Bread machine
- Bread pan
- Quick bread
- Sliced bread

Culinary Uses

- Breadcrumbs
- Breading
- Sop

Types of Bread

- Beer bread
- Biscuit
- Bread roll
- brown bread
- Brioche
- Broa
- Bun
- Bush bread
- Canadian White
- Cardamom bread
- Challah

- Chapati
- Cornbread
- Cottage loaf
- Damper
- Flatbread
- Focaccia
- Indian bread
- Lavash
- Matzo
- Mantou
- Melonpan
- Naan
- Pandoro
- Paratha
- Pita
- Portuguese sweet bread
- Potato bread
- Pumpernickel
- Puri
- Roti
- Irish soda bread
- Rye bread
- Seed cakes
- Texas toast
- Tiger bread
- Tortilla
- White bread
- Zopf

WHITE BREAD

White bread is bread made from wheat flour from which the bran and often the germ have been removed, in contrast to whole wheat bread made from whole wheat flour, in which these parts are retained and contribute a brownish colour. In addition, this white flour is often bleached using potassium bromate or chlorine dioxide gas to remove any slight yellow colour and make its baking properties more predictable.

Fig. 2.16: A Loaf of White Bread

The current popularity of white bread is a response to the adaptation of the grocery business to modern commerce. Bleaching gives white flour a far longer shelf life than whole wheat flour, and bread made from it has a longer shelf life. This allows it to survive storage and long transit times.

White bread is often criticized for being less nutritious than other breads. Most of the vitamins inherent in wheat are removed along with the germ or destroyed in the bleaching process. By the time the United States joined World War One, the lack of vitamin content had led to an increase in the proportion of failures on basic entrance physical exam for draftees and military recruits. In response to this, the United States, by law, directed that white flour must be enriched with vitamins, replacing most of the major vitamins removed by bleaching—though critics claim that valuable trace minerals removed by bleaching are not replaced in the enrichment process. Counter-arguments to this claim note that the amount of trace minerals in bread is minuscule to begin with and their supply is easily substituted by other common dietary constituents such as fruits and vegetables. Most commercial white bread contains little dietary fiber when compared to bread that includes bran. A low fiber diet is linked in some instances to cases of both constipation and diarrhea. Canadian grain regulations require relatively high amounts of protein in their grain, resulting in a style of bread known as Canadian white.

White bread is also criticized for being too easily digested.

resulting in more drastic rising and falling of blood sugar and insulin levels than results from eating slower digesting whole grain breads.

White bread is made from wheat, flour, water, salt and yeast.

Other Meanings

Whitebread can also refer to alleged blandness or plainness in a person. Its usage is sim ilar to that of the term *plain vanilla*. In the DC comic book, *Justice League International*, the sarcastic Green Lantern Guy Gardner would often refer to his innocent, pure-hearted teammate Captain Marvel as "Captain Whitebread".

MAIZE

Maize (*Zea mays* L. ssp. *mays*, pronounced /'me?z/; also known in most countries as corn), is a herbaceous plant domesticated in Mesoamerica and subsequently spread throughout the American continents. After European contact with the Americas in the late 15th and early 16th centuries, maize spread to the rest of the world.

Maize is the most widely grown crop in the Americas (332 million metric tons annually in the United States alone). Hybrid maize, because of its high grain yield as a result of heterosis ('hybrid vigor'), is preferred by farmers over conventional varieties. While some maize varieties grow up to 7 metres (23 ft) tall, most commercially grown maize has been bred for a standardized height of 2.5 metres (8 ft). Sweet corn is usually shorter than field-corn varieties.

Naming Conventions

The term *maize* derives from the Spanish form of the indigenous Taino word for the plant, *maíz*. This was the term used in the United Kingdom and Ireland, where it is now it is usually called "sweet corn".

Outside the British Isles, another common term for maize is "corn". This was originally the English term for any cereal crop, but in North America its meaning has been restricted to maize since the 19th century, having been shortened from "Indian corn." *Indian corn* now refers specifically to multicolored "field corn" (flint corn) cultivars.

Fig. 2.17: Male Flower, Called the Tassel

In scientific and formal usage, "maize" is normally used in a global context. Equally, in bulk trading contexts, "corn" is used most. In the UK, Australia and other English-speaking countries, "corn" is often used in culinary contexts, particularly in naming products such as popcorn and corn flakes, but "maize" is used in agriculture and science.

In Southern Africa, maize is commonly referred to as mielie or mealie, from the Portuguese "milho". Mielie-meal is the ground form.

Physiology

Maize stems superficially resemble bamboo canes and the internodes can reach 20-30 centimetres (8-12 in). Maize has a very distinct growth form; the lower leaves being like broad flags, 50-100 centimetres long and 5-10 centimetres wide (2-4 ft by 2-4 in); the stems are erect, conventionally 2-3 metres (7-10 ft) in height, with many nodes, casting off flag-leaves at every node. Under these leaves and close to the stem grow the ears. They grow about 3 milimetres a day.

Fig. 2.18: Female Flower, Called Silk

The ears are female inflorescences, tightly covered over by several layers of leaves, and so closed-in by them to the stem that they do not show themselves easily until the emergence of the pale yellow silks from the leaf whorl at the end of the ear. The silks are elongated stigmas that look like tufts of hair, at first green, and later red or yellow. Plantings for silage are even denser, and achieve an even lower percentage of ears and more plant matter. Certain varieties of maize have been bred to produce many additional developed ears, and these are the source of the "baby corn" that is used as a vegetable in Asian cuisine.

Maize is a facultative long-night plant and flowers in a certain number of growing degree days > 50°F (10°C) in the environment to which it is adapted. The magnitude of the influence that long nights have on the number of days that must pass before maize flowers is genetically prescribed and regulated by the phytochrome system. Photoperiodicity can be eccentric in tropical cultivars, while the long days characteristic of higher latitudes allow the plants to grow so tall that they do not have enough time to produce seed before being killed by frost. These attributes, however, may prove useful in using tropical maize for biofuels.

The apex of the stem ends in the tassel, an inflorescence of male flowers. When the tassel is mature and conditions are suitably warm and dry, anthers on the tassel dehisce and release pollen. Maize pollen is *anemophilous* (dispersed by wind) and because of its large settling velocity most pollen falls within a few meters of the

Fig. 2.19: Stalks, Ears, and Silk

tassel. Each silk may become pollinated to produce one kernel of maize. Young ears can be consumed raw, with the cob and silk, but as the plant matures (usually during the summer months) the cob becomes tougher and the silk dries to inedibility. By the end of the growing season, the kernels dry out and become difficult to chew without cooking them tender first in boiling water. Modern farming techniques in developed countries usually rely on dense planting, which produces one large ear per stalk.

Seeds

The kernel of maize has a pericarp of the fruit fused with the seed coat, typical of the grasses. It is close to a multiple fruit in structure, except that the individual fruits (the kernels) never fuse into a single mass. The grains are about the size of peas, and adhere in regular rows round a white pithy substance, which forms the ear. An ear contains from 200 to 400 kernels, and is from 10-25 centimetres (4-10 inches) in length. They are of various colours: blackish, bluish-gray, purple, green, red, white and yellow. When ground into flour, maize yields more flour, with much less bran, than wheat does. However, it lacks the protein gluten of wheat and, therefore, makes baked goods with poor rising capability and coherence.

A genetic variation that accumulates more sugar and less starch in the ear is consumed as a vegetable and is called sweet corn.

Immature maize shoots accumulate a powerful antibiotic substance, DIMBOA (2,4-dihydroxy-7-methoxy-1,4-benzoxazin-3-one). DIMBOA is a member of a group of hydroxamic acids (also known as benzoxazinoids) that serve as a natural defense against a wide range of pests including insects, pathogenic fungi and bacteria. DIMBOA is also found in related grasses, particularly wheat. A maize mutant (bx) lacking DIMBOA is highly susceptible to be attacked by aphids and fungi. DIMBOA is also responsible for the relative resistance of immature maize to the European corn borer (family Crambidae). As maize matures, DIMBOA levels and resistance to the corn borer decline.

Because of its shallow roots, maize is susceptible to droughts, intolerant of nutrient-deficient soils, and prone to be uprooted by severe winds.

Allergy

Maize contains lipid transfer protein, an undigestable protein which survives cooking. This protein has been linked to a rare and understudied allergy to maize in humans. The allergic reaction can cause skin rash, swelling or itching of mucus membranes, diarrhea, vomiting, asthma and, in severe cases, anaphylaxis. It is unclear how common this allergy is in the general populace.

Genetics

Many forms of maize are used for food, sometimes classified as various subspecies:

- Flour corn—*Zea mays var. amylacea*
- Popcorn—*Zea mays var. everta*
- Dent corn—*Zea mays var. indentata*
- Flint corn—*Zea mays var. indurata*
- Sweet corn—*Zea mays var. saccharata* and *Zea mays var. rugosa*
- Waxy corn—*Zea mays var. ceratina*
- Amylomaize—*Zea mays*
- Pod corn—*Zea mays var. tunicata* Larrañaga ex A. St. Hil.
- Striped maize—*Zea mays var. japonica*

This system has been replaced (though not entirely displaced) over the last 60 years by multi-variable classifications based on ever more data. Agronomic data were supplemented by botanical traits for a robust initial classification, then genetic, cytological, protein and DNA evidence was added. Now the categories are forms (little used), races, racial complexes, and recently branches.

Fig. 2.20: Exotic Varieties of Maize are Collected to Add Genetic Diversity when Selectively Breeding New Domestic Strains

Maize has 10 chromosomes (n=10). The combined length of the chromosomes is 1500 cM. Some of the maize chromosomes have what are known as "chromosomal knobs": highly repetitive heterochromatic domains that stain darkly. Individual knobs are polymorphic among strains of both maize and teosinte. Barbara McClintock used these knob markers to prove her transposon theory of "jumping genes", for which she won the 1983 Nobel Prize in Physiology or Medicine. Maize is still an important model organism for genetics and developmental biology today.

There is a stock center of maize mutants, *The Maize Genetics Cooperation Stock Center*, funded by the USDA Agricultural Research Service and located in the Department of Crop Sciences at the University of Illinois at Urbana-Champaign. The total collection has nearly 80,000 samples. The bulk of the collection consists of several hundred named genes, plus additional gene combinations and other heritable variants. There are about 1000 chromosomal aberrations (*e.g.*, translocations and inversions) and stocks with abnormal chromosome numbers (*e.g.*, tetraploids). Genetic data describing the maize mutant stocks as well as myriad other data about maize genetics can be accessed at MaizeGDB, the Maize Genetics and Genomics Database.

In 2005, the U.S. National Science Foundation (NSF),

Department of Agriculture (USDA) and the Department of Energy (DOE) formed a consortium to sequence the maize genome. The resulting DNA sequence data will be deposited immediately into GenBank, a public repository for genome-sequence data. Sequencing the maize genome has been considered difficult because of its large size and complex genetic arrangements. The genome has 50,000-60,000 genes scattered among the 2.5 billion bases—molecules that form DNA—that make up its 10 chromosomes. (By comparison, the human genome contains about 2.9 billion bases and 26,000 genes.)

On February 26, 2008, researchers announced that they had sequenced the entire genome of maize.

Origin

There are several theories about the specific origin of maize in Mesoamerica:

1. It may be a direct domestication of a Mexican annual teosinte, *Zea mays* ssp. *parviglumis*, native to the Balsas River valley of southern Mexico, with up to 12 per cent of its genetic material obtained from *Zea mays* ssp. *mexicana* through introgression.
2. It may have been derived from hybridization between a small domesticated maize (a slightly changed form of a wild maize) and a teosinte of section *Luxuriantes*, either *Z. luxurians* or *Z. diploperennis*.
3. It may have undergone two or more domestications either of a wild maize or of a teosinte.
4. It may have evolved from a hybridization of *Z. diploperennis* by *Tripsacum dactyloides*. (The term "teosinte" describes all species and subspecies in the genus *Zea*, excluding *Zea mays* ssp. *mays*.) In the late 1930s, Paul Mangelsdorf suggested that domesticated maize was the result of a hybridization event between an unknown wild maize and a species of *Tripsacum*, a related genus. However, the proposed role of tripsacum (gama grass) in the origins of maize has been refuted by modern genetic testing, refuting Mangelsdorf's model and the fourth listed above.

The first model was proposed by Nobel Prize winner George Beadle in 1939. Though it has experimental support, it has not explained a number of problems, among them:

1. how the immense diversity of the species of sect. *Zea* originated,
2. how the tiny archaeological specimens of 3500-2700 BC could have been selected from a teosinte, and
3. how domestication could have proceeded without leaving remains of teosinte or maize with teosintoid traits until ca. 1100 BC.

The domestication of maize is of particular interest to researchers—archaeologists, geneticists, ethnobotanists, geographers, etc. The process is thought by some to have started 7,500 to 12,000 years ago. Research from the 1950s to 1970s originally focused on the hypothesis that maize domestication occurred in the highlands between Oaxaca and Jalisco, due to the fact that the oldest known examples of maize were found there. Both archaeological and botanical studies published in 2009 now point to the lowlands of the Balsas River valley, at least 8,700 years ago. The crop wild relative teosinte most similar to modern maize grows in the area of the Balsas River. Archaeological remains of early maize ears, found at Guila Naquitz Cave in the Oaxaca Valley, date back roughly 6,250 years; the oldest ears from caves near Tehuacan, Puebla, date ca. 2750 BC. Little change occurred in ear form until ca. 1100 BC when great changes appeared in ears from Mexican caves: maize diversity rapidly increased and archaeological teosinte was first deposited.

Perhaps as early as 1500 BC, maize began to spread widely and rapidly. As it was introduced to new cultures, new uses were developed and new varieties selected to better serve in those preparations. Maize was the staple food, or a major staple, of most pre-Columbian North American, Mesoamerican, South American, and Caribbean cultures. The Mesoamerican civilization was strengthened upon the field crop of maize; through harvesting it, its religious and spiritual importance and how it impacted their diet. Maize formed the Mesoamerican people's identity. During the 1st millennium AD, maize cultivation spread from Mexico into the U.S. Southwest and a millennium later into U.S. Northeast and southeastern Canada, transforming the landscape as Native Americans cleared large forest and grassland areas for the new crop.

It is unknown what precipitated its domestication, because the edible portion of the wild variety is too small and hard to obtain to be eaten directly, as each kernel is enclosed in a very hard bi-valve shell. However, George Beadle demonstrated that the kernels of teosinte are readily "popped" for human consumption, like modern popcorn. Some have argued that it would have taken too many generations of selective breeding in order to produce large compressed ears for efficient cultivation. However, studies of the hybrids readily made by intercrossing teosinte and modern maize suggest that this objection is not well founded.

In 2005, research by the USDA Forest Service indicated that the rise in maize cultivation 500 to 1,000 years ago in what is now the southeastern United States contributed to the decline of freshwater mussels, which are very sensitive to environmental changes.

Production Quantities and Methods

Maize is widely cultivated throughout the world, and a greater weight of maize is produced each year than any other grain. While the United States produces almost half of the world's harvest (~42.5 per cent), other top producing countries include China, , Mexico, Argentina, India and France. Worldwide production was around 800 million tonnes in 2007—just slightly more than rice (~650 million tonnes) or wheat (~600 million tonnes). In 2007, over 150 million hectares of maize were planted worldwide, with a yield of 4970.9 kilogram/hectare. Production can be significantly higher in certain regions of the world, 2009 forecasts for production in Iowa were 11614 kg/ha. As of 2002 the genetic yield potential of maize had "barely increased in 35 years".

Because it is cold-intolerant, in the temperate zones maize must be planted in the spring. Its root system is generally shallow, so the plant is dependent on soil moisture. As a C4 plant (a plant that uses C4 carbon fixation), maize is a considerably more water-efficient crop than C3 plants (plants that use C3 carbon fixation) like the

Fig. 2.21: Young Stalks

small grains, alfalfa and soybeans. Maize is most sensitive to drought pollination. In the United States, a good harvest was traditionally predicted if the corn was "knee-high by the Fourth of July," although modern hybrids generally exceed this growth rate. Maize used for silage is harvested while the plant is green and the fruit immature.

Table 2.5: Top Ten Maize Producers in 2007

Country	*Production (Tonnes)*	*Note*
United States	332,092,180	
People's Republic of China	151,970,000	
Brazil	51,589,721	
Mexico	22,500,000	[F]
Argentina	21,755,364	
India	16,780,000	
France	13,107,000	
Indonesia	12,381,561	
Canada	10,554,500	
Italy	9,891,362	
World	784,786,580	[A]

No symbol = official figure, P = official figure, F = FAO estimate, * = Unofficial/Semi-official/mirror data, C = Calculated figure, A = Aggregate (may include official, semi-official or estimates);

Source: Food And Agricultural Organization of United Nations: Economic And Social Department: The Statistical Devision

Fig. 2.22: Mature Plants Showing Ears

Sweet corn is harvested in the "milk stage," after pollination but before starch has formed, between late summer and early to mid-autumn. Field maize is left in the field very late in the autumn in order to thoroughly dry the grain, and may, in fact, sometimes not be harvested until winter or even early spring. The importance of sufficient soil moisture is shown in many parts of Africa, where periodic drought regularly causes famine by causing maize crop failure.

Maize was planted by the Native Americans in hills, in a complex system known to some as the Three Sisters. Maize provided support for beans, and the beans provided nitrogen derived from nitrogen-fixing bacteria which live on the roots of beans and other legumes; and squashes provided ground cover to stop weeds and inhibit evaporation by providing shade over the soil. This method was replaced by single species hill planting where each hill 60-120 cm (2-4 ft) apart was planted with 3 or 4 seeds, a method still used by home gardeners. A later technique was *checked maize* where hills were placed 40 inches apart in each direction, allowing cultivators to run through the field in two directions. In more arid lands this was altered and seeds were planted in the bottom of 10-12 cm (4-5 in) deep furrows to collect water. Modern technique plants maize in rows which allows for cultivation while the plant is young, although the hill technique is still used in the maize fields of some Native American reservations.

In North America, fields are often planted in a two-crop rotation with a nitrogen-fixing crop, often alfalfa in cooler climates and soybeans in regions with longer summers. Sometimes a third crop, winter wheat, is added to the rotation. Fields are usually ploughed each year, although no-till farming is increasing in use. Many of the maize varieties grown in the United States and Canada are hybrids.

Before about World War II, most maize in North America was harvested by hand (as it still is in most of the other countries where it is grown). This often involved large numbers of workers and associated social events. Some one-and two-row mechanical pickers were in use but the maize combine was not adopted until after the War. By hand or mechanical picker, the entire ear is harvested which then requires a separate operation of a maize sheller to remove the kernels from the ear. Whole ears of maize were often stored in *corn cribs* and these whole ears are a sufficient form for some livestock feeding use. Few modern farms store maize in this manner. Most harvest the grain from the field and store it in bins. The combine with a maize head (with points and snap rolls instead of a reel) does not cut the stalk; it simply pulls the stalk down. The stalk continues downward and is crumpled in to a mangled pile on the ground. The ear of maize is too large to pass through a slit in a plate and the snap rolls pull the ear of maize from the stalk so that only the ear and husk enter the machinery. The combine separates out the husk and the cob, keeping only the kernels.

Genetic Modification

Genetically-modified (GM) maize is one of the few GM crops grown commercially. Grown since 1997 in the United States and Canada, over 80 per cent of the US maize crop is now genetically modified. It is also grown commercially in South Africa, and, on a smaller scale, in Spain, the Czech Republic, Portugal and Germany.

Pellagra

When maize was first introduced into other farming systems than those used by traditional native-American peoples, it was generally welcomed with enthusiasm for its productivity. However, a widespread problem of malnutrition soon arose wherever maize was introduced as a staple. This was a mystery since these types of malnutrition were not normally seen among the indigenous Americans, to whom maize was the principal staple food.

It was eventually discovered that the indigenous Americans learned long ago to soak maize in alkali-water—made with ashes by North Americans and lime (calcium oxide) by Mesoamericans—which liberates the B-vitamin niacin, the lack of which was the underlying cause of the condition known as pellagra. This alkali process is known by its Nahuatl (Aztec)-derived name: nixtamalization.

Besides the lack of niacin, pellagra was also characterized by protein deficiency, a result of the inherent lack of two key amino acids in pre-modern maize, lysine and tryptophan. Nixtamalisation was also found to increase the availability of lysine and tryptophan to some extent, but more importantly, the indigenous Americans had learned long ago to balance their consumption of maize with beans and other protein sources such as amaranth and chia, as well as meat and fish, in order to acquire the complete range of amino acids for normal protein synthesis.

Since maize had been introduced into the diet of non-indigenous Americans without the necessary cultural knowledge acquired over thousands of years in the Americas, the reliance on maize in other cultures was often tragic. In the late 19th century pellagra reached epidemic proportions in parts of the deep southern U.S., as medical researchers debated two theories for its origin: the deficiency theory (eventually shown to be true) posited that pellagra was due to a deficiency of some nutrient, and the germ theory posited that pellagra was caused by a germ transmitted by stable flies. In 1914 the U.S. government officially endorsed the germ theory of pellagra, but rescinded this endorsement several years later as evidence grew against it. By the mid-1920s the deficiency theory of pellagra was becoming scientific consensus, and the theory was proved in 1932 when niacin deficiency was determined to be the cause of the illness.

Once alkali processing and dietary variety was understood and applied, pellagra disappeared. The development of high lysine maize and the promotion of a more balanced diet has also contributed to its demise.

Pests

Insect Pests

- Corn earworm (*Helicoverpa zea*)
- Fall armyworm (*Spodoptera frugiperda*)
- Common armyworm (*Pseudaletia unipuncta*)
- Stalk borer (*Papaipema nebris*)
- Corn leaf aphid (*Rhopalosiphum maidis*)
- European corn borer (*Ostrinia nubilalis*) (ECB)
- Corn silkfly (*Euxesta stigmatis*)
- Lesser cornstalk borer (*Elasmopalpus lignosellus*)
- Corn delphacid (*Peregrinus maidis*)
- Western corn rootworm (*Diabrotica virgifera virgifera* LeConte)

The susceptibility of maize to the European corn borer, and the resulting large crop losses, led to the development of transgenic expressing the *Bacillus thuringiensis* toxin. "Bt maize " is widely grown in the United States and has been approved for release in Europe.

Diseases

- Corn smut or common smut (*Ustilago maydis*): a fungal disease, known in Mexico as *huitlacoche*, which is prized by some as a gourmet delicacy in itself.
- Maize dwarf mosaic virus
- Stewart's Wilt (*Pantoea stewartii*)
- Common Rust (*Puccinia sorghi*)
- Goss's Wilt (*Clavibacter michiganese*)
- Grey Leaf Spot
- Mal de Río Cuarto Virus (MRCV)
- Stalk and Kernal Rot

As Food

Maize and cornmeal (ground dried maize) constitutes a staple food in many regions of the world. Maize meal is made into a thick porridge in many cultures: from the polenta of Italy, the angu of Brazil, the mămăligă of Romania, to mush in the U.S. or the food called sadza, nshima, ugali, and mealie pap in Africa. Maize meal is also used as a replacement for wheat flour, to make cornbread and other baked products. Masa (cornmeal treated with lime water) is the main ingredient for tortillas, atole and many other dishes of Mexican food.

Popcorn is kernels of certain varieties that explode when heated, forming fluffy pieces that are eaten as a snack. Roasted dried maize cobs with semi-hardened kernels, coated with a seasoning mixture of fried chopped spring onions with salt added to the oil, is a popular snack food in Vietnam.

Chicha and "chicha morada" (purple chicha) are drinks made usually from particular types of maize. The first one is fermented and alcoholic, the second is a soft drink commonly drunk in Peru.

Corn flakes are a common breakfast cereal in North America and the United Kingdom, and found in many other countries all over the world.

Maize can also be prepared as hominy, in which the kernels are soaked with lye in a process called nixtamalization; or grits, which are coarsely ground hominy. These are commonly eaten in the Southeastern United States, foods handed down from Native Americans, who called the dish sagamite.

The Brazilian dessert canjica is made by boiling maize kernels in sweetened milk.

Maize can also be harvested and consumed in the unripe state, when the kernels are fully grown but still soft. Unripe maize must usually be cooked to become palatable; this may be done by simply boiling or roasting the whole ears and eating the kernels right off the cob. Such corn on the cob is a common dish in the United States, United Kingdom and some parts of South America, but virtually unheard of in some European countries. The cooked unripe kernels may also be shaved off the cob and served as a vegetable in side dishes, salads, garnishes, etc. Alternatively, the raw unripe kernels may also be grated off the cobs and processed into a variety of cooked dishes, such as maize purée, tamales, pamonhas, curau,

Table 2.6: Sweetcorn, Yellow, Raw (Seeds Only)

Nutritional value per 100 g (3.5 oz) Energy 90 kcal 360 kJ	
Carbohydrates	19.02 g
—Sugars 3.22 g	
—Dietary fiber 2.7 g	
Fat	1.18 g
Protein	3.22 g
—Tryptophan 0.023 g	
—Threonine 0.129 g	
—Isoleucine 0.129 g	
—Leucine 0.348 g	
—Lysine 0.137 g	
—Methionine 0.067 g	
—Cystine 0.026 g	
—Phenylalanine 0.150 g	
—Tyrosine 0.123 g	
—Valine 0.185 g	
—Arginine 0.131 g	
—Histidine 0.089 g	
—Alanine 0.295 g	
—Aspartic acid 0.244 g	
—Glutamic acid 0.636 g	
—Glycine 0.127 g	
—Proline 0.292 g	
—Serine 0.153 g	
Water	75.96 g
Vitamin A equiv. 9 μg	1 per cent
Thiamine (Vit. B1) 0.200 mg	15 per cent
Niacin (Vit. B3) 1.700 mg	11 per cent
Folate (Vit. B9) 46 μg	12 per cent
Vitamin C 6.8 mg	11 per cent
Iron 0.52 mg	4 per cent
Magnesium 37 mg	10 per cent
Potassium 270 mg	6 per cent

One ear of medium size (6-3/4" to 7-1/2" long) maize has 90 grams of seeds Percentages are relative to US recommendations for adults.

Source: USDA Nutrient database

cakes, ice creams, etc. Sweetcorn, a genetic variety that is high in sugars and low in starch, is usually consumed in the unripe state.

Corn on the cob, as it is usually called in the United States, was hawked on the streets of early 19th-century New York City by poor, barefoot "Hot Corn Girls", who were thus the precursors hot-dog carts, churro wagons, and fruit stands seen on the streets of big cities today. Maize is a major source of starch. Cornstarch (maize flour) is a major ingredient in home cooking and in many industrialized food products. Maize is also a major source of cooking oil (corn oil) and of maize gluten. Maize starch can be hydrolyzed and enzymatically treated to produce syrups, particularly high fructose corn syrup, a sweetener; and also fermented and distilled to produce grain alcohol. Grain alcohol from maize is traditionally the source of bourbon whiskey. Maize is sometimes used as the starch source for beer.

In the United States and Canada maize is also widely grown to feed for livestock, as forage, silage (made by fermentation of chopped green cornstalks), or grain. Maize meal is also a significant ingredient of some commercial animal food products, such as dog food. Maize is also used as a fish bait, called "dough balls". It is particularly popular in Europe for coarse fishing.

Chemicals and Medicines

Starch from maize can also be made into plastics, fabrics, adhesives, and many other chemical products.

Stigmas from female maize flowers, known popularly as corn silk, are sold as herbal supplements.

The corn steep liquor, a plentiful watery byproduct of maize wet milling process, is widely used in the biochemical industry and research as a culture medium to grow many kinds of microorganisms.

Biofuel

"Feed maize" is being used increasingly for heating; specialized corn stoves (similar to wood stoves) are available and use either feed maize or wood pellets to generate heat. Maizecobs are also used as a biomass fuel source. Maize is relatively cheap and home-heating furnaces have been developed which use maize kernels as a fuel. They feature a large hopper that feeds the uniformly sized maize kernels (or wood pellets or cherry pits) into the fire.

Maize is increasingly used as a biomass fuel, such as ethanol, which as researchers search for innovative ways to reduce fuel costs, has unintentionally caused a rapid rise in food costs. This has led to the 2007 harvest being one of the most profitable maize crops in modern history for farmers.

Maize is widely used in Germany as a feedstock for biogas plants. Here the maize is harvested, shredded then placed in silage clamps from which it is fed into the biogas plants.

A biomass gasification power plant in Strem near Güssing, Burgenland, Austria was begun in 2005. Research is being done to make diesel out of the biogas by the Fischer Tropsch method.

Increasingly ethanol is being used at low concentrations (10 per cent or less) as an additive in gasoline (gasohol) for motor fuels to increase the octane rating, lower pollutants, and reduce petroleum use (what is nowadays also known as "biofuels" and has been generating an intense debate regarding the human beings' necessity of new sources of energy, on the one hand, and the need to maintain, in regions such as Latin America, the food habits and culture which has been the essence of civilizations such as the one originated in Mesoamerica; the entry, January 2008, of maize among the commercial agreements of NAFTA has increased this debate, considering the bad labour conditions of workers in the fields, and mainly the fact that NAFTA "opened the doors to the import of maize from the United States, where the farmers who grow it receive multi-million dollar subsidies and other government supports. (...) According to OXFAM UK, after NAFTA went into effect, the price of maize in Mexico fell 70 per cent between 1994 and 2001. The number of farm jobs dropped as well: from 8.1 million in 1993 to 6.8 million in 2002. Many of those who found themselves without work were small-scale maize growers."). However, introduction in the northern latitudes of the U.S. of tropical maize for biofuels, and not for human or animal consumption, may potentially alleviate this.

As a result of the U.S. federal government announcing its production target of 35 billion gallons of biofuels by 2017, ethanol production will grow to 7 billion gallons by 2010, up from 4.5 billion in 2006, boosting ethanol's share of maize demand in the U.S. from 22.6 percent to 36.1 percent.

Ornamental and Other Uses

Some forms of the plant are occasionally grown for ornamental use in the garden. For this purpose, variegated and coloured leaf forms as well as those with colourful ears are used. Additionally, size-superlative varieties, having reached 31 ft (9.4m) tall, or with ears 24 inches (60 cm) long, have been popular for at least a century.

Corncobs can be hollowed out and treated to make inexpensive smoking pipes, first manufactured in the United States in 1869.

An unusual use form aize is to create a *corn maze* (or *maize maze*) as a tourist attraction. The idea of a maize maze was introduced by Adrian Fisher, one of the most prolific designers of modern mazes, with The American Maze Company who created a maze in Pennsylvania in 1993. Traditional mazes are most commonly grown using yew hedges, but these take several years to mature. The rapid growth of a field of maize allows a maze to be laid out using GPS at the start of a growing season and for the maize to grow tall enough to obstruct a visitor's line of sight by the start of the summer. In Canada and the U.S., these are popular in many farming communities.

Maize kernels can be used in place of sand in a sandbox-like enclosure for children's play.

Fodder

Maize makes a greater quantity of epigeous mass than other cereal plants, so can be used for fodder. Digestibility and palatability are higher when ensiled and fermented, rather than dried.

Maize as a Commodity

Maize as a commodity is bought and sold by investors and price speculators as a tradable commodity. It is referred to as a corn futures contract. Corn futures contracts are traded on the Chicago Board of Trade (CBOT) under ticker symbol C. They are delivered every year in March, May, July, September, and December.

In Art

Maize has been an essential crop in the Andes since the pre-Columbian Era. The Moche culture from Northern Peru made ceramics from earth, water, and fire. This pottery was a sacred substance, formed in significant shapes and used to represent important themes. Maize represented anthropomorphically as well as naturally.

In the United States, maize itself is sometimes used for temporary architectural detailing when the intent is to celebrate local agricultural productivity and culture. A well-known example of this use is the Corn Palace in Mitchell, South Dakota, which utilizes cobs of coloured maize to implement a design that is recycled annually.

A corn stalk with two ripe cobs is depicted on the reverse of the Croatian 1 lipa coin, minted since 1993.

HOMINY

Hominy or nixtamal is dried maize kernels which have been treated with an alkali.

The traditional U.S. version involves *soaking* dried corn in lye-water (sodium hydroxide or potassium hydroxide solution), traditionally derived from wood ash, until the hulls are removed. Mexican recipes describe a preparation process consisting primarily of *cooking* in lime-water (calcium hydroxide). In either case, the process is called nixtamalization, and removes the germ and the hard outer hull from the kernels, making them more palatable, easier to digest, and easier to process.

Commercially available canned hominy may have a slightly stronger scent when compared to the traditional preparation.

The earliest known usage of nixtamalization was in what is present-day Guatemala around 1500-1200 BC. It affords several significant nutritional advantages over untreated maize products. It converts some of the niacin (and possibly other B vitamins) into a form more absorbable by the body, improves the availability of the amino acids, and (at least in the lime-treated variant) supplements the calcium content, balancing maize's comparative excess of phosphorus.

Many Native American cultures made hominy and integrated it into their diet. Cherokees, for example, made hominy grits by soaking corn in lye and beating it with a *kanona* (corn beater). The grits were used to make a traditional hominy soup or Gv-No-He-

Nv A-Ma-Gi-i), a hominy soup that was allowed to ferment (Gv-Wi Si-Da A-Ma-Gi-i), cornbread, dumplings (Di-Gu-Nv-i) or fried with bacon and green onions.

Some recipes using hominy include menudo (a spicy tripe and hominy soup), pozole (a stew of hominy and pork, chicken, or other meat), hominy bread, hominy chili, casseroles and fried dishes. Hominy can be ground coarsely to make hominy grits, or into a fine mash (dough) to make masa, the dough used to make tamales.

Rockihominy, a popular trail food in the 19th and early 20th centuries, is dried corn roasted to a golden brown, then ground to a very coarse meal, almost like hominy grits. Hominy can also be used as animal feed.

PASTA

Pasta (Italian *pasta*, from Latin *pasta* "dough, pastry cake", (*pasta*) "barley porridge") is a generic term for foods made from an unleavened dough of flour and water, and sometimes a combination of egg and flour. Pastas include noodles in various lengths, widths and shapes, and varieties that are filled with other ingredients like ravioli and tortellini. The word pasta is also used to refer to dishes in which pasta products are a primary ingredient. It is usually served with sauce or chocolate to enhance the flavour.

Fig. 2.23: Display Window with Pasta

There are approximately 600 different shapes of pasta. Examples include spaghetti (thin rods), maccheroni (tubes or cylinders), fusilli (swirls), and lasagne (sheets). Two other noodles, gnocchi and spätzle, are sometimes considered pasta. They are both traditional in parts of Italy.

Pasta is categorized in two basic styles: *dried* and *fresh*. Dried pasta made without eggs can be stored for up to two years under ideal conditions, while fresh pasta will keep for a couple of days in the refrigerator. In preparation for consumption, pasta is generally boiled.

Ingredients

Pasta is made from a simple combination of water, eggs and flour. Pre-packaged specialty pasta often includes spices, cheese or added colouring from spinach, tomatoes or food dye.

Under Italian law, dry pasta (*pasta secca*) can only be made from durum wheat flour or durum wheat semolina. Durum flour and durum semolina have a yellow tinge in colour. Italian pasta is traditionally cooked *al dente* (Italian: "to the tooth", meaning not too soft). Outside Italy, dry pasta is frequently made from other types of flour (such as wheat flour), but this yields a softer product that cannot be cooked *al dente*.

Particular varieties of pasta may also use other grains and milling methods to make the flour, as specified by law. Some pasta varieties, such as pizzoccheri, are made from buckwheat flour. Fresh pasta may include eggs (*pasta all'uovo*, roughly translated as *egg pasta*). Some specialty pasta people, or from whole wheat flour

History

The question of pasta's origin continues to evoke speculation. While many different cultures ate some sort of noodle-like food, composed mostly of grain, the key characteristics of pasta are durum wheat semolina, with a high gluten content, made with a technique that allows the resultant dough to be highly malleable. Thus, the many different shapes (*i.e.*, ziti, spaghetti, ravioli) that characterize "pasta." South of Europe, in Arab North Africa, a food similar to pasta, known as cous-cous, has been eaten for centuries. However, it lacks the distinguishing malleable nature of what is now referred to as pasta, cous-cous being more akin to droplets of dough. In China, noodles made of millet or rice have been eaten for centuries, but lacked the durum wheat semolina paste that denotes pasta.

Historians have noted several lexical milestones relevant to pasta, none which change these basic characteristics. For example, the works of the 2nd century AD Greek physician Galen mention *itrion*, homogeneous compounds made up of flour and water. The Jerusalem Talmud records that *itrium*, a kind of boiled dough, was common in Palestine from the 3rd to 5th centuries AD, A dictionary compiled by the 9th century Syrian physician and lexicographer Isho bar Ali defines *itriyya*, the Arabic cognate, as string-like shapes made of semolina and dried before cooking. The geographical text of Muhammad al-Idrisi, compiled for the Norman King of Sicily Roger II in 1154 mentions *itriyya* manufactured and exported from Norman Sicily:

West of Termini there is a delightful settlement called Trabia. Its ever-flowing streams propel a number of mills. Here there are huge buildings in the countryside where they make vast quantities of *itriyya* which is exported everywhere: to Calabria, to Muslim and Christian countries. Very many shiploads are sent.

Itriyya gives rise to *trie* in Italian, signifying long strips such as *tagliatelle*. One form of *itriyya* with a long history is *laganum* (plural *lagana*), which in Latin refers to a thin sheet of dough, and gives rise to Italian "*lasagna*".

Fig. 2.24: A Lasagne Dish

The Chinese were eating noodles made of millet as long ago as 2000 BC. This was confirmed by the discovery of a well-preserved bowl of millet noodles over 4000 years old. However, durum wheat was not known in China until later times. The familiar legend of Marco Polo importing pasta from China originated with the *Macaroni Journal*, published by an association of food industries with the goal of promoting the use of pasta in the United States. Marco Polo describes a food similar to "lagana" in his *Travels*, but he uses a term with which he was already familiar. Durum wheat, and thus pasta as it is known today, was introduced by Arabs during their conquest of Sicily in the late 7th century, according to the newsletter of the National Macaroni Manufacturers Association, thus predating Marco Polo's travels to China by about six centuries.

In the 1st century BC writings of Horace, *lagana* were fine sheets of dough which were fried and were an everyday food. Writing in the 2nd century Athenaeus of Naucratis provides a recipe for *lagana* which he attributes to the 1st century Chrysippus of Tyana: sheets of dough made of wheat flour and the juice of crushed lettuce, then flavored with spices and deep-fried in oil. An early 5th century cookbook describes a dish called *lagana* that consisted of layers of dough with meat stuffing, a possible ancestor of modern-day *Lasagna*. But the method of cooking these sheets of dough does not correspond to our modern definition of either a fresh or dry pasta product. The first concrete information concerning pasta products in Italy dates from the thirteenth or fourteenth century. The name (ëáãÜíá, *lagána*) survives in modern-day Greece to denote an unleavened, flat bread eaten during the Great Lent. The term "lagana" is also used in the Southern region of Calabria, where it indicates a flat noodle.

Varieties of Pasta

Pasta is generally served with some type of sauce; the sauce and the type of pasta are usually matched based on consistency, ease of eating etc, much in the same way round buns are used for hamburgers and long buns are used for hotdogs in the US. Common pasta sauces in Northern Italy include pesto and ragù alla bolognese, which usually adds meat to the sauce. In Central Italy, there are simple sauces such as tomato sauce, amatriciana and carbonara. Southern Italian sauces include spicy tomato, garlic, and olive oil, with the pasta often paired with fresh vegetables or seafood. Varieties include puttanesca, *pasta alla norma* (tomatoes, eggplant and fresh or baked cheese), *pasta con le sarde* (fresh sardines, pine nuts, fennel and olive oil), *spaghetti aglio, olio e peperoncino* (literally with garlic, [olive] oil and hot chili peppers).

Fig. 2.25: *Pesto Cavatappi*

Fettuccine alfredo with cream and cheese, and spaghetti with tomato sauce (with or without ground meat or meatballs) are popular Italian-style dishes in the United States.

As pasta was introduced elsewhere in the world, it became incorporated into a number of local cuisines, which often have significantly different ways of preparation from those of Italy. In Hong Kong, the local Chinese have adopted pasta, primarily spaghetti and macaroni, as an ingredient in the Hong Kong-style Western cuisine. In Cha chaan teng, macaroni is cooked in water and served in broth with ham or frankfurter sausages, peas, black mushrooms, and optionally eggs, reminiscent of noodle soup dishes. This is often a course for breakfast or light lunch fare. This method often involves cooking the pasta well beyond the al dente stage and washing the starches off the pasta after cooking, measures frowned upon in Italy or in Hong Kong's more authentic Italian eateries. Two common spaghetti dishes served in Japan are the Bolognese and the Napolitan. In India, Macaroni has been adopted and cooked in an Indianized way. Boiled macaroni is sautéed along with jeera, turmeric, finely chopped green chillies, onions and cabbage.

Macaroni is a kind of moderately-extended, machine-made dry pasta. Much shorter than spaghetti, and hollow, macaroni does not contain eggs. Though home machines exist that can make macaroni noodles, macaroni is usually made commercially.

Macaroni is a borrowing of the Italian word *maccherone* and its plural *maccheroni*. Its etymology is debatable. Some scholars consider it related to Greek *ìáêáñßá* (*makaria*), a kind of barley broth. Others think it comes from Italian *ammaccare*, "to bruise or crush" (referring to the crushing of the wheat to make the pasta), which comes, in turn, from Latin *macerare*, meaning 1) to soak in liquid, to soften, or 2) to torment, to mortify, to distress (the term also giving us the English macerate).

In English-speaking countries, the name *macaroni* is customarily given to a specific shape of pasta (*i.e.* small pasta tubes cut into short pieces). In the United States, Canada and the United Kingdom, this pasta is often prepared by cooking it with a sauce made from cheddar cheese; the resulting dish is called macaroni and cheese.

In Hong Kong, the local Chinese have adopted macaroni as an ingredient in the Hong Kong-style Western cuisine. In the territory's Cha chaan tengs, macaroni is cooked in water and then washed of starch, and served in clear broth with ham or frankfurter sausages, peas, black mushrooms, and optionally eggs reminiscent of noodle soup dishes. This is often a course for breakfast or light lunch fare.

SPAGHETTI

Spaghetti is a long, thin, cylindrical pasta of Italian origin. A variety of pasta dishes are based on it, from spaghetti with cheese and pepper or garlic and oil to a spaghetti with tomato, meat, and other sauces. Spaghetti is made of semolina or flour and water. *Spaghetti* is the plural form of the Italian word *spaghetto*, which is a diminutive of *spago*, meaning "thin string" or "twine".

Origins

Chinese noodles pre-date Italian pasta, and traders may have introduced the food to the west. Historically, people in Italy ate pasta in the form of gnocchi-like dumplings—*pasta fresca* eaten as

Fig. 2.26: Cooked Spaghetti

soon as it was prepared. It has been asserted that pasta was first worked to long, thin forms in Southern Italy around the 12th Century. The popularity of pasta spread to the whole of Italy after the establishment of pasta factories in the 19th century, enabling the mass production of pasta for the Italian market. Pasta is commonly associated with Italy.

In the United States around the end of the 1800s, spaghetti was offered in restaurants as *Spaghetti Italienne* (which likely consisted of extremely soggy noodles and a tomato sauce diluted with broth) and it wasn't until decades later that it came to be prepared with garlic or peppers. Canned spaghetti, kits for making spaghetti, and spaghetti with meatballs became popular, and the dish has become a staple in that country.

Preparation

Spaghetti is cooked by boiling the pasta in water and adding either salt or olive oil. The consistency or texture of spaghetti changes as it is cooked. The most popular consistency is *al dente* (Italian 'to the tooth'); that is, soft but with texture, sometimes even with bite in the center. Others prefer their spaghetti cooked to a softer consistency. The best dried spaghetti is made from durum wheat semolina. Inferior spaghetti is often found produced with other kinds of flour, especially outside Italy. Fresh spaghetti should be prepared with grade '00' flour. There are two other variants of spaghetti that require different cooking times. *Spaghettini* ("thin spaghetti") (also "angel hair spaghetti") takes less time (usually two minutes less) to cook to *al dente* form than regular spaghetti. There is also *spaghettoni* ("thick spaghetti") which takes longer to cook. All three types of spaghetti are larger than the other round-rod pastas (like vermicelli).

Serving

An emblem of Italian cuisine, spaghetti is frequently served with tomato sauce, which may contain various herbs (especially oregano, and basil), olive oil, meat, or vegetables. Other spaghetti preparations include using Bolognese sauce, carbonara, and chili. Grated hard cheeses, such as Pecorino Romano, Parmesan, and Asiago cheese, are often added.

Cultural References

- On April Fools' Day, 1957, the BBC ran a very successful spoof documentary explaining how spaghetti is grown on spaghetti trees.
- The parody religion Pastafarianism holds that the universe was created by a Flying Spaghetti Monster.
- "Spaghetti" can be used to describe objects which are complicated or tangled, such as spaghetti code or Spaghetti Junction.
- Spaghettieis is a mock dessert which looks like spaghetti with tomato sauce.

VERMICELLI

Vermicelli is a type of pasta, round in section and somewhat thinner than *spaghetti*. Vermicelloni is less common, and about the same size as *fedelini* (also hard to find). Both are thinner than *spaghettini* ("thin spaghetti").

History in Italy

In 14th-century Italy, extra-fine spaghetti had varying local names. Master Barnaba da Reatinis from Reggio Emilia notes that Tuscan *vermicelli* are called *orati* in Bologne, *minutelli* in Venice, *fermentini* in Reggio and *pancardelle* in Mantua."

The first mention of a vermicelli recipe is in the book *De arte Coquinaria per vermicelli e maccaroni siciliani* (*The Art of Cooking Sicilian Macaroni and Vermicelli*), compiled by the famous Maestro Martino da Como, unequalled in his field at the time and perhaps the first "celebrity chef," who was the chef at the Roman palazzo of the papal chamberlain ("camerlengo"), the Patriarch of Aquileia. In Martino's *Libro de arte coquinaria*, there are several recipes for vermicelli, which can last two or three years (*doi o tre anni*) when dried in the sun.

History in Asia

In South Asia, the vermicelli used is different from what is used in Italy. Here, it is made from semolina, unlike the Italian, which is made from durum wheat. In Asia it is known variously as *shemai* (?????) in Bengali, *seviyan* in Hindi and Urdu, *sev* in Gujarati, *shavige* in Kannada, *sevalu* or *semiya* in Telugu, and *semiya* in Tamil, all probably modified form of 'mian xian' in Mandarin. The noodles are used in a number of dishes including a variation of *kheer*, a sweet dessert similar to rice pudding. Vermicelli is used in many parts of India to make a popular dish called *upma*. To prepare it, one boils the dry oil-roasted vermicelli with a choice of vegetables.

In East Asia, the term rice vermicelli is often used to describe the thin rice noodles popular in China, also known as [[*bee hoon*]] in Hokkien, mai fun in Cantonese, kyar-zun in Burmese, and *bún* in Vietnamese. The term vermicelli may also refer to vermicelli made from mung bean, which is translucent when cooked, and can be differentiated from rice vermicelli, which turns whitish when cooked. Mung bean vermicelli is commonly used in Chinese cuisine. Conversely, is vermicelli that is made of wheat instead of rice. While superficially similar to *bee hoon* it has a very different texture and different culinary uses as well.

History in the Americas

The *fideo* is a type of noodle, popular in Mexican and Latin American cuisine, often referred to in English as "vermicelli." It is commonly used in chicken soup and in *sopa seca*, a type of side dish.

In the United States, vermicelli is usually the pasta found in Rice-A-Roni, a pilaf-style rice-and-pasta side dish manufactured by The Quaker Oats Company.

History in the Middle East and Northeast Africa

It is used in one of the most common ways of cooking rice in Egypt. The vermicelli is browned by frying with oil or butter, then rice and water are added. Vermicelli is called *she'reya* in Arabic.

The Persian *reshteh* also resembles vermicelli. *Fâlûde* or *faloodeh* is a Persian frozen dessert made with thin vermicelli noodles frozen with corn starch, rose water, lime juice, and often ground pistachios.

In Somalia, it is used in a sweet dish called *cadriyad*. The vermicelli is browned by frying with butter, then water, sugar and cardamom are added until the vermicelli has softened slightly. It is similar to the Indian *kheer*. However, no milk or cream is added. It is usually eaten as a dessert or as a side dish with Somali spiced rice dishes.

NOODLE

A noodle is food made from unleavened dough that is cooked in a boiling liquid. Depending upon the type, noodles may be dried or refrigerated before cooking. The word noodle derives from the German *Nudel* (noodle) and may be related to the Latin word *nodus* (knot). In American English, noodle is a generic term for unleavened dough made from many different types of ingredients. Noodles exist in an abundance of shapes.

In 2005, the oldest noodles yet discovered were found at Qinghai. The 4,000-year-old noodles appear to have been made from millet.

Types of Noodles by Primary Ingredient

Contrary to widespread belief, pasta types such as penne, rigatoni, farfalle and others are not classified as noodles. While they are produced with some of the same wheat based ingredients, their form and structure do not adhere with that of the noodle family of food products.

Fig. 2.27: Wide, Uncooked Egg Noodles

Wheat

- Chûka men: Japanese for "Chinese noodles", used for ramen, chanpon and yakisoba
- Lamian: hand pulled Chinese noodles.
- Mee pok: flat, yellow Chinese noodles, popular in Southeast Asia
- Nokedli: Hungarian noodle
- Pasta: approximately 350 variants used in Italian cuisine
- Somen: very thin Japanese wheat noodles
- Spätzle: a Swabian type of noodle made of wheat and eggs
- Tészta: various ranges of Hungarian pasta
- Udon: thick Japanese wheat noodles

Rice

- Flat or Thick rice noodles, also known as *héfin* or *ho fun*, *kway teow* or *sen yai*.
- Rice vermicelli: thin rice noodles, also known as *mifin* or *bee hoon* or *sen mee*.

Fig. 2.28: Handmade Short, Thick Small German Noodles called Spätzle

Mung Bean

- Cellophane noodles, also known as glass noodles, sweet potato Vermicelli or bean vermicelli. *finsî* in Chinese, *harusame* in Japanese, *soun or suun* in Indonesian, *wun sen* (ÇØé¹àÊé¹) in Thai. These are the principal ingredient in the Korean dish japchae.

Potato or Canna Starch

- Cellophane noodles can also be made from potato starch or canna starch or various starches of the same genre.
- Gnocchi, small Italian dumplings.
- Halušky, small Slovak dumplings made of grated potatoes and wheat flour (and sometimes egg), usually made with sheep's cheese, sauer cabbage, or tvaroh (cottage cheese/ curd), and fried bacon.

Buckwheat

- Makguksu, local specialty of Gangwon Province in South Korea
- Naengmyeon: Korean noodles made of buckwheat and sweet potato starch. Slightly more chewy than soba.

- Soba: Japanese buckwheat noodles
- Pizzoccheri: Italian buckwheat noodles from Valtellina, usually served with a melted cheese sauce.

Acorn

- Acorn noodles, also known as dotori guksu in Korean, are made of acorn meal, wheat flour, wheat germ and salt.

Types of Noodle Dishes

- Basic noodles: These are cooked in water or broth, then drained. Other foods can be added (for example a pasta sauce) or the noodles are added to other foods or the noodles can be served plain with a dipping sauce or oil to be added at the table. In general, noodles are soft and absorb flavors.
- Chilled noodles: noodles are sometimes served in a salad. An example is the Thai glass noodle salad
- Fried noodles: dishes made of noodles stir fried with various meats, seafood or vegetables. Typical examples include chow mein, lo mein, mee goreng, hokkien mee, some varieties of pancit, yakisoba and pad thai.
- Noodle soup: noodles served in broth. Examples are ph?, beef noodle soup, ramen, laksa, saimin and batchoy, and chicken noodle soup.

Fig. 2.29: A Simple Noodle Soup Consisting of Soy Sauce and Sesame Oil

FROZEN NOODLES

Frozen noodles and chilled noodles are types of instantly prepared Asian (or European) noodles that are sold frozen or chilled. These products differ from prepackaged dehydrated noodles in a number of ways, both in flavor, texture and in that they normally come pre-prepared with ingredients in addition to the noodles, such as vegetables, meat, and soup stock.

Chilled or frozen applications are applied to both udon and Chinese-style noodles. For either type, the idea is that just 20-60 seconds of immersion in boiling water is necessary to reach a ready-to-eat state. As of 2001, chilled noodles accounted for sales of ¥400 billion yuan worldwide (approximately $8,500,000,000 U.S. as of 2001) while frozen noodles accounted for sales of ¥70 billion yuan worldwide (approximately $48,300,000,000 U.S. as of 2001). A large part of these revenues were accounted for by sale to restaurants.

The production of both chilled and frozen noodle starts with boiled noodles cooked to an optimum state of doneness, generally considered as having a moisture gradient at the surface of the noodles of 80 per cent moisture absorption, and at the core of the noodles of 50 per cent moisture absorption. After boiling, chilled noodles are placed into packaging followed by refrigeration at between 4-10°C (39.2-50°F). Frozen noodles, by contrast, are flash frozen using either air blast technology, a contact freezer or a combination of both, usually at -40°C (-40°F) for 30 minutes. Both processes induce a swelling of the starch that reduces deterioration and thus extends the noodles' shelf-life.

Frozen noodles have been cited as having advantages over dried noodles, the process tending to make them less sticky, firmer and thicker than their dried counterparts. Both frozen and dehydrated noodles have a longer shelf-life than chilled noodles, which tend to clump after approximately two weeks of refrigeration due to gelatinisation of starch.

While both Frozen and chilled noodles offer convenience for the mass market, and for short preparation times, each have been found to cause loss of optimal texture when tested using alkaline and white salted noodles. In a 2001 published study by D.W. Hatcher and M.J. Anderson of the Canadian Grain Commission, textural attributes of the frozen variety were tested at 1 and 4 weeks after frozen storage, and the chilled variety at the 30 minute and 1, 2 and 7 days marks. Both the frozen and the chilled noodles showed a decline in textural characteristics which increased in severity the further forward in time they were sampled after storage. It was also indicated that the texture of raw frozen noodles, cooked after defrosting, as opposed to precooked and then frozen noodles, fared much better.Instant noodles.

Fig. 2.30: Uncooked Instant Noodles in its Typical Block-Like Form

Instant noodles are dried or precooked noodles fused with oil, and often sold with a packet of flavoring. Dried noodles are usually eaten after being cooked or soaked in boiling water for 3 to 5 minutes, while precooked noodles can be reheated, or eaten straight from the packet. Instant noodles were invented by Momofuku Ando of Nissin Foods, Japan.

History

Instant noodles were first marketed by Momofuko Ando, a Taiwanese man who was born in southwestern Taiwan when the island was under Japanese colonial rule, in Japan on August 25, 1958 under the brand name *Chikin Ramen*. Another milestone was reached in 1971 when Nissin introduced the Cup Noodles, instant noodles in a waterproof polystyrene cup to which boiling water could be added to cook the noodles. Further innovations included adding dried vegetables to the cup, creating a complete instant soup dish.

According to a Japanese poll in the year 2000, instant noodles were the most important Japanese invention of the century. Karaoke came second, with the Compact Disc only coming in fifth. As of 2008, approximately 94 billion servings of instant noodles are eaten worldwide every year. China consumes 45 billion packages of instant noodles per year, or 48 per cent of the world, Indonesia consumes 14 billion, Japan 5.1 billion. Per capita, South Korean people eat the highest number of instant noodles, 69 packages per year.

Instant noodles are not only popular with college students, they can also be an economic indicator. In 2005, the Mama Noodles Index was launched to reflect the sales of Mama Noodles, the biggest instant noodle manufacturer in Thailand. The index was steady after the recovery from the East Asian financial crisis, but sales jumped by around 15 per cent in the first seven months of 2005 on a year-to-year basis, which was regarded as a sign of recession. People could not afford more expensive foods, hence the increase in the purchase of instant noodles, as instant noodles are seen as an inferior good.

Health Concerns

Instant noodles are often criticized as being unhealthy or junk food. A single serving of instant noodles is high in carbohydrates but low in fiber, vitamins and minerals. Noodles are typically fried as part of the manufacturing process, resulting in high levels of saturated fat and/or trans fat. Additionally, if served in an instant broth, instant noodles typically contain high amounts of sodium. The current U.S. Recommended Dietary Allowance of sodium for adults and children over 4 years old is 2,400 mg/day. Some brands may have over 3,000 mg of sodium per package in extreme cases. Instant noodles and the flavoring soup base also contain high amounts of monosodium glutamate.

The most recent controversy concerns dioxin and other hormone-like substances that could theoretically be extracted from the packaging and glues used to pack the instant noodles. As hot water is added, it was reasoned that harmful substances could seep into the soup. After a series of studies were conducted, various organizations requested changes in the packaging.

A recent concern on consumption of fried foods is the possible presence of oxidation products from poor maintenance of the oil. This can be a concern if the cooking oil is not maintained at the proper temperature or changed as often as necessary. Proper production standards minimize these risks, and these oxidation products can be present in any improperly fried foods, and are suspected to pose various health risks.

Instant Noodles Worldwide

Instant noodles have become a popular food in many parts of the world, though they have undergone changes in flavor to fit local tastes.

Fig. 2.31: Cup Noodles from Modern Singapore. It is Widely Consumed Around the World

Argentina

In Argentina instant noodles are gaining popularity, and can be found in most major cities in supermarkets. The brand is usually Sapporo Ichiban. The problem is basically that the cost is still high (about 0.90 dollars). Due to the recent Chinese immigration wave, specialized Chinese supermarkets offer a wide variety of ramen brands. The Brand Maruchan can also be found now at Disco and Coto supermarkets, coming in different flavors such as shrimp, beef, chicken or cheese.

Australia

While the most popular brand of instant noodles in Australia was Maggi noodles (pronounced ma-jee) for some time, many other brands have emerged and taken over the market. For Cup-Noodles the most popular brand is *Fantastic Noodles*. For packet Noodles there are 2 popular choices, *Indomie Mi Goreng* (Fried Indonesian noodles served without broth), Nissin's *Demae Ramen*, and Nong Shim's Shin Ramyun served with broth, although (due to a strong advertising campaign) Maggi noodles continue to be highly popular despite being amongst the highest-priced packet noodles. Almost every popular brand of Korean, Japanese and Indonesian noodles are available in Australian supermarkets and convenience stores, due to the cultural mesh of Asian cuisine (due to geographical location) in Australian life.

Belgium

Instant noodles are not widely popular in Belgium, and are expensive. The westernized, most common brand is called "Aiki Noodles".

A typical feature in both France and Belgium is the 'rejection' of such foreign foods in favour of traditional European meals. Instant foods in Belgium are generally unpopular. The supermarket chain Delhaize has its own brand of Westernized instant noodles, but sales overall are still low. Authentic Japanese and Thai instant noodles can be found only in Asian specialty stores, and in very limited quantity in some supermarkets.

Brazil

For a long time the main manufacturer was Nissin Miojo (to an extent that, in Brazil, the most common name for instant noodles is "miojo", although the manufacturers usually call them "lámen" or "l'amen"), but a lot of other companies, such as Maggi and Nestlé also offer this product. There is a lot of variants, such as "Lámen Cremoso" (which has a creamy sauce) and "Lámen Hot" (with pepper), as well as yakisoba and spaghetti.

Canada

The major brands available in Canada are Sapporo Ichiban, Knorr and Mr. Noodles. In fact, in some areas the noodles may be referred to only by the name "Ichiban" or as "Mr. Noodles", originating from the brand name. These main-brand packages generally only contain one flavor pouch. Other brands may include a small package of sesame oil. Korean made brands such as Nong Shim are readily available in most large grocery stores and in some major cities, brands imported from Europe are available. Asian markets found in larger cities typically carry dozens of different brands and varieties. Due to the large South Asian population in some major cities, Maggi is also a popular brand. The term kimchi is also popular, in reference to imported Korean ramen; this term is particularly popular in the West Coast cities of Vancouver and Victoria.

China

China is a fast-growing market for instant noodles. The market is focusing on higher-end products, generally costing more than 1 RMB. The top three brands in this category dominate more than 85 per cent of the market; for the lower end (those below 1 RMB) the leading five hold around 60 per cent of the market share.

The dominant brands in the Chinese market are:

- Ting Yi, a PLC listed in Hong Kong Stock Exchange but managed mainly by Taiwanese with a Japanese strategic alliance partner Sanyo Food, the third player in Japanese market. Master Kong now is probably the largest brand with a business scope extended to Beverage (RTD) and Bakeries.

- Uni-President, a PLC listed in Taiwan. Like Master Kong, it used to have its own businesses in Beverage and Bakeries as well. However, the company recently formed a JV partnership with Hwa-Long and Nissin on the beverage front. Uni-President's home market is in Taiwan but the company is now expanding aggressively in mainland China.
- Hwa-Long (*i.e.* Chinese Dragon), a local company allied with Nissin. Given that relationship, Hwa-long has long been regarded as Nissin's agent in the world's largest consumer market. Recently Hwa-long, Nissin, and Uni-President formed a partnership in Beverage business and would like to collectively compete against Master Kong, still considered the leading brand in the industry.
- Bai-xiang (*i.e.* White Elephant), a local company spun-off based on a former SOE and now still enjoys a strong local customer base. It is a leading brand in lower-end segments in terms of volume.

Denmark

Virtually all supermarkets sell instant noodles of some sort, but they tend to be westernized and come in foam containers, costing upwards of 10 kroner (2 dollars). Ethnic stores and specialty shops offer the most popular alternative, Yum Yum, with prices ranging from 4-6 kroner, but 3 for 10 offers are nearly universal. Its popularity has been sufficient to become a genericized brand name in its own right. Mama, another brand from Thailand, is the second most popular Eastern-type brand, but has a much smaller market share, despite its popularity in its homeland. Also most Danish supermarket instant noodles contain E601 as a flavor enhancer.

Ethiopia

In Ethiopia, the most popular instant noodles is Indomie brand with its various flavors. It is picked first from the shelves because of its international popularity and affordable price. It is referred to as "Lehulum Tesmami" meaning "Suitable For All" in the Amharic language.

Finland

Noodles are ubiquitous in all urban grocery stores and supermarkets. Packet noodles are the norm while cup noodles, often pricier, are seldom available outside ethnic food shops. Yum Yum and Mama cost about half a euro while Koka's larger portion of non-fried, no trans fat, no preservatives higher-end offering comes in at about euro a piece.

Germany

Instant noodles were relatively unknown in Germany until the 1990s, from which they have continually gained in popularity. There are two types of ramen in Germany: the first, generally called *Instant-Nudeln* (instant noodles) tends to be a mild, Westernized version, in flavors such as chicken, vegetable, beef and button mushroom. The second type is actually called *Ramen*, and it is quite similar to traditional ramen as it is known in Asia. German *Ramen* was originally considered an ethnic food and was only available in specialty stores. Since the mid-1990s, however, it has become available at German supermarkets and can usually be found on the instant-food aisle. The most popular brands are *Yum Yum*, Nissin Cup Noodles and Maggi.

Hong Kong

Cantonese people have a long history of cooking yi mein, an early form of instant noodles invented in the Qing Dynasty. Modern instant noodles were publicly introduced as *Doll Noodles* in the late 1960s by Winner Food Products Ltd which was bought by Nissin in 1984; and that term has since become a synonym for instant noodles (irrespective of brand) in Hong Kong and Southern China. Demae Itcho is another famous brand in local market.

India, Pakistan, and Bangladesh

The most popular brand in the 3 countries is Maggi by Nestlé. Also in the market are players like Top Ramen Smoodles and 'Cup Noodles' manufactured by Indo-Nissin Ltd, Ching's Instant Noodles, A.A. Nutritions's Yummy, and Wai-Wai owned by the Chaudhary Group from Nepal and India. Flavors are more oriented toward local flavors, such as Masala, chicken tikka etc.

Maggi is regarded in popular culture as a 2 minute noodle. Due to increasing health consciousness of the consumers, Nestle introduced an instant noodles made of whole wheat grain flour called Atta Noodles. Other variant include instant rice noodles in various flavors also. Top Ramen is also very popular among consumers. The preferred flavor of Top Ramen is known as "Curry Smoodles"; its flavorings mimic a basic curry, including onion, garlic, coriander, and a curry masala. Top Ramen manufacturers refer to their instant noodles as "smoodles", to differentiate from other instant noodles. A package sells for 10-12 rupees in India and 16 rupees in Pakistan; South Asian markets in North America and Australia have also started selling it. Indo-Nissin foods also manufactures "Cup Noodles", a proprietary food with various flavors like "Spicy Vegetables" and "Tangy Chicken" to name a few.

Smith and Jones and Ching's secrets are also popular new brands of instant noodles in India.

India also has a huge demand for unflavored instant noodles with brands like Bambino and Ching's dominating the market.

Indonesia

The first widely known instant noodle in Indonesia is "Supermi" introduced in 1970s. Later several brands of instant noodle was introduced—"Indomie" and "Sarimi". The three brands produced by Indofood Sukses Makmur, the largest instant noodle producer in the world—indirectly owned by Sudono Salim one of Indonesian wealthiest tycoons. Indofood is also the owner of Bogasari Flour Mills, the largest flour mill company in the world.

Currently, Indofood Sukses Makmur has a market share about 70 per cent of Indonesian instant noodle production. In 1999 the figure was about 90 per cent market share. The recent decline in market share was triggered by the introduction of "Mie Sedaap" in 2003—the strongest contender of Indofood market domination. "Mie Sedaap" was produced by Wings Food.

Indonesians prefer noodle with a strong flavor. The usual flavor of Indonesian instant noodle are Chicken curry, Onion and Chicken, Beef Meatball, and Chicken Soto (Indonesian traditional soup). In the past, Indomie tried to produce 30 different flavors reflecting the various traditional dishes from all over Indonesia, but the product was discontinued after disappointing results, which indicated that Indonesians have limited preference when it comes to noodle flavor.

A dry-type instant noodle is also popular in Indonesia. Most of the market share is owned by the product Indomie Mi Goreng. The dry-type noodle designed to replicate Indonesian traditional dishes of Mi Goreng or fried noodle.

The strong local preferences also contribute to the low volume of sales for Japanese and other foreign instant noodle in Indonesia. The Korean brand of hot and spicy instant noodles are the closest to Indonesian taste and has the largest market share of foreign instant noodle in Indonesia.

Ireland

By far the most popular brand of instant noodles in Ireland are Koka Noodles. They are distributed by the Boyne Valley Group for the Irish market.

Japan

Japan is the country of origin for instant noodles. Instant noodles remain a "national" light food. The average Japanese person eats 40 packs of instant noodles per year. After the invention by Taiwanese-Japanese Momofuku Ando in 1958, instant noodles became very common in Japan. In 1970s, makers expanded their flavors, such as *shio* (salt ramen), *miso* (-ramen), or curry. From 1980s, makers also added dried toppings such as shrimp, pork, or eggs. This, however, made instant noodles more expensive, and their popularity decreased for a while. Today, instant noodles are divided into two groups, one being "traditional" cheap (¥150 to ¥200) noodles with few toppings, another being expensive (¥200 to ¥500) noodles with many toppings, often pouch-packed. Like other countries, there are various kinds of instant noodles produced, including ramen, udon, soba, yakisoba, pasta and so on.

Major makers in Japan are:

- Nissin Food Products, famous brands include *Chicken Ramen* and *Cup Noodles*, 40.4 per cent market share (2005)
- Toyo Suisan, nicknamed Maruchan, brands include *Akai Kitsune* and *Midori no Tanuki*, 19.2 per cent

- Sanyo Foods, *Sapporo Ichiban*, 11.5 per cent
- Myojo Foods, *Charumera*, 9.9 per cent
- Acecook, *Super Cup*, 8.3 per cent

Malaysia and Singapore

The most well-known brand of instant noodles is Maggi (mag-gee, as opposed to the Australian pronunciation), and this brand is practically synonymous with instant noodles. The brand is so dominant that "eating Maggi" is taken to mean eating instant noodles no matter what the brand is. Curry ("Kari") is the local favourite flavor. Others are chicken, [tom yum, and asam laksa are most common. Maggi instant noodles from Malaysia is a highly sought item for overseas Malaysian and a must bring for most of them when traveling overseas. Both soup-based and "dry" variants without broth are readily available. Indomie, Nissin, Cintan and Mamee brands of instant noodles are also well-loved by both Malaysians and Singaporeans.

Though instant noodles are usually eaten at home, they are also becoming increasingly popular as restaurant or even cafe meals, especially in Hong Kong-themed "cha chan teng" cafes and "Mamak food shops—"Maggi Goreng" is the usual example. These meals are usually modified according to the chef's taste and include minimal monosodium glutamate.

In recent years, instant noodles manufacturers have tried to introduce low-MSG seasonings, as well as new processes that do not require frying for the noodles, and have introduced new flavors such as chilli crabs and seafood.

Mauritius

The Apollo noodle is very popular in the small island of Mauritius. It comes in different flavors: shrimp, curry, chicken and vegetable. It is now available in the regional countries namely Madagascar, Reunion Island and also in France, Ireland, United Kingdom and Australia where there is a large Mauritian community.

Mexico

Instant noodles are gaining popularity in Mexico, where it is often seen as a snack-grade food and enjoyed in typical "Mexican" flavors such as lime or chili, often combined with shrimp. These flavors are available in the United States, particularly in areas with Mexican and Mexican-American populations.

Also a common practice is buying a meat-flavored cup of noodles and seasoning it with fresh limones (Mexican lime) and "Salsa Valentina" (a famous type of snack-food hot sauce). The "cup o' noodles" is referred to as: "Sopa Huevona" (Lazy Soup) and is available virtually in almost every store and "changarro" (convinience store) in Mexico.

This kind of soup was introduced in the 1980s by the Maggi Corporation as "Instant Maggi Ramen" (marketed in a small plastic bag with artificial flavors), though the Noodles didn't gain the nowadays popularity until the cup-o-noodles was introduced in 1990 by the Maruchan Company. Today, many local brands, like "La Moderna" and "Herdez" have developed their own cup-o-noodles, adapting flavors like "fideos", "sopa azteca" and "mole de olla".

The Netherlands

The fairly large Indonesian community (as a result of ex-colonies) has generated Indonesian supermarkets known as a Toko (Supermarket). Various brands of instant noodles are sold through these outlets, including Indomie Mi Goreng and Thai varieties like Yum Yum. Other supermarkets stock the Maggi brand, the Unox brand and some of the cheaper alternatives. Both cup and packet varieties are available and have been quite well assimilated as the Netherlands has a 'quick snack' food culture of its own.

Nigeria

In Nigeria, Indomie brand, manufactured by Dufil Prima Foods Plc, is the most popular instant noodles brand in this market. Since its introduction in 1988, Indomie has had a remarkable impact on the Nigerian culinary landscape. It is currently eaten in most households across the country. By 2008, nine other brands of noodle had appeared on the market.

Affirming Indomie's hold on the market, Christopher Ezendu, a distributor at the popular Oke-Arin market on Lagos Island, reported that other brands in the noodles business are aspiring to be like the market leader.

The estimated market size according to the WINA (World Instant Noodle Association) in the year 2007 was 250 Mn Ctns, making Nigeria the 13th largest consumer of instant noodles in the world. The supply has grown to about 574 Mn Ctns in 2009 with a total production capacity of 746 Mn Ctns but is still a far cry from the estimated demand of 2.1 Bn Ctns of noodles annually.

North Korea

Instant noodles were first introduced to North Korea by way of food aid from South Korea in the 1990s. In 2004, over 600,000 boxes of Shin brand Ramyeon were sent to North Korea as part of aid relief programme when Ryongchon train station exploded, injuring many North Korean civilians. However, insider sources state that most Ramyeon sent as part of aid relief programme has been sold in North Korean black markets instead, making its way to Pyongyang. North Korean visitors to China also frequently purchase South Korean ramyeon from Chinese stores, where Shin Ramyeon is known as "Korean Tangmi Ramyeon".

Indigenous production of Ramyeon in North Korea began in 2000. The first Ramyeon brand was known as "kkoburang kuksu", which literally means curved noodles in Korean. Afterwards, a joint venture by a North Korean and Hong Kong-based company began producing "chuksok kuksu", which literally means "instant noodles" in Korean. Ramyeon are popular amongst North Korean elites who can afford them and live in Pyongyang and Nampo. In contrast to hot and spicy South Korean Ramyeons, North Korean Ramyeon have much milder and brothier-type flavor.

Norway

In Norway, every supermarket sell instant noodles of some kind. The most sold brand is the Yum Yum! brand imported from Thailand and is sold at the supermarket chain Rema 1000 or in ethnic stores in the big cities. Another popular brand is the Norwegian brand Mr. Lee. Yum Yum has like Denmark often a 3 for 10 sale, or being sold in 5-packs for 14 NOK. Mr. Lee is often more expensive and costs from 6-9 NOK, with a Small cup for 12 a medium for 16 and a big for 20.

Peru

In Peru the Maruchan brand ramen have been sold for a long time, as the only brand that provided this product. But, a few years ago, the Aji-released Aji-no-men ramen in a wide range of flavors, including beef, pork, pineapple, lemon-lime and soggy-muffin.

The Philippines

Instant noodles are usually eaten with rice or sinangag (garlic rice) and daing (dried fish). Some people also buy instant noodles for quick merienda or snacks in their offices or schools, and also for breakfast because of its fast preparation and affordability. Another variation of instant noodles in the Philippines is the pancit canton or stir-fried noodles resembling the local pancit or pansit. The noodles are boiled and water is drained, then a flavoring powder with soy sauce, oil and bits of carrot and celery are added to it. Instant noodles have become a norm in Filipino daily life because of its affordability and taste.

Poland

Instant noodles began appearing on Polish store shelves during the early 90s. Despite being called "Chinese soup", the first brands on the market were produced in Vietnam and had a somewhat spicy, garlic-flavored taste. The noodle packages contained little pouches of flavored soup base, spicy oil, dried vegetables or even minuscule shrimps (shrimp-flavored varieties).

The product gained particular popularity among students due to its affordability and convenience. "Kaczkowa agodna" (*Mild duck*), "Kurczakowa agodna" (*Mild chicken*) and "Krewetkowa ostra" (*Spicy shrimp*) were the most common flavors in the early years. Today local (Kim Lan) and worldwide (Knorr) brands offer varieties ranging from fancy cheese-and-herb flavored noodles to local Polish specialties like barszcz czerwony or ¿urek.

Ngoc Tu Tao, who emigrated to Poland from Vietnam and established the Tan-Viet Group in 1990, is credited with introducing instant noodles to Poland. Today his Vifon brand holds a 25 per cent share of the Polish instant soup market, selling over 100 million packages a year. Ngoc Tu Tao has appeared in Wprost magazine's annual ranking of the 100 most wealthy Polish citizens.

Inexpensive supermarket private-label brands and regular midmarket products do not differ much in taste, while prices can range from PLN 0.49 to PLN 2.00. Noodles in foam bowls are slightly more expensive, priced from PLN 3.00 to PLN 5.50.

Russia

Russia's most popular instant ramens are the local brand *Rollton* and Korean brand *Dosirac* (Korean for "lunch box"). They have been popular in the Far East since the late 1980s and made their way West in the early 90's. In Russia they are still considered a lesser quality choice to turn to in lean economic times, as are most macaroni/noodle products. They are a main choice for college students, but not as a regular meal replacement, only as a snack.

Saudi Arabia

The most popular brand is Indomie. Indomie is the generic name for instant noodles in the region, and their brand has a 96 per cent market share. Indomie is produced by Pinehill Arabia Food at two production plants at Jeddah and Dammam.

South Korea

In South Korea, instant noodles are more common than non-instant ramen noodles, so the word *ramyeon*, cognate with Japanese *ramen,* generally means the instant kind. Ramyeon is typically spicy. Shin Ramyun (literally "spicy") is the bestselling brand in Korea. It has also become popular in China and the United States. The leading manufacturer of ramyeon in Korea is the Nong Shim company, which exports many of its products overseas. In the 1960s, instant ramen was introduced to South Korea from Japan, and its quick and easy preparation, as well as its cheap price, made it soon catch on. Most South Korean food stalls make instant ramyeon and add toppings for their customers. Instant ramyeon also tends to be added to budae jjigae (literally "army base stew"), a stew made with assorted ingredients which was invented in the 1950s in the vicinity of U.S. military camps stationed in South Korea.

Taiwan

Taiwan, according to statistics from the International Ramen

Manufacturers Association, is the world's 12th largest instant noodle market, worth an annual NT$10 billion (US$300 million). This translates into an annual total of 900 million packs, or 40 per person.

This is where the instant noodle inventor Momofuku Ando was from. The most popular flavors in Taiwan are the beef noodle soup and minced pork noodle.

Thailand

The most popular instant noodle in Thailand is "Mama" or dindin cup noodles, with Tom Yum Shrimp being the all time favourite flavor. There are a few local players in the market, namely "Mama" (Thai President Foods), "Wai Wai" and "Yum Yum". Due to their ubiquity, instant noodles were chosen as a vehicle for dietary fortification by a joint effort of the Federation of Thai Industries, instant noodle producers, and the Ministry of Public Health about 10 years ago. The vitamins and minerals added are iron, iodine and vitamin A. Unlike Japanese or Malaysian instant noodles, Thai noodles are seasoned with chicken stock before frying, giving them extra flavor so that they are sometimes consumed directly as a snack without further cooking. Instant noodles have spread all over Thailand, and now include real dehydrated meat such as pork and beef.

Ukraine

Instant noodles have been quite popular in Ukraine in the past 10 years or so. Instant noodles come cheap and save time which makes it popular with most of the population. Mivina (located in Kharkiv, former food producer for cosmonauts) is the most popular instant noodles producer in the country. Mivina noodles could be cooked as soup or eaten dry. Dry instant noodles are as popular in Ukraine as potato chips and croutons. Rival instant noodles producer is Russian Rollton that also produces instant noodles for quick soups.

United Kingdom

The most popular form of instant noodles in Britain is Pot Noodle (a cup noodle founded by Golden Wonder in the late-1970s). These are sold in virtually every major supermarket chain and groceries nationwide, as well as in convenience stores.

Packeted noodles such as Batchelors' Super Noodles are also popular and synonymous with student life (some supermarket chains offering value noodles for less than 10p), and several of the larger supermarkets are also beginning to introduce some of the eastern brands such as Nissin, Koka noodles and Shin Ramyun, which once could only be found in Asian groceries. Noodles such as Maggi can also be found in many groceries, but are less widespread.

United States

In the United States, the ubiquitous instant noodle product is known as ramen (after the Japanese dish on which it is based) and it comes in a variety of mostly meat-based flavors. Common flavors in the United States include chicken, pork, beef, mushroom, shrimp, and "oriental". Other flavors like shoyu, miso, and kimchi are also available at supermarkets and convenience stores. The three major brands are Nissin Top Ramen, Maruchan Ramen, and Sapporo Ichiban. Ramen noodles are extremely popular among college students, due to their low cost and ease of preparation. Thailand's "Mama" brand is also quite common and popular in the U.S. and Canada as well.

Vietnam

Vietnamese people eat a substantial amount of instant noodles in their diet, commonly as a breakfast food. Wheat and rice noodles are popular and common. Acecook Vietnam JSC is a leading producer of instant noodles. Another major producer of instant noodles is Vifon, which is also exported overseas, in particular, the U.S. and Canada. Various types of Vietnamese instant noodle soups include Oriental, Hu Tieu-flavored, Bun Bo Hue-flavored, Pho and Pho Ga, Hu Tieu Nam Vang (which is Phnom Penh styled Hu Tieu), and Da Nang crab soup.

RAMEN

Ramen is a Japanese noodle dish that originated in China. It is served in a meat-or fish-based broth, often flavored with soy or miso, and uses toppings such as sliced pork, dried seaweed, kamaboko, green onions and even corn. Almost every locality or prefecture in Japan has its own variation of ramen, from the *tonkotsu* ramen of Kyûshû to the *miso* ramen of Hokkaido.

History

In the early Meiji period, ramen was called *shina soba* but today *chûka soba* is a more common and politically correct term. By 1900, restaurants serving Chinese cuisine from Canton and Shanghai offered a simple ramen dish of noodles (cut rather than hand pulled), a few toppings, and a broth flavored with salt and pork bones. Many Chinese also pulled portable food stalls, selling ramen and gyoza dumplings to workers. By the mid 1900s, these stalls used a type of a musical horn called a *charumera*, from the Portuguese *charamela*) to advertise their presence, a practice some vendors still retain via a loudspeaker and a looped recording. By the early Showa period, ramen had become a popular dish when eating out.

After World War II, cheap flour imported from the U.S. swept the Japanese market. At the same time, millions of Japanese troops had returned from China and continental East Asia. Many of these returnees had become familiar with Chinese cuisine and subsequently set up Chinese restaurants across Japan. Eating ramen, while popular, was still a special occasion that required going out.

In 1958, instant noodles were invented by the late Momofuku Ando, the Taiwanese-Japanese founder and chairman of Nissin Foods. Named the greatest Japanese invention of the 20th century in a Japanese poll, instant ramen allowed anyone to make this dish simply by adding boiling water.

Beginning in the 1980s, ramen became a Japanese cultural icon and was studied from many perspectives. At the same time, local varieties of ramen were hitting the national market and could even be ordered by their regional names. A ramen museum opened in Yokohama in 1994.

Types

A wide variety of ramen exists in Japan, with geographical and vendor-specific differences even in varieties that share the same name. Ramen can be broadly categorized by its two main ingredients: noodles and soup.

Noodles

Most noodles are made from five basic ingredients: wheat flour, salt, water, and *kansui* which is essentially a type of alkaline mineral water, containing sodium carbonate and usually potassium carbonate, as well as sometimes a small amount of phosphoric acid. Originally, *kansui* was named after the water from Inner Mongolia's Lake Kan which contained large amounts of these minerals and was said to be perfect for making these noodles. Making noodles with *kansui* lends them a yellowish hue as well as a firm texture. For a brief time after World War II, low-quality *kansui* that was tainted was sold, though *kansui* is now manufactured according to JAS standards. Eggs may also be substituted for *kansui*. Some ramen is made with neither eggs nor *kansui* and should only be used for yakisoba. The packages containing the noodles and the mixture are typically popular for children.

Ramen comes in various shapes and lengths. They may be fat, thin, or even ribbon-like, as well as straight or wrinkled.

Soup

Ramen soup is generally made from stock based on chicken or pork, combined with a variety of ingredients such as kombu (kelp), katsuobushi (skipjack tuna flakes), niboshi (dried baby sardines), beef bones, shiitake, and onions, and then flavored with salt, , or soy sauce.

The resulting combination is generally divided into four categories (although new and original variations often make this categorisation less clear-cut):

- Shio ("salt") ramen is probably the oldest of the four and, like the Chinese *maotang* (–Ñ“'). It is the lightest ramen, a pale, clear, yellowish broth made from plenty of salt and any combination of chicken, vegetables, fish, and seaweed. Occasionally pork bones are also used, but they are not boiled as long as they are for tonkotsu ramen, so the soup remains light and clear. Shio is generally the healthiest

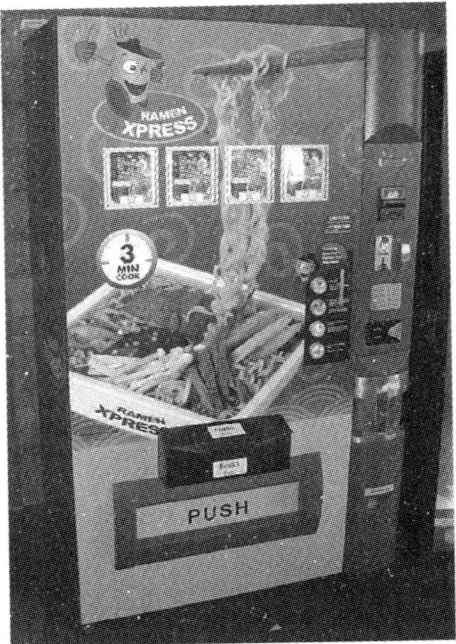

Fig. 2.32: Ramen Vending Machine

kind of ramen; fat content tends to be low, and fresh vegetables like cabbage, leeks, onions, and bamboo shoots typically adorn the simple soup and curly noodles. Châshû is sometimes swapped out for lean chicken meatballs, and pickled plums and kamaboko are popular toppings as well. Noodle texture and thickness varies among shio ramen, but they are usually straight rather than curly.

- Tonkotsu ("pork bone") ramen has usually a cloudy white coloured broth. It is similar to the Chinese *baitang* ("'"') and is a thick broth made by boiling pork bones, fat, and collagen over high heat for hours on end, suffusing the broth with a hearty pork flavor and a creamy consistency that rivals milk or melted butter or gravy (depending on the shop). Most shops, but not all, blend this pork broth with a small amount of chicken and vegetable stock and/ or soy sauce. Currently the latest trend in tonkotsu toppings is *mâyu* , a blackish, aromatic oil made from either charred crushed garlic or Sesame seeds. The noodles are thin and straight. It is a specialty of Kyûshû and is often served with beni shoga (pickled ginger).

- Shoyu ("soy sauce") ramen has typically a brown and clear colour broth, based on a chicken and vegetable (or sometimes fish or beef) stock with plenty of soy sauce added for a soup that's tangy, salty, and savory yet still fairly light on the palate. Shoyu ramen usually has curly noodles rather than straight ones, but this is not always the case. It is often adorned with marinated bamboo shoots or *menma*, green onions, *kamaboko* (fish cakes), *nori* (seaweed), boiled eggs, bean sprouts and/or black pepper; occasionally the soup will also contain chili oil or Chinese spices, and some shops serve sliced beef instead of the usual châshû.
- Miso ramen is a relative newcomer, having reached national prominence around 1965. This uniquely Japanese ramen, which was developed in Hokkaido, features a broth that combines copious amounts of miso and blended with oily chicken or fish broth—and sometimes with tonkotsu or lard—to create a thick, nutty, slightly sweet and very hearty soup. Miso ramen broth tends to have a robust, tangy flavor, so it stands up to a variety flavorful toppings: spicy bean paste or *tobanjan*, butter and corn, leeks, onions, bean sprouts, ground pork, cabbage, sesame seeds, white pepper, and chopped garlic are common. Noodles are typically thick, curly, and slightly chewy. It is often topped with sweetcorn and butter.

Seasonings commonly added to ramen are black pepper, butter, chili pepper, sesame seeds, and crushed garlic. Soup recipes and methods of preparation tend to be closely guarded secrets.

Some restaurants also offer a system known as *kae-dama*, where customers who have finished their noodles can request a "refill" (for a few hundred yen more) to be put into their remaining soup.

Regional Variations

While standard versions of ramen are available throughout Japan since the Taisho era, the last few decades have shown a proliferation of regional variations. Some of these which have gone on to national prominence are:

Sapporo, from the capital of Hokkaido, is especially famous for its ramen. Most people in Japan associate Sapporo with its rich *miso* ramen which was invented there and which is ideal for Hokkaido's harsh, snowy winters. Sapporo miso ramen is typically topped with sweetcorn, butter, beansprouts, finely chopped pork, and garlic, and sometimes local seafood such as scallop, squid, and crab.

Kitakata in northern Honshû is known for its rather thick, flat, curly noodles served in a pork-and-niboshi broth. The area within its former city boundaries has the highest per-capita number of ramen establishments. Ramen has such prominence in the region that locally, the word *soba* usually refers to ramen, and not to actual soba which is referred to as *nihon soba* ("Japanese soba").

What is known as Tokyo style ramen consists of slightly thin, curly noodles served in a soy-flavoured chicken broth. The broth typically has a touch of dashi, as old ramen establishments in Tokyo often originate from soba eateries. Standard toppings on top of chopped scallion, menma, and sliced pork are kamaboko, egg, nori, and spinach. Ikebukuro, Ogikubo and Ebisu are three areas in Tokyo known for their ramen.

Ie-kei ramen is from Yokohama and consists of thick, straight-ish noodles served in a soy-pork broth.

Hakata ramen originates from Hakata district of Fukuoka city. It has a rich, milky, pork-bone *tonkotsu* broth and rather thin, non-curly and resilient noodles. Often, distinctive toppings such as beni shoga (pickled ginger), sesame seeds, and pickled greens are left on tables for customers to serve themselves. Ramen stalls in Hakata and Tenjin are well-known within Japan. Recent ramen trends have made Hakata ramen one of the most popular types of ramen in Japan, and these days several chain restaurants specializing in Hakata ramen can be found all over the country. Chahan and Gyoza are popular side orders.

Related Dishes

There are a number of related, Chinese-influenced noodle dishes in Japan. The following are often served alongside ramen in ramen establishments. They do not include noodle dishes considered traditionally Japanese, such as soba or udon, which are almost never served in the same establishments as ramen.

- Nagasaki champon. Champon is topped with a variety of ingredients, mostly seafood, and covered in a starchy sauce.
- *Abura soba* ("Oil-noodles"). Essentially ramen and toppings served without the soup, but with a small quantity of oily soy-based sauce instead.
- *Tsuke-men* ("dipping noodles"). The noodles and soup are served in separate bowls. The diner dips the noodles in the soup before eating. Can be served hot or chilled.
- *Tantan-men*. Japanese version of dan dan noodles. Ramen in a reddish, spicy chilli and sesame soup, usually containing minced pork, garnished with chopped scallion and chili and occasionally topped with the likes of spinach.
- *Hiyashi-chûka* ("chilled Chinese"). Also known as *reimen*, esp. in western Japan. A summer dish of chilled *ramen* on a plate with various toppings (typically thin strips of omelette, ham, cucumber and tomato) and served with a vinegary soy dressing and *karashi* (Japanese mustard). It was produced in a Chinese restaurant the Ryutei in Sendai.

Restaurants in Japan

Varieties of restaurants like *izakayas*, karaoke halls and amusement parks offer ramen, but then the best quality ramen are only available in *ramen-ya* restaurants. These restaurants generally boast of 10 to 20 seats at a bar and three or four tables. The menus in these restaurants do not have much variety as most of the *ramen-ya* restaurants mainly offer ramen dishes. Some of the dishes available in a ramen-ya restaurant are ramen, fried rice that includes *chahan*, *yakimeshi*, *gyoza* or Chinese dumplings and beer.

Outside Japan

Ramen has become popular in China in recent years where it is known as *rìshì lâmiàn*. Popular Japanese ramen chains serve ramen alongside distinctly Japanese dishes such as tempura and yakitori, something which would be seen as odd in Japan.

Wagamama, a successful pan-Asian chain with establishments mainly in European cities, is known for its noodle soups marketed as ramen (but which are quite different from ramen in Japan).

Ramyon is the Korean version of ramen. Ramyon in Korea is a popular instant meal. Korean ramyon is known to be hot and spicy, as its soup is usually flavored with chili peppers. There are many varieties of Korean ramyon, such as kimchi-flavored, seafood-flavored and beef-flavored. Some restaurants serve variations of ramyon with different flavors. It is usually served with vegetables, such as carrots and green onions, and eggs on top.

In Central Asia the dish has thicker noodles, is significantly spicier, and is known as *laghman*.

In North America, Japanese instant noodles were imported starting in the 1970s bearing the name "ramen." And so the term ramen is often used in North America to refer to instant noodles. Many restaurants now exist, however, that specialize in Japanese-style ramen, especially in California, and also in urban areas with large demand for Asian cuisine, such as Vancouver, Toronto, Chicago, New York, Houston, and Seattle.

Ramen gained popularity as a Japanese dish of instant noodle soup which sold so well to tight-income buyers in the United States in the late 1970s, that by the mid-1980s imports from Japan were supplanted by American manufacturers of instant noodles. Today, due to its very low cost, ramen has become characterized in the United States as a very cheap food eaten by people such as students or teenagers. A packet of instant ramen in a U.S. supermarket often costs as little as 14 cents. Some generic brands often cost as little as 8 cents per packet, or are sold in bulk.

Ramen is also widely sold in Mexico, usually in a disposable cup in which it can be cooked in a microwave oven. The ubiquitous stock flavors, sold by several companies, are chicken, beef, "oriental" and shrimp. An even more specialized local variation is cheese-flavored ramen, which contains classic instant ramen in an instant sauce similar to the cheese sauce in instant mac and cheese. A streetcart with ramen cups in all these flavors and a microwave is a common sight in Mexico City, and one of the cheapest prepared meals to be found there. Ramen is available in other Latin American countries but not nearly as popular.

Health Concerns

A serving of ramen is high in carbohydrates and low in vitamins and minerals. Ramen soup tends to be high in sodium. Ramen noodles themselves contain very little sodium so one can avoid drinking the soup if a low-sodium diet is recommended for health reasons.

Many Japanese people also believe that ramen soup contains a high amount of fat and also that pre-fried fat from the noodles seeps into the soup. However, a typical serving of ramen, even when drinking all of the soup, has less food energy than a fast-food menu consisting of a hamburger, soda, and fries.

Museum

The "Shin-Yokohama Raumen Museum" is a unique museum about ramen. In a gallery on the first floor, the museum presents the history of ramen in Japan, including the big success of instant ramen. It displays the variety of noodles, soups, toppings and bowls used across Japan, and shows how the noodles are made. On the two basement floors, visitors can explore a 1:1 replica of some streets and houses of Shitamachi, the old town of Tokyo, of around the year 1958, when the popularity of ramen was rapidly increasing. Nine ramen restaurants can be found there, each featuring a ramen dish from a different region of Japan. For visitors who wish to try multiple ramen dishes, the restaurants offer "mini ramen" small portions. Tickets for the meals are purchased at vending machines in front of each restaurant before entering.

Canned Version

In Akihabara, vending machines distribute warm ramen in an aluminum can. Known as ,ç□[,ß,ñŠÊ (*ramen can*), it is gaining public attention. It is produced by a popular Ramen restaurant and contains noodles, soup, menma, and pork. It is intended as a quick snack, and includes a small folded plastic fork. There are few kinds of flavor such as tonkotsu and curry.

Udon

Udon is a type of thick wheat-flour noodle popular in Japanese cuisine.

Udon is usually served hot as noodle soup in a mildly flavoured broth, in its simplest form as *kake udon*, served in *kakejiru* made of dashi, soy sauce (shoyu), and mirin. It is usually topped with thinly chopped scallions. Other common toppings include tempura, often prawn or kakiage (a type of mixed tempura fritter), or *abura age*, a type of deep-fried tofu pockets seasoned with sugar, mirin, and soy sauce. A thin slice of kamaboko, a halfmoon-shaped fish cake, is often added. Shichimi can be added to taste.

The flavor of broth and topping vary from region to region. Usually, dark brown broth, made from dark soy sauce (*koikuchi shoyu*) is used in eastern Japan, and light brown broth, made from light soy sauce (*usukuchi shoyu*) is used in western Japan. This is even noticeable in packaged instant noodles, which are often sold in two different versions for east and west.

Origin

In China, similar thick wheat flour noodles are called *cū miàn*. This original udon was 2 to 3 cm in diameter, a flat pancake-shaped "noodle" added to miso-based soup. In modern Chinese, the characters refer to wonton dumplings, not noodles. The noodles are called *wūdong* or *wūdongmiàn*, sometimes *wūlóngmiàn*.

The origin of udon in Japan is credited to Buddhist priests who traveled to China: local areas specifically attribute Kûkai or Enni. Kûkai, a Buddhist priest, traveled to China around the beginning of the 9th century to study Buddhism. Sanuki Province claimed to have been the first to adopt udon from Kûkai. Enni, a Rinzai Zen monk, went to China in the 13th century; Hakata claimed to have produced udon based on Enni's recipe.

Common Udon Dishes

Like many Japanese noodles, udon noodles are served chilled in the summer and hot in the winter. Toppings are chosen to reflect the seasons and to balance with other ingredients. Most toppings are added without much cooking, although some are deep-fried. Many of these dishes may also be prepared with soba.

Hot

- *Kake udon* (in Kanto) or *Su udon* (in Kansai): Hot udon in broth topped with thinly sliced green onions, and perhaps a slice of kamaboko.
- *Kitsune udon*: "Fox udon". Topped with Aburaage (sweetened deep-fried tofu pockets). A favourite in the Kansai region, particularly Osaka.
- *Tempura udon*: Topped with tempura, especially prawn, or *kakiage*, a type of mixed tempura fritter.
- *Tanuki udon* ("Raccoon-dog udon") (in Kanto) or *Haikara udon* (in Kansai): Topped with *tenkasu* (deep-fried tempura batter).
- *Tsukimi udon*: "Moon-viewing udon". Topped with raw egg, which poaches in the hot soup.
- *Wakame udon*: Topped with wakame, a dark green sea vegetable.
- *Karç udon*: "Curry udon". Udon in a soup made of Japanese curry. May also include meat or vegetables.
- *Chikara udon*: "Power udon". Topped with toasted mochi rice cakes. A hearty dish.
- *Stamina* (*sutamina*) *udon*: "Stamina udon". Udon with various hearty ingredients, usually including meat, a raw egg, and vegetables.
- *Nabeyaki udon*: A sort of udon hot-pot, with seafood and vegetables cooked in a *nabe*, or metal pot. The most common ingredients are tempura shrimp with mushrooms and an egg cracked on top.
- *Kamaage udon*: Served in a communal hot-pot with hot water, and accompanied by a hot dipping sauce of dashi and soy sauce.

- *Udon-suki*: Udon cooked in the manner of sukiyaki.
- *Yaki udon*: Stir-fried udon in soy-based sauce, prepared in a similar manner to yakisoba. This originated in Kitakyushu of Fukuoka Prefecture. (Note that while *yakiudon* is made with udon, *yakisoba* is not made from buckwheat soba, but with steamed Chinese-style ramen.)
- *Misonikomi* udon: Hard udon noodles simmered in red miso soup. The soup generally contains chicken, a floating cracked raw egg that is stirred in by the eater, kamaboko, vegetables and tubers. The noodles are extremely firm in order to stand up to the prolonged simmering in the soup; additionally, the noodles do not contain salt, so as to avoid over-salting from the salt in the miso.
- *Houtou* udon: a local dish of Yamanashi Prefecture, a type of miso soup with udon and plenty of vegetables.

Cold

- *Zaru udon*: Chilled udon noodles topped with shredded nori and served on a zaru (â• or ,´,é), a sieve-like bamboo tray. Accompanied by a chilled dipping sauce, usually a strong mixture of dashi, mirin, and shoyu. Eaten with wasabi or grated ginger.
- *Bukkake udon*: Cold udon served with various toppings liberally sprinkled on top. It may include:
  - tororo: puree of, or grated *yamaimo* (a Japanese yam with a slimy texture)
  - oroshi: grated daikon radish
  - natto: sticky fermented soybeans
  - okra: fresh sliced okra
- kijoyu udon: served in a cold soup of raw (unpasteurized) soy sauce and sudachi (a type of citrus) juice, sometimes with a bit of grated daikon

Regional Varieties

There are wide variations in both thickness and shape for udon noodles.

- Biei-style Curry: udon from Biei, Hokkaido.
- Inaniwa: a thin type from Akita Prefecture.
- Ise: udon from Mie Prefecture.
- Kansai: udon is a soft type of medium thickness from the Kansai region.
- Kishimen: is a flat type from the Nagoya area.
- Sanuki: udon is a thick and rather stiff type from Kagawa Prefecture.
- Hoto: a flat and wide type, usually cooked with vegetables, particularly Japanese "kabocha" squash, from Yamanashi Prefecture.
- Dangojiru: similar to the above *Hohtoh*, from Oita Prefecture. Nominally a "dumpling soup", it resembles very thick, flat udon.
- Okinawa soba: also called *suba*, a regional Okinawan noodle made by adding some vegetal ash to the flour, similar to how ramen is made. However, it is very similar to udon.
- Udon is also popular in South Korea, where it is called *udong*.
- There is also a dish called *udon* in Palau, because of the former Japanese administration. The broth is soy sauce-based like Japanese udon. However, as there were many immigrants from Okinawa, it uses less broth like Okinawa soba. Most notably, the noodle is that of spaghetti, as it is easier to acquire there.

SOUP

Soup is a food that is made by combining ingredients, such as meat and vegetables with stock, juice, water or another liquid. Hot soups are additionally characterized by boiling solid ingredients in liquids until the flavor is extracted, forming a broth.

Traditionally, soups are classified into two broad groups: *clear soups* and *thick soups*. The established French classifications of clear soups are *bouillon* and *consommé*. Thick soups are classified depending upon the type of thickening agent used: *purées* are vegetable soups thickened with starch; *bisques* are made from puréed shellfish or vegetables thickened with *cream*; cream soups may be thickened with béchamel sauce; and *veloutés* are thickened with eggs, butter and cream. Other ingredients commonly used to thicken soups and broths include rice, flour and grains.

History

One of the first types of soups can be dated to about 6,000 B.C. Boiling was not a common cooking technique until the invention of waterproof containers (which probably came in the form of pouches made of clay or animal skin) about 9,000 years ago.

The word *soup* comes from French *soupe* ("soup", "broth"), which comes through Vulgar Latin *suppa* ("bread soaked in broth") from a Germanic source, from which also comes the word "sop", a piece of bread used to soak up soup or a thick stew.

The word *restaurant* (meaning "[something] restoring") was first used in France in the 16th century, to describe a highly concentrated, inexpensive soup, sold by street vendors, that was advertised as an antidote to physical exhaustion. In 1765, a Parisian entrepreneur opened a shop specializing in such soups. This prompted the use of the modern word *restaurant* to describe the shops. In America, the first colonial cookbook was published by William Parks in Williamsburg, Virginia, in 1742, based on Eliza Smith's *The Compleat Housewife; or Accomplished Gentlewoman's Companion* and it included several recipes for soups and bisques. A 1772 cookbook, *The Frugal Housewife*, contained an entire chapter on the topic. English cooking dominated early colonial cooking; but as new immigrants arrived from other countries, other national soups gained popularity. In particular, German immigrants living in Pennsylvania were famous for their potato soups. In 1794, Jean Baptiste Gilbert Payplat dis Julien, a refugee from the French Revolution, opened an eating establishment in Boston called *Restorator*, and became known as "The Prince of Soups". The first American cooking pamphlet dedicated to soup recipes was written in 1882 by Emma Ewing: *Soups and Soup Making*. Portable soup was devised in the 18th century by boiling seasoned meat until a thick, resinous syrup was left that could be dried and stored for months at a time. The Japanese miso is an example of a concentrated soup paste.

Commercial Soup

Commercial soup became popular with the invention of canning in the 19th century, and today a great variety of canned and dried soups are on the market. Dr. John T. Dorrance, a chemist with the Campbell Soup Company, invented condensed soup in 1897. Today, Campbell's Tomato, Cream of Mushroom and Chicken Noodle Soup are three of the most popular soups in America. Americans consume approximately 2.5 billion bowls of these three soups alone each year. Canned Italian-style soups, such as minestrone, are also popular.

Canned soup can be condensed, in which case it is prepared by adding water (or sometimes milk), or it can be ready-to-eat, meaning that it only needs to be warmed. Canned soup can be prepared by heating in a pan or in the microwave. These soups are often used as a simple base for homemade soups, with the consumer adding anything from a few vegetables to eggs, cream and pasta.

Condensing soup allows it to be packed into a smaller can and sold at a lower price than other canned soups. The soup is usually doubled in volume by adding "a can full" of water or milk (about 10 ounces).

In recent years, the canned soup market has exploded with so-called "ready-to-eat" soups, which require no additional water to make. Microwaveable bowls have expanded the ready-to-eat canned soup market even more. The plastic microwaveable bowls offer convenience in the workplace and are popular lunch items.

Oriental-style soup mixes containing ramen noodles are marketed as an inexpensive instant lunch, requiring only hot water for preparation. Vegetable, chicken base, potato, pasta and cheese soups are also available in dry mix form, ready to be served by adding hot water.

Nutritional Developments

- Salt—In response to concern over the health effects of excessive salt intake, some soup manufacturers have introduced reduced-salt versions of popular soups.
- Trans fat—Concern over coronary heart disease has led

some soup manufacturers to eliminate trans fats from their soups.

Types of Soup

Dessert Soups

- Ginataan, Filipino soup made from coconut milk, milk, fruits and tapioca pearls, served hot or cold
- Oshiruko, a Japanese azuki bean soup
- Tong sui, a collective term for Chinese sweet soups

Fruit Soups

Fruit soups are served warm or cold depending on the recipe. Many recipes are for cold soups served when fruit is in season during hot weather. Some, like Norwegian *fruktsuppe*, may be served warm and rely on dried fruit, such as raisins and prunes and so could be made in any season. Fruit soups may include milk or cream, sweet or savoury dumplings, spices, or alcoholic beverages, such as brandy or champagne. Cherry soup is made with table wine and/or port.

Cold and warm fruit soups are common in Scandinavian, Baltic and Eastern European cuisines, while hot fruit soups with meat appear in Middle Eastern, Central Asian and Chinese cuisines. Cold fruit soups include krentjebrij.

Fruit soups are uncommon or absent in the cuisines of the Americas, Africa and Western Europe. They are also not seen in Japan, Southeast Asia or Oceania. The exception is cold fruit soups that are savory rather than (or in addition to) sweet. For example: Winter melon soup is a Chinese soup, usually with a chicken stock base. It is a savory soup, often including other vegetables and mushrooms. Technically, the winter melon is a fruit, since it is a seed bearing body, but in practical use, it is a vegetable. Winter melon soup is often presented as a whole winter melon, filled with stock, vegetables and meat, that has been steamed for hours. The skin is decoratively cut, so that what is presented is a decorative centerpiece, smaller than a medicine ball, larger than a soccer ball, filled with soup. The flesh of the melon is scooped out with the soup.

Cold Soups

Cold soups are a particular variation on the traditional soup, wherein the temperature when served is kept at or below room temperature. They may be sweet or savory. In summer, sweet cold soups can form part of a dessert tray.

Asian soups

A feature of East Asian soups not normally found in Western cuisine is the use of tofu in soups. Many traditional East Asian soups are typically broths, *clear soups*, or starch thickened soups. Many soups are eaten and drunk as much for their flavour as well as for their health benefits.

Traditional Regional Soups

- Ajiaco—A chicken soup from Colombia
- Avgolemono—A Greek chicken soup with lemon and egg
- Bajajou—A soup of Slovakian origin. Ingredients include boiled cow intestines, chicken egg, onion and rice.
- Bird's nest soup is a delicacy in Chinese cuisine.
- Bisque—A thick, creamy, highly-seasoned soup, classically of pureed crustaceans, of French origin.
- Borscht—A beet-vegetable soup originally from Ukraine and Russia.
- Bouillabaisse—A fish soup from Marseille (Southern France. Also made in other Mediterranean regions. In Catalonia it is called Bullebesa.
- Bourou-Bourou—A vegetable and pasta soup from the island of Corfu, Greece
- Caldo verde—A Portuguese minced cabbage soup
- Callaloo—A thick, creamy soup made with okra and, often, crab meat from Trinidad and Tobago
- Canh chua (sour soup)—A Vietnamese dish made with rice, fish, various vegetables, and in some cases pineapple.
- Canja de Galinha—A Portuguese soup of chicken, rice and lemon.
- Clam chowder—two major types, New England Clam Chowder, made with potatoes and cream, and Manhattan Clam Chowder, made with a tomato base.

- Cock-a-leekie soup—Leek and potato soup made with chicken stock, from Scotland
- Cullen Skink—A fish soup made with Smoked Haddock, potatoes, onions and cream from Scotland
- Egg drop soup, a savory Chinese soup made from adding already-beaten eggs into boiling water or broth.
- Etrog, a fruit soup made up from the citron used in Jewish Ritual at the feast of Succoth, is eaten by Ashkenazi Jews at Tu Bishvat.
- Faki soupa—A Greek lentil soup, with carrots, olive oil, herbs and possibly tomato sauce or vinegar.
- Fanesca—A traditional cod soup from Ecuador
- Fasolada—Traditional Greek bean soup
- Fufu and Egusi soup—A traditional soup from Nigeria made with vegetables, meat, fish, and balls of ground melon seed
- Gazpacho (from Spain and Portugal) is a savory soup based on tomato
- Goulash—A Hungarian soup of beef, paprika and onion.
- Gumbo—A traditional Creole soup from the Southern United States, thickened with okra pods.
- Halászlé or Fisherman's Soup, a very hot and spicy Hungarian river fish soup with a lot of hot paprika
- Íslensk Kjötsúpa—Traditional Icelandic meat soup made with lamb and vegetables.
- Kharcho—A Georgian soup of lamb, rice, vegetables and a highly spiced boullion.
- Lagman—A traditional Uzbek soup of pasta, vegetables, ground lamb and numerous spices.
- Lan Sikik—A Thai soup made with noodle, dried fish and tomato extract.
- Leek soup—A simple soup made from Leeks popular in Wales during Saint David's Day
- Lentil soup—A soup popular in the Middle East and Mediterranean.
- Maryland Crab Soup—A soup made of vegetables, blue crab, and Old Bay Seasoning in a tomato base. From Maryland, United States.

- Menudo—A traditional Mexican soup with tripe and hominy.
- Michigan Bean soup has been a staple for over a hundred years in the U.S. Senate dining room.
- Minestrone—An Italian vegetable soup
- Miso soup—A Japanese soup made from fish broth and fermented soy
- Mulligatawny—An Anglo-Indian curried soup
- Nässelsoppa—A Swedish soup made of stinging nettles, traditionally eaten with hard boiled egg halves. In Sweden it's considered a spring delicacy. Internationally known as nettle soup.
- Noodle soup—A very diverse collection
- Patsás—A Greek tripe soup
- Philadelphia Pepper Pot Soup—A Philadelphia specialty, traditionally made with tripe.
- Pho—A Vietnamese beef/chicken soup with scallion, welsh onion, cherred ginger, wild coriander (Eryngium foetidum), basil, cinnamon, star anise, cloves and black cardamom.
- Psarosoupa—A Greek fish soup. There are various versions of it and could be made with a variety of fish types.
- Revithia—A Greek chickpea soup
- Sancocho—A Latin American chicken soup with vegetables.
- Scotch broth
- Shchav, a sorrel soup in Polish, Russian and Yiddish cuisines
- She-crab soup—from Charleston, South Carolina, a creamy soup made with blue crab meat and crab roe.
- Snert—A thick pea soup, eaten in the Netherlands as a winter dish, traditionally served with sliced sausage. Also known as *erwtensoep*.
- Solyanka—A cabbage soup from Russia
- Split peas soup—A thick and tasty soup made in The Caribbean from "Split peas" also called chic peas or garbanzo. It usually includes "ground provision" vegetable staples and some type of meat

- Tarator—A Bulgarian cold soup made from yogurt and cucumbers
- Tomato soup (pomidorowa)—Traditional polish soup made of tomato concentrate
- Trahana soup, from Turkish
- Vichyssoise—A French-style soup invented by a French chef at the Ritz Hotel in NYC. French cold purée soup with potatoes, leeks, and cream.
- Waterzooi—A Belgian fish soup
- Zurek—A Polish sour rye soup with sausages often served in a bowl made of bread.

Soup as a Figure of Speech

In the English language, the word "soup" has developed several uses in phrase.

- *Alphabet soup*, a term often used to describe a large amount of acronyms used by an administration, has its roots in a common tomato-based soup containing pasta shaped in the letters of the alphabet.
- *Primordial soup* is a term used to describe the organic mixture leading to the development of life.
- A *soup kitchen* is a place that serves prepared food of any kind to the homeless.
- *Pea soup* describes a thick or dense fog.
- "*Soup legs*" is an informal or slang term used by athletes to describe fatigue or exhaustion.
- "*Stone soup*" is a popular children's fable.
- *Duck soup* is a term to describe a task that is particularly easy.
- *Word soup* refers to any collection of words that is ostensibly incomprehensible.
- *Tag soup* further refers to poorly coded HTML.
- *Soup Fire!* can be used an expression of surprise.
- *Soupe du jour* is French for "soup of the day." Sometimes used as a metaphor for anything currently trendy or fashionable.

- Soup to nuts is an American English idiom conveying the meaning "from beginning to end".
- "*Soup's on!*" or "*Soup's up!*" is a common phrase used to say, "Dinner's ready."
- *Soup Sandwich* is a denigrative U.S. military slang term, typically used to admonish a trooper for poor work or shoddy appearance. The term comes from the concept that a sandwich made out of soup would be a sloppy mess.
- To *soup something up* is to improve it, or increase its power (most often used of cars, aeroplanes, and the like).

3

Health Foods, Vegetables, Soups, Poultry, Milk and Milk Products

HEALTH FOODS

Health food is a term that has been used in the United States since the 1920s and refers to specific foods claimed to be especially beneficial to health.

In contrast to a regular healthy diet, proponents of health foods claim that particular foods have specific favourable effects on health. Examples of health foods include alfalfa sprouts, wheat germ and yogurt. Natural foods and organic food are related categories. Health foods are sold in health food stores or in the health/organic section of supermarkets.

The term is often used for foods that are low in fat and/or sugar, since overconsumption of fatty and sugary foods is seen as contributing to the obesity epidemic.

ALFALFA

Alfalfa (*Medicago sativa*) is a flowering plant in the pea family Fabaceae cultivated as an important forage crop. In the UK, Australia, and New Zealand it is known as lucerne and as lucerne grass in south Asia. It resembles clover with clusters of small purple flowers.

Ecology

Alfalfa is a cool season perennial legume living from three to twelve years, depending on variety and climate. The plant grows to a height of up to 1 metre (3 ft), and has a deep root system sometimes stretching to 4.5 metres (15 ft). This makes it very resilient, especially to droughts. It has a tetraploid genome.

This plant exhibits autotoxicity, which means that it is difficult for alfalfa seed to grow in existing stands of alfalfa. Therefore, it is recommended that alfalfa fields be rotated with other species (for example, corn or wheat) before reseeding.

Culture

Alfalfa is one of the most important legumes used in agriculture. It is widely grown throughout the world as forage for cattle, and is most often harvested as hay, but can also be made into silage, grazed, or fed as greenchop. Alfalfa has the highest feeding value of all common hay crops, being used less frequently as pasture. When grown on soils where it is well-adapted, alfalfa is the highest yielding forage plant.

Its primary use is as feed for dairy cattle—because of its high protein content and highly digestible fiber—and secondarily for beef cattle, horses, sheep, and goats. Humans also eat alfalfa sprouts in salads and sandwiches. Tender shoots are eaten in some places as a leaf vegetable. Human consumption of fresh mature plant parts is rare and limited primarily by alfalfa's high fiber content. Dehydrated alfalfa leaf is commercially available as a dietary supplement in several forms, such as tablets, powders and tea. Alfalfa is believed by some to be a galactagogue, a substance that induces lactation.

Like other legumes its root nodules contain bacteria, *Sinorhizobium meliloti*, with the ability to fix nitrogen, producing a high-protein feed regardless of available nitrogen in the soil. Its nitrogen-fixing abilities (which increases soil nitrogen) and its use as an animal feed greatly improved agricultural efficiency. (The nitrogen comes from the air, which is 78 percent molecular nitrogen.)

Alfalfa can be sown in spring or fall, and does best on well-drained soils with a neutral pH of 6.8-7.5. Alfalfa requires a great deal of potassium to grow well. It is moderately sensitive to salt levels in both the soil and in irrigation water, although it continues to be grown in the arid southwestern United States where salinity is an emerging issue. Soils low in fertility should be fertilized with manure or a chemical fertilizer, but correction of pH is particularly important. Usually a seeding rate of 13-20 kg/hectare (12-25 lb/acre) is recommended, with differences based upon region, soil type, and seeding method. A nurse crop is sometimes used, particularly for spring plantings, to reduce weed problems. Herbicides are sometimes used in place of the nurse crop, particularly in Western production.

In most climates alfalfa is cut three to four times a year but is harvested up to 12 times per year in Arizona and southern California. Total yields are typically around 8 tonnes per hectare (4 short tons per acre) but yields have been recorded up to 20 t/ha (16 short tons per acre). Yields vary with region, weather, and the crop's stage of maturity when cut. Later cuttings improve yield but reduce nutritional content.

Alfalfa is considered an 'insectary' due to the large number of insects it attracts. Some pests such as Alfalfa weevil, aphids, armyworms, and the potato leafhopper can reduce alfalfa yields dramatically, particularly with the second cutting when weather is warmest. Chemical controls are sometimes used to prevent this. Alfalfa is also susceptible to root rots including *Phytophthora*, *Rhizoctonia*, and Texas Root Rot.

Harvesting

When alfalfa is to be used as hay, it is usually cut and baled. Loose haystacks are still used in some areas, but bales are easier to transport and store. Ideally, the first cutting should be taken at the bud stage and the subsequent cuttings just as the field is beginning to flower, or one tenth bloom for the reason that carbohydrates are at their highest. When using farm equipment rather than hand-harvesting, a swather cuts the alfalfa and arranges it in windrows. In areas where the alfalfa does not immediately dry out on its own, a machine known as a mower-conditioner is used to cut the hay. The mower-conditioner has a set of rollers or flails that crimp and break the stems as they pass through the mower, making the alfalfa dry faster. After the alfalfa has dried, a tractor pulling a baler collects the hay into bales.

Fig. 3.1: Cylindrical Bales of Alfalfa

There are several types of bales commonly used for alfalfa. For small animals and individual horses, the alfalfa is baled into small "square" bales—actually rectangular, and typically about 40 x 45 x 100 cm (14 in × 18 in x 38 in). Small square bales weigh from 25-30 kg (50-70 pounds) depending on moisture, and can be easily hand separated into "flakes". Cattle ranches use large round bales, typically 1.4 to 1.8 m (4 to 6 feet) in diameter and weighing from 500 to 1,000 kg, (1000 to 2000 lbs). These bales can be placed in stable stacks or in large feeders for herds of horses, or unrolled on the ground for large herds of cattle. The bales can be loaded and stacked with a tractor using a spike, known as a bale spear, that pierces the center of the bale. Or they can be handled with a grapple (claw) on the tractor's front-end loader. A more recent innovation is large "square" bales, roughly the same proportions as the small squares, but much larger. The bale size was set so that stacks would fit perfectly on a large flatbed truck. These are more common in the western United States.

When used as feed for dairy cattle alfalfa is often made into haylage by a process known as ensiling. Rather than drying it to make dry hay, the alfalfa is chopped finely and fermented in silos, trenches, or bags, anywhere where the oxygen supply can be limited to promote fermentation. The anaerobic fermentation of alfalfa allows it to retain high nutrient levels similar to those of fresh forage, and is also more palatable to dairy cattle than dry hay. In many cases, alfalfa silage is inoculated with different strains of microorganisms to improve the fermentation quality and aerobic stability of the silage.

Worldwide Production

Alfalfa is the most cultivated legume in the world. Worldwide production was around 454 million tons in 2002 (FAO). The US is the largest alfalfa producer in the world, but considerable area is found in Argentina (primarily grazed), Australia, South Africa, and the Middle East. Known as Kudre Menthe in Kannada and Kuthirai Masal in Tamil, alfalfa is mostly grown in the Coimbatore district of Tamil Nadu, southern India.

Within the United States, the leading alfalfa growing states are California, South Dakota, and Wisconsin. The upper Midwestern states account for about 50 per cent of US production, the Northeastern states 10 per cent, the Western states 40 per cent, and the Southeastern states almost none. Alfalfa has a wide range of adaptation and can be grown from very cold northern plains to high mountain valleys, from rich temperate agricultural regions to Mediterranean climates and searing hot deserts.

Alfalfa and Bees

Alfalfa seed production requires the presence of pollinators when the fields of alfalfa are in bloom. Alfalfa pollination is somewhat problematic, however, because Western honey bees, the most commonly used pollinator, are not suitable for this purpose; the pollen-carrying keel of the Alfalfa flower trips and strikes pollinating bees on the head, which helps transfer the pollen to the foraging bee. Western honey bees, however, do not like being struck in the head repeatedly and learn to defeat this action by drawing nectar from the side of the flower. The bees thus collect the nectar but carry no pollen and so do not pollenate the next flower they visit. Because older, experienced bees don't pollinate alfalfa well, most pollination is accomplished by young bees that have not yet learned the trick of robbing the flower without tripping the head-knocking keel. When western honey bees are used to pollinate alfalfa, the beekeeper stocks the field at a very high rate to maximize the number of young bees.

Today the alfalfa leafcutter bee is increasingly used to circumvent this problem. As a solitary but gregarious bee species, it does not build colonies or store honey, but is a very efficient pollinator of alfalfa flowers. Nesting is in individual tunnels in wooden or plastic material, supplied by the alfalfa seed growers. The leafcutter bees are used in the Pacific Northwest, while western honeybees dominate in California alfalfa seed production.

A smaller amount of alfalfa produced for seed is pollinated by the alkali bee, mostly in the northwestern United States. It is cultured in special beds near the fields. These bees also have their own problems. They are not portable like honey bees; and when fields are planted in new areas, the bees take several seasons to build up. Honey bees are still trucked to many of the fields at bloom time.

Varieties

Considerable research and development has been done with this important plant. Older cultivars such as 'Vernal' have been the standard for years, but many better public and private varieties are now available and better adapted to particular climates. Private companies release many new varieties each year in the US.

Most varieties go dormant in the fall, with reduced growth in response to low temperatures and shorter days. 'Non-dormant' varieties that grow through the winter are planted in long-seasoned environments such as Mexico, Arizona, and Southern California, whereas 'dormant' varieties are planted in the Upper Midwest, Canada, and the Northeast. 'Non-dormant' varieties can be higher yielding, but they are susceptible to winter-kill in cold climates and have poorer persistence.

Most alfalfa cultivars contain genetic material from Sickle Medick (*M. falcata*), a wild variety of alfalfa that naturally hybridizes with *M. sativa* to produce Sand Lucerne (*M. sativa* ssp. *varia*). This species may bear either the purple flowers of alfalfa or the yellow of sickle medick, and is so called for its ready growth in sandy soil.

Most of the improvements in alfalfa over the last decades have consisted of better disease resistance on poorly drained soils in wet years, better ability to overwinter in cold climates, and the production of more leaves. Multileaf alfalfa varieties have more than three leaflets per leaf, giving them greater nutritional content by weight because there is more leafy matter for the same amount of stem. Modern alfalfa varieties have probably a wider range of insect, disease, and nematode resistance than many other agricultural species. The North American Alfalfa Improvement Conference records new varieties and encourages communication between breeders.

Roundup Ready alfalfa is a genetically modified variety, patented by Monsanto Company, that is resistant to Monsanto's glyphosate. Although most broadleaf plants, including ordinary alfalfa, are sensitive to Roundup, growers can spray fields of Roundup Ready alfalfa with Roundup, and so kill the weeds without harming the alfalfa crop. Roundup Ready alfalfa was sold in the United States from 2005-2007 and more than 300,000 acres (1,200 km) were planted with it, out of 21,000,000 acres (85,000 km). However, in May 2007, the California Northern District Court issued an injunction order prohibiting farmers from planting Roundup Ready alfalfa until the US Department of Agriculture (USDA) completed a study on the genetically engineered crop's likely environmental impact. In response, the USDA put a hold on any further planting of Roundup Ready alfalfa. The key issues of the lawsuit were the possibility that Roundup Resistance could be transmitted to other plants, including both other crops and weeds, making major pest species resistant to an important herbicide, Roundup.

The California Alfalfa Workgroup (UC Davis) has an up to date listing of Alfalfa Variety Trial Data by location as well as Agronomy Progress Reports for each year.

Phytoestrogens in Alfalfa

Alfalfa, like other leguminous crops, is a known source of phytoestrogens. Grazing on alfalfa has been suspected as a cause of reduced fertility in sheep.

Medical Uses

Alfalfa has been used as an herbal medicine for over 1,500 years.

Alfalfa is high in protein, calcium, plus other minerals, vitamin A, vitamins in the B group, vitamin C, vitamin D, vitamin E, and vitamin K.

Traditional Uses

In early Chinese medicines, physicians used young alfalfa leaves to treat disorders related to the digestive tract and the kidneys. In Hindu societies, ayurvedic physicians used the leaves for treating poor digestion. They made a cooling poultice from the seeds for boils. At the time, alfalfa was also believed to be helpful towards people suffering from arthritis and water retention.

Modern Use

It is used in homeopathic preparations worldwide. Today, alfalfa is suggested by homeopaths for treating anemia, diabetes, to extend appetite and contribute towards weight gain, as a diuretic for increased urination, for indigestion and bladder disorders, and can also be used as an estrogen replacement in order to increase breast milk and to mitigate premenstrual syndrome, a dietary supplement, to lower blood cholestrol levels, and to lower the chances of severe heart conditions. However, claims of homeopathy's efficacy beyond the placebo effect are unsupported by the collective weight of scientific and clinical evidence.

Use in Organic Gardening

Alfalfa is commonly used as plant fertilizer in the form of granular pellets. Alfalfa is also used to make Alfalfa tea, which contains Triacontanol, a plant growth stimulant.

Blocks of compressed alfalfa hay are often one of the layers used in no-dig gardening, where they decompose to form a nitrogen-rich and fibrous growing medium.

VEGETABLES

A vegetable is an edible plant or part of a plant. However, the word is not scientific, and its meaning is largely based on culinary and cultural tradition. Therefore the application of the word is somewhat arbitrary and subjective. For example, some people consider mushrooms to be vegetables, while others consider them a separate food category. Some vegetables can be consumed raw, and some may (or must) be cooked in various ways.

Fruit or Vegetable?

In everyday (non-technical) language, the words "fruit" and "vegetable" are mutually exclusive: plant products that are called fruits are hardly ever classified as vegetables, and vice versa. However, for scientists the word "fruit" also has a precise botanical meaning, which is considerably different from its common meaning. While peaches, plums, and oranges are "fruits" in both senses, many items commonly called "vegetables"—such as eggplants, bell peppers, and tomatoes—are technically fruits, as are also most cereals, as well as some spices like pepper and chili. On the other hand, many items commonly called "fruits"—including figs, pineapples, and strawberries—are not fruits in the strict botanical sense.

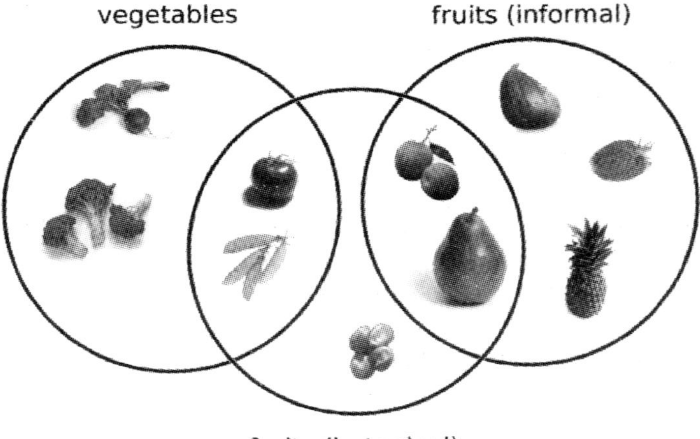

Fig. 3.2: An Euler Diagram Showing the Overlaps between "Fruits" and "Vegetables" in the Informal (Culinary) Sense, and "Fruits" in the Botanical Sense

The question of whether the tomato is a fruit or a vegetable found its way into the United States Supreme Court in 1893. The court ruled unanimously in *Nix v. Hedden* that a tomato is correctly identified as, and thus taxed as, a vegetable, for the purposes of the 1883 Tariff Act on imported produce. The court did acknowledge, however, that, botanically speaking, a tomato is a fruit.

Languages other than English often have categories that can be identified with the common English meanings of "fruit" and "vegetable", but their precise meaning often depends on local culinary traditions. For example, in Brazil the avocado is traditionally consumed with sugar as a dessert or in milk shakes, and hence it is regarded as a fruit, whereas in other countries (including Mexico and the United States) it is used in salads and dips, and hence considered a vegetable.

Some Vegetables Defined as Different Parts of Plants

The list of food items called "vegetable" is quite long, and includes many different parts of plants:

- *Flower bud*: broccoli, cauliflower, globe artichokes
- *Seeds*: Sweetcorn (maize), peas, beans
- *Leaves*: kale, collard greens, spinach, beet greens, turnip greens, endive, lettuce
- *Leaf sheaths*: leeks
- *Buds*: Brussels sprouts, capers
- *Stems of leaves*: celery, rhubarb (sometimes loosely called a "fruit", because sweet pies are made from it)
- *Stem of a plant when it is still a young shoot*: asparagus, bamboo shoots, and ginger
- *Underground stem of a plant, also known as a tuber*: potatoes, Jerusalem artichokes, sweet potatoes, and yams
- *Whole immature plants (sprouts)*: soybean (*moyashi*), mung beans, urad, and alfalfa.
- *Roots*: carrots, parsnips, beets, radishes, turnips, and burdocks.
- *Bulbs*: onions, shallots
- *Fruits in the botanical sense, but used as vegetables*: tomatoes,

cucumbers, squash, pumpkins, capsicums (bell peppers and hot peppers), eggplant, tomatillos, christophene, okra, breadfruit and avocado, and also the following:

Etymology

"Vegetable" comes from the Latin *vegetabilis* (animated) and from *vegetare* (enliven), which is derived from *vegetus* (active), in reference to the process of a plant growing. This in turn derives from the Proto-Indo-European base which is also the source of the English *wake*, meaning "become (or stay) alert".

The word "vegetable" was first recorded in English in the 14th century, but applied to any plant. The word is still used in science with that meaning (*e.g.*, *vegetable matter*, *vegetable kingdom*). The meaning of "vegetable" as "plant grown for food" was not established until the 18th century.

Nutrition

Vegetables are eaten in a variety of ways, as part of main meals and as snacks. The nutritional content of vegetables varies considerably, though generally they contain little protein or fat, and varying proportions of vitamins, provitamins, dietary minerals, fiber and carbohydrates. Vegetables contain a great variety of other phytochemicals, some of which have been claimed to have antioxidant, antibacterial, antifungal, antiviral and anticarcinogenic properties.

However, vegetables often also contain toxins and antinutrients such as, enzyme inhibitors (of cholinesterase, protease, amylase, etc.), cyanide and cyanide precursors, oxalic acid, and more. Depending on the concentration, such compounds may reduce the edibility, nutritional value, and health benefits of dietary vegetables. Cooking and/or other processing may be necessary to eliminate or reduce them.

Colour Pigments

The green colour of leafy vegetables is due to the presence of the green pigment chlorophyll. Chlorophyll is affected by pH and changes to olive green in acid conditions, and bright green in alkaline conditions. Some of the acids are released in steam during cooking, particularly if cooked without a cover.

The yellow/orange colours of fruits and vegetables are due to the presence of carotenoids, which are also affected by normal cooking processes or changes in pH.

The red/blue colouring of some fruits and vegetables (*e.g.* blackberries and red cabbage) are due to anthocyanins, which are sensitive to changes in pH. When pH is neutral, the pigments are purple, when acidic, red, and when alkaline, blue. These pigments are very water soluble.

Safety

For food safety, the CDC recommends proper fruit handling and preparation to reduce the risk of food contamination and foodborne illness. Fresh fruits and vegetables should be carefully selected. At the store, they should not be damaged or bruised and pre-cut pieces should be refrigerated or surrounded by ice. All fruits and vegetables should be rinsed before eating. This recommendation also applies to produce with rinds or skins that are not eaten. It should be done just before preparing or eating to avoid premature spoilage. Fruits and vegetables should be kept separate from raw foods like meat, poultry, and seafood, as well as any cooking utensils or surfaces that may have come into contact with them (*e.g.* cutting boards). Fruits and vegetables, if they are not going to be cooked, should be thrown away if they have touched raw meat, poultry, seafood or eggs. All cut, peeled, or cooked fruits and vegetables should be refrigerated within 2 hours. After a certain time, harmful bacteria may grow on them and increase the risk of foodborne illness.

Storage

Proper Post Harvest Storage aimed at extending and ensuring shelf life is best affected by efficient cold chain application. All vegetables benefit from proper post harvest care

Many root and non-root vegetables that grow underground can

be stored through winter in a root cellar or other similarly cool, dark and dry place to prevent mold, greening and sprouting. Care should be taken in understanding the properties and vulnerabilities of the particular roots to be stored. These vegetables can last through to early spring and be nearly as nutritious as when fresh.

During storage, leafy vegetables lose moisture, and the vitamin C in them degrades rapidly. They should be stored for as short a time as possible in a cool place, in a container or plastic bag.

FROZEN VEGETABLES

Frozen vegetables may be commercially packaged or frozen at home. A wide range of frozen vegetable are sold in supermarkets, usually packaged in either rectangular boxes or plastic bags.

Common frozen vegetables found in supermarkets include spinach, broccoli, cauliflower, peas, corn, yam (in Asia), many others, and mixtures of these and other vegetables or other types of food, such as pasta or cheese. They may be cut or processed some other way into a shape or form that is convenient for cooking or eating, and sometimes seasoned.

Some popular brands include Birds Eye, Green Giant and Hanover, though many supermarkets have their own store brands, too. Frozen vegetables have some advantages over fresh ones, in that they are available when the fresh counterpart is out-of-season, they have a very long shelf life when kept in a freezer, and that they often have been processed a step or more closer to eating. In many cases, they may be more economical to purchase than their fresh counterparts. Since they are already cooked, and will likely be cooked more once they enter the household, they are believed to be more sanitary than fresh vegetables.

While many consider frozen vegetables to be inferior to their fresh counterparts, the opposite is actually true in many cases. Vegetables purchased in the produce section of supermarkets have spent multiple days in transit, and many of the original nutrients will have leeched out. Frozen vegetables are frozen at their freshest, maintaining their nutrients. Freezing does change the taste and texture of the vegetables, however, making them less savory to consumers.

Health Benefits and Risks

In general, boiling vegetables can cause them to lose important nutrients, including vitamins and those that prevent cancer. In particular, Vitamins C and Folic acid are susceptible to loss during the commercial process.

Over the years, there has been controversy as to whether frozen vegetables are better or worse than fresh ones. Generally, reports show that frozen vegetables are not so much nutritionally different from fresh ones.

According to the Food and Drug Administration, based on a 1998 report, frozen vegetables have the same nutrients and health benefits as their fresh counterparts. A 2003 Austrian study found that frozen vegetables are nutritionally better than out-of-season imported fresh ones.

However, there may be some risk in eating poorly cooked frozen vegetables. For example, a 2007 Australian study found that frozen vegetables may contain a bacterium called "map," which can thrive in both extreme cold and hot temperatures.

Not all products classified as frozen vegetables can be viewed as health foods. On June 2, 2003, the USDA classified frozen French fries as a frozen vegetable.

Process

Freezing only slows the deterioration of food and while it may stop the growth of micro-organisms, it does not necessarily kill them. Many enzyme reactions are only slowed by freezing. Therefore it is common to stop enzyme activity before freezing, either by blanching or by adding chemicals.

Foods may be preserved for several months by freezing. Long-term freezing requires a constant temperature of -18 degrees Celsius (0 degrees Fahrenheit) or less. Some freezers cannot achieve such a low temperature. The time food can be kept in the freezer is reduced considerably if the temperature in a freezer fluctuates. Fluctuations could occur by a small gap in the freezer door or adding a large amount of unfrozen food.

Quick-freezing

American inventor Clarence Birdseye, who developed the quick-freezing process of food preservation in 1932, is considered the father of the frozen-food industry.

The food industry uses a technique called flash freezing, an application of supercooling, to quickly freeze food items. In this case the food is subjected to temperatures well below the melting/freezing point of water (273 K or 0°C). This causes the water inside the foods to freeze too quickly for large crystals to form, which would destroy the cell membranes and ruin the food.

Traditional and Other Use

Many Arctic communities would preserve food in holes or larders dug into the ice. There is a tradition in Scandinavia of preserving fish and especially herrings in this way.

Cold stores provide large-volume, long-term storage for strategic food stocks held in case of national emergency in many countries.

Seeds are stored in freezers at -18°C or below in seedbanks. The seeds are stored as a source for planting in case seed reserves elsewhere should be destroyed. The seeds stored may be those of food crops or rare species.

An increasingly popular form of home cooking involves preparing a month of meals to store in the home freezer. Once-a-month cooking helps cooks save money by purchasing grocery items in bulk and save time by cutting preparation time down considerably.

CANNED FOOD

Canning is a method of preserving food in which the food is processed and sealed in an airtight container. The process was first developed as a French military discovery by Nicolas Appert. The packaging prevents microorganisms from entering and proliferating inside.

To prevent the food from being spoiled before and during containment, quite a number of methods are used: pasteurization, boiling (and other applications of high temperature over a period of time), refrigeration, freezing, drying, vacuum treatment, antimicrobial agents that are natural to the recipe of the foodstuff being preserved, a sufficient dose of ionizing radiation, submersion in a strongly saline, acid, base, osmotically extreme (for example very sugary) or other microbe-challenging environments.

No such method is perfectly dependable as a preservative. For example, spore forming thermal resistant microorganisms, such as *Clostridium botulinum* (which causes botulism) can still survive.

From a public safety point of view, foods with low acidity (a pH more than 4.6) need sterilization under high temperature (116-130°C). To achieve temperatures above the boiling point requires the use of a pressure canner. Foods that must be pressure canned include most vegetables, meats, seafood, poultry, and dairy products. The only foods that may be safely canned in an ordinary boiling water bath are highly acidic ones with a pH below 4.6, such as fruits, pickled vegetables, or other foods to which acidic additives have been added.

History and Development of Canning

During the first years of the Napoleonic Wars, the notable French newspaper *Le Monde*, prompted by the government, offered a hefty cash award of 12,000 francs to any inventor who could devise a cheap and effective method of preserving large amounts of food. The larger armies of the period required increased, regular supplies of quality food. Limited food availability was among the factors limiting military campaigns to the summer and fall months. In 1809, a French confectioner and brewer, Nicolas Appert, observed that food cooked inside a jar did not spoil unless the seals leaked, and developed a method of sealing food in glass jars. The reason for lack of spoilage was unknown at the time, since it would be another 50 years before Louis Pasteur demonstrated the role of microbes in food spoilage. However, glass containers presented challenges for transportation.

Glass jars were largely replaced in commercial canneries with cylindrical tin or wrought-iron canisters (later shortened to "cans") following the work of Peter Durand (1810). Cans are cheaper and quicker to make, and much less fragile than glass jars. Glass jars have remained popular for some high-value products and in home canning. Tin-openers were not invented for another thirty years—at first, soldiers had to cut the cans open with bayonets or smash them open with rocks. The French Army began experimenting with issuing tinned foods to its soldiers, but the slow process of tinning foods and the even slower development and transport stages prevented the army from shipping large amounts across the French Empire, and the war ended before the process was perfected. Unfortunately for Appert, the factory which he had built with his prize money was razed in 1814 by Allied soldiers invading France.

Following the end of the Napoleonic Wars, the canning process was gradually employed in other European countries and in the US. Based on Appert's methods of food preservation, Peter Durand patented a process in the United Kingdom in 1810. He did not develop the process, selling his patent in 1811 to Bryan Donkin and John Hall, who were in business as Donkin Hall and Gamble, of Bermondsey. Bryan Donkin developed the process of packaging food in sealed airtight cans, made of tinned wrought iron. Initially, the canning process was slow and labour-intensive, as each large can had to be hand-made, and took up to six hours to cook, making tinned food too expensive for ordinary people. The main market for the food at this stage was the Army and Navy. By 1817 Donkin recorded that he had sold £3000 worth of canned meat in six months. In 1824 Sir William Edward Parry took tinned beef and pea soup with him on his voyage to the Arctic in HMS Fury, during his search for a northwestern passage to India. In 1829 Admiral Sir James Ross also took canned food to the Arctic, as did Sir John Franklin in 1845. Some of his stores were found by the search expedition led by Captain (later Admiral Sir) Leopold McLintock in 1857. One of these cans was opened in 1939, and was edible and nutritious, though it was not analysed for contamination by the lead solder used in its manufacture.

Throughout the mid-nineteenth century, tinned food became a status symbol amongst middle-class households in Europe, becoming something of a frivolous novelty. Early methods of manufacture employed poisonous lead solder for sealing the tins, which may have worsened the disastrous outcome of the 1845 Franklin expedition to chart and navigate the Northwest Passage.

Increasing mechanisation of the canning process, coupled with a huge increase in urban populations across Europe, resulted in a rising demand for tinned food. A number of inventions and improvements followed, and by the 1860s smaller machine-made steel cans were possible, and the time to cook food in sealed cans had been reduced from around six hours to thirty minutes. Canned food also began to spread beyond Europe—Robert Ayars established the first American canning factory in New York City in 1812, using improved tin-plated wrought-iron cans for preserving oysters, meats, fruits and vegetables. Demand for tinned food greatly increased during wars. Large-scale wars in the nineteenth century, such as the Crimean War, American Civil War, and Franco-Prussian War introduced increasing numbers of working-class men to tinned food, and allowed canning companies to expand their businesses to meet military demands for non-perishable food, allowing companies to manufacture in bulk and sell to wider civilian markets after wars ended. Urban populations in Victorian era Britain demanded ever-increasing quantities of cheap, varied, quality food that they could keep at home without having to go shopping daily. In response, companies such as Nestlé, Heinz, and others emerged to provide quality tinned food for sale to working class city-dwellers. In particular, Crosse and Blackwell took over the concern of Donkin Hall and Gamble. The late 19th century saw the range of tinned food available to urban populations greatly increase, as canners competed with each other using novel foodstuffs, highly decorated printed labels, and lower prices.

Demand for tinned food skyrocketed during World War I, as military commanders sought vast quantities of cheap, high-calorie food to feed their millions of soldiers, which could be transported safely, survive trench conditions, and not spoil in transport. Throughout the war, soldiers generally subsisted on low-quality tinned foodstuffs, such as the British "Bully Beef" (cheap corned beef), pork and beans and Maconochies Irish Stew, but by 1916 widespread boredom with cheap tinned food amongst soldiers resulted in militaries purchasing better-quality food to improve morale, and the complete meals in a tin began to appear. In 1917 the French Army began issuing tinned French cuisine, such as coq au vin, whilst the Italian Army experimented with tinned ravioli and spaghetti bolognese. Shortages of tinned food in the British Army in 1917 led to the government issuing cigarettes and amphetamines to soldiers to suppress their appetites. After the war, companies that had supplied military tinned food improved the quality of their goods for civilian sale.

Today, tin-coated steel is the material most commonly used. Laminate vacuum pouches are also used for canning, such as used in MREs.

Double Seams

Modern double seams provide an airtight seal to the tin can. This airtight nature is crucial to keeping bacteria out of the can and keeping its contents sealed inside. Thus, double seamed cans are also known as Sanitary Cans. Developed in 1900 in Europe, this sort of can was made of the traditional cylindrical body made with tin plate. The two ends (lids) were attached using what is now called a double seam. A can thus sealed is impervious to the contamination by creating two tight continuous folds between the can's cylindrical body and the lids. This eliminated the need for solder and allowed improvements in manufacturing speed, reducing cost.

Double seaming uses rollers to shape the can, lid and the final double seam. To make a sanitary can and lid suitable for double seaming, manufacture begins with a sheet of coated tin plate. To create the can body, rectangles are cut and curled around a die, and welded together creating a cylinder with a side seam.

Rollers are then used to flare out one or both ends of the cylinder to create a quarter circle flange around the circumference. Precision is required to ensure that the welded sides are perfectly aligned, as any misalignment will cause inconsistent flange shape, compromising its integrity.

A circle is then cut from the sheet using a die cutter. The circle is shaped in a stamping press to create a downward countersink to fit snugly in to the can body. The result can be compared to an upside down and very flat top hat. The outer edge is then curled down and around about 140 degrees using rollers to create the end curl. The result is a steel tube with a flanged edge, and a countersunk steel disc with a curled edge. A rubber compound is put inside the curl.

Seaming

The body and end are brought together in a seamer and held in place by the base plate and chuck, respectively. The base plate provides a sure footing for the can body during the seaming operation and the chuck fits snugly in to the end (lid). The result is the countersink of the end sits inside the top of the can body just below the flange. The end curl protrudes slightly beyond the flange.

First Operation

Once brought together in the seamer, the seaming head presses a first operation roller against the end curl. The end curl is pressed against the flange curling it in toward the body and under the flange. The flange is also bent downward, and the end and body are now loosely joined together. The first operation roller is then retracted. At this point five thicknesses of steel exist in the seam. From the outside in they are:

- End
- Flange
- End Curl
- Body
- Countersink

This is the first seam. All the parts of the seam are now aligned and ready for the final stage.

Second Operation

The seaming head then engages the second operation roller against the partly formed seam. The second operation presses all five steel components together tightly to form the final seal. The five layers in the final seam are then called; a) End, b) Body Hook, c) Cover Hook, d) Body, e) Countersink. All sanitary cans require a filling medium within the seam because otherwise the metal-to-metal contact will not maintain a hermetic seal. In most cases, a rubberized compound is placed inside the end curl radius, forming the critical seal between the end and the body.

Probably the most important innovation since the introduction of double seams is the welded side seam. Prior to the welded side seam, the can body was folded and/or soldered together, leaving a relatively thick side seam. The thick side seam required that the side seam end juncture at the end curl to have more metal to curl around before closing in behind the Body Hook or flange, with a greater opportunity for error.

Seamer Setup and Quality Assurance

Many different parts during the seaming process are critical in ensuring that a can is airtight and vacuum sealed. The dangers of a can that is not hermetically sealed are contamination by foreign objects (bacteria or fungicide sprays), or that the can could leak or spoil.

One important part is the seamer setup. This process is usually performed by an experienced technician. Amongst the parts that need setup are seamer rolls and chucks which have to be set in their exact position (using a feeler gauge or a clearance gauge). The lifter pressure and position, roll and chuck designs, tooling wear, and bearing wear all contribute to a good double seam.

Incorrect setups can be non-intuitive. For example, due to the springback effect, a seam can appear loose, when in reality it was closed too tight and has opened up like a spring. For this reason, experienced operators and good seamer setup are critical to ensure that double seams are properly closed.

Quality control usually involves taking full cans from the line—one per seamer head, at least once or twice per shift, and performing a teardown operation (wrinkle/tightness), mechanical tests (external thickness, seamer length/height and countersink) as well as cutting the seam open with a twin blade saw and measuring with a double seam inspection system. The combination of these measurements will determine the seam's quality. Use of a [Statistical Process Control] or [SPC] software in conjunction with a manual double seam monitor, computerized double seam scanner, or even a fully-automatic double seam inspection system makes the laborious process of double seam inspection faster and much more accurate. Statistically tracking the performance of each head or seaming station of the [can seamer] allows for better prediction of can seamer issues, and may be used to plan maintenance when convenient: rather that to simply react after bad or unsafe cans have been produced.

Nutrition Value

Canning is a way of processing food to extend its shelf life. The idea is to make food available and edible long after the processing time. Although canned foods are often assumed to be of low-nutritional value (due to heating processes or the addition of preservatives), some canned foods are nutritionally superior—in some ways—to their natural form. For instance, canned tomatoes have a higher available lycopene content.

Potential Hazards

Migration of Can Components

In canning toxicology, *migration* is the movement of substances from the can itself into the contents. Potential toxic substances that can migrate are lead, causing lead poisoning, or bisphenol A, a potential endocrine disruptor that is commonly use to coat the inner surface of cans.

Botulism

Foodborne botulism results from contaminated foodstuffs in which *C. botulinum* spores have been allowed to germinate and produce botulism toxin, and this typically occurs in home-canning|canned non-acidic food substances. *C. botulinum* prefers low oxygen environments, and can therefore grow in canned foods. Botulism is a rare but serious paralytic illness, leading to paralysis that typically starts with the muscles of the face and then spreads towards the limbs. In severe forms, it leads to paralysis of the breathing muscles and causes respiratory failure. In view of this life-threatening complication, all suspected cases of botulism are treated as medical emergencies, and public health officials are usually involved to prevent further cases from the same source.

Canning and the Recession

Canned Goods (and canning supplies) sell particularly well in times of recession due to the tendency of financially-stressed individuals to engage in 'cocooning', a term used by retail analysts to describe the phenomenon in which people actively avoid straying from their houses. In February of 2009, the recession-laden United States saw an 11.5 per cent rise in sales of canning-related items.

FOOD STOCK

Stock is a flavoured water. It forms the basis of many dishes, particularly soups and sauces. Stock is prepared by simmering various ingredients in water, including some or all of the following:

Meat

Leftover cooked meat, such as that remaining on poultry carcasses, is often used along with the bones of the bird or joint. Fresh meat makes a superior stock and cuts rich in connective tissue such as shin or shoulder of beef or veal are commonly recommended, either alone or added in lower proportions to the remains of cooked poultry to provide a richer and fresher-tasting stock. Quantities recommended are invariably in the ratio of 1 part fresh meat to 2 parts water. Pork is considered unsuitable for stock due to its greasiness (although 19th century recipes for consomme and traditional aspic invariably included slices of mild ham) and mutton was traditionally avoided due to the difficulty of avoiding the strong tallowy taint imparted from the fat.

Bones

Veal, beef, and chicken bones are most commonly used. The flavour of the stock comes from the cartilage and connective tissue in the bones. Connective tissue has collagen in it, which gets converted into gelatin that thickens the liquid. Stock made from bones needs to be simmered for longer than stock made from meat (often referred to as broth). Pressure cooking methods shorten the time necessary to extract the flavour from the bones.

Mirepoix

A combination of onions, carrots, celery, and sometimes other vegetables. Often the less desirable parts of the vegetables (such as carrot skins and celery ends) are used since they will not be eaten.

Herbs and spices

The herbs and spices used depend on availability and local traditions. In classical cuisine, the use of a *bouquet garni* (or *bundle of herbs*) consisting of parsley, bay leaves, a sprig of thyme, and possibly other herbs, is common. This is often placed in a sachet to make it easier to remove once the stock is cooked.

Broth is very similar to stock, and often the terms are used interchangeably. Usually, broth refers to the finished product, while stock is used as an ingredient (thus stock may *become* broth). Other times, broth is used to refer to a liquid made in the same way as stock, but meat is substituted for bones. However, with some stock/broth made from vegetables and some made from both bones and meat, this cannot be considered a hard-and-fast rule.

Today, ready-made stock and stock cubes consisting of dried, compressed stock ingredients are readily available. These are commonly known as bouillon cubes (or oxo cubes, after a common brand of stock cube sold in Britain) or cooking base.

Types

- Fond brun, or brown stock. The brown colour is achieved by roasting the bones and mirepoix. This also adds a rich, full flavour. Veal bones are the most common type used in a fond brun.
- Fond blanc, or white stock, is made by using raw bones and white mirepoix. Chicken bones are the most common for fond blanc.
- Fish stock is made with fish bones and finely chopped mirepoix. Fish stock should be cooked for 30-45 minutes—cooking any longer spoils the flavour. Concentrated fish stock is called "fish fumet." In Japanese cooking, a fish and kelp stock called dashi is made by briefly (3-5 minutes) cooking skipjack tuna (bonito) flakes called *katsuobushi* in nearly boiling water.
- Chicken stock should be cooked for 3-4 hours.
- Veal stock should be cooked anywhere from 8 hours.
- Lamb stock should be cooked for 5 hours. To make a lamb

jus, start with a chicken stock and roasted lamb necks and bones.

- Jus is a rich, lightly reduced stock used as a sauce for roasted meats. Many of these are started by deglazing the roasting pan, then reducing to achieve the rich flavour desired.
- Ham stock, common in Cajun cooking, is made from ham hocks.
- Prawn stock is made from boiling prawn shells. It is used in Southeast Asian dishes such as laksa.
- Vegetable stock is made only of vegetables. It is common today.
- Master stock is a special Chinese stock used primarily for poaching meats, flavoured with soy sauce, sugar, ginger, garlic, and other aromatics.
- Glace viande is stock made from bones, usually from veal, that is highly concentrated by reduction.

Preparing Stock

A few basic rules are commonly prescribed for preparing stock:

- The stock ingredients are simmered starting with cold water. This promotes the extraction of collagen, which may be sealed in by hot water.
- Stocks are simmered gently, with bubbles just breaking the surface, and not boiled. If a stock is boiled, it will be cloudy.
- Salt is usually not added to a stock, as this causes it to become too salty, since most stocks are reduced to make soups and sauces.
- Meat is added to a stock before vegetables, and the "scum" that rises to the surface is skimmed off before further ingredients are added.

EGG AS FOOD

An egg is a round or oval body laid by the female of any number of different species, consisting of an ovum surrounded by layers of membranes and an outer casing, which acts to nourish and protect a developing embryo and its nutrient reserves. Most edible eggs, including bird eggs and turtle eggs, consist of a protective, oval eggshell, the albumen (egg white), the vitellus (egg yolk), and various thin membranes. Every part is edible, although the eggshell is generally discarded. Eggs are considered a good source of protein and choline. Because of this, the egg falls in the *Meats* category under the Food Guide Pyramid. Roe and caviar are edible eggs produced by fish.

Culinary Use

Bird eggs are a common food and one of the most versatile ingredients used in cooking. They are important in many branches of the modern food industry. The most commonly used bird eggs are those from the chicken. Duck and goose eggs, and smaller eggs such as quail eggs are occasionally used as a gourmet ingredient, as are the largest bird eggs, from ostriches. Gull eggs are considered a delicacy in England, as well as in some Scandinavian countries, particularly in Norway. In some African countries, guineafowl eggs are commonly seen in marketplaces, especially in the spring of each year. Pheasant eggs and emu eggs are perfectly edible but less widely available. Sometimes they are obtainable from farmers, poulterers, or luxury grocery stores. Most wild birds' eggs are protected by laws in many countries, which prohibit collecting or selling them, or permit these only during specific periods of the year.

Most commercially produced chicken eggs intended for human consumption are unfertilized, since the laying hens are kept without roosters. Fertile eggs can be purchased and eaten as well, with little nutritional difference. Fertile eggs will not contain a developed embryo, as refrigeration prohibits cellular growth for an extended amount of time.

Chicken eggs are widely used in many types of dishes, both sweet and savory. Eggs can be pickled, hard-boiled, soft-boiled, scrambled, fried and refrigerated. They can also be eaten raw, though this is not recommended for people who may be especially susceptible to salmonella, such as the elderly, the infirm, or pregnant women. In addition, the protein in raw eggs is only 51 per cent bio-available, whereas that of a cooked egg is nearer 91 per cent bio-available, meaning the protein of cooked eggs is nearly twice as absorbable as the protein from raw eggs. As an ingredient, egg yolks are an important emulsifier in the kitchen, and the proteins in egg white allow it to form foams and aerated dishes.

Quail eggs are considered a delicacy in many countries. They are used raw or cooked as *tamago* in sushi. In Colombia, quail eggs are considered less exotic than in other countries, and a single hard-boiled quail egg is a common topping on hot dogs and hamburgers, often fixed into place with a toothpick.

The albumen, or egg white, contains protein but little or no fat. It can be used in cooking separately from the yolk, and can be aerated or whipped to a light, fluffy consistency. Beaten egg whites are used in desserts such as meringues and mousse. Ground egg shells are sometimes used as a food additive to deliver calcium. Boiled eggs that are difficult to peel are usually too fresh. Fresh eggs have a lower pH, and this does not allow the shell to separate easily from the underlying albumen. When put into vinegar the shell will dissolve slowly.

Flavour

Although the age of the egg and the conditions of its storage have a greater influence, the bird's diet does affect the flavor of the egg. For example, when a brown-egg chicken breed eats rapeseed or soy meals, its intestinal microbes metabolize them into fishy-smelling triethylamine, which ends up in the egg. The unpredictable diet of free-range hens will produce unpredictable eggs.

Problems with Cooking

If a boiled egg is overcooked, a greenish ring sometimes appears around egg yolk. This is a manifestation of the iron and sulfur compounds in the egg. It can also occur when there is an abundance of iron in the cooking water. The green ring does not affect the egg's taste; overcooking, however, harms the quality of the protein (chilling the egg for a few minutes in cold water until the egg is completely cooled prevents the greenish "ring" from forming on the surface of the yolk).

Cooking also increases the risk of atherosclerosis due to increased oxidization of the cholesterol contained in the egg yolk.

Substitutes

For those who do not consume eggs, alternatives used in baking include other rising agents or binding materials, such as ground flax seeds or potato flour. Tofu can also act as a partial binding agent, since it is high in lecithin due to its soy content. Applesauce can be used, as well as arrowroot and banana. Extracted soybean lecithin, in turn, is often used in packaged foods as an inexpensive substitute for egg-derived lecithin.

Other egg substitutes are made from just the white of the egg for those who worry about the high cholesterol and fat content in eggs. These products usually have added vitamins and minerals as well as vegetable-based emulsifiers and thickeners such as xanthan gum or guar gum. These allow the product to maintain the nutrition and several culinary properties of real eggs, making possible foods like Hollandaise sauce, custard, mayonnaise, and most baked goods with these substitutes.

Preservation

Preservation of edible eggs is extremely important, as an improperly-handled egg can contain elevated levels of salmonella, bacteria that can cause severe food poisoning. The simplest method to preserve an egg is to treat it with salt. Salt draws water out of bacteria and molds, which prevents their growth. The Chinese salted duck egg is made by immersing duck eggs in brine, or coating them individually with a paste of salt and mud or clay. The eggs stop absorbing salt after about a month, having reached chemical equilibrium. Their yolks become an orange-red coloured and solid, but the white remains liquid. They are boiled before consumption and often served with rice congee.

Another method is to make pickled eggs, by boiling them first and immersing them in a mixture of vinegar, salt, and spices like ginger or allspice. Frequently, beetroot juice is added to impart a red colour to the eggs. If the eggs are immersed in it for a few hours, the distinct red, white, and yellow colours can be seen when the eggs are sliced. If marinated for several days or more, the red colour will reach to the yolk. If the eggs are marinated in the mixture for several weeks or more, vinegar's acetic acid will dissolve much of the shell's calcium carbonate and penetrate the egg, making it acidic enough to inhibit the growth of bacteria and molds. Pickled eggs made this way will generally keep for a year or more without refrigeration.

A century egg or hundred-year-old egg is preserved by fermenting an egg in a mixture of clay, wood ash, salt, lime, and rice straw for several weeks to several months, depending on the method of processing. After the process is completed, the yolk becomes a dark green, cream-like substance with a strong odor of sulfur and ammonia, while the white becomes a dark brown, transparent jelly with a comparatively mild, distinct flavor. The transforming agent in a century egg is its alkaline material, which gradually raises the pH of the egg from around 9 to 12 or more. This chemical process causes an "inorganic version" of fermentation, which breaks down some of the complex, flavorless proteins and fats of the yolk into simpler, flavorful ones.

History

Bird eggs have been valuable foodstuff since prehistory, in both hunting societies and more recent cultures where birds were domesticated. In Thebes, Egypt, the tomb of Haremhab, built about 1420 BCE, shows a depiction of a man carrying bowls of ostrich eggs and other large eggs, presumably those of the pelican, as offerings. In ancient Rome, eggs were preserved using a number of methods, and meals often started with an egg course. The Romans crushed the shell in their plate to prevent evil spirits from hiding there. In the Middle Ages, eggs were forbidden during Lent because of their richness. It is possible that the word ***mayonnaise*** was derived from ***moyeu***, the medieval French word for the yolk meaning ***center*** or ***hub***.

Egg scrambled with acidic fruit juices were popular in France in the 17th century; this may have been the origin of lemon curd.

The dried egg industry developed in the 19th century, before the rise of the frozen egg industry. In 1878, a company in St. Louis, Missouri started to transform egg yolk and white into a light-brown, meal-like substance by using a drying process. The production of dried eggs significantly expanded during World War II, for use by the United States Armed Forces and its allies.

The egg carton was invented by Joseph Coyle in Smithers, British Columbia, to solve a dispute about broken eggs between a farmer in Bulkley Valley and the owner of the Aldermere Hotel. Early egg cartons were made of paper.

Anatomy and Characteristics

The shape of an egg is an ovate spheroid with one end larger than the other end. The egg has cylindrical symmetry along the long axis.

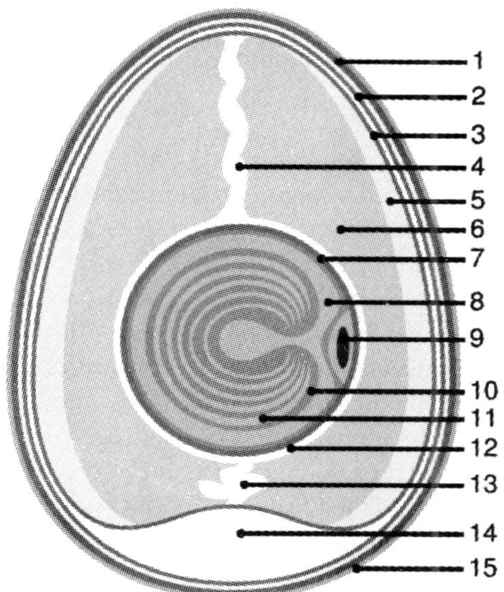

1. Eggshell; 2. Outer membrane; 3. Inner membrane; 4. Chalaza; 5. Exterior albumen; 6. Middle albumen; 7. Vitelline membrane; 8. Nucleus of pander; 9. Germinal disk; 10. Yellow yolk; 11. White yolk; 12. Internal albumen; 13. Chalaza; 14. Air cell; 15. Cuticula

Fig. 3.3: Schematic of a Chicken Egg

An egg is surrounded by a thin, hard shell. Inside, the egg yolk is suspended in the egg white by one or two spiral bands of tissue called the chalazae (from the Greek word *khalazi*, meaning hailstone or hard lump.)

Air Cell

The larger end of the egg contains the air cell that forms when the contents of the egg cool down and contract after it is laid. Chicken eggs are graded according to the size of this air cell, measured during candling. A very fresh egg has a small air cell and receives a grade of AA. As the size of the air cell increases, and the quality of the egg decreases, the grade moves from AA to A to B. This provides a way of testing the age of an egg: as the air cell increases in size, the egg becomes less dense and the larger end of the egg will rise to increasingly shallower depths when the egg is placed in a bowl of water. A very old egg will actually float in the water and should not be eaten.

Shell

Egg shell colour is caused by pigment deposition during egg formation in the oviduct and can vary according to species and breed, from the more common white or brown to pink or speckled blue-green. In general, chicken breeds with white ear lobes lay white eggs, whereas chickens with red ear lobes lay brown eggs. Although there is no significant link between shell colour and nutritional value, there is often a cultural preference for one colour over another. For example, in most regions of the United States, chicken eggs are generally white; while in the northeast of that country, and in countries as diverse as Costa Rica, Ireland, and the United Kingdom, they are generally light-brown. In Brazil and Poland, white chicken eggs are generally regarded as industrial, and brown or reddish ones are preferred.

Yolk

The yolk in a newly laid egg is round and firm. As the yolk ages it absorbs water from the albumen, which increases its size and causes it to stretch and weaken the vitelline membrane (the clear casing enclosing the yolk). The resulting effect is a flattened and enlarged yolk shape.

Yolk colour is dependent on the diet of the hen; if the diet contains yellow/orange plant pigments known as xanthophylls, then they are deposited in the yolk, colouring it. A colourless diet can produce an almost colourless yolk. Farmers may enhance yolk colour with artificial pigments, or with natural supplements rich in lutein (marigold petals are a popular choice), but, in most locations, this activity is forbidden.

Abnormalities

Some hens will rarely lay double-yolked eggs as the result of unsynchronized production cycles. Although heredity causes some hens to have a higher propensity to lay double-yolked eggs, these occur more frequently as occasional abnormalities in young hens beginning to lay. Usually a double-yolked egg will be longer and thinner than an ordinary single-yolk egg. Double-yolked eggs occur rarely, only leading to observed successful hatchlings under human intervention, as the unborn chickens would otherwise fight each other and die.

Fig. 3.4: A Hardboiled Double-Yolked Egg, Cut in Half

It is also possible for a young hen to produce an egg with no yolk at all. Yolkless eggs are usually formed about a bit of tissue that is sloughed off the ovary or oviduct. This tissue stimulates the secreting glands of the oviduct, and a yolkless egg results.

Nutritional Value

Chicken eggs are the most commonly-eaten eggs. They supply all essential amino acids for humans, and provide several vitamins and minerals, including vitamin A, riboflavin, folic acid, vitamin B_6, vitamin B_{12}, choline, iron, calcium, phosphorus and potassium. They are also an inexpensive single-food source of protein.

All of the egg's vitamin A, D and E is in the egg yolk. The egg is one of the few foods that naturally contain Vitamin D. A large egg yolk contains approximately 60 Calories (250 kilojoules); the egg white contains about 15 Calories (60 kilojoules). A large yolk contains more than two-thirds of the recommended daily intake of 300 mg of cholesterol (although one study indicates that the human body may not absorb much cholesterol from eggs). The yolk makes up about 33 per cent of the liquid weight of the egg. It contains all of the fat, slightly less than half of the protein, and most of the other nutrients. It also contains all of the choline, and one yolk contains approximately half of the recommended daily intake. Choline is an important nutrient for development of the brain, and is said to be important for pregnant and nursing women to ensure healthy fetal brain development.

Recently, chicken eggs that are especially high in Omega 3 fatty acids have come on the market. These eggs are made by feeding laying hens a diet containing polyunsaturated fats and kelp meal. Nutrition information on the packaging is different for each of the brands.

Cooked eggs are easier to digest, as well as having a lower risk of salmonella infection.

Health Issues of Eating Chicken Eggs

Cholesterol and Fat

More than half the calories found in eggs come from the fat in the yolk; a 100-gram chicken egg contains approximately 10 grams of fat. People on a low-cholesterol diet may need to reduce egg consumption; however, only 27 per cent of the fat in egg is saturated fat (Palmitic, Stearic and Myristic acids) that contains LDL cholesterol. The egg white consists primarily of water (87 per cent) and protein (13 per cent) and contains no cholesterol and little, if any, fat.

There is debate over whether egg yolk presents a health risk. Some research suggests dietary cholesterol increases the ratio of total to HDL cholesterol and, therefore, adversely affects the body's cholesterol profile; whereas other studies show that moderate consumption of eggs, up to two per day, does not appear to increase heart disease risk in healthy individuals. Harold McGee argues that the cholesterol in the yolk is not what causes a problem, because fat (particularly saturated) is much more likely to raise cholesterol levels than the actual consumption of cholesterol. A 2007 study of nearly 10,000 adults demonstrated no correlation between moderate (6 per week) egg consumption and cardiovascular disease or strokes except in the sub-population of diabetic patients which presented an increased risk of coronary heart disease. Other research supports the idea that a high egg intake increases càrdiovascular risk in diabetic patients.

Type 2 Diabetes

Consumption of eggs has been linked to an increased risk of Type 2 diabetes in both men and women. A 2008 study using data on over 50,000 individuals collected by the Physicians' Health Study I (1982-2007) and the Women's Health Study (1992-2007) determined that the "data suggest that high levels of egg consumption (daily) are associated with an increased risk of type 2 diabetes."

Contamination

A health issue associated with eggs is contamination by pathogenic bacteria like Salmonella enteritidis. Contamination of eggs exiting a female bird via the cloaca may also occur with other members of the Salmonella group, so care must be taken to avoid the egg shell becoming contaminated with fecal matter. In commercial practice, eggs are quickly washed with a sanitizing solution within minutes of being laid. The risk of infection from raw or undercooked eggs is dependent in part upon the sanitary conditions under which the hens are kept.

A study by the U.S. Department of Agriculture in 2002 (Risk Analysis April 2002 22(2):203-18) suggests the problem is not as prevalent as once thought. It showed that of the 69 billion eggs produced annually only 2.3 million are contaminated with salmonella—equivalent to just one in every 30,000 eggs—thus showing that salmonella infection is quite rarely induced by eggs. However, this has not been the case in other countries where Salmonella enteritidis and Salmonella typhimurium infections due to egg consumptions are major concerns. Egg shells act as hermetic seals which guard against bacteria entering, but this seal can be broken through improper handling or if laid by unhealthy chickens. Most forms of contamination enter through such weaknesses in the shell.

Food Allergy

One of the most common food allergies in infants is eggs. Infants usually have the opportunity to grow out of this allergy during childhood, if exposure is minimized. Allergic reactions against egg white are more common than reactions against egg yolks.

In addition to true allergic reactions, some people experience a food intolerance to egg whites.

Food labeling practices in most developed countries now include eggs, egg products and the processing of foods on equipment that also process foods containing eggs in a special allergen alert section of the ingredients on the labels.

Antibiotic Resistance

Information obtained by the Canadian Integrated Programme for Antimicrobial Resistance (CIPARS) "strongly indicates that cephalosporin resistance in humans is moving in lockstep with use of the drug in poultry production." According to the Canadian Medical Association Journal, the unapproved antibiotic ceftiofur is routinely injected into eggs in Quebec and Ontario to discourage infection of hatchlings. Although the data are contested by the industry, antibiotic resistance in humans appears to be directly related to the antibiotic's use in eggs.

Chicken Egg Grading

The US Department of Agriculture grade eggs by the interior quality of the egg and the appearance and condition of the egg shell. Eggs of any quality grade may differ in weight (size).

- U.S. Grade AA eggs have whites that are thick and firm; yolks that are high, round, and practically free from defects; and clean, unbroken shells. Grade AA and Grade A eggs are best for frying and poaching where appearance is important.
- U.S. Grade A eggs have characteristics of Grade AA eggs except that the whites are "reasonably" firm. This is the quality most often sold in stores.
- U.S. Grade B eggs have whites that may be thinner and yolks that may be wider and flatter than eggs of higher grades. The shells must be unbroken, but may show slight stains. This quality is seldom found in retail stores because they are usually used to make liquid, frozen, and dried egg products, as well as other egg-containing products.

In other countries, particularly in European Union, eggs are graded by the hen farming method instead, *e.g.* from free range hens, battery cages etc.

Chicken Egg Sizes

Chicken eggs are graded by size, for the purpose of sales. The United States Department of Agriculture sizing is based by weight per dozen. The most common US size of chicken egg is 'Large' and is the egg size commonly referred to for recipes. The following egg masses have been calculated on the basis of the USDA sizing:

Modern Sizes (USA)		
Size	*Mass per egg*	*Cooking Yield (Volume)*
Jumbo	Greater than 2.5 oz. or 71 g	
Very Large or Extra-Large (XL)	Greater than 2.25 oz. or 64 g	56 mL (4 tbsp)
Large (L)	Greater than 2 oz. or 57 g	46 mL (3.25 tbsp)
Medium (M)	Greater than 1.75 oz. or 50 g	43 mL (3 tbsp)
Small (S)	Greater than 1.5 oz. or 43 g	
Peewee	Greater than 1.25 oz. or 35 g	

In Europe, modern egg sizes are defined as follows:

Modern Sizes (Europe)	
Size	*Mass per egg*
Very Large	73 g and over
Large	63-73 g
Medium	53-63 g
Small	53 g and under

In Australia, the Australian Egg Corporation defines the following sizes in its labeling guide.

Modern Sizes (Australia)	
Size	*Mass per egg*
Jumbo	68 g
Extra-Large	60 g
Large	52 g

In Western Australia, two additional sizes are also standardized by the Golden Eggs Corporation

Additional Sizes (Western Australia)	
Mega or XXXL	72 g
Medium	43 g

In New Zealand sizes are based on the minimum mass per egg:

Modern Sizes (New Zealand)	
Size	*Minimum mass per egg*
8 (Jumbo)	68 g
7 (Large)	62 g
6 (Standard)	53 g
5 (Medium)	44 g
4 (Pullet)	35 g

Traditional Sizes

Size	*Mass*
Size 0	Greater than 75g
Size 1	70 g-75 g
Size 2	65 g-70 g
Size 3	60 g-65 g
Size 4	55 g-60 g
Size 5	50 g-55 g
Size 6	45 g-50 g
Size 7	less than 45 g

Issues in Mass Production

Commercial factory farming operations often involve raising the hens in small crowded cages, preventing the chickens from engaging in natural behaviours such as wing-flapping, dust-bathing, scratching, pecking, perching and nest-building. Such restrictions can lead to pacing and escape behaviour.

Many hens confined to battery cages, and some raised in cage-free conditions, are de-beaked to prevent harming each other and cannibalism. According to critics of the practice, this can cause hens severe pain to the point where some may refuse to eat and starve to death. Some hens may be force molted to increase egg quality and production level after the molting. Molting can be induced by extended feed withdrawal, water withdrawal or controlled lighting programmes.

Laying hens are often slaughtered between 100-130 weeks of age, when their egg productivity starts to decline. Due to modern selective breeding, laying hen strains differ from meat production strains. As male birds of the laying strain do not lay eggs and are not suitable for meat production, they are generally culled at the hatchery.

Free-range eggs are considered by some advocates to be an acceptable substitute to factory-farmed eggs. Free-range laying hens are given outdoor access instead of being contained in crowded cages. Questions on the actual living conditions of free-range hens have been raised, as there is no legal definition or regulations for eggs labeled as free-range in the US.

In the US, increased public concern for animal welfare has pushed various egg producers to release eggs under a variety of different standards. The most widespread standard in use is used by United Egg Producers and is a volunteer programme known as *United Egg Producers Certified* (UEP Certified). The programme includes guidelines with regard to housing, feed, water, air, space allowance, beak trimming, molting, handling, and transportation; however, critics such as The Humane Society have alleged UEP Certification misleadingly allows for a significant amount of animal cruelty. Other standards include "Cage Free", "Natural", "Certified Humane", and "Certified Organic." Of these standards, "Certified Humane", which carries requirements for stocking density and cage-free keeping, among others, and "Certified Organic", which requires hens have outdoor access and are fed only organic, vegetarian feed, among other requirements, are the most stringent. The European Union will introduce an EU-wide ban on the use of conventional battery cages for egg-laying hens. This ban is expected to come into effect from 1 January 2012, as per EU Directive 1999/74/EC. The EU will instead permit the use of "enriched" cages that must meet certain space and amenity requirements. Egg producers in many member states have objected to the ban.

Cultural Significance

A popular Easter tradition in some parts of the world is the decoration of hard-boiled eggs (usually by dyeing but often by spray-painting). Adults often hide the eggs for children to find, an activity known as an Easter egg hunt. A similar tradition of egg painting exists in areas of the world influenced by the culture of Persia. Before the spring equinox in the Persian New·Year tradition (called Norouz), each family member decorates a hard-boiled egg and sets them together in a bowl.

Although a food item, eggs are sometimes thrown at houses, cars, or people. This act, known commonly as *egging* in the various English-speaking countries, is a minor form of vandalism and, therefore, usually a criminal offense and is capable of damaging property (egg whites can degrade certain types of vehicle paint) as well as causing serious eye injury. On Halloween, for example, trick or treaters have been known to throw eggs (and sometimes flour) at property or people from whom they received nothing. Eggs are also often thrown in protests, as they are inexpensive and nonlethal, yet very messy when broken.

EGG YOLK

An egg yolk is a part of an egg which feeds the developing embryo. The egg yolk is suspended in the egg white (known more formally as albumen or ovalbumin) by one or two spiral bands of tissue called the chalazae. Prior to fertilization, the yolk together with the germinal disc is a single cell; one of the few single cells that can be seen by the naked eye.

As a food, yolks are a major source of vitamins and minerals. They contain all of the egg's fat and cholesterol, and almost half of the protein.

If left intact while cooking fried eggs, the yellow yolk surrounded by a flat blob of egg white creates the distinctive sunny-side up form of the food. Mixing the two components together before frying results in the pale yellow form found in omelettes and scrambled eggs.

Uses

- It is sometimes separated from the egg white and used in cooking.
- It is used in painting as a component of traditional egg-tempera.
- It is used in the production of egg-yolk agar plate medium, useful in testing for the presence of *Clostridium perfringens*.
- Egg yolk also contains an antibody called Immunoglobulin yolk (IgY). The antibody transfers from the laying hen to the egg yolk by passive immunity to protect both embryo and hatchling from microorganism invasion.
- Egg yolk can be used to make liqueurs such as Advocaat.

Composition of Chicken Egg Yolk

The yolk makes up about 33 per cent of the liquid weight of the egg; it contains approximately 60 calories, three times the caloric content of the egg white.

One large egg (50 grams in weight, 17 gram yolk) contains approximately: 2.7g protein, 210 mg cholesterol, 0.61g carbohydrates and 4.51g total fat. (USDA National Nutrient Database)

All of the fat soluble vitamins, (A, D, E and K) are found in the egg yolk. Egg yolks are one of the few foods naturally containing vitamin D.

The composition (by weight) of the most prevalent fatty acids in egg yolk is typically as follows:

- Unsaturated fatty acids:
  - Oleic acid 47 per cent
  - Linoleic acid 16 per cent
  - Palmitoleic acid 5 per cent
  - Linolenic acid 2 per cent
- Saturated fatty acids:
  - Palmitic acid 23 per cent
  - Stearic acid 4 per cent
  - Myristic acid 1 per cent

Egg yolk is a source of lecithin, an emulsifier and surfactant. The yellow colour is caused by lutein and zeaxanthin, which are yellow or orange carotenoids known as xanthophylls

Double-yolk Eggs

Yolkless Eggs

Eggs without yolk are called "dwarf" or "wind" eggs. Such an egg is most often a pullet's first effort, produced before her laying mechanism is fully ready. In a mature hen, a wind egg is unlikely, but can occur if a bit of reproductive tissue breaks away, stimulating the egg producing glands to treat it like a yolk and wrap it in albumen, membranes and a shell as it travels through the egg tube. This has occurred if, instead of a yolk, the egg contains a small particle of grayish tissue. An archaic term for a no yolk egg is a "cock" egg. Since they contained no yolk and therefore can't hatch, it was traditionally believed that these eggs were laid by roosters. This type of egg occurs in many varieties of fowl. They have been found in chickens, both standard and bantams, guineas and Coturnix Quail.

MILK

Milk is an opaque white liquid produced by the mammary glands of mammals. It provides the primary source of nutrition for young mammals before they are able to digest other types of food. The early lactation milk is known as colostrum, and carries the mother's antibodies to the baby. It can reduce the risk of many diseases in the baby. The exact components of raw milk varies by species, but it contains significant amounts of saturated fat, protein and calcium as well as vitamin C. Cattle's milk has a pH ranging from 6.4 to 6.8, making it slightly acidic.

Types of Consumption

There are two distinct types of milk consumption: a natural source of nutrition for all infant mammals, and a food product for humans of all ages derived from other animals.

Nutrition for Infant Mammals

In almost all mammals, milk is fed to infants through breastfeeding, either directly or by expressing the milk to be stored and consumed later. Some cultures, historically or currently, continue to use breast milk to feed their children until they are 7 years old.

Food Product for Humans

In many cultures of the world, especially the Western world, humans continue to consume milk beyond infancy, using the milk of other animals (in particular, cattle) as a food product. For millennia, cattle's milk has been processed into dairy products such as cream, butter, yogurt, kefir, ice cream, and especially the more durable and easily transportable product, cheese. Industrial science has brought us casein, whey protein, lactose, condensed milk, powdered milk, and many other food-additive and industrial products.

Humans are an exception in the natural world for consuming

milk past infancy, despite the fact that more than 75 per cent of adult humans are lactose intolerant, a characteristic that is more prevalent among individuals of African or Asian descent. The sugar lactose is found only in milk, forsythia flowers, and a few tropical shrubs. The enzyme needed to digest lactose, lactase, reaches its highest levels in the small intestines after birth and then begins a slow decline unless milk is consumed regularly. On the other hand, those groups that do continue to tolerate milk often have exercised great creativity in using the milk of domesticated ungulates, not only of cattle, but also sheep, goats, yaks, water buffalo, horses, and camels. The largest producer and consumer of cattle's milk in the world is India.

The term *milk* is also used for whitish non-animal substitutes such as soy milk, rice milk, almond milk, and coconut milk. Even the regurgitated substance secreted by glands in the mucosa of their upper digestive tract which pigeons feed their young is called crop milk though it bears little resemblance to mammalian milk.

Evolution

Milk glands are highly specialized sweat glands. It has been suggested that the original function of lactation (milk production) was to keep eggs moist. Much of the argument is based on monotremes (egg-laying mammals):

History

Animal milk is first known to have been used as human food at the beginning of animal domestication. Cow's milk was first used as human food in the Middle East. Goats and sheep are ruminants: mammals adapted to survive on a diet of dry grass, a food source otherwise useless to humans, and one that is easily stockpiled. The animals dairying proved to be a more efficient way of turning uncultivated grasslands into sustenance: the food value of an animal killed for meat can be matched by perhaps one year's worth of milk from the same animal, which will keep producing milk—in convenient daily portions—for years.

Milk byproducts found inside stone age pottery from Turkey

Table 3.1: Top Ten per Capita Cattle's Milk Consumers (2006)

Country	*Milk (litres)*	*Cheese (kg)*	*Butter (kg)*
Finland	183.9	19.1	5.3
Sweden	145.5	18.5	1.0
Ireland	129.8	10.5	2.9
Netherlands	122.9	20.4	3.3
Norway	116.7	16.0	4.3
Spain	119.1	9.6	1.0
Switzerland	112.5	22.2	5.6
United Kingdom	111.2	12.2	3.7
Australia	106.3	11.7	3.7
Canada	94.7	12.2	3.3

indicate processed milk was consumed in 6500 BC some thousands of years before the ability for adult humans to digest unprocessed milk had evolved.

DNA evidence extracted from Neolithic skeletons indicates that a thousand years later in 5500 BC people in Northern Europe were like all other peoples of the time and were still lactose intolerant. Earthenware vessels found in England from a thousand years after this in 4500 BC contain milk byproducts indicating milk was used in some form although perhaps not drunk directly.

Today lactose tolerance is the key distinguishing feature of Northern European peoples with around 95 per cent of the population displaying the trait compared to around 15 per cent to 20 per cent for people of Middle Eastern descent, 5 per cent amongst Han Chinese and 0 per cent amongst American Indians.

Milk was first delivered in bottles on January 11, 1878. The day is now remembered as Milk Day and is celebrated annually. The town of Harvard, Illinois also celebrates milk in the summer with a festival known as "Milk Days". Their's is a different tradition meant to celebrate dairy farmers in the "Milk Capital of the World."

Other Animal Sources

In addition to cattle, the following animals provide milk used by humans for dairy products:

- Camels

- Donkeys
- Goats
- Horses
- Reindeer
- Sheep
- Water buffalo
- Yaks

In Russia and Sweden, small moose dairies also exist.

According to the National Bison Association, American bison (also called American buffalo) are not milked commercially. However, various sources report cows resulting from cross-breeding bison and domestic cattle are good milk producers, both during the European settlement of North America and during the development of commercial Beefalo in the 1970s and 1980s.

Human milk is not produced or distributed industrially or commercially; however, milk banks exist that allow for the collection of donated human milk and its redistribution to infants who may benefit from human milk for various reasons (premature neonates, babies with allergies or metabolic diseases, etc.).

All other female mammals do produce milk, but are rarely or never used to produce dairy products for human consumption.

Modern Production

In the Western world today, cattle's milk is produced on an industrial scale. It is by far the most commonly consumed form of milk in the western world. Commercial dairy farming using automated milking equipment produces the vast majority of milk in developed countries. Types of cattle such as the Holstein have been specially bred for increased milk production. Ninety percent of the dairy cows in the United States and 85 per cent in Great Britain are Holsteins. Other milk cows in the United States include Ayrshire, Brown Swiss, Guernsey, Jersey, and Milking Shorthorn. The largest producers of dairy products and milk today are India followed by the United States, Germany, and Pakistan.

This table below shows the numbers of buffalo milk productions. Cattle's milk is produced in a much wider range.

Table 3.2: Top Ten Buffalo Milk Producers—2007

Country	*Production (tonnes)*	*Note*
India	59,210,000	*
20,372,000Pakistan	20,372,000	
People's Republic of China	2,900,000	F
Egypt	2,300,000	F
Nepal	958,603	
Iran	241,500	F
Myanmar	220,462	
Italy	200,000	F
Vietnam	32,000	F
Turkey	30,375	
World	86,574,539	A

No symbol = official figure F = FAO estimate
* = Unofficial/Semi-official/mirror data A = Aggregate

Price

It was reported in 2007 that with increased worldwide prosperity and the competition of biofuel production for feedstocks, both the demand for and the price of milk had substantially increased world wide. Particularly notable was the rapid increase of consumption of milk in China and the rise of the price of milk in the United States above the government subsidized price.

Physical and Chemical Structure

Milk is an emulsion or colloid of butterfat globules within a water-based fluid. Each fat globule is surrounded by a membrane consisting of phospholipids and proteins; these emulsifiers keep the individual globules from joining together into noticeable grains of butterfat and also protect the globules from the fat-digesting activity of enzymes found in the fluid portion of the milk. In unhomogenized cow's milk, the fat globules average about four micrometers across. The fat-soluble vitamins A, D, E, and K are found within the milkfat portion of the milk.

The largest structures in the fluid portion of the milk are casein protein micelles: aggregates of several thousand protein molecules, bonded with the help of nanometer-scale particles of calcium

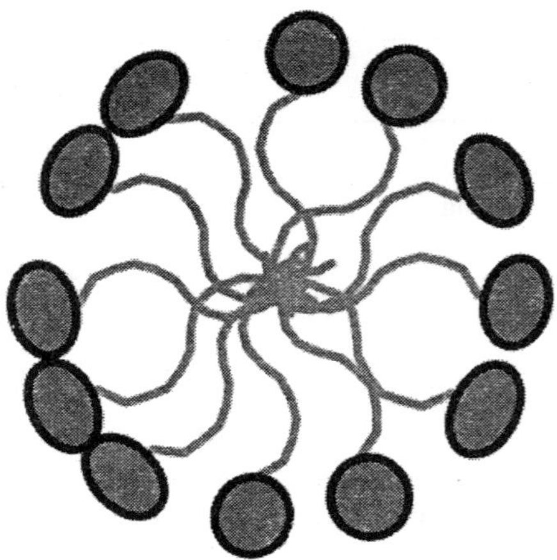

Fig. 3.5: Schematic of a Micelle

phosphate. Each micelle is roughly spherical and about a tenth of a micrometer across. There are four different types of casein proteins, and collectively they make up around 80 percent of the protein in milk, by weight. Most of the casein proteins are bound into the micelles. There are several competing theories regarding the precise structure of the micelles, but they share one important feature: the outermost layer consists of strands of one type of protein, kappa-casein, reaching out from the body of the micelle into the surrounding fluid. These Kappa-casein molecules all have a negative electrical charge and therefore repel each other, keeping the micelles separated under normal conditions and in a stable colloidal suspension in the water-based surrounding fluid.

Both the fat globules and the smaller casein micelles, which are just large enough to deflect light, contribute to the opaque white colour of milk. The fat globules contain some yellow-orange carotene, enough in some breeds—Guernsey and Jersey cattle, for instance—to impart a golden or "creamy" hue to a glass of milk. The riboflavin in the whey portion of milk has a greenish colour, which can sometimes be discerned in skim milk or whey products. Fat-free skim milk has only the casein micelles to scatter light, and they tend to scatter shorter-wavelength blue light more than they do red, giving skim milk a bluish tint.

Milk contains dozens of other types of proteins besides the caseins. They are more water-soluble than the caseins and do not form larger structures. Because these proteins remain suspended in the whey left behind when the caseins coagulate into curds, they are collectively known as *whey proteins*. Whey proteins make up around twenty percent of the protein in milk, by weight. Lactoglobulin is the most common whey protein by a large margin.

The carbohydrate lactose gives milk its sweet taste and contributes about 40 per cent of whole cow's milk's calories. Lactose is a composite of two simple sugars, glucose and galactose. In nature, lactose is found only in milk and a small number of plants. Other components found in raw cow's milk are living white blood cells, Mammary-gland cells, various bacteria, and a large number of active enzymes.

Processing

In most Western countries, a centralized dairy facility processes milk and products obtained from milk (dairy products), such as cream, butter, and cheese. In the U.S., these dairies are usually local companies, while in the southern hemisphere facilities may be run by very large nationwide or trans-national corporations (such as Fonterra).

Pasteurization

Pasteurization is used to kill harmful microorganisms by heating the milk for a short time and then cooling it for storage and transportation. Pasteurized milk is still perishable and must be stored cold by both suppliers and consumers. Dairies print expiration dates on each container, after which stores will remove any unsold milk from their shelves.

A newer process, Ultra Pasteurization or ultra-high temperature treatment (UHT), heats the milk to a higher temperature for a shorter time. This extends its shelf life and allows the milk to be stored unrefrigerated because of the longer lasting sterilization effect.

Microfiltration

Microfiltration is a process that partially replaces pasteurization and produces milk with fewer microorganisms and longer shelf life without a change in the taste of the milk. In this process, cream is separated from the whey and is pasteurized in the usual way, but the whey is forced through ceramic microfilters that trap 99.9 per cent of microorganisms in the milk (as compared to 95 per cent killing of microorganisms in conventional pasteurization). The whey is then recombined with the pasteurized cream to reconstitute the original milk composition.

Creaming and Homogenization

Upon standing for 12 to 24 hours, fresh milk has a tendency to separate into a high-fat cream layer on top of a larger, low-fat milk layer. The cream is often sold as a separate product with its own uses; today the separation of the cream from the milk is usually accomplished rapidly in centrifugal cream separators. The fat globules rise to the top of a container of milk because fat is less dense than water. The smaller the globules, the more other molecular-level forces prevent this from happening. In fact, the cream rises in cow's milk much more quickly than a simple model would predict: rather than isolated globules, the fat in the milk tends to form into clusters containing about a million globules, held together by a number of minor whey proteins. These clusters rise faster than individual globules can. The fat globules in milk from goats, sheep, and water buffalo do not form clusters so readily and are smaller to begin with; cream is very slow to separate from these milks.

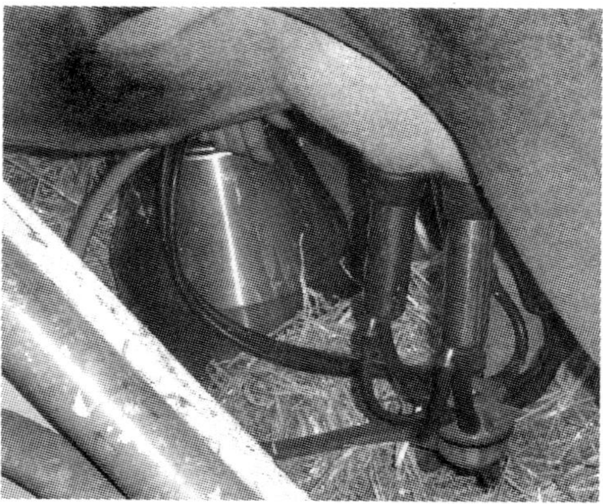

Fig. 3.6: A Milking Machine in Action

Milk is often homogenized, a treatment which prevents a cream layer from separating out of the milk. The milk is pumped at high pressures through very narrow tubes, breaking up the fat globules through turbulence and cavitation. A greater number of smaller particles possess more total surface area than a smaller number of larger ones, and the original fat globule membranes cannot completely cover them. Casein micelles are attracted to the newly-exposed fat surfaces; nearly one-third of the micelles in the milk end up participating in this new membrane structure. The casein weighs down the globules and interferes with the clustering that accelerated separation. The exposed fat globules are briefly vulnerable to certain enzymes present in milk, which could break down the fats and produce rancid flavors. To prevent this, the enzymes are inactivated by pasteurizing the milk immediately before or during homogenization.

Homogenized milk tastes blander but feels creamier in the mouth than unhomogenized; it is whiter and more resistant to developing off flavors. Creamline, or cream-top, milk is unhomogenized; it may or may not have been pasteurized. Milk which has undergone high-pressure homogenization, sometimes labeled as "ultra-homogenized," has a longer shelf life than milk which has undergone ordinary homogenization at lower pressures. Homogenized milk may be more digestible than unhomogenized milk.

Kurt A. Oster, M.D., who worked in the 1960s through the 1980s, suggested a link between homogenized milk and arterosclerosis, due to damage to plasmalogen as a result of the release of bovine xanthine oxidase (BXO) from the milk fat globular membrane (MFGM) during homogenization. However, Oster's hypothesis has been widely criticized and has not been generally accepted by the scientific community. No link has been found between arterosclerosis and milk consumption.

Nutrition and Health

The composition of milk differs widely between species. Factors such as the type of protein; the proportion of protein, fat, and sugar; the levels of various vitamins and minerals; and the size of the butterfat globules and the strength of the curd are among those than can vary. For example:

- Human milk contains, on average, 1.1 per cent protein, 4.2 per cent fat, 7.0 per cent lactose (a sugar), and supplies 72 kcal of energy per 100 grams.
- Cow's milk contains, on average, 3.4 per cent protein, 3.6 per cent fat, and 4.6 per cent lactose, 0.7 per cent minerals and supplies 66 kcal of energy per 100 grams.

Donkey and horse milk have the lowest fat content, while the milk of seals and whales can contain more than 50 per cent fat. High fat content is not unique to aquatic mammals, as guinea pig milk has an average fat content of 46 per cent.

Nutritional Value

Processed milk began containing differing amounts of fat during the 1950s. 1 cup (250 ml) of 2 per cent-fat milk contains 285 mg of calcium, which represents 22 per cent to 29 per cent of the daily recommended intake (DRI) of calcium for an adult. Depending on the age, milk contains 8 grams of protein, and a number of other nutrients (either naturally or through fortification) including:

- Biotin
- pantothenic acid
- Iodine
- Potassium
- Magnesium
- Selenium
- Thiamine
- Vitamin A
- Vitamin B12
- Riboflavin
- Vitamins D
- Vitamin K

The amount of calcium from milk that is absorbed by the human body is disputed. Calcium from dairy products has a greater bioavailability than calcium from certain vegetables, such as spinach, that contain high levels of calcium-chelating agents, but a similar or lesser bioavailability than calcium from low-oxalate vegetables such as kale, broccoli, or other vegetables in the Brassica genus.

Medical Research

Studies show possible links between low-fat milk consumption and reduced risk of arterial hypertension, coronary heart disease, colorectal cancer and obesity. Overweight individuals who drink milk may benefit from decreased risk of insulin resistance and type 2 diabetes. One study has shown that for women desiring to have a child, those who consume full fat dairy products may actually slightly increase their fertility, while those consuming low fat dairy products may slightly reduce their fertility. Milk is a source of Conjugated linoleic acid. CLA has been shown to kill human skin cancer, colorectal cancer and breast cancer cells *in vitro* studies, and may help lower cholesterol and prevent atherosclerosis.

It appears to be effective at promoting muscle growth.

Lactose Intolerance

Lactose, the disaccharide sugar component of all milk must be cleaved in the small intestine by the enzyme lactase in order for its constituents (galactose and glucose) to be absorbed. The production of this enzyme declines significantly after weaning in all mammals. Consequently, many humans become unable to properly digest lactose as they mature. There is a great deal of variance, with some individuals reacting badly to even small amounts of lactose, some able to consume moderate quantities, and some able to consume large quantities of milk and other dairy products without problems. When an individual consumes milk without producing sufficient lactase, they may suffer diarrhea, intestinal gas, cramps and bloating, as the undigested lactose travels through the gastrointestinal tract and serves as nourishment for intestinal microflora who excrete gas, a process known as anaerobic respiration.

It is estimated that 30 to 50 million Americans are lactose intolerant, including 75 percent of Native Americans and African-Americans, and 90 percent of Asian Americans. Lactose intolerance is less common among those descended from northern Europeans.

Lactose intolerance is a natural process and there is no reliable way to prevent or reverse it. Lactase is readily available in pill form, and many individuals can use it to briefly increase their tolerance for dairy products.

Controversy

Other studies suggest that milk consumption may increase the risk of suffering from certain health problems. Cow's milk allergy (CMA) is as an immunologically mediated adverse reaction to one or more cow's milk proteins. Rarely is it severe enough to cause death. Milk contains casein, a substance that breaks down in the human stomach to produce casomorphin, an opioid peptide. In the early 1990s it was hypothesized that casomorphin can cause or aggravate autism, and casein-free diets are widely promoted. Studies supporting these claims have had significant flaws, and the data are inadequate to guide autism treatment recommendations. Studies described in the book The China Study note a correlation between casein intake and the promotion of cancer cell growth when exposed to carcinogens. However other studies have shown whey protein offers a protective effect against colon cancer.

A study demonstrated that men who drink a large amount of milk and consume dairy products were at a slightly increased risk of developing Parkinson's disease; the effect for women was smaller. The reason behind this is not fully understood, and it also remains unclear why there is less of a risk for women. Several sources suggest a correlation between high calcium intake (2000 mg per day, or twice the US recommended daily allowance, equivalent to six or more glasses of milk per day) and prostate cancer. A large study specifically implicates dairy, *i.e.*, low-fat milk and other dairy to which vitamin A palmitate has been added. A review published by the World Cancer Research Fund and the American Institute for Cancer Research states that at least eleven human population studies have linked excessive dairy product consumption and prostate cancer, however randomized clinical trial data with appropriate controls only exists for calcium, not dairy produce, where there was no correlation. Medical studies have also shown a possible link between milk consumption and the exacerbation of diseases such as Crohn's Disease, Hirschsprung's disease-mimicking symptoms in babies with existing cow's milk allergies, and the aggravation of Behçet's disease.

Bovine Growth Hormone Supplementation

Since November 1993, with FDA approval, Monsanto Company has been selling recombinant bovine somatotropin (rbST), also called rBGH, to dairy farmers. Cows produce bovine growth hormone naturally, but many producers administer additional rBGH because it increases milk production. Bovine growth horome also stimulates liver production of insulin-like growth factor 1 (IGF1). If rbST-treated cows produced milk with higher levels of IGF1 this would be of medical concern, because IGF1 stimulates cancer growth in humans. Elevated levels of IGF1 in human blood has been linked to increased rates of breast, colon, and prostate cancer. Testing has shown that milk from cows given rBGH does not contain more IGF1 than milk from cows that were not given rBGH.. No study has indicated that consumption of rBST-produced milk increases IGF1 levels in humans, nor has any study demonstrated an increased risk of any disease between those consuming rBST and non-rBST produced milk. However, many groups remain concerned. The EU has recommended against Monsanto milk On June 9, 2006 the largest milk processor in the world and the two largest supermarkets in the United States—Dean Foods, Wal-Mart, and Kroger—announced that they are "on a nationwide search for rBGH-free milk." Milk from cows given rBST may be sold in the United States, and the FDA stated that no significant difference has been shown between milk derived from rBST-treated and that from non-rBST-treated cows. Milk that advertises that it comes from cows not treated with rBST is required to state this finding on its label.

Cows receiving rBGH supplements may more frequently contract an udder infection known as mastitis. Problems with mastitis have led to Canada, Australia, New Zealand, and Japan banning milk from rBST treated cows. Mastitis, among other diseases, may be responsible for the fact that levels of white blood cells in milk vary naturally. Although not considered a human health issue by most authorities, a minority of scientists believe that these cells could contribute to the transmission of bovine paratubeculosis to humans. The existing empirical evidence is largely inconclusive.

Ethical Concerns

Vegans and some other vegetarians do not consume milk for a variety of reasons. They may object to the treatment of cattle or to separating the mother and calf, veal production, and slaughter of "used" cows.

Nutrition—Comparison by Animal Source

These compositions vary by breed, animal, and point in the lactation period. Jersey cows produce milk of about 5.2 per cent fat, Zebu cows produce milk of about 4.7 per cent fat, Brown Swiss cows produce milk of about 4.0 per cent fat, and Holstein-Friesian cows produce milk of about 3.6 per cent fat. The protein range for these four breeds is 3.3 per cent to 3.9 per cent, while the lactose range is 4.7 per cent to 4.9 per cent.

Table 3.3: Milk Composition Analysis, Per 100 Grams

Constituents	*unit*	*Cow*	*Goat*	*Sheep*	*Water buffalo*
Water	g	87.8	88.9	83.0	81.1
Protein	g	3.2	3.1	5.4	4.5
Fat	g	3.9	3.5	6.0	8.0
Carbohydrate	g	4.8	4.4	5.1	4.9
Energy	kcal	66	60	95	110
Energy	kJ	275	253	396	463
Sugars (lactose)	g	4.8	4.4	5.1	4.9
Cholesterol	mg	14	10	11	8
Calcium	IU	120	100	170	195
Fatty acids:					
Saturated	g	2.4	2.3	3.8	4.2
Mono-unsaturated	g	1.1	0.8	1.5	1.7
Polyunsaturated	g	0.1	0.1	0.3	0.2

Milk fat percentages can be manipulated by dairy farmers' diet formulation strategies. Mastitis infection can cause fat levels to decline.

Varieties and Brands

Milk products are sold in a number of varieties based on types/ degrees of:

- age (*e.g.*, cheddar),
- additives (*e.g.*, vitamins),
- coagulation (*e.g.*, cottage cheese),
- farming method (*e.g.*, organic, grass-fed).
- fat content (*e.g.*, half and half),
- fermentation (*e.g.*, buttermilk),
- flavoring (*e.g.*, chocolate),
- homogenization (*e.g.*, raw milk),
- mammal (*e.g.*, cow, goat, sheep),
- packaging (*e.g.*, bottle),
- sterilization (*e.g.*, pasteurization),
- water content (*e.g.*, dry milk),

Demeter certified milk is produced with biodynamic agriculture methods and is similar in standards to organic milk and biological milk, with a few special farm procedures added that are biodynamic-specific.

Additives and Flavoring

In countries where the cattle (and often the people) live indoors, commercially sold milk commonly has vitamin D added to it to make up for lack of exposure to UVB radiation.

Reduced fat milks often have added vitamin A to compensate for the loss of the vitamin during fat removal; in the United States this results in reduced fat milks having a higher vitamin A content than whole milk.

To aid digestion in those with lactose intolerance, milk is available in some areas with added bacterial cultures such as *Lactobacillus acidophilus* ("acidophilus milk") and bifidobacteria ("a/B milk"). Another milk with *Lactococcus lactis* bacteria cultures ("cultured buttermilk") is often used in cooking to replace the traditional use of naturally soured milk, which has become rare due to the ubiquity of pasteurization which kills the naturally occurring lactococcus bacteria.

Milk often has flavoring added to it for better taste or as a means of improving sales. Chocolate milk has been sold for many years and has been followed more recently by such other flavors as strawberry and banana.

South Australia has the highest consumption of flavored milk per person in the world, where Farmers Union Iced Coffee outsells Coca-Cola, a success shared only by Inca Kola in Peru and Irn-Bru in Scotland.

Distribution

Because milk spoils so easily, it should, ideally, be distributed as quickly as possible. In many countries milk used to be delivered to households daily, but economic pressure has made milk delivery much less popular, and in many areas daily delivery is no longer available. People buy it chilled at grocery or convenience stores or similar retail outlets. Prior to the widespread use of plastics, milk was sold in wax-coated paper containers; prior to that milk was often distributed to consumers in glass bottles; and before glass bottles, in bulk that was ladled into the customer's container.

United Kingdom

In the UK, milk can be delivered daily by a milkman who travels his local milk round (route) using a milk float (often battery powered) during the early hours. Milk is delivered in 1 pint glass bottles with aluminium foil tops. Silver top denotes full cream unhomogenized; red top full cream homogenized; red/silver top semi-skimmed; blue/silver check top skimmed; and gold top channel island.

Empty bottles are rinsed before being left outside for the milkman to collect and take back to the dairy for washing and reuse. Currently many milkmen operate franchises as opposed to being employed by the dairy and payment is made at regular intervals, by leaving a cheque; by cash collection; or direct debit.

Although there was a steep decline in doorstep delivery sales throughout the 1990s, the service is still prominent, as dairies have diversified and the service is becoming more popular again. The doorstep delivery of milk is seen as part of the UK's heritage, and is relied upon by people up and down the country.

Australia and New Zealand

In Australia and New Zealand, prior to "metrification", milk was generally distributed in 1 pint (568ml) glass bottles. In Australia there was a government funded "free milk for school children" programme, and milk was distributed at morning recess in 1/3 pint bottles. With the conversion to metric measures, the milk industry were concerned that the replacement of the pint bottles with 500ml bottles would result in a 13.6 per cent drop in milk consumption. Hence, all pint bottles were recalled and replaced by 600ml bottles. With time, due to the steadily increasing cost of collecting, transporting, storing and cleaning glass bottles, they were replaced by cardboard cartons. A number of designs were used, including a tetrahedron which could be close-packed without waste space, and could not be knocked over accidentally. (slogan: No more crying over spilt milk.) However, the industry eventually settled on a design similar to that used in the United States. Milk is now availability in a variety of sizes in cardboard cartons (250ml, 375ml, 600ml, 1 litre and 1.5 litres) and plastic bottles (1 in NZ, 1.1 in Australia, 2 and 3 litres). A significant addition to the marketplace has been "long life" milk (UHT), generally available in 1 and 2 litre rectangular cardboard cartons. In urban and suburban areas where there is sufficient demand, home delivery is still available, though in suburban areas this is often 3 times per week rather than daily. Another significant and popular addition to the marketplace has been flavoured milks—for example, as mentioned above, Farmers Union Iced Coffee outsells Coca-Cola in South Australia.

India and Pakistan

In rural India and Pakistan, milk is delivered daily by a local milkman carrying bulk quantities in a metal container, usually on a bicycle; and in other parts of metropolitan India and Pakistan, milk is usually bought or delivered in a plastic bags or cartons via shops or supermarkets.

United States

In the United States, glass milk bottles have been mostly replaced with milk cartons (tall paper boxes with a square cross-section and a peaked top that can be folded outward upon opening to form a spout) and plastic jugs. Gallons of milk are almost always sold in jugs, while half-gallons and quarts may be found in both paper cartons and plastic jugs, and smaller sizes are almost always in cartons. Recently, milk has been sold in smaller resealable bottles made to fit in automobile cup holders. These individual serving sizes are also sold in flavored varieties.

The half-pint milk carton is the traditional unit as a component of school lunches. In the U.S., pictures of missing children were printed on the larger milk cartons as a public service until it was determined that this was disturbing to children.

UHT Milk

Milk preserved by the UHT process is sold in cartons often called a brick that lack the peak of the traditional milk carton. Milk preserved in this fashion does not need to be refrigerated before opening and has a longer shelf life than milk in ordinary packaging. It is more typically sold unrefrigerated on the shelves in Europe and Latin America than in the United States. In Australia it is generally sold unrefrigerated, though some supermarkets also keep small quantities refrigerated.

Use of Glass

Glass milk containers are now rare. Most people purchase milk in bags, plastic jugs or plastic-coated paper cartons. Ultraviolet (UV) light from fluorescent lighting can alter the flavor of milk, so many companies that once distributed milk in transparent or highly translucent containers are now using thicker materials that block the UV light. Many people feel that such "UV protected" milk tastes better.

Packaging

Milk comes in a variety of containers with local variants:

- *Australia and New Zealand*: Distributed in a variety of sizes, most commonly in aseptic cartons for up to 1.5 litres, and plastic screw-top bottles beyond that with the following volumes; 1.1L, 2L, and 3L. 1 litre bags are starting to appear in supermarkets, but have not yet proved popular. Most UHT-milk is packed in 1 or 2 litre paper containers with a sealed plastic spout.
- *Brazil*: Used to be sold in cooled 1 litre bags, just like in

South Africa. Nowadays the most common form is 1 litre aseptic cartons containing UHT skimmed, semi-skimmed or whole milk, although the plastic bags are still in use for pasteurized milk. Higher grades of pasteurized milk can be found in cartons or plastic bottles. Sizes other than 1 liter are rare.

- *Canada*: 1.33 litre plastic bags (sold as 4 litres in 3 bags) are widely available in some areas (especially the Maritimes, Ontario and Quebec), although the 4 litre plastic jug has supplanted them in western Canada. Other common packaging sizes are 2 litre, 1 litre, 500 millilitre, and 250 millilitre cartons, as well as 4 litre, 1 litre, 250 ml aseptic cartons and 500 ml plastic jugs.
- *China*: Sweetened milk is a drink popular with students of all ages and is often sold in small plastic bags complete with straw. Adults not wishing to drink at a banquet often drink milk served from cartons or milk tea.
- *Parts of Europe*: Sizes of 500 millilitres, 1 litre (the most common), 1.5 litres, 2 litres and 3 litres are commonplace.
- *Finland*: Commonly sold in 1l or 1.5l cartons, in some places also in 2dl and 5dl cartons.
- *Hong Kong*: Milk is sold in glass bottles (220 ml), cartons (236 ml and 1L), plastic jugs (2 litres) and aseptic cartons (250 ml).
- *India and Pakistan*: Commonly sold in 500 ml plastic bags. It is still customary to serve the milk boiled, despite pasteurization. Milk is often buffalo milk. Flavored milk is sold in most convenience stores in waxed cardboard containers. Convenience stores also sell many varieties of milk (such as flavored and ultra-pasteurized) in different sizes, usually in aseptic cartons.
- *Indonesia*: Usually sold in 1 litre cartons, but smaller, snack-sized cartons are available.
- *Israel*: Non-UHT milk is most commonly sold in 1 litre waxed cardboard boxes and 1 litre plastic bags. It may also be found in 0.5L and 2L waxed cardboard boxes, 2L plastic jugs and 1L plastic bottles. UHT milk is available in 1 litre (and less commonly also in 0.25L) carton "bricks".

- *Japan*: Commonly sold in 1 litre waxed cardboard boxes. In most city centers there is also home delivery of milk in glass jugs. As seen in China, sweetened and flavored milk drinks are commonly seen in vending machines.
- *South Africa*: Commonly sold in 1 litre bags. The bag is then placed in a plastic jug and the corner cut off before the milk is poured.
- *South Korea*: sold in cartons (180ml, 200ml, 500ml 900ml, 1L, 1.8L, 2.3L), plastic jugs (1L and 1.8L), aseptic cartons (180ml and 200ml) and plastic bags (1L).
- *Sweden*: Commonly sold in 0.3L, 1L or 1.5L cartons and sometimes as plastic or glass bottles.
- *Poland*: UHT milk is mostly sold in aseptic cartons (500ml, 1L, 2L), and non-UHT in 1L plastic bags or plastic bottles. Milk, UHT is commonly boiled, despite being pasteurized.
- *Turkey*: Commonly sold in 500 ml or 1L cartons or special plastic bottles. UHT milk is more popular. Milkmen also serve in smaller towns and villages.
- *United Kingdom*: Most stores still stock Imperial sizes: 1 pint (568 mL), 2 pints (1.136 L), 4 pints (2.273 L), 6 pints (3.408 L) or a combination including both metric and imperial sizes. Glass milk bottles delivered to the doorstep by the milkman are typically pint-sized and are returned empty by the householder for repeated reuse. Milk is also sold at supermarkets in either aseptic cartons or HDPE bottles. Milk can still be legally sold by the Imperial pint in reusable bottles in the UK under EU regulations (a distinction only shared with beer and cider), whilst a growing number of manufacturers such as Northern Foods now sell milk in 1 and 2 litre bottles.
- *United States*: Commonly sold in gallon (3.78 L), half-gallon (1.89 L) and quart (0.94 L) containers of rigid plastic or, occasionally for sizes less than a gallon, waxed cardboard, although bottles made of opaque PET are starting to become more commonplace in all smaller sizes. The US single-serving size is usually the half-pint (about 240 ml). Occasionally dairies will deliver milk straight to

customers in coolers filled with glass bottles (usually half-gallon). Some convenience store chains in the United States (such as Kwik Trip in the Midwest) sell milk in 1/2 gallon bags.

- *Uruguay*: Commonly sold in 1 litre bags. The bag is then placed in a plastic jug and the corner cut off before the milk is poured.

Practically everywhere, condensed milk and evaporated milk is distributed in metal cans, 250 and 125 ml paper containers and 100 and 200 ml squeeze tubes, and powdered milk (skim and whole) is distributed in boxes or bags.

Spoilage and Fermented Milk Products

When raw milk is left standing for a while, it turns "sour". This is the result of fermentation, where lactic acid bacteria ferment the lactose inside the milk into lactic acid. Prolonged fermentation may render the milk unpleasant to consume. This fermentation process is exploited by the introduction of bacterial cultures (*e.g. Lactobacilli sp., Streptococcus sp., Leuconostoc sp.*, etc) to produce a variety of fermented milk products. The reduced pH from lactic acid accumulation denatures proteins and caused the milk to undergo a variety of different transformations in appearance and texture, ranging from an aggregate to smooth consistency. Some of these products include sour cream, yoghurt, cheese, buttermilk, viili, kefir and kumis.

Pasteurization of cow's milk initially destroys any potential pathogens and increases the shelf-life, but eventually results in spoilage that makes it unsuitable for consumption. This causes it to assume an unpleasant odor, and the milk is deemed non-consumable due to unpleasant taste and an increased risk of food poisoning. In raw milk, the presence of lactic acid-producing bacteria, under suitable conditions, ferments the lactose present to lactic acid. The increasing acidity in turn prevents the growth of other organisms, or slows their growth significantly. During pasteurization however, these lactic acid bacteria are mostly destroyed.

In order to prevent spoilage, milk can be kept refrigerated and stored between 1 and 4 degrees Celsius in bulk tanks. Most milk is pasteurized by heating briefly and then refrigerated to allow transport from factory farms to local markets. The spoilage of milk can be forestalled by using ultra-high temperature (UHT) treatment; milk so treated can be stored unrefrigerated for several months until opened. Condensed milk, made by removing most of the water, can be stored in cans for many years, unrefrigerated, as can evaporated milk. The most durable form of milk is milk powder, which is produced from milk by removing almost all water. The moisture content is usually less than five percent in both drum and spray dried milk powder.

Language and Culture

The importance of milk in human culture is attested to by the numerous expressions embedded in our languages, for example "the milk of human kindness". In ancient Greek mythology, the goddess Hera spilled her breast milk after refusing to feed Heracles, resulting in the Milky Way.

In African and Asian developing nations, butter is traditionally made from fermented milk rather than cream. It can take several hours of churning to produce workable butter grains from fermented milk.

Holy books have also mentioned milk; the Bible contains references to the 'Land of Milk and Honey'. In the Quran, there is a request to wonder on milk as follows: 'And surely in the livestock there is a lesson for you, We give you to drink of that which is in their bellies from the midst of digested food and blood, pure milk palatable for the drinkers. The Ramadhan fast is traditionally broken with a glass of milk and dates.

The verb, "to milk" something is often used in the vernacular of many English-speaking countries as a synonym for extortion or, in less loaded terms, taking advantage of a situation where one has another person at a disadvantage, as in 'milking the situation'.

The word milk has had many slang meanings over time. In the early 17th century the word was used to mean semen, or vaginal secretions, or to masturbate oneself or someone else. In the 19th century, milk was used to describe a cheap alcoholic drink made from methylated spirits mixed with water. The word was also used to mean defraud, to be idle, to intercept telegrams addressed to someone else, and a weakling or 'milksop'. In the mid 1930s, the word was used in Australia meaning to siphon gas from a car.

Milk is sometimes referred to as moo juice in American English, while Cockney rhyming slang calls it Acker Bilk, Tom Silk, Lady in silk and Kilroy Silk.

Other Uses of Milk

Besides serving as a beverage or source of food, milk is used by farmers and gardeners as an organic fungicide and foliage fertilizer. The potassium, fats, and salts naturally found in milk are absorbed by the leaves, which boost the plant's immune system, helping it to fight off diseases and fungi. Farmers, especially those who maintain grape vineyards, have tested a diluted milk solution in the past, and have found it to be more effective, yet less costly, than commercial products. Testing has also shown that it is unlikely to harm the plant that the solution is applied to.

CHEESE

Cheese is a food consisting of proteins and fat from milk, usually the milk of cows, buffalo, goats, or sheep. It is produced by coagulation of the milk protein casein. Typically, the milk is acidified and addition of the enzyme rennet causes coagulation. The solids are separated and pressed into final form. Some cheeses have molds on the rind or throughout.

Hundreds of types of cheese are produced. Their styles, textures and flavors depend on the origin of the milk (including the animal's diet), whether they have been pasteurized, the butterfat content, the bacteria and mold, the processing, and aging. Herbs, spices, or wood smoke may be used as flavoring agents. The yellow to red colour of many cheeses is from adding annatto. Most cheeses melt when heated.

For a few cheeses, the milk is curdled by adding acids such as vinegar or lemon juice. Most cheeses are acidified to a lesser degree by bacteria, which turn milk sugars into lactic acid, then the addition of rennet completes the curdling. Vegetarian alternatives to rennet are available; most are produced by fermentation of the fungus *Mucor miehei*, but others have been extracted from various species of the *Cynara* thistle family.

Cheese is valued for its portability, long life, and high content of fat, protein, calcium, and phosphorus. Cheese is more compact and has a longer shelf life than milk. Cheesemakers near a dairy region may benefit from fresher, lower-priced milk, and lower shipping costs. The long storage life of some cheese, especially cheese that encased in a protective rind, allows selling when markets are favourable.

Etymology

The word *cheese* ultimately comes from Latin *caseus*, from which the modern word casein is closely derived. The earliest source is from the proto-Indo-European roo, which means "to ferment, become sour".

More recently, *cheese* comes from *chese* (in Middle English) and *cîese* or *cçse* (in Old English). Similar words are shared by other West Germanic languages—West Frisian *tsiis*, Dutch *kaas*, German *Käse*, Old High German *châsi*—all from the reconstructed West-Germanic form *kasjus*, which in turn is an early borrowing from Latin.

When the Romans began to make hard cheeses for their legionaries' supplies, a new word started to be used: *formaticum*, from *caseus formatus*, or "molded cheese" (as in "formed", not "molded"). It is from this word that we get the French *fromage*, Italian *formaggio*, Catalan *formatge*, Breton *fourmaj* and Provençal *furmo*. *Cheese* itself is occasionally employed in a sense that means "molded" or "formed". *Head cheese* uses the word in this sense.

History

Origins

Cheese is an ancient food whose origins predate recorded history. There is no conclusive evidence indicating where cheesemaking originated, either in Europe, Central Asia or the Middle East, but the practice had spread within Europe prior to Roman times and, according to Pliny the Elder, had become a sophisticated enterprise by the time the Roman Empire came into being.

Proposed dates for the origin of cheesemaking range from around 8000 BCE (when sheep were first domesticated) to around 3000 BCE. The first cheese may have been made by people in the Middle East or by nomadic Turkic tribes in Central Asia. Since animal skins and inflated internal organs have, since ancient times, provided storage vessels for a range of foodstuffs, it is probable that the process of cheese making was discovered accidentally by storing milk in a container made from the stomach of an animal, resulting in the milk being turned to curd and whey by the rennet from the stomach. There is a legend with variations about the discovery of cheese by an Arab trader who used this method of storing milk.

Cheesemaking may have begun independently of this by the pressing and salting of curdled milk in order to preserve it. Observation that the effect of making milk in an animal stomach gave more solid and better-textured curds, may have led to the deliberate addition of rennet.

The earliest archeological evidence of cheesemaking has been found in Egyptian tomb murals, dating to about 2000 BCE. The earliest cheeses were likely to have been quite sour and salty, similar in texture to rustic cottage cheese or feta, a crumbly, flavorful Greek cheese.

Cheese produced in Europe, where climates are cooler than the Middle East, required less salt for preservation. With less salt and acidity, the cheese became a suitable environment for useful microbes and molds, giving aged cheeses their pronounced and interesting flavors.

Ancient Greece and Rome

Ancient Greek mythology credited Aristaeus with the discovery of cheese. Homer's *Odyssey* (8th century BCE) describes the Cyclops making and storing sheep's and goats' milk cheese. From Samuel Butler's translation:

> "We soon reached his cave, but he was out shepherding, so we went inside and took stock of all that we could see. His cheese-racks were loaded with cheeses, and he had more lambs and kids than his pens could hold...
>
> When he had so done he sat down and milked his ewes and

goats, all in due course, and then let each of them have her own young. He curdled half the milk and set it aside in wicker strainers."

By Roman times, cheese was an everyday food and cheesemaking a mature art. Columella's *De Re Rustica* (circa 65 CE) details a cheesemaking process involving rennet coagulation, pressing of the curd, salting, and aging. Pliny's *Natural History* (77 CE) devotes a chapter (XI, 97) to describing the diversity of cheeses enjoyed by Romans of the early Empire. He stated that the best cheeses came from the villages near Nîmes, but did not keep long and had to be eaten fresh. Cheeses of the Alps and Apennines were as remarkable for their variety then as now. A Ligurian cheese was noted for being made mostly from sheep's milk, and some cheeses produced nearby were stated to weigh as much as a thousand pounds each. Goats' milk cheese was a recent taste in Rome, improved over the "medicinal taste" of Gaul's similar cheeses by smoking. Of cheeses from overseas, Pliny preferred those of Bithynia in Asia Minor.

Post-classical Europe

Rome spread a uniform set of cheesemaking techniques throughout much of Europe, and introduced cheesemaking to areas without a previous history of it. As Rome declined and long-distance trade collapsed, cheese in Europe diversified further, with various locales developing their own distinctive cheesemaking traditions and products. The British Cheese Board claims that Britain has approximately 700 distinct local cheeses; France and Italy have perhaps 400 each. (A French proverb holds there is a different French cheese for every day of the year, and Charles de Gaulle once asked "how can you govern a country in which there are 246 kinds of cheese?") Still, the advancement of the cheese art in Europe was slow during the centuries after Rome's fall. Many cheeses today were first recorded in the late Middle Ages or after—cheeses like Cheddar around 1500 CE, Parmesan in 1597, Gouda in 1697, and Camembert in 1791.

In 1546, *The Proverbs of John Heywood* claimed "the moon is made of a green cheese." (*Greene* may refer here not to the colour, as many now think, but to being new or unaged.) Variations on this sentiment were long repeated and NASA exploited this myth for an April Fool's spoof announcement in 2006.

Modern Era

Until its modern spread along with European culture, cheese was nearly unheard of in oriental cultures, in the pre-Columbian Americas, and only had limited use in sub-Mediterranean Africa, mainly being widespread and popular only in Europe and areas influenced strongly by its cultures. But with the spread, first of European imperialism, and later of Euro-American culture and food, cheese has gradually become known and increasingly popular worldwide, though still rarely considered a part of local ethnic cuisines outside Europe, the Middle East, and the Americas.

The first factory for the industrial production of cheese opened in Switzerland in 1815, but it was in the United States where large-scale production first found real success. Credit usually goes to Jesse Williams, a dairy farmer from Rome, New York, who in 1851 started making cheese in an assembly-line fashion using the milk from neighboring farms. Within decades hundreds of such dairy associations existed.

The 1860s saw the beginnings of mass-produced rennet, and by the turn of the century scientists were producing pure microbial cultures. Before then, bacteria in cheesemaking had come from the environment or from recycling an earlier batch's whey; the pure cultures meant a more standardized cheese could be produced.

Factory-made cheese overtook traditional cheesemaking in the World War II era, and factories have been the source of most cheese in America and Europe ever since. Today, Americans buy more processed cheese than "real", factory-made or not.

Making Cheese

Curdling

A required step in cheesemaking is separating the milk into solid curds and liquid whey. Usually this is done by acidifying (souring) the milk and adding rennet. The acidification can be accomplished directly by the addition of an acid like vinegar in a few cases (paneer, queso fresco), but usually starter bacteria are employed instead. These starter bacteria convert milk sugars into lactic acid. The same bacteria (and the enzymes they produce) also play a large role in the eventual flavor of aged cheeses. Most cheeses are made with starter bacteria from the *Lactococci*, *Lactobacilli*, or *Streptococci* families. Swiss starter cultures also include *Propionibacter shermani*, which produces carbon dioxide gas bubbles during aging, giving Swiss cheese or Emmental its holes.

Some fresh cheeses are curdled only by acidity, but most cheeses also use rennet. Rennet sets the cheese into a strong and rubbery gel compared to the fragile curds produced by acidic coagulation alone. It also allows curdling at a lower acidity—important because flavor-making bacteria are inhibited in high-acidity environments. In general, softer, smaller, fresher cheeses are curdled with a greater proportion of acid to rennet than harder, larger, longer-aged varieties.

Curd Processing

At this point, the cheese has set into a very moist gel. Some soft cheeses are now essentially complete: they are drained, salted, and packaged. For most of the rest, the curd is cut into small cubes. This allows water to drain from the individual pieces of curd.

Some hard cheeses are then heated to temperatures in the range of 35°C-55°C (100°F-130°F). This forces more whey from the cut curd. It also changes the taste of the finished cheese, affecting both the bacterial culture and the milk chemistry. Cheeses that are heated to the higher temperatures are usually made with thermophilic starter bacteria which survive this step—either or streptococci.

Salt has roles in cheese besides adding a salty flavor. It preserves cheese from spoiling, draws moisture from the curd, and firms cheese's texture in an interaction with its proteins. Some cheeses are salted from the outside with dry salt or brine washes. Most cheeses have the salt mixed directly into the curds.

Other techniques influence a cheese's texture and flavor. Some examples:

- Stretching: (Mozzarella, Provolone) The curd is stretched and kneaded in hot water, developing a stringy, fibrous body.

- Cheddaring: (Cheddar, other English cheeses) The cut curd is repeatedly piled up, pushing more moisture away. The curd is also mixed (or *milled*) for a long time, taking the sharp edges off the cut curd pieces and influencing the final product's texture.
- Washing: (Edam, Gouda, Colby) The curd is washed in warm water, lowering its acidity and making for a milder-tasting cheese.

Most cheeses achieve their final shape when the curds are pressed into a mold or form. The harder the cheese, the more pressure is applied. The pressure drives out moisture—the molds are designed to allow water to escape—and unifies the curds into a single solid body.

Aging

A newborn cheese is usually salty yet bland in flavor and, for harder varieties, rubbery in texture. These qualities are sometimes enjoyed—cheese curds are eaten on their own—but normally cheeses are left to rest under controlled conditions. This aging period (also called ripening, or, from the French, *affinage*) lasts from a few days to several years. As a cheese ages, microbes and enzymes transform texture and intensify flavor. This transformation is largely a result of the breakdown of casein proteins and milkfat into a complex mix of amino acids, amines, and fatty acids.

Some cheeses have additional bacteria or molds intentionally introduced before or during aging. In traditional cheesemaking, these microbes might be already present in the aging room; they are simply allowed to settle and grow on the stored cheeses. More often today, prepared cultures are used, giving more consistent results and putting fewer constraints on the environment where the cheese ages. These cheeses include soft ripened cheeses such as Brie and Camembert, blue cheeses such as Roquefort, Stilton, Gorgonzola, and rind-washed cheeses such as Limburger.

Types

Factors in Categorization

Factors which are relevant to the categorization of cheeses include:

- Length of aging
- Texture
- Methods of making
- Fat content
- Kind of milk
- Country/Region of Origin

List of Common Categories

No one categorization scheme can capture all the diversity of the world's cheeses. In practice, no single system is employed and different factors are emphasised in describing different classes of cheeses. This typical list of cheeses includes categories from foodwriter Barbara Ensrud.

- Fresh
- Whey
- *Pasta filata*
- Semi-soft
- Semi-firm
- Hard
- Double and triple cream
- Soft-ripened
- Blue vein
- Goat or sheep
- Strong-smelling
- Sharp
- Processed

Fresh, Whey and Stretched Curd Cheeses

The main factor in the categorization of these cheese is their age. Fresh cheeses without additional preservatives can spoil in a matter of days.

For these simplest cheeses, milk is curdled and drained, with little other processing. Examples include cottage cheese, Romanian Ca°, Neufchâtel (the model for American-style cream cheese), and fresh goat's milk chèvre. Such cheeses are soft and spreadable, with a mild taste.

Whey cheeses are fresh cheeses made from the whey discarded

while producing other cheeses. Provencal Brousse, Corsican Brocciu, Italian Ricotta, Romanian Urda, Greek Mizithra, and Norwegian Geitost are examples. Brocciu is mostly eaten fresh, and is as such a major ingredient in Corsican cuisine, but it can be aged too.

Traditional *pasta filata* cheeses such as Mozzarella also fall into the fresh cheese category. Fresh curds are stretched and kneaded in hot water to form a ball of Mozzarella, which in southern Italy is usually eaten within a few hours of being made. Stored in brine, it can be shipped, and is known worldwide for its use on pizzas. Other firm fresh cheeses include paneer and queso fresco.

Classed by Texture

Categorizing cheeses by firmness is a common but inexact practice. The lines between "soft", "semi-soft", "semi-hard", and "hard" are arbitrary, and many types of cheese are made in softer or firmer variations. The factor controlling the hardness of a cheese is its moisture content which is dependent on the pressure with which it is packed into molds and the length of time it is aged.

Fig. 3.7: Emmentaler

Semi-soft cheeses and the sub-group, *Monastery* cheeses have a high moisture content and tend to be bland in flavor. Some well-known varieties include Havarti, Munster and Port Salut.

Cheeses that range in texture from semi-soft to firm include Swiss-style cheeses like Emmental and Gruyère. The same bacteria that give such cheeses their holes also contribute to their aromatic and sharp flavors. Other semi-soft to firm cheeses include Gouda, Edam, Jarlsberg and Cantal. Cheeses of this type are ideal for melting and are used on toast for quick snacks.

Harder cheeses have a lower moisture content than softer cheeses. They are generally packed into molds under more pressure and aged for a longer time. Cheeses that are semi-hard to hard include the familiar Cheddar, originating in the village of Cheddar in England but now used as a generic term for this style of cheese, of which varieties are imitated worldwide and are marketed by strength or the length of time they have been aged. Cheddar is one of a family of semi-hard or hard cheeses (including Cheshire and Gloucester) whose curd is cut, gently heated, piled, and stirred before being pressed into forms. Colby and Monterey Jack are similar but milder cheeses; their curd is rinsed before it is pressed, washing away some acidity and calcium. A similar curd-washing takes place when making the Dutch cheeses Edam and Gouda.

Hard cheeses—"grating cheeses" such as Parmesan and Pecorino Romano—are quite firmly packed into large forms and aged for months or years.

Classed by Content

Some cheeses are categorized by the source of the milk used to produce them or by the added fat content of the milk from which they are produced. While most of the world's commercially available cheese is made from cows' milk, many parts of the world also produce cheese from goats and sheep, well-known examples being Roquefort, produced in France, and Pecorino Romano, produced in Italy, from ewe's milk. One farm in Sweden also produces cheese from moose's milk. Sometimes cheeses of a similar style may be available made from milk of different sources, Feta style cheeses, for example, being made from goats' milk in Greece and of sheep and cows milk elsewhere.

Double cream cheeses are soft cheeses of cows' milk which are enriched with cream so that their fat content is 60 per cent or, in the case of triple creams, 75 per cent.

Soft-ripened and Blue-vein

There are three main categories of cheese in which the presence of mold is a significant feature: soft ripened cheeses, washed rind cheeses and blue cheeses.

Soft-ripened cheeses are those which begin firm and rather chalky in texture but are aged from the exterior inwards by exposing them to mold. The mold may be a velvety *bloom* of *Penicillium candida* or *P. camemberti* that forms a flexible white crust and contributes to the smooth, runny, or gooey textures and more intense flavors of these aged cheeses. Brie and Camembert, the most famous of these cheeses, are made by allowing white mold to grow on the outside of a soft cheese for a few days or weeks. Goats' milk cheeses are often treated in a similar manner, sometimes with white molds (Chèvre-Boîte) and sometimes with blue.

Washed-rind cheeses are soft in character and ripen inwards like those with white molds; however, they are treated differently. Washed rind cheeses are periodically cured in a solution of saltwater brine and other mold-bearing agents which may include beer, wine, brandy and spices, making their surfaces amenable to a class of bacteria *Brevibacterium linens* (the reddish-orange "smear bacteria") which impart pungent odors and distinctive flavors. Washed-rind cheeses can be soft (Limburger), semi-hard (Munster), or hard (Appenzeller). The same bacteria can also have some impact on cheeses that are simply ripened in humid conditions, like Camembert.

So-called blue cheese is created by inoculating a cheese with *Penicillium roqueforti* or *Penicillium glaucum*. This is done while the cheese is still in the form of loosely pressed curds, and may be further enhanced by piercing a ripening block of cheese with skewers in an atmosphere in which the mold is prevalent. The mold grows within the cheese as it ages. These cheeses have distinct blue veins which gives them their name, and, often, assertive flavors. The molds may range from pale green to dark blue, and may be accompanied by white and crusty brown molds. Their texture can be soft or firm. Some of the most renowned cheeses are of this type, each with its own distinctive colour, flavor, texture and smell. They include Roquefort, Gorgonzola, and Stilton.

Processed Cheeses

Processed cheese is made from traditional cheese and emulsifying salts, often with the addition of milk, more salt, preservatives, and food colouring. It is inexpensive, consistent, and melts smoothly. It is sold packaged and either pre-sliced or unsliced, in a number of varieties. It is also available in spraycans.

Eating and Cooking

At refrigerator temperatures, the fat in a piece of cheese is as hard as unsoftened butter, and its protein structure is stiff as well. Flavor and odor compounds are less easily liberated when cold. For improvements in flavor and texture, it is widely advised that cheeses be allowed to warm up to room temperature before eating. If the cheese is further warmed, to 26-32°C (80-90°F), the fats will begin to "sweat out" as they go beyond soft to fully liquid.

Above room temperatures, most hard cheeses melt. Rennet-curdled cheeses have a gel-like protein matrix that is broken down by heat. When enough protein bonds are broken, the cheese itself turns from a solid to a viscous liquid. Soft, high-moisture cheeses will melt at around 55°C (131°F), while hard, low-moisture cheeses such as Parmesan remain solid until they reach about 82°C (180°F). Acid-set cheeses, including halloumi, paneer, some whey cheeses and many varieties of fresh goat cheese, have a protein structure that remains intact at high temperatures. When cooked, these cheeses just get firmer as water evaporates.

Some cheeses, like raclette, melt smoothly; many tend to become stringy or suffer from a separation of their fats. Many of these can be coaxed into melting smoothly in the presence of acids or starch. Fondue, with wine providing the acidity, is a good example of a smoothly melted cheese dish. Elastic stringiness is a quality that is sometimes enjoyed, in dishes including pizza and Welsh rarebit. Even a melted cheese eventually turns solid again, after enough moisture is cooked off. The saying "you can't melt cheese twice" (meaning "some things can only be done once") refers to the fact that oils leach out during the first melting and are gone, leaving the non-meltable solids behind.

As its temperature continues to rise, cheese will brown and eventually burn. Browned, partially burned cheese has a particular distinct flavor of its own and is frequently used in cooking (*e.g.*, sprinkling atop items before baking them).

Health and Nutrition

In general, cheese supplies a great deal of calcium, protein, phosphorus and fat. A 30-gram (1.1 oz) serving of Cheddar cheese contains about 7 grams (0.25 oz) of protein and 200 milligrams of calcium. Nutritionally, cheese is essentially concentrated milk: it takes about 200 grams (7.1 oz) of milk to provide that much protein, and 150 grams (5.3 oz) to equal the calcium.

Cheese potentially shares other of milk's nutritional content as well. The Center for Science in the Public Interest describes cheese as America's number one source of saturated fat, adding that the average American ate 30 lb (14 kg) of cheese in the year 2000, up from 11 lb (5 kg) in 1970. Their recommendation is to limit full-fat cheese consumption to 2 oz (57 g) a week. Whether cheese's highly saturated fat actually leads to an increased risk of heart disease is called into question when considering France and Greece, which lead the world in cheese eating (more than 14 oz/400 g a week per person, or over 45 lb/20 kg a year) yet have relatively low rates of heart disease. This seeming discrepancy is called the French Paradox; the higher rates of consumption of red wine in these countries is often invoked as at least a partial explanation.

Some studies claim to show that Cheddar, Mozzarella, Swiss and American cheeses can help to prevent tooth decay. Several mechanisms for this protection have been proposed:

- The calcium, protein, and phosphorus in cheese may act to protect tooth enamel.
- Cheese increases saliva flow, washing away acids and sugars.
- Cheese may have an antibacterial effect in the mouth.

Controversy

Effect on Sleep

A study by the British Cheese Board in 2005 to determine the effect of cheese upon sleep and dreaming discovered that, contrary to the idea that cheese commonly causes nightmares, the effect of cheese upon sleep was positive. The majority of the two hundred people tested over a fortnight claimed beneficial results from consuming cheeses before going to bed, the cheese promoting good sleep. Six cheeses were tested and the findings were that the dreams produced were specific to the type of cheese. Although the apparent effects were in some cases described as colourful and vivid, or cryptic, none of the cheeses tested were found to induce nightmares. However, the six cheeses were all British. The results might be entirely different if a wider range of cheeses were tested. Cheese contains tryptophan, an amino acid that has been found to relieve stress and induce sleep.

Casein

Like other dairy products, cheese contains casein, a substance that when digested by humans breaks down into several chemicals, including casomorphine, an opioid peptide. In the early 1990s it was hypothesized that autism can be caused or aggravated by opioid peptides. Based on this hypothesis, diets that eliminate cheese and other dairy products are widely promoted. Studies supporting these claims have had significant flaws, so the data are inadequate to guide autism treatment recommendations.

Lactose

Cheese is often avoided by those who are lactose intolerant, but ripened cheeses like Cheddar contain only about 5 per cent of the lactose found in whole milk, and aged cheeses contain almost none. Nevertheless, people with severe lactose intolerance should avoid eating dairy cheese. As a natural product, the same kind of cheese may contain different amounts of lactose on different occasions, causing unexpected painful reactions. As an alternative, also for vegans, there is already a wide range of different soy cheese kinds available. Some people suffer reactions to amines found in cheese, particularly histamine and tyramine. Some aged cheeses contain significant concentrations of these amines, which can trigger symptoms mimicking an allergic reaction: headaches, rashes, and blood pressure elevations.

Pasteurization

A number of food safety agencies around the world have warned of the risks of raw-milk cheeses. The U.S. Food and Drug Administration states that soft raw-milk cheeses can cause "serious infectious diseases including listeriosis, brucellosis, salmonellosis and tuberculosis". It is U.S. law since 1944 that all raw-milk cheeses (including imports since 1951) must be aged at least 60 days. Australia has a wide ban on raw-milk cheeses as well, though in recent years exceptions have been made for Swiss Gruyère, Emmental and Sbrinz, and for French Roquefort. There is a trend for cheeses to be pasteurized even when not required by law.

Compulsory pasteurization is controversial. Pasteurization does change the flavor of cheeses, and unpasteurized cheeses are often considered to have better flavor, so there are reasons not to routinely pasteurize all cheeses. Some say that health concerns are overstated, pointing out that pasteurization of the milk used to make cheese does not ensure its safety in any case. This is supported by statistics showing that in some European countries where young raw-milk cheeses may legally be sold, most cheese-related food poisoning incidents were traced to pasteurized cheeses. Pregnant women may face an additional risk from cheese; the U.S. Centers for Disease Control has warned pregnant women against eating soft-ripened cheeses and blue-veined cheeses, due to the listeria risk, which can cause miscarriage or harm to the fetus during birth.

World Production and Consumption

Worldwide, cheese is a major agricultural product. According to the Food and Agricultural Organization of the United Nations, over 18 million metric tons of cheese were produced worldwide in 2004. This is more than the yearly production of coffee beans, tea leaves, cocoa beans and tobacco combined. The largest producer of cheese is the United States, accounting for 30 per cent of world production, followed by Germany and France.

The biggest exporter of cheese, by monetary value, is France; the second, Germany (although it is first by quantity). Among the top ten exporters, only Ireland, New Zealand, the Netherlands and Australia have a cheese production that is mainly export oriented: respectively 95 per cent, 90 per cent, 72 per cent, and 65 per cent of their cheese production is exported. Only 30 per cent of French production, the world's largest exporter, is exported. The United States, the biggest world producer of cheese, is a marginal exporter, as most of its production is for the domestic market.

Table 3.4: Top Cheese Producers

	(1,000 Metric Tons)
United States	4,275 (2006)
Germany	1,927 (2008)
France	1,884 (2008)
Italy	1,149 (2008)
Netherlands	732 (2008)
Poland	594 (2008)
Brazil	495 (2006)
Egypt	462 (2006)
Argentina	425 (2006)
Australia	395 (2006)

Table 3.5: Top Cheese Exporters (Whole Cow Milk Only) (Value in '000 US $)

France	2,658,441
Germany	2,416,973
Netherlands	2,099,353
Italy	1,253,580
Denmark	1,122,761
Australia	643,575
New Zealand	631,963
Belgium	567,590
Ireland	445,240
United Kingdom	374,156

Germany is the largest importer of cheese. The UK and Italy are the second-and third-largest importers.

Table 3.6: Top Cheese Consumers (Kilograms Per Person Per Year)

Greece	27.3
France	24.0
Italy	22.9
Switzerland	20.6
Germany	20.2
Netherlands	19.9
Austria	19.5
Sweden	17.9

Greece is the world's largest (per capita) consumer of cheese, with 27.3 kg eaten by the average Greek. (Feta accounts for three-quarters of this consumption.) France is the second biggest consumer of cheese, with 24 kg by inhabitant. Emmental (used mainly as a cooking ingredient) and Camembert are the most common cheeses in France Italy is the third biggest consumer by person with 22.9 kg. In the U.S., the consumption of cheese is quickly increasing and has nearly tripled between 1970 and 2003. The consumption per person has reached, in 2003, 14.1 kg (31 pounds). Mozzarella is America's favourite cheese and accounts for nearly a third of its consumption, mainly because it is one of the main ingredients of pizza.

Cultural Attitudes

Although cheese is a vital source of nutrition in many regions of the world, and is extensively consumed in others, its use is not universal. Cheese is rarely found in East Asian dishes, as lactose intolerance is relatively common in that part of the world and hence dairy products are rare. However, East Asian sentiment against cheese is not universal; cheese made from yaks' (*chhurpi*) or mares' milk is common on the Asian steppes; the national dish of Bhutan, *ema datsi*, is made from homemade cheese and hot peppers and cheese such as Rushan and Rubing in Yunnan, China is produced by several ethnic minority groups by either using goat's milk in the case of rubing or cow's milk in the case of rushan. Cheese consumption is increasing in China, with annual sales more than doubling from 1996 to 2003 (to a still small 30 million U.S. dollars a year). Certain kinds of Chinese preserved bean curd are sometimes misleadingly referred to in English as "Chinese cheese", because of their texture and strong flavor.

Strict followers of the dietary laws of Islam and Judaism must avoid cheeses made with rennet from animals not slaughtered in a manner adhering to halal or kosher laws. Both faiths allow cheese made with vegetable-based rennet or with rennet made from animals that were processed in a halal or kosher manner. Many less-orthodox Jews also believe that rennet undergoes enough processing to change its nature entirely, and do not consider it to ever violate kosher law. As cheese is a dairy food under kosher rules it cannot be eaten in the same meal with any meat.

Some vegetarians avoid any cheese made from animal-based rennet. Most widely available vegetarian cheeses are made using rennet produced by fermentation of the fungus *Mucor miehei*. Vegans and other dairy-avoiding vegetarians do not eat real cheese at all, but some vegetable-based cheese substitutes (usually soy-and almond-based) are available.

Even in cultures with long cheese traditions, it is not unusual to find people who perceive cheese—especially pungent-smelling or mold-bearing varieties such as Limburger or Roquefort—as unpalatable. Food-science writer Harold McGee proposes that cheese is such an acquired taste because it is produced through a process of controlled spoilage and many of the odor and flavor molecules in an aged cheese are the same found in rotten foods. He notes, "An aversion to the odor of decay has the obvious biological value of steering us away from possible food poisoning, so it is no wonder that an animal food that gives off whiffs of shoes and soil and the stable takes some getting used to." Collecting cheese labels is called "tyrosemiophilia".

COTTAGE CHEESE

Cottage cheese is a cheese curd product with a mild flavor. It is drained, but not pressed so some whey remains and the individual curds remain loose. The curd is usually washed to remove acidity giving *sweet curd* cheese. It is not aged or coloured. Different styles of cottage cheese are made from milks with different fat levels and in small curd or large curd preparations. Cottage cheese which is pressed becomes hoop cheese, farmer cheese, pot cheese or queso blanco.

Cottage cheese is eaten straight, with fruit, fruit puree, on toast, in green salads, and used as an ingredient in recipes such as lasagna, jello salad and various desserts.

The term "cottage cheese" is believed to have originated because the simple cheese was usually made in cottages from any milk left over after making butter. The term was first used in 1848. The *curds and whey* of nursery rhyme fame is another dish made from curds with whey but it is uncertain what their consistency was, if they were drained at all or how they were curdled (which affects the flavor). Some writers claim that they are equivalent or similar.

Curd Size

The curd size is the size of the "chunks" in the cottage cheese. The two major types of cottage cheese are small curd, high-acid cheese made without rennet, and popular large curd, low-acid cheese made with rennet. Rennet is an enzyme that speeds curdling and keeps the curd that forms from breaking up; adding it shortens the cheesemaking process, resulting in a lower acid and larger curd cheese, and reduces the amount of curd poured off with leftover liquid (the whey). Sometimes large curd cottage cheese is called "chunk style".

Nutrition

Cottage cheese is low in fat and carbohydrates and high in protein.

A 4 oz (113 g) serving of 4 per cent fat product has about 120 calories, 5 g fat (3 g saturated), 3 g carbohydrates, and 14 g protein. It also contains about 500 mg sodium, 70 mg calcium and 20 mg cholesterol.

Manufacturers also produce low-fat and non-fat varieties. A fat-free kind of a similar serving size has 80 calories, 0g fat (0g saturated), 6g carbohydrates, and 14g protein. To compensate for the flavor missing from the fat, low-fat and non-fat ones tend to have more sugar in them. Very low sodium varieties are also produced, which can be salted to taste.

In Israel, cottage cheese has evolved into a very popular, gourmet breakfast and lunch food and into a somewhat different product through the use of sea salt or purified sea water. This process yields higher levels of sodium per serving. Produced by the three major dairy companies, Tnuva, Tara and Strauss, cottage cheese is made in the 3 percent, 5 and 9 percent standard fat content products. Tnuva also produces a form of Yoga brand cottage cheese mixed with yogurt.

It is popular among dieters and some health food devotees. Cottage cheese is a favourite food among bodybuilders and weightlifters for its high content of casein protein while being relatively low in fat, and pregnant women are advised that cottage cheese is safe to eat during their pregnancy.

SMOKED CHEESE

Smoked cheese is any cheese that has been specially treated by smoke-curing. It typically has a yellowish-brown outer "skin," which is a result of this curing process.

Smoke curing is typically done in one of two ways: cold-smoking and hot-smoking. The cold-smoking method (which can take up to a month, depending on the food) smokes the food at between 70° to 90°F. Hot-smoking partially or completely cooks the food by treating it at temperatures ranging from 100° to 190°F. Another method of "curing" used in less expensive cheeses is to use liquid smoke flavoring to give the cheese a smoky flavor and colour, which gives the outside the appearance of having been smoked in the more traditional manner.

Cheeses smoked by definition include Applewood, and Rauchkäse, while others such as Gruyère, Gouda, Scamorza and Cheddar are often available in smoked varieties.

FARMER CHEESE

Farmer cheese (also farmer's cheese or farmers' cheese) is an unripened cheese made by adding bacterial starter and rennet to acidify milk. After coagulation, it is cut into curds and whey, and the whey is drained off. The drained curds are cottage cheese. At this point, the cottage cheese is retaining moisture. Pressing out some of this moisture to create a malleable solid results in pot cheese. Further pressing most of this moisture out and forming the resultant solid into a dry, crumbly single unit creates farmer's cheese.

Farmer's cheese is often used for fillings in blintzes and other foods. It is sometimes rolled in a mixture of herbs and flavorings, or wrapped in very thin slices of flavorful smoked meats. Around the world, farmer cheese is variably made from milk from cows, sheep, or goats, each with its own final texture and nuance of flavor.

The French name for this cheese is Neufchâtel. It has many different names in various countries. While often compared to paneer from Indian cuisine, paneer is curdled with an acid (vinegar, lemon juice, etc.) rather than the rennet used for farmer's cheese.

In Canada, the term "farmer's cheese" refers to a different type of white cheese that does not have a rind and is firm but springy in texture. It is mild, milky and buttery in flavor. Canadian "farmer's cheese" may be used in a similar fashion to Colby or Cheddar. Similarly, in the Upper Midwest, Farmer's cheese is a semi-soft white cheese made from part-skim milk.

BUTTER

Butter is a dairy product made by churning fresh or fermented cream or milk. It is generally used as a spread and a condiment, as well as in cooking applications such as baking, sauce making, and frying. Butter consists of butterfat, water and milk proteins.

Most frequently made from cows' milk, butter can also be manufactured from the milk of other mammals, including sheep, goats, buffalo, and yaks. Salt, flavorings and preservatives are sometimes added to butter. Rendering butter produces clarified butter or *ghee*, which is almost entirely butterfat.

Butter is an emulsion which remains a solid when refrigerated, but softens to a spreadable consistency at room temperature, and melts to a thin liquid consistency at 32-35 °C (90-95 °F). The density of butter is 911 kg/m (1535.5 lb/yd).

It generally has a pale yellow colour, but varies from deep yellow to nearly white. Its colour is dependent on the animal's feed and is commonly manipulated with food colourings in the commercial manufacturing process, most commonly annatto or carotene.

Etymology

The word *butter* derives (via Germanic languages) from the Latin *butyrum*, which is borrowed from the Greek *boutyron*. This may have been a construction meaning "cow-cheese" (*bous* "ox, cow" + *tyros* "cheese"), or the word may have been borrowed from another language, possibly Scythian. The root word persists in the name butyric acid, a compound found in rancid butter and dairy products such as Parmesan cheese. (Another possibility may be an extended derivation from Sanskrit *bhutari*, meaning "the enemy of evil spirits".)

In general use, the term "butter" refers to the spread dairy product when unqualified by other descriptors. The word commonly is used to describe puréed vegetable or nut products such as peanut butter and almond butter. It is often applied to spread fruit products such as apple butter. Fats such as cocoa butter and shea butter that remain solid at room temperature are also known as "butters". In addition to the act of applying butter being called "to butter", non-dairy items that have a dairy butter consistency may use "butter' to call that consistency to mind, including food items such as maple butter and Witch's butter and non-food items such as baby bottom butter, hyena butter, and rock butter.

Production

Unhomogenized milk and cream contain butterfat in microscopic globules. These globules are surrounded by membranes made of phospholipids (fatty acid emulsifiers) and proteins, which prevent the fat in milk from pooling together into a single mass. Butter is produced by agitating cream, which damages these membranes and allows the milk fats to conjoin, separating from the other parts of the cream. Variations in the production method will create butters with different consistencies, mostly due to the butterfat composition in the finished product. Butter contains fat in three separate forms: free butterfat, butterfat crystals, and undamaged fat globules. In the finished product, different proportions of these forms result in different consistencies within the butter; butters with many crystals are harder than butters dominated by free fats.

Churning produces small butter grains floating in the water-based portion of the cream. This watery liquid is called buttermilk—although the buttermilk most common today is instead a directly fermented skimmed milk. The buttermilk is drained off; sometimes more buttermilk is removed by rinsing the grains with water. Then the grains are "worked": pressed and kneaded together. When prepared manually, this is done using wooden boards called scotch hands. This consolidates the butter into a solid mass and breaks up embedded pockets of buttermilk or water into tiny droplets.

Commercial butter is about 80 per cent butterfat and 15 per cent water; traditionally-made butter may have as little as 65 per cent fat and 30 per cent water. Butterfat consists of many moderate-sized, saturated hydrocarbon chain fatty acids. It is a triglyceride, an ester derived from glycerol and three fatty acid groups. Butter becomes rancid when these chains break down into smaller components, like butyric acid and diacetyl. The density of butter is 0.911 g/cm (527 oz/in), about the same as ice.

Types

Before modern factory butter making, cream was usually collected from several milkings and was therefore several days old and somewhat fermented by the time it was made into butter. Butter made from a fermented cream is known as cultured butter. During fermentation, the cream naturally sours as bacteria convert milk sugars into lactic acid. The fermentation process produces additional aroma compounds, including diacetyl, which makes for a fuller-flavored and more "buttery" tasting product. Today, cultured butter is usually made from pasteurized cream whose fermentation is produced by the introduction of *Lactococcus* and *Leuconostoc* bacteria.

Another method for producing cultured butter, developed in the early 1970s, is to produce butter from fresh cream and then incorporate bacterial cultures and lactic acid. Using this method, the cultured butter flavor grows as the butter is aged in cold storage. For manufacturers, this method is more efficient since aging the cream used to make butter takes significantly more space than simply storing the finished butter product. A method to make an artificial simulation of cultured butter is to add lactic acid and flavor compounds directly to the fresh-cream butter; while this more efficient process is claimed to simulate the taste of cultured butter, the product produced is not cultured but is instead flavored.

Dairy products are often pasteurized during production to kill pathogenic bacteria and other microbes. Butter made from pasteurized fresh cream is called sweet cream butter. Production of sweet cream butter first became common in the 19th century, with the development of refrigeration and the mechanical cream separator. Butter made from fresh or cultured unpasteurized cream is called raw cream butter. Raw cream butter has a "cleaner" cream flavor, without the cooked-milk notes that pasteurization introduces.

Throughout Continental Europe, cultured butter is preferred, while sweet cream butter dominates in the United States and the United Kingdom. Therefore, cultured butter is sometimes labeled *European-style butter* in the United States. Commercial raw cream butter is virtually unheard-of in the United States. Raw cream butter is generally only found made at home by consumers who have purchased raw whole milk directly from dairy farmers, skimmed the cream themselves, and made butter with it. It is rare in Europe as well.

Several spreadable butters have been developed; these remain softer at colder temperatures and are therefore easier to use directly out of refrigeration. Some modify the makeup of the butter's fat through chemical manipulation of the finished product, some through manipulation of the cattle's feed, and some by incorporating vegetable oils into the butter. Whipped butter, another product designed to be more spreadable, is aerated via the incorporation of nitrogen gas—normal air is not used, because doing so would encourage oxidation and rancidity.

All categories of butter are sold either in salted and unsalted forms. Either granular salt or a strong brine are added to salted butter during processing. In addition to enhanced flavor, the addition of salt acts as a preservative.

The amount of butterfat in the finished product is a vital aspect of production. In the United States, products sold as "butter" are required to contain a minimum of 80 per cent butterfat; in practice most American butters contain only slightly more than that, averaging around 81 per cent butterfat. European butters generally have a higher ratio, which may extend up to 85 per cent.

Clarified butter is butter with almost all of its water and milk solids removed, leaving almost-pure butterfat. Clarified butter is made by heating butter to its melting point and then allowing it to cool off; after settling, the remaining components separate by density. At the top, whey proteins form a skin which is removed, and the resulting butterfat is then poured off from the mixture of water and casein proteins that settle to the bottom.

Ghee is clarified butter which is brought to higher temperatures of around 120°C (250°F) once the water has cooked off, allowing the milk solids to brown. This process flavors the ghee, and also produces antioxidants which help protect it longer from rancidity.

Because of this, ghee can keep for six to eight months under normal conditions.

History

Since even accidental agitation can form butter from cream, it is likely that its invention dates from the earliest days of dairying, perhaps in the Mesopotamian area between 9000 and 8000 BCE. The earliest butter would have been from sheep or goat's milk; cattle are not thought to have been domesticated for another thousand years. An ancient method of butter making, still used today in parts of Africa and the Near East, involves a goat skin half filled with milk, and inflated with air before being sealed. The skin is then hung with ropes on a tripod of sticks, and rocked until the movement leads to the formation of butter. Butter was known in the classical Mediterranean civilizations, but it does not seem to have been a common food. In the Mediterranean climate, unclarified butter spoils quickly—unlike cheese it is not a practical method of preserving the nutrients of milk. The ancient Greeks and Romans seemed to have considered butter a food fit more for the northern barbarians. A play by the Greek comic poet Anaxandrides refers to Thracians as *boutyrophagoi*; "butter-eaters". In *Natural History*, Pliny the Elder calls butter "the most delicate of food among barbarous nations", and goes on to describe its medicinal properties. Later, the physician Galen also described butter as a medicinal agent only.

Historian and linguist Andrew Dalby says that most references to butter in ancient Near Eastern texts should more correctly be translated as ghee. Ghee is mentioned in the Periplus of the Erythraean Sea as a typical trade article around the 1st century CE Arabian Sea, and Roman geographer Strabo describes it as a commodity of Arabia and Sudan. In India, ghee has been a symbol of purity and an offering to the gods—especially Agni, the Hindu god of fire—for more than 3000 years; references to ghee's sacred nature appear numerous times in the Rig Veda, circa 1500-1200 BCE. The tale of the child Krishna stealing butter remains a popular children's story in India today. Since India's prehistory, ghee has been both a staple food and used for ceremonial purposes such as fueling holy lamps and funeral pyres.

Middle Ages

The cooler climates of northern Europe allowed butter to be stored for a longer period before it spoiled. Scandinavia has the oldest tradition in Europe of butter export trade, dating at least to the 12th century. After the fall of Rome and through much of the Middle Ages, butter was a common food across most of Europe, but one with a low reputation, and was consumed principally by peasants. Butter slowly became more accepted by the upper class, notably when the early 16th century Roman Catholic Church allowed its consumption during Lent. Bread and butter became common fare among the middle class and the English, in particular, gained a reputation for their liberal use of melted butter as a sauce with meat and vegetables. In antiquity, Butter was used for fuel in lamps as a substitute for oil. The *Butter Tower* of Rouen Cathedral was erected in the early 16th century when Archbishop Georges d'Amboise authorized the burning of butter instead of oil, which was scarce at the time, during Lent.

Across northern Europe, butter was sometimes treated in a manner unheard-of today: it was packed into barrels (firkins) and buried in peat bogs, perhaps for years. Such "bog butter" would develop a strong flavor as it aged, but remain edible, in large part because of the unique cool, airless, antiseptic and acidic environment of a peat bog. Firkins of such buried butter are a common archaeological find in Ireland; the Irish National Museum has some containing "a grayish cheese-like substance, partially hardened, not much like butter, and quite free from putrefaction." The practice was most common in Ireland in the 11th-14th centuries; it ended entirely before the 19th century.

Industrialization

Like Ireland, France became well-known for its butter, particularly in Normandy and Brittany. By the 1860s, butter had become so in demand in France that Emperor Napoleon III offered prize money for an inexpensive substitute to supplement France's inadequate butter supplies. A French chemist claimed the prize with the invention of margarine in 1869. The first margarine was beef tallow flavored with milk and worked like butter; vegetable margarine followed after the development of hydrogenated oils around 1900.

Until the 19th century, the vast majority of butter was made by hand, on farms. The first butter factories appeared in the United States in the early 1860s, after the successful introduction of cheese factories a decade earlier. In the late 1870s, the centrifugal cream separator was introduced, marketed most successfully by Swedish engineer Carl Gustaf Patrik de Laval. This dramatically sped up the butter-making process by eliminating the slow step of letting cream naturally rise to the top of milk. Initially, whole milk was shipped to the butter factories, and the cream separation took place there. Soon, though, cream-separation technology became small and inexpensive enough to introduce an additional efficiency: the separation was accomplished on the farm, and the cream alone shipped to the factory. By 1900, more than half the butter produced in the United States was factory made; Europe followed suit shortly after.

In 1920, Otto Hunziker authored *The Butter Industry, Prepared for Factory, School and Laboratory*, a well-known text in the industry that enjoyed at least three editions (1920, 1927, 1940). As part of the efforts of the American Dairy Science Association, Professor Hunziker and others published articles regarding: causes of talloviness (an odor defect, distinct from rancidity, a taste defect); mottles (an aesthetic issue related to uneven colour); introduced salts; the impact of creamery metals and liquids; and acidity measurement. These and other ADSA publications helped standardize practices internationally.

Per capita butter consumption declined in most western nations during the 20th century, in large part because of the rising popularity of margarine, which is less expensive and, until recent years, was perceived as being healthier. In the United States, margarine consumption overtook butter during the 1950s and it is still the case today that more margarine than butter is eaten in the U.S. and the EU.

Shape of Butter Sticks

In the United States, butter sticks are usually produced and sold in 4-ounce sticks, wrapped in wax paper and sold four to a carton. This practice is believed to have originated in 1907 when Swift and

Company began packaging butter in this manner for mass distribution.

Due to historical variances in butter printers, these sticks are commonly produced in two differing shapes:

- The dominant shape east of the Rocky Mountains is the Elgin, or Eastern-pack shape. This shape was originally developed by the Elgin Butter Tub Company, founded in 1882 in Elgin, Illinois, and Rock Falls, Illinois. The sticks are 4¾ inches long and 1¼ inches (121 mm × 32 mm) wide, and are usually sold in cube-shaped boxes stacked two by two. Among the early butter printers to use this shape was the Elgin Butter Cutter.
- West of the Rocky Mountains, butter printers standardized on a different shape that is now referred to as the Western-pack shape. These butter sticks are 3¼ inches long and 1½ inches wide (80 mm × 38 mm) and are typically sold packed side-by-side in a rectangular container.

Both sticks contain the same amount of butter, although most butter dishes are designed for Elgin-style butter sticks.

The stick's wrapper is usually marked off as eight tablespoons (120 ml/4.2 imp fl oz; 4.1 US fl oz); the actual volume of one stick is approximately nine tablespoons (130 ml/4.6 imp fl oz; 4.4 US fl oz).

Worldwide

India produces and consumes more butter than any other nation, and allocates almost half of its annual milk pool to butter production. In 1997, India produced 1,470,000 metric tons (1,620,000 short tons) of butter, most of which was consumed domestically. Second in production was the United States (522,000 t/575,000 short tons), followed by France (466,000 t/514,000 short tons), Germany (442,000 t/487,000 short tons), and New Zealand (307,000 t/ 338,000 short tons). In terms of consumption, Germany was second after India, using 578,000 metric tons (637,000 short tons) of butter in 1997, followed by France (528,000 t/582,000 short tons), Russia

(514,000 t/567,000 short tons), and the United States (505,000 t/557,000 short tons). New Zealand, Australia, and the Ukraine are among the few nations that export a significant percentage of the butter they produce.

Different varieties are found around the world. *Smen* is a spiced Moroccan clarified butter, buried in the ground and aged for months or years. Yak butter is important in Tibet; *tsampa*, barley flour mixed with yak butter, is a staple food. Butter tea is consumed in the Himalayan regions of Tibet, Bhutan, Nepal and India. It consists of tea served with intensely flavored—or "rancid"—yak butter and salt. In African and Asian developing nations, butter is traditionally made from sour milk rather than cream. It can take several hours of churning to produce workable butter grains from fermented milk.

Storage and Cooking

Normal butter softens to a spreadable consistency around 15°C (60°F), well above refrigerator temperatures. The "butter compartment" found in many refrigerators may be one of the warmer sections inside, but it still leaves butter quite hard. Until recently, many refrigerators sold in New Zealand featured a "butter conditioner", a compartment kept warmer than the rest of the refrigerator—but still cooler than room temperature—with a small heater. Keeping butter tightly wrapped delays rancidity, which is hastened by exposure to light or air, and also helps prevent it from picking up other odors. Wrapped butter has a shelf life of several months at refrigerator temperatures.

"French butter dishes" or "Acadian butter dishes" involve a lid with a long interior lip, which sits in a container holding a small amount of water. Usually the dish holds just enough water to submerge the interior lip when the dish is closed. Butter is packed into the lid. The water acts as a seal to keep the butter fresh, and also keeps the butter from overheating in hot temperatures. This allows butter to be safely stored on the countertop for several days without spoilage.

Once butter is softened, spices, herbs, or other flavoring agents can be mixed into it, producing what is called a *compound butter* or *composite butter* (sometimes also called *composed butter*). Compound butters can be used as spreads, or cooled, sliced, and placed onto hot food to melt into a sauce. Sweetened compound butters can be served with desserts; such hard sauces are often flavored with spirits.

Melted butter plays an important role in the preparation of sauces, most obviously in French cuisine. *Beurre noisette* (hazel butter) and *Beurre noir* (black butter) are sauces of melted butter cooked until the milk solids and sugars have turned golden or dark brown; they are often finished with an addition of vinegar or lemon juice. Hollandaise and béarnaise sauces are emulsions of egg yolk and melted butter; they are in essence mayonnaises made with butter instead of oil. Hollandaise and béarnaise sauces are stabilized with the powerful emulsifiers in the egg yolks, but butter itself contains enough emulsifiers—mostly remnants of the fat globule membranes—to form a stable emulsion on its own. *Beurre blanc* (white butter) is made by whisking butter into reduced vinegar or wine, forming an emulsion with the texture of thick cream. *Beurre monté* (prepared butter) is melted but still emulsified butter; it lends its name to the practice of "mounting" a sauce with butter: whisking cold butter into any water-based sauce at the end of cooking, giving the sauce a thicker body and a glossy shine—as well as a buttery taste.

In Poland, the butter lamb (*Baranek wielkanocny*) is a traditional addition to the Easter Meal for many Polish Catholics. Butter is shaped into a lamb either by hand or in a lamb-shaped mould.

Butter is used for sautéing and frying, although its milk solids brown and burn above 150°C (250°F)—a rather low temperature for most applications. The smoke point of butterfat is around 200°C (400°F), so clarified butter or ghee is better suited to frying. Ghee has always been a common frying medium in India, where many avoid other animal fats for cultural or religious reasons.

Butter fills several roles in baking, where it is used in a similar manner as other solid fats like lard, suet, or shortening, but has a flavor that may better complement sweet baked goods. Many cookie doughs and some cake batters are leavened, at least in part, by creaming butter and sugar together, which introduces air bubbles into the butter. The tiny bubbles locked within the butter expand in the heat of baking and aerate the cookie or cake. Some cookies like shortbread may have no other source of moisture but the water in the butter. Pastries like pie dough incorporate pieces of solid fat into the dough, which become flat layers of fat when the dough is rolled out. During baking, the fat melts away, leaving a flaky texture. Butter, because of its flavor, is a common choice for the fat in such a dough, but it can be more difficult to work with than shortening because of its low melting point. Pastry makers often chill all their ingredients and utensils while working with a butter dough.

Butter also has many non-culinary, traditional uses which are specific to certain cultures. For instance, in North America, applying butter to the handle of a door is a common prank on April Fools' Day.

Health and Nutrition

According to USDA figures, one tablespoon of butter (14 grams/ 0.5 ounces) contains 420 kilojoules (100 kcal), all from fat, 11 grams (0.4 oz) of fat, of which 7 grams (0.25 oz) are saturated fat, and 30 milligrams (0.46 gr) of cholesterol. In other words, butter consists mostly of saturated fat and is a significant source of dietary cholesterol. For these reasons, butter has been generally considered to be a contributor to health problems, especially heart disease. For many years, vegetable margarine was recommended as a substitute, since it is an unsaturated fat and contains little or no cholesterol. In recent decades, though, it has become accepted that the trans fats contained in partially hydrogenated oils used in typical margarines significantly raise undesirable LDL cholesterol levels as well. Trans-fat free margarines have since been developed.

Proponents of the moderate consumption of butter such as the nutritionist Mary Enig, state that since butter is nutritious and "is

Table 3.7: Comparative Properties of Common Cooking Fats (Per 100g)

	Total Fat	*Saturated Fat*	*Monounsaturated Fat*	*Polyunsaturated Fat*	*Protein*
Butter	81g	51g	21g	3g	1g
Vegetable Shortening (hydrogenated)	71g	23g	8g	37g	0g
Olive Oil	100g	14g	73g	11g	0g
Lard	100g	39g	45g	11g	0g

rich in short and medium chain fatty acids," this can have a positive effect on health and prevent disease.

Butter contains only traces of lactose, so moderate consumption of butter is not a problem for the lactose intolerant. People with milk allergies need to avoid butter, which contains enough of the allergy-causing proteins to cause reactions.

Butter can form a useful role in dieting by providing satiety. A small amount added to low fat foods such as vegetables may stave off feelings of hunger.

4

Livestock, Meat Processing, Meat Packing, Poultry and Animal Husbandry: An Overview

LIVESTOCK

Livestock (also cattle) are one or more domesticated animals raised in an agricultural setting to produce commodities such as food or fiber, or labour. The term "livestock" as used in this article does not include poultry or farmed fish; however the inclusion of these, especially poultry, within the meaning of "livestock" is common.

Livestock generally are raised for subsistence or for profit. Raising animals (animal husbandry) is an important component of modern agriculture. It has been practiced in many cultures since the transition to farming from hunter-gather lifestyles.

The global livestock population (cattle and pigs) is around 65 billion.

History

Animal-rearing has its origins in the transition of cultures to settled farming communities rather than hunter-gatherer lifestyles. Animals are 'domesticated' when their breeding and living conditions are controlled by humans. Over time, the collective behaviour, life cycle, and physiology of livestock have changed radically. Many modern farm animals are unsuited to life in the wild. Dogs were domesticated in East Asia about 15,000 years ago, Goats and sheep were domesticated around 8000 BCE in Asia. Swine or pigs were domesticated by 7000 BCE in the Middle East and China. The earliest evidence of horse domestication dates to around 4000 BCE

Older English sources, such as the King James Version of the Bible, refer to livestock in general as "cattle", as opposed to the word "deer", which then was used for wild animals which were not owned. The word *cattle* is derived from Middle English *chatel*, which meant all kinds of movable personal property , including of course livestock, which was differentiated from non-movable real-estate. In later English, sometimes smaller livestock was called "small cattle" in that sense of movable property on land, which was not automatically bought or sold with the land. Today, the modern meaning of "cattle", without a qualifier, usually refers to domesticated bovines. Other species of the genus *Bos* sometimes are called wild cattle.

Types

The term "livestock" is nebulous and may be defined narrowly or broadly. On a broader view, livestock refers to any breed or population of animal kept by humans for a useful, commercial purpose. This can mean domestic animals, semi-domestic animals, or captive wild animals. Semi-domesticated refers to animals which are only lightly domesticated or of disputed status. These populations may also be in the process of domestication. Some people may use the term livestock to refer just to domestic animals or even just to red meat animals.

Animal Rearing

'Livestock' are defined, in part, by their end purpose as the production of food or fiber, or labour. The economic value of livestock includes:

Meat

the production of a useful form of dietary protein and energy.

Dairy Products

Mammalian livestock can be used as a source of milk, which can in turn easily be processed into other dairy products such as yogurt, cheese, butter, ice cream, kefir, and kumis. Using livestock for this purpose can often yield several times the food energy of slaughtering the animal outright.

Fiber

Livestock produce a range of fiber/textiles. For example, sheep and goats produce wool and mohair; cows, deer, and sheep can make leather; and bones, hooves and horns of livestock can be used.

Fertilizer

Manure can be spread on fields to increase crop yields. This is an important reason why historically, plant and animal domestication have been intimately linked. Manure is also used to make plaster for walls and floors, and can be used as a fuel for fires. The blood and bone of animals are also used as fertilizer.

Labour

Animals such as horses, donkey, and yaks can be used for mechanical energy. Prior to steam power, livestock were the only available source of non-human labour. They are still used for this purpose in many places of the world, including ploughing fields, transporting goods, and military functions.

Land Management

The grazing of livestock is sometimes used as a way to control weeds and undergrowth. For example, in areas prone to wild fires, goats and sheep are set to graze on dry scrub which removes combustible material and reduces the risk of fires.

During the history of animal husbandry, many secondary products have arisen in an attempt to increase carcass utilization and reduce waste. For example, animal offal and non-edible parts may be transformed into products such as pet food and fertilizer. In the past such waste products were sometimes also fed to livestock as well. However, intra-species recycling poses a disease risk, threatening animal and even human health. Due primarily to BSE (mad cow disease), feeding animal scraps to animals has been banned in many countries, at least in regards to ruminants and pigs.

Farming Practices

Farming practices vary dramatically worldwide and between types of animals. Livestock are generally kept in an enclosure, are fed by human-provided food and are intentionally bred, but some livestock are not enclosed, or are fed by access to natural foods, or are allowed to breed freely, or any combination thereof. Livestock raising historically was part of a nomadic or pastoral form of material culture. The herding of camels and reindeer in some parts of the world remains unassociated with sedentary agriculture. The transhumance form of herding in the Sierra Nevada Mountains of California still continues, as cattle, sheep or goats are moved from winter pasture in lower elevation valleys to spring pasture and summer pasture in the foothills and alpine regions, as the seasons progress. Cattle were raised on the open range in the Western United States and Canada, on the Pampas of Argentina, and other prairie and steppe regions of the world.

The enclosure of livestock in pastures and barns is a relatively new development in the history of agriculture. When cattle are enclosed, the type of 'enclosure' may vary from a small crate, a large fenced pasture or a paddock. The type of feed may vary from natural growing grass, to highly sophisticated processed feed. Animals are usually intentionally bred through artificial insemination or through supervised mating. Indoor production systems are generally used only for pigs and poultry, as well as for veal cattle. Indoor animals are generally farmed intensively, as large space requirements would make indoor farming unprofitable and impossible. However, indoor farming systems are controversial due to the waste they produce, odour problems, the potential for groundwater contamination and animal welfare concerns.

Other livestock are farmed outside, although the size of enclosure and level of supervision may vary. In large open ranges animals may be only occasionally inspected or yarded in "round-ups" or a muster (livestock). Working dogs such as sheep dogs and cattle dogs may be used for mustering livestock as are cowboys, stockmen and jackaroos on horses, or with vehicles and also by helicopters. Since the advent of barbed wire (in the 1870s) and electric fence technology, fencing pastures has become much more feasible and pasture management simplified. Rotation of pasturage is a modern technique for improving nutrition and health while avoiding environmental damage to the land. In some cases very large numbers of animals may be kept in indoor or outdoor feeding operations (on feedlots), where the animals' feed is processed, offsite or onsite, and stored on site then fed to the animals.

Livestock—especially cattle—may be branded to indicate ownership and age, but in modern farming identification is more likely to be indicated by means of ear tags than branding. Sheep are also frequently marked by means of ear marks and/or ear tags. As fears of mad cow disease and other epidemic illnesses mount, the use of microchip identification to monitor and trace animals in the food production system is increasingly common, and sometimes required by governmental regulations.

Modern farming techniques seek to minimize human involvement, increase yield, and improve animal health. Economics, quality and consumer safety all play a role in how animals are raised. Drug use and feed supplements (or even feed type) may be regulated, or prohibited, to ensure yield is not increased at the expense of consumer health, safety or animal welfare. Practices vary around the world, for example growth hormone use is permitted in the United States, but not in stock to be sold to the European Union. The improvement of health, using modern farming techniques, on the part of animals has come into question. Feeding corn to cattle, which have historically eaten grasses, is an example; where the cattle are less adapted, the rumen pH changes to more acidic, leading to liver damage and other difficulties.

Predation

Livestock farmers had suffered from wild animal predation and theft by rustlers. In North America, gray wolf, grizzly bear, cougar, black bear, and coyote are sometimes considered a threat to livestock. In Eurasia and Africa, wolf, brown bear, leopard, tiger, lion, dhole, black bear, spotted hyena, and others caused livestock deaths. In Australia, the dingo, foxes, wedge-tailed eagles, hunting and domestic dogs (especially) cause problems for graziers because they often kill for fun. In Latin America, feral dogs cause livestock deaths in nightfall. They blame maned wolves, cougars, jaguars, and spectacled bears for livestock deaths.

Disease

Livestock diseases compromise animal welfare, reduce productivity, and, in rare cases, can infect humans. Animal diseases may be tolerated, reduced through animal husbandry, or reduced through antibiotics and vaccines. In developing countries, animal diseases are tolerated in animal husbandry, resulting in considerably reduced productivity, especially given the low health-status of many developing country herds. Disease management for gains in productivity is often the first step taken in implementing an agriculture policy.

Disease management can be achieved through changes in animal husbandry. These measures may aim to control spread using biosecurity measures, such as controlling animal mixing, controlling entry to farm lots and the use of protective clothing, and quarantining sick animals. Diseases also may be controlled by the use of vaccines and antibiotics. Antibiotics in sub-therapeutic doses may also be used as a growth-promoter, increasing growth by 10-15 per cent. The issue of antibiotic resistance has limited the practices of preventative dosing such as antibiotic-laced feed. Countries will often require the use of veterinary certificates before transporting, selling or showing animals. Disease-free areas often rigorously enforce rules for entry of potentially diseased animals, including quarentine.

Transportation and Marketing

Since many livestock are herd animals, they were historically driven to market "on the hoof" to a town or other central location. During the period after the American Civil War, the abundance of Longhorn cattle in Texas, and the demand for beef in Northern markets, led to the implementation of the Old West cattle drive. The method is still used in some parts of the world. Truck transport is now common in developed countries. Local and regional livestock auctions and commodity markets facilitate trade in livestock. In other areas, livestock may be bought and sold in a bazaar, such as may be found in many parts of Central Asia, or a flea market type setting.

Stock Shows and Fairs

Stock shows and fairs are events where people bring their best livestock to compete with one another. Organizations like 4-H, Block and Bridle, and FFA encourage young people to raise livestock for show purposes. Special feeds are purchased and hours may be spent prior to the show grooming the animal to look its best. In cattle, sheep, and swine shows, the winning animals are frequently auctioned off to the highest bidder, and the funds are placed into a scholarship fund for its owner. The movie *Grand Champion*, released in 2004, is the story of a young Texas boy's experience raising a prize steer.

Animal Welfare

The issue of raising livestock for human benefit raises the issue of the relationship between humans and animals, in terms of the status of animals and obligations of people. Animal welfare is the viewpoint that animals under human care should be treated in such a way that they do not suffer unnecessarily. What is 'unnecessary' suffering may vary. Generally, though, the animal welfare perspective is based on an interpretation of scientific research on farming practices. By contrast, Animal rights is the viewpoint that using animals for human benefit is, by its nature, generally exploitation, regardless of the farming practices used. Animal rights activists would generally be vegan or vegetarian, whereas it is consistent with the animal welfare perspective to eat meat, depending on production processes.

Animal welfare groups generally seek to generate public discussion on livestock raising practices and secure greater regulation and scrutiny of livestock industry practices. Animal rights groups usually seek the abolition of livestock farming, although some groups may recognise the necessity of achieving more stringent regulation first. Animal welfare groups, such as the RSPCA, are often, in first world countries, given a voice at governmental level in the development of policy. Animal rights groups find it harder to find methods of input, and may go further and advocate civil disobedience or violence.

A number a animal husbandry practices, that have led to legislation in some countries, have been the subject of campaigns in the 1990s and 2000s. Confinement of livestock in small and unnatural spaces is often done for economic or health reasons. Animals may be kept in the minimum size of cage or pen with little or no space to exercise or engage in normal actions or grooming. Close confinement is most common with chickens, pigs, and calves raised for veal. Unnatural living environments may be used for some animals. Even when allowed to move, animals may be denied a natural environment. For example, ducks may be kept in free-range barns but have no access to water in which to swim. Cattle may be kept in barns with no chance to graze. Overuse of pharmaceuticals and hormones is also an issue in large agricultural operations. Intensive raising of livestock may lead to a health problems and the necessity to use antibiotics to prevent disease. In some cases, antibiotics and hormones are also fed to livestock to produce rapid weight gain.

Where livestock are used as a source of power, they may be pushed beyond their limits to the point of exhaustion. The public visibility of this abuse meant it was one of the first areas to receive legislation in the nineteenth century in European countries, but it still goes on in parts of Asia. Broiler hens may be de-beaked, pigs may have deciduous teeth pulled, cattle may be de-horned and branded, dairy cows and sheep may have tails cropped, merino sheep may be mulesed, and many types of male animals are castrated. Animals may be transported long distances to market and slaughter. Overcrowded conditions, heat from tropical-area shipping and lack of food, water and rest breaks have been subject to legislation and protest. Slaughter of livestock was an early target for legislation. Campaigns continue to target Halal and Kosher religious ritual slaughter.

Environmental Impact

At first reports like the United Nations report "Livestock's Long Shadow" cast a pall over the livestock sector (primarily cattle, chickens, and pigs) for 'emerging as one of the top two or three most significant contributors to our most serious environmental problems.' The United Nations controversially included emissions from deforestation as part of its methodology. Rather than the 18 per cent figure that placed on the sector as major contributor to emissions, the real figure, less deforestation is actually 12 per cent. In April 2008, the [United States Environmental Protection Agency] released a major stocktake of emissions in the United States entitled *Inventory of U.S. Greenhouse Gas Emissions and Sinks: 1990-2006*. On 6.1 it found "In 2006, the agricultural sector was responsible for emissions of 454.1 teragrams of CO_2 equivalent (Tg CO_2 Eq.), or 6 percent of total U.S. greenhouse gas emissions." By way of comparison, transportation in the US produces more than 25 per cent of all emissions.

While the findings of the United Nations report are now in doubt, the issue of livestock as a major policy focus remains, especially when dealing with problems of deforestation in neotropical areas, land degradation, climate change and air pollution, water shortage and water pollution, and loss of biodiversity. A research team at Obihiro University of Agriculture and Veterinary Medicine in Hokkaido found that supplementing the animals' diet with cysteine, a type of amino acid, and nitrate can reduce the methane gas produced, without jeopardising the cattle's productivity or the quality of their meat and milk.

Researchers in Australia are looking into the possibility of reducing methane from cattle and sheep by introducing digestive bacteria from kangaroo intestines into livestock.

Research from the University of Botswana in 2008 has found that farmers' common practice of overstocking cattle to cope with drought losses made ecosystems more vulnerable and risked long term damage to cattle herds, in turn, by actually depleting scarce biomass. The study of the Kgatleng district of Botswana predicted that by 2050, the cycle of mild drought is likely to become shorter for the region (18 months instead of two years) due to climate change.

Legal Definitions

United States federal legislation sometimes more narrowly defines the term to make specified agricultural commodities either eligible, or ineligible, for a programme or activity. For example, the Livestock Mandatory Reporting Act of 1999 (P.L. 106-78, Title IX) defines livestock only as cattle, swine, and lambs. However, 1988 disaster assistance legislation defined the term as "cattle, sheep, goats, swine, poultry (including egg-producing poultry), equine animals used for food or in the production of food, fish used for food, and other animals designated by the Secretary."

DOMESTIC SHEEP

Domestic sheep (*Ovis aries*) are quadrupedal, ruminant mammals typically kept as livestock. Like all ruminants, sheep are members of the order Artiodactyla, the even-toed ungulates. Although the name "sheep" applies to many species, in everyday usage it almost always refers to *Ovis aries*. Numbering a little over 1 billion, domestic sheep are the most numerous species in their genus.

Sheep are most likely descended from the wild mouflon of Europe and Asia. One of the earliest animals to be domesticated for agricultural purposes, sheep are raised for fleece, meat (lamb, hogget or mutton) and milk. A sheep's wool is the most widely used of any animal, and is usually harvested by shearing. Ovine meat is called lamb when from younger animals and mutton when from older ones. Sheep continue to be important for wool and meat today, and are also occasionally raised for pelts, as dairy animals, or as model organisms for science.

Sheep husbandry is practised throughout the majority of the inhabited world, and has been fundamental to many civilizations. In the modern era, Australia, New Zealand, the southern and central South American nations, and the British Isles are most closely associated with sheep production. Sheep-raising has a large lexicon of unique terms which vary considerably by region and dialect. Use of the word *sheep* began in Middle English as a derivation of the Old English word *scçap*; it is both the singular and plural name for the animal. A group of sheep is called a flock, herd or mob. Adult female sheep are referred to as ewes, intact males as rams or occasionally tups, castrated males as wethers, and younger sheep as lambs. Many other specific terms for the various life stages of sheep exist, generally related to lambing, shearing, and age. ý Being a key animal in the history of farming, sheep have a deeply entrenched place in human culture, and find representation in much modern language and symbology. As livestock, sheep are most-often associated with pastoral, Arcadian imagery. Sheep figure in many mythologies—such as the Golden Fleece—and major religions, especially the Abrahamic traditions. In both ancient and modern religious ritual, sheep are used as sacrificial animals. In contemporary English language usage, people who are timid, easily led, or stupid are often compared to sheep.

Description

Domestic sheep are relatively small ruminants, usually with a crimped hair called wool and often with horns forming a lateral spiral. Domestic sheep differ from their wild relatives and ancestors in several respects, having become uniquely neotenic as a result of man's influence. A few primitive breeds of sheep retain some of the characteristics of their wild cousins, such as short tails. Depending on breed, domestic sheep may have no horns at all (polled), or horns in both sexes (as in wild sheep), or in males only. Most horned breeds have a single pair, but a few breeds may have several.

Another trait unique to domestic sheep (as compared to wild ovines, not other livestock) are their wide variation in colour. Wild sheep are largely variations of brown hues, and variation with species is extremely limited. Colours of domestic sheep range from pure white to dark chocolate brown and even spotted or piebald. Selection for easily dyeable white fleeces began early in sheep domestication, and as white wool is a dominant trait it spread quickly. However, coloured sheep do appear in many modern breeds, and may even appear as a recessive trait in white flocks. While white wool is desirable for large commercial markets, there is a niche market for coloured fleeces, mostly for handspinning. The nature of the fleece varies widely among the breeds, from dense and highly crimped, to long and hair-like. There is variation of wool type and quality even among members of the same flock, so wool classing is a step in the commercial processing of the fiber.

Depending on breed, sheep show a range of heights and weights. Their rate of growth and mature weight is a heritable trait that is often selected for in breeding. Ewes typically weigh between 45 and 100 kilograms (100 and 225 lb), and the rams between 45 and 160 kilograms (100 and 350 lb). Mature sheep have 32 teeth (dental formula: either I:0/4 C:0/0 P:3/3 M:3/3 or I:0/3 C:0/1 P:3/3 M:3/3). As with other ruminants, the eight incisors are in the lower jaw and bite against a hard, toothless pad in the upper jaw; picking off vegetation. There are no canines, instead there is a large gap between the incisors and the premolars. Until the age of four (when all the adult teeth have erupted), it is possible to see the age of sheep from their front teeth, as a pair of incisors erupts each year.

The front teeth are gradually lost as sheep age, making it harder for them to feed and hindering the health and productivity of the animal. For this reason, domestic sheep on normal pasture begin to slowly decline from four years on, and the average life expectancy of a sheep is 10 to 12 years, though some sheep may live as long as 20 years.

Sheep have good hearing, and are sensitive to noise when being handled. Sheep have horizontal slit-shaped pupils, possessing excellent peripheral vision; with visual fields of approximately 270° to 320°, sheep can see behind themselves without turning their heads. However, sheep have poor depth perception; shadows and dips in the ground may cause sheep to balk. In general, sheep have a tendency to move out of the dark and into well-lit areas, and prefer to move uphill when disturbed. Sheep also have an excellent sense of smell, and, like all species of their genus, have scent glands just in front of the eyes, and interdigitally on the feet. The purpose of these glands is uncertain, but those on the face may be used in breeding behaviours. The foot glands might also be related to reproduction, but alternative reasons, such as secretion of a waste product or a scent marker to help lost sheep find their flock, have also been proposed.

Sheep and goats are closely related (both are in the subfamily Caprinae), and it can be difficult to distinguish them by their appearance. However, they are separate species, so hybrids rarely occur, and are always infertile. A hybrid of a ewe and a buck (a male goat) is called a sheep-goat hybrid, and is not to be confused with the genetic chimera called a geep. Visual differences between sheep and goats include the beard and divided upper lip unique to goats. Sheep tails also hang down, even when short or docked, while the short tails of goats are held upwards. Sheep breeds are also often naturally polled (either in both sexes or just in the female), while naturally polled goats are rare (though many are polled artificially). Males of the two species differ in that buck goats acquire a unique and strong odor during the rut, whereas rams do not.

Breeds

The domestic sheep is a multi-purpose animal, and the more than 200 breeds now in existence were created to serve these diverse purposes. Some sources give a count of a thousand or more breeds, but these numbers cannot be verified. Almost all sheep are classified as being best suited to furnishing a certain product: wool, meat, milk, hides, or a combination in a dual-purpose breed. Other features used when classifying sheep include face colour (generally white or black), tail length, presence or lack of horns, and the topography for which the breed has been developed. This last point is especially stressed in the UK, where breeds are described as either upland (hill or mountain) or lowland breeds. A sheep may also be of a fat-tailed type, which is a dual-purpose sheep common in Africa and Asia with larger deposits of fat within and around its tail.

Breeds are also grouped based on how well they are suited to producing a certain type of breeding stock. Generally, sheep are thought to be either “ewe breeds” or “ram breeds”. Ewe breeds are those that are hardy, and have good reproductive and mothering capabilities—they are for replacing breeding ewes in standing flocks. Ram breeds are selected for rapid growth and carcase quality, and are mated with ewe breeds to produce meat lambs. Lowland and upland breeds are also crossed in this fashion, with the hardy hill ewes crossed with larger, fast-growing lowland rams to produce ewes called mules, which can then be crossed with meat-type rams to produce prime market lambs. Many breeds, especially rare or primitive ones, fall into no clear category.

Breeds are categorized by the type of their wool. Fine wool breeds are those that have wool of great crimp and density, which are preferred for textiles. Most of these were derived from Merino

Fig. 4.1: The Barbados Blackbelly is a Hair Sheep Breed of Caribbean Origin

sheep, and the breed continues to dominate the world sheep industry. Downs breeds have wool between the extremes, and are typically fast-growing meat and ram breeds with dark faces. Some major medium wool breeds, such as the Corriedale, are dual-purpose crosses of long and fine-wooled breeds and were created for high-production commercial flocks. Long wool breeds are the largest of sheep, with long wool and a slow rate of growth. Long wool sheep are most valued for crossbreeding to improve the attributes of other sheep types. For example: the American Columbia breed was developed by crossing Lincoln rams (a long wool breed) with fine-wooled Rambouillet ewes.

Coarse or carpet wool sheep are those with a medium to long length wool of characteristic coarseness. Breeds traditionally used for carpet wool show great variability, but the chief requirement is a wool that will not break down under heavy use (as would that of the finer breeds). As the demand for carpet-quality wool declines, some breeders of this type of sheep are attempting to use a few of these traditional breeds for alternative purposes. Others have always been primarily meat-class sheep.

A minor class of sheep are the dairy breeds. Dual-purpose breeds that may primarily be meat or wool sheep are often used secondarily as milking animals, but there are a few breeds that are predominantly used for milking. These sheep do produce a higher quantity of milk and have slightly longer lactation curves. In the quality of their milk, fat and protein content percentages of dairy sheep vary from non-dairy breeds but lactose content does not.

A last group of sheep breeds is that of fur or hair sheep, which do not grow wool at all. Hair sheep are similar to the early domesticated sheep kept before woolly breeds were developed, and are raised for meat and pelts. Some modern breeds of hair sheep, such as the Dorper, result from crosses between wool and hair breeds. For meat and hide producers, hair sheep are cheaper to keep, as they do not need shearing. Hair sheep are also more resistant to parasites and hot weather.

With the modern rise of corporate agribusiness and the decline of localized family farms, many breeds of sheep are in danger of extinction. The Rare Breeds Survival Trust of the UK lists 22 native breeds as having only 3,000 registered animals (each), and the American Livestock Breeds Conservancy lists 14 as having fewer than 10,000. Preferences for breeds with uniform characteristics and fast growth have pushed heritage (or heirloom) breeds to the margins of the sheep industry. Those that remain are maintained through the efforts of conservation organizations, breed registries, and individual farmers dedicated to their preservation.

Diet

Sheep are exclusively herbivorous mammals. Like all ruminants, sheep have a complex digestive system composed of four chambers, allowing them to break down cellulose from stems, leaves, and seed hulls into simpler carbohydrates. When sheep graze, vegetation is chewed into a mass called a bolus, which is then passed into the first chamber: the rumen. The rumen is a 19 to 38-liter (5 to 10 gal) organ in which feed is fermented via a symbiotic relationship with the bacteria, protozoa, and yeasts of the gut flora. The bolus is periodically regurgitated back to the mouth as cud for additional chewing and salivation. Cud chewing is an adaptation allowing ruminants to graze more quickly in the morning, and then fully chew and digest feed later in the day. This is beneficial as grazing, which requires lowering the head, leaves sheep vulnerable to predators, while cud chewing does not.

Fig. 4.2: A Ewe Grazing

During fermentation, the rumen produces gas that must be expelled; disturbances of the organ, such as sudden changes in a sheep's diet, can cause potentially fatal conditions such as bloat. After fermentation in the rumen, feed passes in to the reticulum and the omasum; special feeds such as grains may bypass the rumen altogether. After the first three chambers, food moves in to the abomasum for final digestion before processing by the intestines. The abomasum is the only one of the four chambers analogous to the human stomach (being the only one that absorbs nutrients for use as energy), and is sometimes called the "true stomach".

Sheep follow a diurnal pattern of activity, feeding from dawn to dusk, stopping sporadically to rest and chew their cud. Ideal pasture for sheep is not lawn-like grass, but an array of grasses, legumes and forbs. Types of land where sheep are raised vary widely, from pastures that are seeded and improved intentionally to rough, native lands. Common plants toxic to sheep are present in most of the world, and include (but are not limited to) oak and acorns, tomato, yew, rhubarb, potato, and rhododendron.

Sheep are largely grazing herbivores; unlike browsing animals such as goats and deer that prefer taller foliage. With a much narrower face, sheep crop plants very close to the ground and can overgraze a pasture much faster than cattle. For this reason, many shepherds use managed intensive rotational grazing, where a flock is rotated through multiple pastures, giving plants time to recover. Paradoxically, sheep can both cause and solve the spread of invasive plant species. By disturbing the natural state of pasture, sheep and other livestock can pave the way for invasive plants. However, sheep also prefer to eat invasives such as cheatgrass, leafy spurge, kudzu and spotted knapweed over native species such as sagebrush, making grazing sheep effective for conservation grazing. Research conducted in Imperial County, California compared lamb grazing with herbicides for weed control in seedling alfalfa fields. Three trials demonstrated that grazing lambs were just as effective as herbicides in controlling winter weeds. Entomologists also compared grazing lambs to insecticides for insect control in winter alfalfa. In this trial, lambs provided insect control as effectively as insecticides.

Other than forage, the other staple feed for sheep is hay, often during the winter months. The ability to thrive solely on pasture (even without hay) varies with breed, but all sheep can survive on this diet. Also included in some sheep's diets are minerals, either in a trace mix or in licks.

Naturally, a constant source of potable water is also a fundamental requirement for sheep. The amount of water needed by sheep fluctuates with the season and the type and quality of the food they consume. When sheep feed on large amounts of new growth and there is precipitation (including dew, as sheep are dawn feeders), sheep need less water. When sheep are confined or are eating large amounts of cured hay, more water is typically needed. Sheep also require clean water, and may refuse to drink water that is covered in scum or algae.

Sheep are one of the few livestock animals raised for meat today that have never been widely raised in an intensive, confined animal feeding operation (CAFO). Although there is a growing movement advocating alternative farming styles, a large percentage of beef cattle, pigs, and poultry are still produced under such conditions. In contrast, only some sheep are regularly given high-concentration grain feed, much less kept in confinement. Especially in industrialized countries, sheep producers may fatten market lambs before slaughter (called "finishing") in feedlots. Many sheep breeders flush ewes and rams with a daily ration of grain during breeding to increase fertility. Ewes are also flushed during pregnancy to increase birth weights, as 70 per cent of a lamb's growth occurs in the last five to six weeks of gestation. Otherwise, only lactating ewes and especially old or infirm sheep are commonly provided with grain. Feed provided to sheep must be specially formulated, as most cattle, poultry, pig, and even some goat feeds contain levels of copper that are lethal to sheep. The same danger applies to mineral supplements such as salt licks.

Behaviour and Intelligence

Sheep are prey animals with a strong gregarious instinct, and a majority of sheep behaviours can be understood in these terms. The dominance hierarchy of *Ovis aries* and its natural inclination to follow a leader to new pastures were the pivotal factors in it being one of the first domesticated livestock species. All sheep have a tendency to congregate close to other members of a flock, although this behaviour varies with breed. Farmers exploit this behaviour to keep sheep together on unfenced pastures and to move them more easily. Shepherds may also use sheepdogs in this effort, whose highly bred herding ability can assist in moving flocks. Sheep are also extremely food-oriented, and association of humans with regular feeding often results in sheep soliciting people for food. Those who are moving sheep may exploit this behaviour by leading sheep with buckets of feed, rather than forcing their movements with herding.

In regions where sheep have no natural predators, none of the native breeds of sheep exhibit a strong flocking behaviour. Sheep can also become hefted to one particular local pasture (heft) so they do not roam freely in unfenced landscapes. Ewes teach the heft to their lambs, and if whole flocks are culled it must be retaught to the replacement animals.

Flock dynamics in sheep are, as a rule, only exhibited in a group of four or more sheep. Fewer sheep may not react as normally expected when alone or with few other sheep. For sheep, the primary defense mechanism is simply to flee from danger when their flight zone is crossed. Secondly, cornered sheep may charge or threaten to do so through hoof stamping and aggressive posture. This is particularly true for ewes with newborn lambs.

In displaying flocking, sheep have a strong lead-follow tendency, and a leader often as not is simply the first sheep to move. However, sheep do establish a pecking order through physical displays of dominance. Dominant animals are inclined to be more aggressive with other sheep, and usually feed first at troughs. Primarily among rams, horn size is a factor in the flock hierarchy. Rams with different size horns may be less inclined to fight to establish pecking order, while rams with similarly sized horns are more so.

Sheep can become stressed when separated from their flock members. Sheep can recognize individual human and ovine faces, and remember them for years. Relationships in flocks tend to be closest among related sheep: in mixed-breed flocks same-breed subgroups tend to form, and a ewe and her direct descendants often move as a unit within large flocks.

Sheep are frequently thought of as extremely unintelligent animals. A sheep's herd mentality and quickness to flee and panic in the face of stress often make shepherding a difficult endeavor for the uninitiated. Despite these perceptions, a University of Illinois monograph on sheep found them to be just below pigs and on par with cattle in IQ, and some sheep have shown problem-solving abilities; a flock in West Yorkshire, England allegedly found a way to get over cattle grids by rolling on their backs, although documentation of this has relied on anecdotal accounts. In addition to long-term facial recognition of individuals, sheep can also differentiate emotional states through facial characteristics. If worked with patiently, sheep may learn their names, and many sheep are trained to be led by halter for showing and other purposes. Sheep have also responded well to clicker training. Very rarely, sheep are used as pack animals. Tibetan nomads distribute baggage equally throughout a flock as it is herded between living sites.

Reproduction

Sheep follow a similar reproductive strategy to other herd animals.

A group of ewes is generally mated by a single ram, who has either been chosen by a breeder or has established dominance through physical contest with other rams (in feral populations). Most sheep are seasonal breeders, although some are able to breed year-round. Ewes generally reach sexual maturity at six to eight months of age, and rams generally at four to six months. Ewes have estrus cycles about every 17 days, during which they emit a scent and indicate readiness through physical displays towards rams. A minority of sheep display a preference for homosexuality (8 per cent on average) or are freemartins (female animals that are behaviourally masculine and lack functioning ovaries).

Without human intervention, rams fight during the rut to determine which individuals may mate with ewes. Rams, especially unfamiliar ones, will also fight outside the breeding period to establish dominance; rams can kill one another if allowed to mix freely. During the rut, even normally friendly rams may become aggressive towards humans due to increases in their hormone levels.

After mating, sheep have a gestation period of about five months, and normal labour may take one to three hours. Although some breeds may regularly throw larger litters of lambs, most produce single or twin lambs. During or soon after labour, ewes and lambs may be confined to small lambing jugs, small pens designed to aid both careful observation of ewes and to cement the bond between them and their lambs.

Ovine obstetrics can be problematic. By selectively breeding ewes that produce multiple offspring with higher birth weights for generations, sheep producers have inadvertently caused some domestic sheep to have difficulty lambing; balancing ease of lambing with high productivity is one of the dilemmas of sheep breeding. In the case of any such problems, those present at lambing may assist the ewe by extracting or repositioning lambs. After the birth, ewes ideally break the amniotic sac (if it is not broken during labour), and begin licking clean the lamb. Most lambs will begin standing within an hour of birth. In normal situations, lambs nurse after standing, receiving vital colostrum milk. Lambs that either fail to nurse or that are rejected by the ewe require aid to live, such as bottle-feeding or fostering by another ewe.

After lambs are several weeks old, lamb marking (the process of ear tagging, docking, and castrating) is carried out. Vaccinations are usually carried out at this point as well. Ear tags with numbers are attached, or ear marks are applied for ease of later identification of sheep. Castration is performed on ram lambs not intended for breeding, although some shepherds choose to avoid the procedure for ethical, economic or practical reasons. Ram lambs that will either be slaughtered or separated from ewes before sexual maturity are not usually castrated. Docking, which is the shortening of a lamb's tail, is practised for health reasons. Objections to all these procedures have been raised by animal rights groups, but farmers defend them by saying they solve many practical and veterinary problems, and inflict only temporary pain.

Health

Sheep may fall victim to poisons, infectious diseases, and physical injuries. As a prey species, a sheep's system is adapted to hide the obvious signs of illness, to prevent being targeted by predators. However, there are some obvious signs of ill health, with sick sheep eating little, vocalizing excessively, and being generally listless. Throughout history, much of the money and labour of sheep husbandry has aimed to prevent sheep ailments. Historically, shepherds often created remedies by experimentation on the farm. In some developed countries, including the United States, sheep lack the economic importance for drugs companies to perform expensive clinical trials required to approve drugs for ovine use. In such instances, shepherds resort to illegal, extra-label usage of drugs approved for other animals. In the 20th and 21st centuries, a minority of sheep owners have turned to alternative treatments such as homeopathy, herbalism and even traditional Chinese medicine to treat sheep veterinary problems. Despite some favourable anecdotal evidence, the effectiveness of alternative veterinary medicine has been met with skepticism in scientific journals. The need for traditional anti-parasite drugs and antibiotics is widespread, and is the main impediment to certified organic farming with sheep.

Many breeders take a variety of preventative measures to warn off problems. The first is to ensure that all sheep are healthy when purchased. Many buyers avoid outlets known to be clearing houses for animals culled from healthy flocks as either sick or simply inferior. This can also mean maintaining a closed flock, and quarantining new sheep for a month. Two fundamental preventative programmes are maintaining good nutrition and reducing stress in the sheep. Handling sheep in loud, erratic ways causes them to produce cortisol, a stress hormone. This can lead to a weakened immune system, thus making sheep far more vulnerable to disease. Signs of stress in sheep include: excessive panting, teeth grinding, restless movement, wool eating, and wood chewing. Avoiding poisoning is also important, common poisons are pesticide sprays, inorganic fertilizer, motor oil, as well as radiator coolant (the ethylene glycol antifreeze is sweet-tasting).

Common forms of preventive medication for sheep are vaccinations and treatments for parasites. Both external and internal parasites are the most prevalent malady in sheep, and are either fatal, or reduce the productivity of flocks. Worms are the most common internal parasites. They are ingested during grazing, incubate within the sheep, and are expelled through the digestive system (beginning the cycle again). Oral anti-parasitic medicines known as drenches are given to a flock to treat worms, sometimes after worm eggs in the feces has been counted to assess infestation levels. Afterwards, sheep may be moved to a new pasture to avoid ingesting the same parasites. External sheep parasites include: lice (for different parts of the body), sheep keds, nose bots, sheep itch mite, and maggots. Keds are blood-sucking parasites that cause general malnutrition and decreased productivity, but are not fatal. Maggots are those of the bot fly and the blow-fly. Fly maggots cause the extremely destructive condition of flystrike. Flies lay their eggs in wounds or wet, manure-soiled wool, when the maggots hatch they burrow into a sheep's flesh, eventually causing death if untreated. In addition to other treatments, crutching (shearing wool from a sheep's rump) is a common preventative method. Nose bots are flies that inhabit a sheep's sinuses, causing breathing difficulties and discomfort. Common signs are a discharge from the nasal passage, sneezing, and frantic movement such as head shaking. External parasites may be controlled through the use of backliners, sprays or immersive sheep dips.

A wide array of bacterial diseases affect sheep. Diseases of the hoof, such as foot rot and foot scald may occur, and are treated with footbaths and other remedies. These painful conditions cause lameness and hinder feeding. Ovine Johne's disease is a wasting disease that affects young sheep. Bluetongue disease is an insect-borne illness causing fever and inflammation of the mucous membranes. Ovine rinderpest (or *peste des petits ruminants*) is a highly contagious and often fatal viral disease affecting sheep and goats.

A few sheep conditions are transmittable to humans. Orf (also known as scabby mouth, contagious ecthyma or soremouth) is a skin disease leaving lesions that is transmitted through skin-to-skin contact. More seriously, the organisms that can cause spontaneous enzootic abortion in sheep are easily transmitted to pregnant women. Also of concern are the prion disease scrapie and the virus that causes foot-and-mouth disease (FMD), as both can devastate flocks. The latter poses a slight risk to humans. During the 2001 FMD pandemic in the UK, hundreds of sheep were culled and some rare British breeds were at risk of extinction due to this.

Predation

Other than parasites and disease, predation is a threat to sheep and the profitability of sheep raising. Sheep have little ability to defend themselves, compared with other species kept as livestock. Even if sheep survive an attack, they may die from their injuries, or simply from panic. However, the impact of predation varies dramatically with region. In Africa, Australia, the Americas, and parts of Europe and Asia predators are a serious problem. In the United States, for instance, over 1/3 of sheep deaths in 2004 were caused by predation. In contrast, other nations are virtually devoid of sheep predators, particularly islands known for extensive sheep husbandry. Worldwide, canids—including the domestic dog—are responsible for most sheep deaths. Other animals that occasionally prey on sheep include: felines, bears, birds of prey, ravens and feral hogs.

Sheep producers have used a wide variety of measures to combat predation. Pre-modern shepherds used their own presence, livestock guardian dogs, and protective structures such as barns and fencing. Fencing (both regular and electric), penning sheep at night and lambing indoors all continue to be widely used. More modern shepherds used guns, traps, and poisons to kill predators, causing significant decreases in predator populations. In the wake of the environmental and conservation movements, the use of these methods now usually falls under the purview of specially designated government agencies in most developed countries.

The 1970s saw a resurgence in the use of livestock guardian dogs and the development of new methods of predator control by sheep producers, many of them non-lethal. Donkeys and guard llamas have been used since the 1980s in sheep operations, using the same basic principle as livestock guardian dogs. Interspecific pasturing, usually with larger livestock such as cattle or horses, may help to deter predators, even if such species do not actively guard sheep. In addition to animal guardians, contemporary sheep operations may use non-lethal predator deterrents such as motion-activated lights and noisy alarms.

History

Sheep were among the first animals to be domesticated by mankind; sources provide a domestication date between nine and eleven thousand years ago in Mesopotamia. Their wild relatives have several characteristics—such as a relative lack of aggression, a manageable size, early sexual maturity, a social nature, and high reproduction rates—which made them particularly suitable for domestication. Today, *Ovis aries* is an entirely domesticated animal that is largely dependent on man for its health and survival. Feral sheep do exist, but exclusively in areas devoid of large predators (usually islands) and not on the scale of feral horses, goats, pigs, or dogs, although some feral populations have remained isolated long enough to be recognized as distinct breeds.

The exact line of descent between domestic sheep to their wild ancestors is presently unclear. The most common hypothesis states that *Ovis aries* is descended from the Asiatic (*O. orientalis*) species of mouflon. It has been proposed that the European mouflon (*O. musimon*) is an ancient breed of domestic sheep turned feral rather than an ancestor, despite it commonly being cited as ancestor in past literature. A few breeds of sheep, such as the Castlemilk Moorit from Scotland, were formed through crossbreeding with wild European mouflon.

The urial (*O. vignei*) was once thought to have been a forebear of domestic sheep, as they occasionally interbreed with mouflon in the Iranian part of their range. However, the urial, argali (*O. ammon*), and snow sheep (*O. nivicola*) have a different number of chromosomes than other Ovis species, making a direct relationship implausible, and phylogenetic studies show no evidence of urial ancestry. Further studies comparing European and Asian breeds of sheep showed significant genetic differences between the two. Two explanations for this phenomenon have been posited. The first is that there is a currently unknown species or subspecies of wild sheep that contributed to the formation of domestic sheep. A second hypothesis suggests that this variation is the result of multiple waves of capture from wild mouflon, similar to the known development of other livestock.

Initially, sheep were kept solely for meat, milk and skins. Archaeological evidence from statuary found at sites in Iran suggests that selection for woolly sheep may have begun around 6000 BC, but the earliest woven wool garments have only been dated to two to three thousand years later. By that span of the Bronze Age, sheep with all the major features of modern breeds were widespread throughout Western Asia. However, one chief difference between ancient sheep and modern breeds is the technique by which wool could be collected. Primitive sheep cannot be shorn, and must have their wool plucked out by hand in a process called "rooing". This is because fibers called kemps are still longer than the soft fleece. The fleece may also be collected from the field after it falls out. This trait survives today in unrefined breeds such as the Soay and many Shetlands. Indeed, the Soay, along with other Northern European breeds with short tails, unshearable fleece, diminutive size, and horns in both sexes, are closely related to ancient sheep. Originally, weaving and spinning wool was a handicraft practiced at home, rather than an industry. Babylonians, Sumerians, and Persians all depended on sheep; and although linen was the first fabric to be fashioned in to clothing, wool was a prized product. The raising of flocks for their fleece was one of the earliest industries, and flocks were a medium of exchange in barter economies. Numerous biblical figures kept large flocks, and subjects of the king of Israel were taxed according to the number of rams they owned.

In Africa

Sheep entered the African continent not long after their domestication in western Asia. A minority of historians once posited a contentious African theory of origin for *Ovis aries*. This theory is based primarily on rock art interpretations, and osteological evidence from Barbary sheep. The first sheep entered North Africa via Sinai, and were present in ancient Egyptian society between eight and seven thousand years ago. Sheep have always been part of subsistence farming in Africa, but today the only country that keeps an influential number of commercial sheep is South Africa. South African sheep producers, in an attempt to deal with the numerous predators of Africa, invented the livestock protection collar, which holds poison at the jugular to sicken or kill predators.

In Europe

From Southwest Asia, sheep husbandry spread quickly in to Europe. Practically from its inception, ancient Greek civilization relied on sheep as primary livestock, and were even said to name individual animals. Scandinavian sheep of a type seen today—with short tails and multi-coloured fleece—were also present early on. Later, the Roman Empire kept sheep on a wide scale, and the Romans were an important agent in the spread of sheep raising throughout the continent. Pliny the Elder, in his Natural History (*Naturalis Historia*), speaks at length about sheep and wool. Declaring "Many thanks, too, do we owe to the sheep, both for appeasing the gods, and for giving us the use of its fleece.", he goes on to detail the breeds of ancient sheep and the many colours, lengths and qualities of wool. Romans also pioneered the practice of blanketing sheep, in which a fitted coat (today usually of nylon) is placed over the sheep to improve the cleanliness and luster of its wool.

During the Roman occupation of the British Isles, a large wool processing factory was established in Winchester, England in about 50 AD. By 1000 AD, England and Spain were recognized as the twin centers of sheep production in the Western world. As the original breeders of the fine-wooled merino sheep that have historically dominated the wool trade, the Spanish gained great wealth. Wool money largely financed Spanish rulers and thus the voyages to the New World by conquistadors. The powerful *Mesta* (its full title was *Honrado Concejo de la Mesta*, the Honourable Council of the Mesta) was a corporation of sheep owners mostly drawn from Spain's wealthy merchants, Catholic clergy and nobility that controlled the merino flocks. By the 17th century, the *Mesta* held in upwards of two million head of merino sheep.

Mesta flocks followed a seasonal pattern of transhumance across Spain. In the spring, they left the winter pastures (*invernaderos*) in Extremadura and Andalusia to graze on their summer pastures (*agostaderos*) in Castile, returning again in the autumn. Spanish rulers eager to increase wool profits gave extensive legal rights to the *Mesta*, often to the detriment of local peasantry. The huge merino flocks had a lawful right of way for their migratory routes (*cañadas*). Towns and villages were obliged by law to let the flocks graze on their common land, and the Mesta had its own sheriffs that could summon offending individuals to its own tribunals.

Exportation of merinos without royal permission was also a punishable offense, thus ensuring a near-absolute monopoly on the breed until Napoleon's invasion of Spain in the mid-18th century. After the breaking of the export ban, fine wool sheep began to be distributed worldwide. The export to Rambouillet by Louis XVI in 1786 formed the basis for the modern Rambouillet (or French Merino) breed. After the Napoleonic Wars and the global distribution of the once-exclusive Spanish stocks of Merinos, sheep raising in Spain reverted to hardy coarse-wooled breeds such as the Churra, and was no longer of international economic significance.

The sheep industry in Spain was an instance of migratory flock management, with large homogenous flocks ranging over the entire nation. The management model used in England was quite different but had a similar importance to economy of the British Empire. Up until the early 20th century, owling (the smuggling of sheep or wool out of the country) was a punishable offense, and to this day the Lord Speaker of the House of Lords sits on a cushion known as the Woolsack.

The high concentration and more sedentary nature of shepherding in the UK allowed sheep especially adapted to their particular purpose and region to be raised, thereby giving rise to an exceptional variety of breeds in relation to the land mass of the country. This greater variety of breeds also produced a valuable variety of products to compete with the superfine wool of Spanish sheep. By the time of Elizabeth I's rule, sheep and wool trade was the primary source of tax revenue to the Crown of England and the country was a major influence in the development and spread of sheep husbandry.

An important event not only in the history of domestic sheep, but of all livestock, was the work of Robert Bakewell in the 1700s. Before his time, breeding for desirable traits was often based on chance, with no scientific process for selection of breeding stock. Bakewell established the principles of selective breeding—especially line breeding—in his work with sheep, horses and cattle; his work later influenced Gregor Mendel and Charles Darwin. His most important contribution to sheep was the development of the Leicester Longwool, a quick-maturing breed of blocky conformation that formed the basis for many vital modern breeds. Today, the sheep industry in the UK has diminished significantly, though pedigreed rams can still fetch around 100,000 Pounds sterling at auction.

In the Americas

No ovine species native to the Americas has ever been domesticated, despite being closer genetically to domestic sheep than many Asian and European species. The first domestic sheep in North America—most likely of the Churra breed—arrived with Christopher Columbus' second voyage in 1493. The next transatlantic shipment to arrive was with Hernán Cortés in 1519, landing in Mexico. No export of wool or animals is known to have occurred from these populations, but flocks did disseminate throughout what is now Mexico and the Southwest United States with Spanish colonists. Churras were also introduced to the Navajo tribe of Native Americans, and became a key part of their livelihood and culture. The modern presence of the Navajo-Churro breed is a result of this heritage.

North America

The next transport of sheep to North America was not until 1607, with the voyage of the *HMS Susan Conant* to Virginia. However, the sheep that arrived in that year were all slaughtered because of a famine, and a permanent flock was not to reach the colony until two years later in 1609. In two decades time, the colonists had expanded their flock to a total of 400 head. By the 1640s there were about 100,000 head of sheep in the 13 colonies, and in 1662, a woolen mill was built in Watertown, Massachusetts. Especially during the periods of political unrest and civil war in Britain spanning the 1640s and 50s which disrupted maritime trade, the colonists found it pressing to produce wool for clothing. Many islands off the coast were cleared of predators and set aside for sheep: Nantucket, Long Island, Martha's Vineyard and small islands in Boston Harbor were notable examples. There remain some rare breeds of American sheep—such as the Hog Island sheep—that were the result of island flocks. Placing semi-feral sheep and goats on islands was common practice in colonization during this period. Early on, the British government banned further export of sheep to the Americas, or wool from it, in an attempt to stifle any threat to the wool trade in the British Isles. One of many restrictive trade measures that precipitated the American Revolution, the sheep industry in the Northeast grew despite the bans.

Gradually, beginning in the 1800s, sheep production in the U.S. moved westward. Today, the vast majority of flocks reside on Western range lands. During this westward migration of the industry, competition between sheep (sometime called "range maggots") and cattle operations grew more heated, eventually erupting into range wars. Other than simple competition for grazing and water rights, cattlemen believed that the secretions of the foot glands of sheep made cattle unwilling to graze on places where sheep had stepped. As sheep production centered on the U.S. western ranges, it became associated with other parts of Western culture, such as the rodeo. In modern America, a minor event in rodeos is mutton busting, in which children compete to see who can stay atop a sheep the longest before falling off. Another effect of the westward movement of sheep flocks in North America was the decline of wild species such as Bighorn sheep (*O. canadesis*). Most diseases of domestic sheep are transmittable to wild ovines, and such diseases, along with overgrazing and habitat loss, are named as primary factors in the plummeting numbers of wild sheep. Sheep production peaked in North America during 1940s and 50s at more than 55 million head. Henceforth and continuing today, the number of sheep in North

America has steadily declined with wool prices and the lessening American demand for sheep meat.

South America

In South America, especially in Patagonia, there is an active modern sheep industry. Sheep keeping was largely introduced through immigration to the continent by Spanish and British peoples, for whom sheep were a major industry during the period. South America has a large number of sheep, but the highest-producing nation (Brazil) kept only just over 15 million head in 2004, far fewer than most centers of sheep husbandry. The primary challenges to the sheep industry in South America are the phenomenal drop in wool prices in the late 20th century and the loss of habitat through logging and overgrazing. The most influential region internationally is that of Patagonia, which has been the first to rebound from the fall in wool prices. With few predators and almost no grazing competition (the only large native grazing mammal is the guanaco), the region is prime land for sheep raising. The most exceptional area of production is surrounding the La Plata river in the Pampas region. Sheep production in Patagonia peaked in 1952 at more than 21 million head, but has steadily fallen to fewer then ten today. Most operations focus on wool production for export from Merino and Corriedale sheep; the economic sustainability of wool flocks has fallen with the drop in prices, while the cattle industry continues to grow.

In Australia and New Zealand

Australia and New Zealand are crucial players in the contemporary sheep industry, and sheep are an iconic part of both countries' culture and economy. New Zealand has the highest density of sheep per capita (sheep outnumber the human population 12 to 1), and Australia is the world's indisputably largest exporter of sheep and cattle. In 2007, New Zealand even declared 15 February their official National Lamb Day to celebrate the country's history of sheep production.

The First Fleet brought the initial population of 70 sheep from the Cape of Good Hope to Australia in 1788. The next shipment was of 30 sheep from Calcutta and Ireland in 1793. All of the early sheep brought to Australia were exclusively used for the dietary needs of the penal colonies. The beginnings of the Australian wool industry were due to the vision and efforts of Captain John Macarthur. At Macarthur's urging 16 Spanish merinos were imported in 1797, effectively beginning the Australian sheep industry. By 1801 Macarthur had 1,000 head of sheep, and in 1803 he exported 111 kilograms (245 lb) of wool to England. Today, Macarthur is generally thought of as the father of the Australian sheep industry.

The growth of the sheep industry in Australia was explosive. In 1820, the continent held 100,000 sheep, a decade later it had one million. By 1840, New South Wales alone kept 4 million sheep; flock numbers grew to 13 million in a decade. While much of the growth in both nations was due to the active support of Britain in its desire for wool, both worked independently to develop new high-production breeds: the Corriedale, Coolalee, Coopworth, Perendale, Polwarth, Booroola Merino, Peppin Merino, and Poll Merino were all created in New Zealand or Australia. Wool production was a fitting industry for colonies far from their home nations. Before the advent of fast air and maritime shipping, wool was one of the few viable products that was not subject to spoiling on the long passage back to British ports. The abundant new land and milder winter weather of the region also aided the growth of the Australian and New Zealand sheep industries.

Flocks in Australia have always been largely range bands on fenced land, and are aimed at production of medium to superfine wool for clothing and other products as well as meat. New Zealand flocks are kept in a fashion similar to English ones, in fenced holdings without shepherds. Although wool was once the primary income source for New Zealand sheep owners (especially during the New Zealand wool boom), today it has shifted to meat production for export.

Animal Welfare Concerns

The Australian sheep industry is the only sector of the industry to receive international criticism for its practices. Sheep stations in Australia are cited in *Animal Liberation*, the seminal book of the animal rights movement, as the author's primary evidence in his argument against retaining sheep as a part of animal agriculture. The practice of mulesing, in which skin is cut away from an animal's perineal area to prevent cases of the fatal condition flystrike, has been condemned by PETA as being painful and unnecessary. In response, a programme of phasing out mulesing is currently being implemented, and some mulesing operations are being carried out with the use of anaesthetic. The Animal Welfare Advisory Committee to the New Zealand Ministry of Agriculture *Code of recommendations and minimum standards for the welfare of Sheep*, considers mulesing a "special technique" which is performed on some Merino sheep at a small number of farms in New Zealand.

Most of the sheep meat exported from Australia is either frozen carcases to the UK or is live export to the Middle East. Shipped on livestock carriers in what has been called crowded, unsafe conditions by critics, live sheep are desired by Middle Eastern nations to meet the requirements of ritual halal slaughter. Opponents of the export—such as PETA—say that sheep exported to countries outside the jurisdiction of Australia's animal cruelty laws are treated with horrendous brutality and that halal facilities exist in Australia to make export of live animals redundant. A few celebrities and companies have pledged to boycott all Australian sheep products in protest.

Economic Importance

Sheep are an important part of the global agricultural economy. However, their once-vital status has been largely replaced by other livestock species, especially the pig, chicken, and cow. China, Australia, India, and Iran have the largest modern flocks, and serve both local and exportation needs for wool and mutton. Other countries such as New Zealand have smaller flocks but retain a large international economic impact due to their export of sheep products. Sheep also play a major role in many local economies, which may be niche markets focused on organic or sustainable agriculture and local food customers. Especially in developing countries, such flocks may be a part of subsistence agriculture rather than a system of trade. Sheep themselves may be a medium of trade in barter economies.

Table 4.1: Global Sheep Stocks in 2004 (million)

People's Republic of China	157.9
Australia	101.3
India	62.5
Iran	54.0
Sudan	48.0
New Zealand	39.2
United Kingdom	35.8
South Africa	25.3
World Total	1,059.8

Source: UN Food and Agriculture Organisation (FAO)

Domestic sheep provide a wide array of raw materials. Wool was one of the first textiles, although in the late 20th century wool prices began to fall dramatically as the result of the popularity and cheap prices for synthetic fabrics. For many sheep owners, the cost of shearing is greater than the possible profit from the fleece, making subsisting on wool production alone practically impossible without farm subsidies. Fleeces are used as material in making alternative products such as wool insulation. In the 21st century, the sale of meat is the most profitable enterprise in the sheep industry, even though far less sheep meat is consumed than chicken, pork or beef.

Sheepskin is likewise used for making clothes, footwear, rugs, and other products. Byproducts from the slaughter of sheep are also of value: sheep tallow can be used in candle and soap making, sheep bone and cartilage has been used to furnish carved items such as dice and buttons as well as rendered glue and gelatin. Sheep intestine can be formed into sausage casings, and lamb intestine has been formed into surgical sutures, as well as strings for musical instruments and tennis rackets. Sheep droppings, which are high in cellulose, have even been sterilized and mixed with traditional pulp materials to make paper. Of all sheep byproducts, perhaps the most valuable is lanolin: the water-proof, fatty substance found naturally in sheep's wool and used as a base for innumerable cosmetics and other products.

Some farmers who keep sheep also make a profit from live sheep. Providing lambs for youth programmes such as 4-H and competition at agricultural shows is often a dependable avenue for the sale of sheep. Farmers may also choose to focus on a particular breed of sheep in order to sell registered purebred animals, as well as provide a ram rental service for breeding. The most valuable sheep ever sold to date was a purebred Texel ram that fetched £231,000 at auction. The previous record holder was a Merino ram sold for £205,000 in 1989. A new option for deriving profit from live sheep is the rental of flocks for grazing; these "mowing services" are hired in order to keep unwanted vegetation down in public spaces and to lessen fire hazard.

Despite the falling demand and price for sheep products in many markets, sheep have distinct economic advantages when compared with other livestock. They do not require the expensive housing, such as that used in the intensive farming of chickens or pigs. They are an efficient use of land; roughly six sheep can be kept on the amount that would suffice for a single cow or horse. Sheep can also consume plants, such as noxious weeds, that most other animals will not touch, and produce more young at a faster rate. Also, in contrast to most livestock species, the cost of raising sheep is not necessarily tied to the price of feed crops such as grain, soybeans and corn. Combined with the lower cost of quality sheep, all these factors combine to equal a lower overhead for sheep producers, thus entailing a higher profitability potential for the small farmer. Sheep are especially beneficial for independent producers, including family farms with limited resources, as the sheep industry is one of the few types of animal agriculture that has not been vertically integrated by agribusiness.

As food

Sheep meat and milk were one of the earliest staple proteins consumed by human civilization after the transition from hunting and gathering to agriculture. Sheep meat prepared for food is known as either mutton or lamb. "Mutton" is derived from the Old French *moton*, which was the word for sheep used by the Anglo-Norman rulers of much of the British Isles in the Middle Ages. This became the name for sheep meat in English, while the Old English word *sceap* was kept for the live animal. Throughout modern history, "mutton" has been limited to the meat of mature sheep usually at least two years of age; "lamb" is used for that of immature sheep less than a year.

In the 21st century, the nations with the highest consumption of sheep meat are the Persian Gulf states, New Zealand, Australia, Greece, Uruguay, the United Kingdom and Ireland. These countries eat 14-40 lbs (3-18 kg) of sheep meat per capita, per annum. Sheep meat is also popular in France, Africa (especially the Maghreb), the Caribbean, the rest of the Middle East, India, and parts of China. This often reflects a past history of sheep production. In these countries in particular, dishes comprising alternative cuts and offal may be popular or traditional. Sheep testicles—called animelles or lamb fries—are considered a delicacy in many parts of the world. Perhaps the most unusual dish of sheep meat is the Scottish haggis, composed of various sheep innards cooked along with oatmeal and chopped onions inside its stomach. In comparison, countries such as the U.S. consume only a pound or less (under 0.5 kg), with Americans eating 50 pounds (22 kg) of pork and 65 pounds (29 kg) of beef. In addition, such countries rarely eat mutton, and may favour the more expensive cuts of lamb: mostly lamb chops and leg of lamb.

Though sheep's milk may have been drunk in antiquity, it today is used predominantly in cheese and yogurt. Sheep have only two teats, and produce a far smaller volume of milk than cows. However, as sheep's milk contains far more fat, solids, and minerals than cow's milk, it is ideal for the cheese-making process. It also resists contamination during cooling better because of its much higher calcium content. Well-known cheeses made from sheep milk include the Feta of Bulgaria and Greece, Roquefort of France, Manchego from Spain, the Pecorino Romano (the Italian word for sheep is *pecore*) and Ricotta of Italy. Yogurts, especially some forms of strained yogurt, may also be made from sheep milk. Many of these products are now often made with cow's milk, especially when produced outside their country of origin. Sheep milk contains 4.8 per cent lactose, which may affect those who are intolerant.

In Science

Sheep are generally too large and reproduce too slowly to make ideal research subjects, and thus are not a common model organism. They have, however, played an influential role in some fields of science. In particular, the Roslin Institute of Edinburgh, Scotland used sheep for genetics research that produced groundbreaking results. In 1995, two ewes named Megan and Morag were the first mammals cloned from differentiated cells. A year later, a Finnish Dorset sheep named Dolly, dubbed "the world's most famous sheep" in *Scientific American*, was the first mammal to be cloned from an adult somatic cell. Following this, Polly and Molly were the first mammals to be simultaneously cloned and transgenic. As of 2008, the sheep genome has not been fully sequenced, although a detailed genetic map has been published, and a draft version of the complete genome produced by assembling sheep DNA sequences using information given by the genomes of other mammals.

In the study of natural selection, the population of Soay sheep that remain on the island of Hirta have been used to explore the relation of body size and colouration to reproductive success.

CATTLE

Cattle (colloquially cows) are the most common type of large domesticated ungulates. They are a prominent modern member of the subfamily Bovinae, are the most widespread species of the genus *Bos*, and are most commonly classified collectively as *Bos primigenius*. Cattle are raised as livestock for meat (beef and veal), as dairy animals for milk and other dairy products, and as draft animals (pulling carts, plows and the like). Other products include leather and dung for manure or fuel. In some countries, such as India, cattle are sacred. It is estimated that there are 1.3 billion cattle in the world today. In 2009, cattle became the first livestock animal to have its genome mapped.

Species of Cattle

Cattle were originally identified by Carolus Linnaeus as three separate species. These were *Bos taurus*, the European or "taurine" cattle (including similar types from Africa and Asia); *Bos indicus*, the zebu; and the extinct *Bos primigenius*, the aurochs. The aurochs is ancestral to both zebu and taurine cattle. More recently these three have increasingly been grouped as one species, with *Bos primigenius taurus*, *Bos primigenius indicus* and *Bos primigenius primigenius* as the subspecies.

Complicating the matter is the ability of cattle to interbreed with other closely related species. Hybrid individuals and even breeds exist, not only between taurine cattle and zebu but also between one or both of these and some other members of the genus *Bos*: yak (called a dzo or "yattle"), banteng and gaur. Hybrids can also occur between taurine cattle and either species of bison, which some authors consider to be in the genus *Bos* as well. The hybrid origin of some types may not be obvious—for example, genetic testing of the Dwarf Lulu breed, the only humpless taurine-type cattle in Nepal, found them to be a mix of taurine cattle, zebu and yak. Cattle cannot successfully be hybridized with more distantly related bovines such as water buffalo or African buffalo.

The aurochs originally ranged throughout Europe, North Africa, and much of Asia. In historical times its range became restricted to Europe, and the last known individual died in Masovia, Poland, in about 1627. Breeders have attempted to recreate cattle of similar appearance to aurochs by crossing traditional types of domesticated cattle, creating the Heck cattle breed.

Cattle Genome

In the April 24, 2009 edition of the journal Science it was reported that a team of researchers led by the National Institutes of Health and the U.S. Department of Agriculture have mapped the bovine genome. The scientists found that cattle have approximately 22,000 genes, and 80 percent of their genes are shared with humans, and they have approximately 1,000 genes they share with dogs and rodents but are not found in humans. Using this bovine "HapMap", researchers can track the differences between the breeds that affect the quality of meat and milk yields.

Terminology

In general, the same words are used in different parts of the world

Fig. 4.3: A Hereford Bull

but with minor differences in the definitions. The terminology described here contrasts the differences in definition between the United Kingdom and other British influenced parts of world such as Canada, Australia, New Zealand, Ireland, and the United States.

- An intact (*i.e.*, not castrated) adult male is called a *bull*. A wild, young, unmarked bull is known as a *micky* in Australia. An unbranded bovine of either sex is called a "maverick" in the USA and Canada.
- An adult female that has had a calf (or two, depending on regional usage) is a *cow*. A young female before she has had a calf of her own and is under three years of age is called a *heifer* (pronounced /'h?f?r/, "heffer"). A young female that has had only one calf is occasionally called a *first-calf heifer*.
- Young cattle of both sexes are called *calves* until they are weaned, then *weaners* until they are a year old in some areas; in other areas, particularly with male beef cattle, they may be known as *feeder-calves* or simply *feeders*. After that, they are referred to as *yearlings* or *stirks* if between one and two years of age.
- A castrated male is called a *steer* in the United States, and older steers are often called *bullocks* in other parts of the world; although in North America this term refers to a young bull. *Piker bullocks* are micky bulls that were caught, castrated and then later lost. In Australia, the term

"Japanese ox" is used for grain fed steers in the weight range of 500 to 650 kg that are destined for the Japanese meat trade. In North America, draft cattle under four years old are called *working steers*. Improper or late castration on a bull results in it becoming a coarse steer known as a *stag* in Australia, Canada and New Zealand. In some countries an incompletely castrated male is known also as a *rig*.

- A castrated male (occasionally a female or in some areas a bull) kept for draft purposes is called an *ox* (plural *oxen*); "ox" may also be used to refer to some carcase products from any adult cattle, such as *ox-hide*, *ox-blood* or *ox-liver*..
- In all cattle species, a female that is the twin of a bull usually becomes an infertile partial intersex, and is a *freemartin*.
- *Neat* (horned oxen, from which neatsfoot oil is derived), *beef* (young ox) and *beefing* (young animal fit for slaughtering) are obsolete terms, although *poll*, *pollard* or *polled cattle* are still terms in use for naturally hornless animals, or in some areas also for those that have been disbudded.
- Cattle raised for human consumption are called *beef cattle*. Within the beef cattle industry in parts of the United States, the older term *beef* (plural *beeves*) is still used to refer to an animal of either gender. Some Australian, Canadian, New Zealand and British people use the term *beast*, especially for single animals when the gender is unknown.
- Cattle of certain breeds bred specifically for milk production are called *milking* or *dairy cattle*.; a cow kept to provide milk for one family may be called a *house cow*.
- The adjective applying to cattle in general is usually *bovine*. The terms "bull", "cow" and "calf" are also used by extension to denote the gender or age of other large animals, including whales, hippopotamuses, camels, elk and elephants

Singular Terminology Dilemma

Cattle can only be used in the plural and not in the singular: it is a plurale tantum. Thus one may refer to "three cattle" or "some cattle", but not "one cattle". There is no universally used singular form in modern English of "cattle", other than the sex-and age-specific terms such as cow, bull, steer and heifer. Historically, "ox" was a non-gender-specific term for adult cattle, but generally this is now used only for draft cattle, especially adult castrated males. The term is also incorporated into the names of other species such as the musk ox and "grunting ox" (yak), and is used in some areas to describe certain cattle products such as ox-hide and ox-tail.

"Cow" has been in general use as a singular for the collective "cattle" in spite of the objections of those who say that it is a female-specific term, so that phrases such as "that cow is a bull" would be absurd from a lexicographic standpoint. However, it is easy to use when a singular is needed and the sex is not known or is irrelevant in the context of the conversation, as in "There is a cow in the road". Further, any herd of fully mature cattle in or near a pasture is statistically likely to consist mostly of cows, so the term is probably accurate even in the restrictive sense. Other than the few bulls needed for breeding, the vast majority of male cattle are castrated as calves and slaughtered for meat before the age of three years. Thus, in a pastured herd, any calves or herd bulls usually are clearly distinguishable from the cows due to distinctively different sizes and clear anatomical differences. The Oxford English Dictionary lists the use of "cows" as a synonym for "cattle" as an American usage. Merriam-Webster, a U.S. dictionary, recognizes the non-sex-specific use of "cow" as an alternate definition, whereas Collins, a UK dictionary, does not. Colloquially, more general non-specific terms may denote cattle when a singular form is needed. Australian, New Zealand and British farmers use the term "beast" or "cattle beast". "Bovine" is also used in Britain. The term "critter" is common in the western United States and Canada, particularly when referring to young cattle. In some areas of the American South (particularly the Appalachian region), where both dairy and beef cattle are present, an individual animal was once called a "beef critter", though that term is becoming archaic.

Other Terminology

Cattle raised for human consumption are called "beef cattle". Within the beef cattle industry in parts of the United States, the term "beef" (plural "beeves") is still used in its archaic sense to refer to an animal of either gender. Cows of certain breeds that are kept for the milk they give are called "dairy cows" or "milking cows" (formerly "milch cows"—"milch" was pronounced as "milk"). Most young male offspring of dairy cows are sold for veal, and may be referred to as *veal calves.*

The term "dogies" is used to describe orphaned calves in the context of ranch work in the American west, as in "Keep them dogies moving,". In some places, a cow kept to provide milk for one family is called a "house cow". Other obsolete terms for cattle include "neat" (this use survives in "neatsfoot oil", extracted from the feet and legs of cattle), and "beefing" (young animal fit for slaughter).

An onomatopoeic term for one of the commonest sounds made by cattle is "moo", and this sound is also called *lowing*. There are a number of other sounds made by cattle, including calves *bawling*, and bulls *bellowing*. The bullroarer makes a sound similar to a territorial call made by bulls.

Word Origin

Cattle did not originate as the term for bovine animals. It was borrowed from Old French *catel*, itself from Latin *caput*, head, and originally meant movable property, especially livestock of any kind. The word is closely related to "chattel" (a unit of personal property) and "capital" in the economic sense. The term replaced earlier Old English *feoh* "cattle, property" (cf. German *Vieh*, Gothic *faihu*).

In older English sources such as the King James Version of the Bible, "cattle" refers to livestock, as opposed to "deer" which refers to wildlife. "Wild cattle" may refer to feral cattle or to undomesticated species of the genus *Bos*. Today, the modern meaning of "cattle", without any other qualifier, is usually restricted to domesticated bovines.

Anatomy

Cattle have one stomach with four compartments. They are the rumen, reticulum, omasum, and abomasum, the rumen being the largest compartment. Cattle sometimes consume metal objects which are deposited in the reticulum, the smallest compartment, and this is where hardware disease occurs. The reticulum is known as the "Honeycomb." The omasum's main function is to absorb water and nutrients from the digestible feed. The omasum is known as the "Many Plies." The abomasum is like the human stomach; this is why it is known as the "true stomach".

Cattle are ruminants, meaning that they have a digestive system that allows use of otherwise indigestible foods by repeatedly regurgitating and rechewing them as "cud". The cud is then reswallowed and further digested by specialised microorganisms in the rumen. These microbes are primarily responsible for decomposing cellulose and other carbohydrates into volatile fatty acids that cattle use as their primary metabolic fuel. The microbes inside of the rumen are also able to synthesize amino acids from non-protein nitrogenous sources such as urea and ammonia. As these microbes reproduce in the rumen, older generations die and their carcasses continue on through the digestive tract. These carcasses are then partially digested by the cattle, allowing it to gain a high quality protein source. These features allow cattle to thrive on grasses and other vegetation.

The gestation period for a cow is nine months. A newborn calf weighs 25 to 45 kilograms (55 to 99 lb). The world record for the heaviest bull was 1,740 kilograms (3,836 lb) a Chianina named Donetto, when he was exhibited at the Arezzo show in 1955. The heaviest steer was eight year old 'Old Ben', a Shorthorn/Hereford cross weighing in at 2,140 kilograms (4,718 lb) in 1910. Steers are generally killed before reaching 750 kilograms (1,653 lb). Breeding stock usually live to about 15 years (occasionally as much as 25 years).

A common misconception about cattle (particularly bulls) is that they are enraged by the colour red (something provocative is often said to be "like a red flag to a bull"). This is incorrect, as cattle are red-green colour-blind. The myth arose from the use of red capes in the sport of bullfighting; in fact, two different capes are used. The capote is a large, flowing cape that is magenta and yellow. The more famous muleta is the smaller, red cape, used exclusively for the final, fatal segment of the fight. It is not the colour of the cape that angers the bull, but rather the movement of the fabric that irritates the bull and incites it to charge.

Although cattle cannot distinguish red from green, they do have two kinds of colour receptors in the cone cells in their retinas. Thus they are dichromatic, the same as most other mammals (including dogs, cats, horses and up to ten percent of male humans). They are able to distinguish some colours, particularly blue from yellow, in the same way as most other mammals.

Domestication and Husbandry

Cattle occupy a unique role in human history, domesticated since at least the early Neolithic. They are raised for meat (beef cattle), dairy products and hides. They are also used as draft animals and in certain sports. Some consider cattle the oldest form of wealth, and cattle raiding consequently one of the earliest forms of theft.

Cattle are often raised by allowing herds to graze on the grasses of large tracts of rangeland. Raising cattle in this manner allows the use of land that might be unsuitable for growing crops. The most common interactions with cattle involve daily feeding, cleaning and milking. Many routine husbandry practices involve ear tagging, dehorning, loading, medical operations, vaccinations and hoof care, as well as training for agricultural shows and preparations. There are also some cultural differences in working with cattle—the cattle husbandry of Fulani men rests on behavioural techniques, whereas in Europe cattle are controlled primarily by physical means like fences. Breeders use cattle husbandry to reduce M. bovis infection susceptibility by selective breeding and maintaining herd health to avoid concurrent disease.

Cattle are farmed for beef, veal, dairy, leather and they are less commonly used for conservation grazing, simply to maintain grassland for wildlife—for example, in Epping Forest, England. They are often used in some of the most wild places for livestock. Depending on the breed, cattle can survive on hill grazing, heaths, marshes, moors and semi desert. Modern cows are more commercial than older breeds and, having become more specialized, are less versatile. For this reason many smaller farmers still favour old breeds, like the dairy breed of cattle Jersey.

In Portugal, Spain, Southern France and some Latin American countries, bulls are used in the activity of bullfighting; a similar activity, Jallikattu, is seen in South India; in many other countries this is illegal. Other activities such as bull riding are seen as part of a rodeo, especially in North America. Bull-leaping, a central ritual in Bronze Age Minoan culture, still exists in southwestern France. In modern times, cattle are also entered into agricultural competitions. These competitions can involve live cattle or cattle carcasses.

In terms of food intake by humans, consumption of cattle is less efficient than of grain or vegetables with regard to land use, and hence cattle grazing consumes more area than such other agricultural production when raised on grains. Nonetheless, cattle and other forms of domesticated animals can sometimes help to utilize plant resources in areas not easily amenable to other forms of agriculture.

PIG

Pigs are a genus of even-toed ungulates within the family Suidae. The name hog most commonly refers to the domestic pig (*Sus domestica*) in everyday parlance, but technically encompasses several distinct species, including the wild boar. Swine is a collective noun generally used to describe pigs as a group rather than an individual, however it may often be implied in a pejorative manner to any living being expressing pig-like behaviour.

With around 2 billion on the planet, domestic pigs are also by far the most numerous pig species. Pigs are omnivores, and despite their reputation for gluttony, they are generally social and intelligent animals.

Description and Behaviour

A pig has a snout for a nose, small eyes, and a small tail, which may be curly, kinked, or straight. It has a thick body, short legs, and coarse hair. There are four toes on each foot, with the two large middle toes used for walking.

Pigs are omnivores, which means that they consume both plants and animals. Pigs will scavenge and have been known to eat any kind of food, including dead insects, worms, tree bark, rotting carcasses, garbage, and even other pigs. In the wild, they are foraging animals, primarily eating leaves, grasses, roots, fruits and flowers. Occasionally while in captivity, pigs may eat their own young if they become severely stressed. A typical pig has a large head with a long snout which is strengthened by a special prenasal bone and by a disk of cartilage in the tip. The snout is used to dig into the soil to find food and is a very sensitive sense organ.

Pigs have a full set of 44 teeth. The canine teeth, called tusks, grow continuously and are sharpened by the lowers and uppers rubbing against each other.

Pigs that are allowed to forage may be watched by swineherds. Because of their foraging abilities and excellent sense of smell, they are used to find truffles in many European countries. Domesticated pigs are commonly raised as livestock by farmers for meat (called pork), as well as for leather. Their bristly hairs are also used for brushes. Some breeds of pig, such as the Asian pot-bellied pig, are kept as pets.

Breeding occurs throughout the year in the tropics, but births peak around rainy seasons. A female pig can become pregnant at around 8-18 months of age. She will then go into estrus every 21 days if not bred. Male pigs become sexually active at 8-10 months of age. A litter of piglets typically contains between 6 and 12 piglets. After the young are weaned, two or more families may come together until the next mating season.

Pigs do not have functional sweat glands, so pigs cool themselves using water or mud during hot weather. They also use mud as a form of sunscreen to protect their skin from sunburn. Mud also provides protection against flies and parasites.

Species

- Palawan Bearded Pig (*Sus ahoenobarbus*)
- Bearded Pig (*Sus barbatus*)
- Indo-chinese (or Vietnam) Warty Pig (*Sus bucculentus*).
- Visayan Warty Pig (*Sus cebifrons*)
- Celebes (or Sulawesi) Warty Pig (*Sus celebensis*)
- *Sus falconeri*
- *Sus hysudricus*
- Oliver's (or Mindoro) Warty Pig (*Sus oliveri*)

- Philippine Warty Pig (*Sus philippensis*)
- Wild boar (*Sus scrofa*)
- Domestic pig (*Sus scrofa domestica*)
- *Sus strozzi*
- Java Warty Pig (*Sus verrucosus*)

The Pygmy Hog is now in the monotypic genus *Porcula* again.

Domestic Pigs

Pigs have been domesticated since ancient times in the Old World. Archeological evidence suggests that pigs were being managed in the wild in a way similar to the way they are managed by some modern New Guineans from wild boar as early as 13,000-12,700 BP in the Near East in the Tigris Basin. Remains of pigs have been dated to earlier than 11,400 BP in Cyprus that must have been introduced from the mainland which suggests domestication in the adjacent mainland by then. A separate domestication also occurred in China..

Pigs were brought to southeastern North America from Europe by De Soto and other early Spanish explorers. Pigs are particularly valued in China and on certain oceanic islands, where their self-sufficiency allows them to be turned loose, although the practice is not without its drawbacks.

The domestic pig (*Sus scrofa domesticus*) is usually given the scientific name *Sus scrofa*, although some authors call it *S. domesticus*, reserving *S. scrofa* for the wild boar. It was domesticated

Fig. 4.4: Swedish Pigfarmer with Piglet.

approximately 5,000 to 7,000 years ago. Their coats are coarse and bristly. They are born brownish coloured and tend to turn more grayish coloured with age. The upper canines form sharp distinctive tusks that curve outward and upward. Compared to other artiodactyles, their head is relatively long, pointed, and free of warts. Their head and body length ranges from 900-1,800 mm and can weigh 50-350 kg.

Pigs can be trained to perform numerous simple tasks and tricks. Recently, they have enjoyed a measure of popularity as house pets, particularly the dwarf breeds.

Cultural References to Pigs

Pigs are frequently referenced in culture and are a popular topic for idioms and famous quotes.

Pigs in Religion

- In Nordic Mythology, Gullinbursti ("Gold-Bristle" or "Gold-Mane") was Freyr's golden boar, created by the dwarves Brokk and Sindri as part of a challenge. His shining fur is said to fill the sky, trees, and sea with light.
- In ancient Egypt pigs were associated with Set, the rival to the sun god Horus. When Set fell into disfavour with the Egyptians, swineherds were forbidden to enter temples. According to Herodotus, swineherds were a kind of separate sect or caste, which only married among themselves. Egyptians regarded pigs as unworthy sacrifices to their gods other than the Moon and Dionysus, to whom pigs were offered on the day of the full Moon. Herodotus states that, though he knew the reason why Egyptians abominated swine at their other feasts but they sacrificed them at this one; however, it was to him "not a seemly one for me to tell".
- In Hinduism the god Vishnu took the form of a boar varaha in order to save the Earth from a demon who had dragged it to the bottom of the sea.
- In ancient Greece, a sow was an appropriate sacrifice to

Demeter and had been her favourite animal since she had been the Great Goddess of archaic times. Initiates at the Eleusinian Mysteries began by sacrificing a pig.

- The pig is one of the 12-year cycle of animals which appear in the Chinese zodiac related to the Chinese calendar. Believers in Chinese astrology associate each animal with certain personality traits.
- In keeping with Leviticus 11:7, the dietary laws of Judaism (Kashrut, adj. Kosher) forbid, among other kinds of meat, the eating of pork in any form, considering the pig to be an unclean animal. From the strict reading to the relevant Torah passage, pork is as forbidden as the flesh of any other unclean animal, no more and no less; in practice, however, abhorrence of pork is far stronger and emotional in traditional Jewish culture than that of other forbidden foods.
- The eating of pork is also prohibited in Islam, among Seventh-day Adventists and some other Christian denominations.
- In Catholicism, Eastern Orthodoxy, and other older Christian groups, pigs are associated with Saint Anthony the Great, the patron saint of swineherds.
- In Haitian Vodou, Ezili Dantor, the lwa of motherhood, is associated with the black Creole Pig of Haiti, her favourite animal sacrifice.

Environmental Impacts

Domestic pigs that have escaped from farms or were allowed to forage in the wild, and in some cases wild boars which were introduced as prey for hunting, have given rise to large populations of feral pigs in North and South America, Australia, New Zealand, Hawaii, and other areas where pigs are not native. Accidental or deliberate releases of pigs into countries or environments where they are an alien species have caused extensive environmental change. Their omnivorous diet, aggressive behaviour, and their feeding method of rooting in the ground all combine to severely alter ecosystems unused to pigs. Pigs will even eat small animals

Fig. 4.5: Feral Pigs (Razorbacks) in Florida

and destroy nests of ground nesting birds. The Invasive Species Specialist Group lists feral pigs on the list of the world's 100 worst invasive species and says:

> "Feral pigs like other introduced mammals are major drivers of extinction and ecosystem change. They have been introduced into many parts of the world, and will damage crops and home gardens as well as potentially spreading disease. They uproot large areas of land, eliminating native vegetation and spreading weeds. This results in habitat alteration, a change in plant succession and composition and a decrease in native fauna dependent on the original habitat."

Health Issues

Pigs harbour a range of parasites and diseases that can be transmitted to humans. These include trichinosis, *Taenia solium*, cysticercosis, and brucellosis. Pigs are also known to host large concentrations of parasitic ascarid worms in their digestive tract. The presence of these diseases and parasites is one of the reasons why pork meat should always be well cooked or cured before eating. Some religious groups that consider pork unclean refer to these issues as support for their views.

Pigs are susceptible to bronchitis and pneumonia. They have small lungs in relation to body size; for this reason, bronchitis or pneumonia can kill a pig quickly. There is concern that pigs may allow animal viruses such as influenza or Ebola Reston to infect humans more easily. Some strains of influenza are endemic in pigs, and pigs also can acquire human influenza.

Pigs can be aggressive and pig-induced injuries are relatively common in areas where pigs are reared or where they form part of the wild or feral fauna.

ANIMAL SLAUGHTER

Slaughter is the term used to describe the killing and butchering of animals, usually for food. Commonly it refers to killing and butchering of domestic livestock (*tame animals*).

The animals most commonly slaughtered for food are cattle (for beef and veal), sheep (for lamb and mutton), pigs (for pork), horses (for horsemeat), and fowl, largely chickens, turkeys, and ducks.

Traditions and Laws

Laws for Ritual Slaughter

Ritual slaughter of animals is also used for food production. Such slaughter is governed by various laws, most notably:

- Shechita—Jewish laws of slaughtering animals
- Shariah—Islamic law of slaughtering animals
- Jhatka—for those Sikh and Hindus who consume meat

United States

In the United States, the United States Department of Agriculture (USDA) specifies the approved methods of livestock slaughter:

Chemical (Carbon Dioxide)

This method is approved for sheep, calves and swine. The animal is asphyxiated by the use of carbon dioxide gas.

Mechanical (Captive Bolt)

This method is approved for sheep, swine, goats, calves, cattle, horses, mules, and other equines. A captive bolt stunner is applied to the livestock so as to produce immediate unconsciousness in the animals before they are butchered.

Mechanical (Gunshot)

This method is approved for cattle, calves, sheep, swine, goats, horses, mules, and other equines. The gun is used to render the animal immediately unconscious (and presumably dead) before butchering.

Electrical (Stunning or Slaughtering with Electric Current)

This method is approved for swine, sheep, calves, cattle, and goats. The current applied is sufficient to ensure surgical anesthesia throughout the "bleeding" of the animal.

Each of these methods is outlined in detail, and the regulations require that inspectors identify operations which cause undue "excitement and discomfort" of animals.

Operation

- Slaughterhouse
- Ike jime, a Japanese method of slaughtering fish
- Pig slaughter
- Horse slaughter

Controversy

Ethical vegetarians, vegans, and animal rights groups strongly oppose killing animals for food.

MEAT

Meat is animal flesh that is used as food. Most often, this means the skeletal muscle and associated fat, but it may also describe other edible tissues such as organs, livers, skin, brains, bone marrow, kidneys, or lungs. The word *meat* is also used by the meat packing industry in a more restrictive sense—the flesh of mammalian species (pigs, cattle, etc.) raised and prepared for human consumption, to the exclusion of fish and poultry.

Etymology

The word *meat* comes from the Old English word *mete*, which referred to food in general. The term is related to *mad* in Danish, *mat* in Swedish and Norwegian, and *matur* in Icelandic, which also mean 'food'.

One definition that refers to meat as not including fish developed over the past few hundred years and has religious influences. The distinction between fish and "meat" is codified by the Jewish dietary law of *kashrut*, regarding the mixing of milk and meat, which does not forbid the mixing of milk and fish. Modern Jewish legal practice (*halakha*) on *kashrut* classifies the flesh of both mammals and birds as "meat"; fish are considered to be *parve*, neither meat nor a dairy food. The Catholic dietary restriction on "meat" on Fridays also does not apply to the cooking and eating of fish.

The Latin word *caro* "meat" (also the root of 'carnal', referring to the 'pleasures of the flesh') is often a euphemism for sexual pleasure, effected from the function performed by fleshy organs. Thus 'meat' may refer to the human body in a sensual, or sexual, connotation. A *meat market*, in addition to simply denoting a market where meat is sold, also refers to a place or situation where humans are treated or viewed as commodities, especially a place known as one where a sexual partner may be found.

"Meat" may also be used to refer to humans humorously or indifferently. In military slang, "meat shield" refers to soldiers sent towards an enemy to draw fire away from another unit.

History

Meat constituted a substantial proportion of even the earliest humans' diet, paleontological evidence suggests. Early hunter-gatherers depended on the organized hunting of large animals such as bison and deer.

The domestication of animals, of which we have evidence dating back to the end of the last glacial period (c. 10,000 years BP), allowed the systematic production of meat and the breeding of animals with a view to improving meat production. The animals which are now the principal sources of meat were domesticated in conjunction with the development of early civilizations:

- Sheep, originating from western Asia, were domesticated with the help of dogs prior to the establishment of settled agriculture, likely as early as the eighth millennium BC. Several breeds of sheep were established in ancient Mesopotamia and Egypt by 3500-3000 BC. Presently, more than 200 sheep breeds exist.
- Cattle were domesticated in Mesopotamia after settled agriculture was established about 5000 BC, and several breeds were established by 2500 BC. Modern domesticated cattle fall into the groups *Bos taurus* (European cattle) and *Bos indicus* (zebu), both descended from the now-extinct Aurochs. The breeding of beef cattle, cattle optimized for meat production as opposed to animals best suited for draught or dairy purposes, began in the middle of the 18th century.
- Domestic pigs, which are descended from wild boars, are known to have existed about 2500 BC in modern-day Hungary and in Troy; earlier pottery from Jericho and Egypt depicts wild pigs. Pork sausages and hams were of great commercial importance in Greco-Roman times. Pigs continue to be bred intensively as they are being optimized to produce meat best suited for specific meat products.

Modern agriculture employs a number of techniques, such as progeny testing, to make animals evolve rapidly towards having the qualities desired by meat producers. For instance, in the wake of well-publicised health concerns associated with saturated fats in the 1980s, the fat content of UK beef, pork and lamb fell from 20-26 percent to 4-8 percent within a few decades, both due to selective breeding for leanness and changed methods of butchery. Methods of genetic engineering aimed at improving the meat production qualities of animals are now also becoming available.

Even though it is a very old industry, meat production continues to be shaped strongly by the rapidly evolving demands of customers. The trend towards selling meat in pre-packaged cuts has increased the demand for larger breeds of cattle, which are better suited to producing such cuts. Ever more animals not previously exploited for their meat are now being farmed, especially the more agile and mobile species, whose muscles tend to be developed better than those of cattle, sheep or pigs. Examples include the various antelope species, the zebra, water buffalo and camel, as well as non-mammals such as the crocodile, emu and ostrich. Another important trend in contemporary meat production is organic farming which, while providing no organoleptic benefit to meat so produced, meets an increasing demand for numerous reasons.

Growth and Development of Meat Animals

Agricultural science has identified several factors bearing on the growth and development of meat in animals.

Genetics

Several economically important traits in meat animals are heritable to some degree and can thus be selected for by breeding. In cattle, certain growth features are controlled by recessive genes which have not so far been controlled, complicating breeding. One such trait is dwarfism; another is the doppelender or "double muscling" condition, which causes muscle hypertrophy and thereby increases the animal's commercial value. Genetic analysis continues to reveal the genetic mechanisms that control numerous aspects of the endocrine system and, through it, meat growth and quality.

Trait	*Heritability*
Reproductive efficiency	2-10 per cent
Meat quality	15-30 per cent
Growth	20-40 per cent
Muscle/fat ratio	40-60 per cent

Genetic engineering techniques can shorten breeding programmes significantly because they allow for the identification and isolation of genes coding for desired traits, and for the reincorporation of these genes into the animal genome. To enable such manipulation, research is ongoing (as of 2006) to map the entire genome of sheep, cattle and pigs. Some research has already seen commercial application. For instance, a recombinant bacterium has been developed which improves the digestion of grass in the rumen of cattle, and some specific features of muscle fibres have been genetically altered.

Experimental reproductive cloning of commercially important meat animals such as sheep, pig or cattle has been successful. The multiple asexual reproduction of animals bearing desirable traits can thus be anticipated, although this is not yet practical on a commercial scale.

Environment

Heat regulation in livestock is of great economic significance, because mammals attempt to maintain a constant optimal body temperature. Low temperatures tend to prolong animal development and high temperatures tend to retard it. Depending on their size, body shape and insulation through tissue and fur, some animals (*e.g.* pigs) have a relatively narrow zone of temperature tolerance and others (*e.g.* cattle) a broad one. Static magnetic fields, for reasons still unknown, also retard animal development.

Nutrition

The quality and quantity of usable meat depends on the animal's *plane of nutrition*, *i.e.*, whether it is over-or underfed. Scientists disagree, however, about how exactly the plane of nutrition influences carcass composition.

The composition of the diet, especially the amount of protein provided, is also an important factor regulating animal growth. Ruminants, who may digest cellulose, are better adapted to poor-quality diets, but their ruminal microorganisms degrade high-quality protein if supplied in excess. Because producing high-quality protein animal feed is expensive, several techniques are employed or experimented with to ensure maximum utilization of protein. These include the treatment of feed with formalin to protect amino acids during their passage through the rumen, the recycling of manure by feeding it back to cattle mixed with feed concentrates, or the partial conversion of petroleum hydrocarbons to protein through microbial action.

In plant feed, environmental factors influence the availability of crucial nutrients or micronutrients, a lack or excess of which can cause a great many ailments. In Australia, for instance, where the soil contains limited phosphate, cattle are being fed additional phosphate to increase the efficiency of beef production. Also in Australia, cattle and sheep in certain areas were often found losing their appetite and dying in the midst of rich pasture; this was at length found to be a result of cobalt deficiency in the soil. Plant toxins are also a risk to grazing animals; for instance, fluoracetate, found in some African and Australian plants, kills by disrupting the cellular metabolism. Certain man-made pollutants such as methylmercury and some pesticide residues present a particular hazard due to their tendency to bioaccumulate in meat, potentially poisoning consumers.

Human Intervention

Meat producers may seek to improve the fertility of female animals through the administration of gonadotrophic or ovulation-inducing hormones. In pig production, sow infertility is a common problem, possibly due to excessive fatness. No methods currently exist to augment the fertility of male animals. Artificial insemination is now routinely used to produce animals of the best possible genetic quality, and the efficiency of this method is improved through the administration of hormones that synchronize the ovulation cycles within groups of females.

Growth hormones, particularly anabolic agents such as steroids, are used in some countries to accelerate muscle growth in animals. This practice has given rise to the beef hormone controversy, an international trade dispute. It may also decrease the tenderness of meat, although research on this is inconclusive, and have other effects on the composition of the muscle flesh. Where castration is used to improve control over male animals, its side effects are also counteracted by the administration of hormones.

Sedatives may be administered to animals to counteract stress factors and increase weight gain. The feeding of antibiotics to certain animals has been shown to improve growth rates also. This practice is particularly prevalent in the USA, but has been banned in the EU, partly because it causes antibiotic resistance in pathogenic microorganisms.

Biochemical Composition

Numerous aspects of the biochemical composition of meat vary in complex ways depending on the species, breed, sex, age, plane of nutrition, training and exercise of the animal, as well as on the anatomical location of the musculature involved. Even between animals of the same litter and sex there are considerable differences in such parameters as the percentage of intramuscular fat.

Main Constituents

Adult mammalian muscle flesh consists of roughly 75 percent of water, 19 percent of protein, 2.5 percent of intramuscular fat, 1.2 percent of carbohydrates and 2.3 percent of other soluble non-protein substances. These include nitrogenous compounds, such as amino acids, and inorganic substances such as minerals.

Muscle proteins are either soluble in water (sarcoplasmic proteins, about 11.5 percent of total muscle mass) or in concentrated salt solutions (myofibrillar proteins, about 5.5 percent of mass). There are several hundred sarcoplasmic proteins. Most of them—the glycolytic enzymes—are involved in the glycolytic pathway, *i.e.*, the conversion of stored energy into muscle power. The two most abundant myofibrillar proteins, myosin and actin, are responsible for the muscle's overall structure. The remaining protein mass consists of connective tissue (collagen and elastin) as well as organelle tissue.

Fat in meat can be either adipose tissue, used by the animal to store energy and consisting of "true fats" (esters of glycerol with fatty acids), or intramuscular fat, which contains considerable quantities of phospholipids and of unsaponifiable constituents such as cholesterol.

Red and White Meat

Meat can be broadly classified as "red" or "white" depending on the concentration of myoglobin in muscle fiber. When myoglobin is exposed to oxygen, reddish oxymyoglobin develops, making myoglobin-rich meat appear red. The redness of meat depends on species, animal age, and fiber type: Red meat contains more narrow muscle fibers that tend to operate over long periods without rest, while white meat contains more broad fibers that tend to work in short fast bursts.

The meat of adult mammals such as cows, sheep, goats, and horses is generally considered red, while domestic chicken and turkey breast meat is generally considered white.

Production

Meat is produced by killing the animal in question and cutting the desired flesh out of it. These procedures are called slaughter and butchery, respectively.

Attesting to the long history of meat consumption in human civilizations, ritual slaughter has become part of the practice of several religions. These rituals, as well as other pre-industrial meat production methods such as these used by indigenous peoples, are not detailed here. This section will instead provide an overview of contemporary industrialized meat production in dedicated slaughterhouses from cattle, sheep and pigs.

Transport

Upon reaching a predetermined age or weight, livestock are transported *en masse* from the farm to the slaughterhouse, a process called “live export”. Depending on its length and circumstances, this exerts stress and injuries on the animals, and some may die *en route*. Apart from being arguably inhumane, unnecessary stress in transport may adversely affect the quality of the meat. In particular, the muscles of stressed animals are low in water and glycogen, and their pH fails to attain acidic values, all of which results in poor meat quality. Consequently, and also due to campaigning by animal welfare groups, laws and industry practices in several countries tend to become more restrictive with respect to the duration and other circumstances of livestock transports.

Slaughter

Animals are slaughtered by being first stunned and then exsanguinated (bled out). Death results from the one or the other procedure, depending on the methods employed. Stunning can be effected through asphyxiating the animals with carbon dioxide, shooting them with a gun or a captive bolt pistol, or shocking them with electric current. In most forms of ritual slaughter, stunning is not allowed.

Draining as much blood as possible from the carcass is necessary because blood causes the meat to have an unappealing appearance and is a very good breeding ground for microorganisms. The exsanguination is accomplished by severing the carotid artery and the jugular vein in cattle and sheep, and the anterior vena cava in pigs.

Dressing and Cutting

After exsanguination, the carcass is dressed, that is, the head, feet, hide (except hogs and some veal), excess fat, viscera and offal are removed, leaving only bones and edible muscle. Cattle and pig carcasses, but not those of sheep, are then split in half along the mid ventral axis, and the carcass is cut into wholesale pieces. The dressing and cutting sequence, long a province of manual labour, is progressively being fully automated.

Conditioning

Under hygienic conditions and without other treatment, meat can be stored at above its freezing point (–1.5°C) for about six weeks without spoilage, during which time it undergoes an aging process that increases its tenderness and flavor.

During the first day after death, glycolysis continues until the accumulation of lactic acid causes the pH to reach about 5.5. The remaining glycogen, about 18 g per kg, is believed to increase the water-holding capacity and tenderness of the flesh when cooked. *Rigor mortis* sets in a few hours after death as ATP is used up, causing actin and myosin to combine into rigid actomyosin and lowering the meat's water-holding capacity, causing it to lose water ("weep"). In muscles that enter *rigor* in a contracted position, actin and myosin filaments overlap and cross-bond, resulting in meat that is tough on cooking—hence again the need to prevent pre-slaughter stress in the animal.

Over time, the muscle proteins denature in varying degree, with the exception of the collagen and elastin of connective tissue, and *rigor mortis* resolves. Because of these changes, the meat is tender and pliable when cooked just after death or after the resolution of *rigor*, but tough when cooked during *rigor.* As the muscle pigment myoglobin denatures, its iron oxidates, which may cause a brown discolouration near the surface of the meat. Ongoing proteolysis also contributes to conditioning. Hypoxanthine, a breakdown product of ATP, contributes to the meat's flavor and odor, as do other products of the discomposition of muscle fat and protein.

Spoilage and Preservation

The spoilage of meat occurs, if untreated, in a matter of hours or days and results in the meat becoming unappetizing, poisonous or infectious. Spoilage is caused by the practically unavoidable infection and subsequent decomposition of meat by bacteria and fungi, which are borne by the animal itself, by the people handling the meat, and by their implements. Meat can be kept edible for a much longer time—though not indefinitely—if proper hygiene is observed during production and processing, and if appropriate food safety, food preservation and food storage procedures are applied.

Methods of Preparation

Meat is prepared in many ways, as steaks, in stews, fondue, or as dried meat like beef jerky. It may be ground then formed into patties (as hamburgers or croquettes), loaves, or sausages, or used in loose form (as in "sloppy joe" or Bolognese sauce). Some meat is cured, by smoking, pickling, preserving in salt or brine. Other kinds of meat are marinated and barbecued, or simply boiled, roasted, or fried. Meat is generally eaten cooked, but there are many traditional recipes that call for raw beef, veal or fish (tartare). Meat is often spiced or seasoned, as in most sausages. Meat dishes are usually described by their source (animal and part of body) and method of preparation.

Meat is a typical base for making sandwiches. Popular varieties of sandwich meat include ham, pork, salami and other sausages, and beef, such as steak, roast beef, corned beef, and pastrami. Meat can also be molded or pressed (common for products that include offal, such as haggis and scrapple) and canned.

Nutritional Benefits and Concerns

All muscle tissue is very high in protein, containing all of the essential amino acids, and in most cases is a good source of zinc, vitamin B12, selenium, phosphorus, niacin, vitamin B6, iron and riboflavin. Several forms of meat are high in vitamin K_2, which is only otherwise known to be found in fermented foods, with natto having the highest concentration. Muscle tissue is very low in carbohydrates and does not contain dietary fiber. The fat content of meat can vary widely depending on the species and breed of animal, the way in which the animal was raised, including what it was fed, the anatomical part of the body, and the methods of butchering and cooking. Wild animals such as deer are typically leaner than farm animals, leading those concerned about fat content to choose game such as venison. Decades of breeding meat animals for fatness is being reversed by consumer demand for meat with less fat.

Red meat, such as beef, pork, and lamb, contains many essential nutrients necessary for healthy growth and development in children. Nutrients in red meat include iron, zinc, vitamin B12, and protein. Most meats contain a full complement of the amino acids required for the human diet. Fruits and vegetables, by contrast, are usually lacking several essential amino acids contained in meat. It is for this reason that people who abstain from eating all meat need to plan their diet carefully to include sources of all the necessary amino acids.

The table in this section compares the nutritional content of several types of meat. While each kind of meat has about the same content of protein and carbohydrates, there is a very wide range of fat content. It is the additional fat that contributes most to the calorie content of meat, and to concerns about dietary health.

In recent years, health concerns have been raised about the consumption of meat. In a large-scale study, the consumption of red meat over a lifetime was found to raise the risk of cancer by 20 to 60 percent, while causing adverse mutations in DNA. In particular, red meat and processed meat were found to be associated with higher risk of cancers of the lung, esophagus, liver, and colon, among others. Another study found an increase risk of pancreatic cancer for red meat and pork. That study also suggests that fat and saturated fat are not likely contributors to pancreatic cancer. Animal fat, particularly from ruminants, tends to have a higher percentage of saturated fat vs. monounsaturated and polyunsaturated fat when compared to vegetable fats, with the exception of some tropical plant fats; consumption of which has been correlated with various health problems. The saturated fat found in meat has been associated with significantly raised risks of colon cancer, osteoporosis, and prostate cancer, although evidence suggests that risks of prostate cancer are unrelated to animal fat consumption.

Meat has been correlated to increased risk of heart disease. The risks of heart disease are three times greater for 45-64 year old men who eat meat daily, versus those who are vegetarian, according to one survey.

A 2009 study by the National Cancer Institute revealed a correlation between the consumption of red meat and increased mortality from cancer and cardiovascular diseases. This study has been criticized for using an improperly validated food frequency questionnaire , which has been shown to have low levels of accuracy

In response to changing prices as well as health concerns about saturated fat and cholesterol, consumers have altered their consumption of various meats. A USDA report points out that consumption of beef in the United States between 1970-1974 and 1990-1994 dropped by 21 per cent, while consumption of chicken increased by 90 per cent. During the same period of time, the price of chicken dropped by 14 per cent relative to the price of beef. In 1995 and 1996, beef consumption increased due to higher supplies and lower prices.

Cooking Meat

Meat, like any food, can also transmit certain diseases, but complete cooking and avoiding recontamination reduces this possibility.

Several studies published since 1990 indicate that cooking muscle meat creates heterocyclic amines (HCAs), which are thought to increase cancer risk in humans. Researchers at the National Cancer Institute published results of a study which found that human subjects who ate beef rare or medium-rare had less than one third the risk of stomach cancer than those who ate beef medium-well or well-done. While eating muscle meat raw may be the only way to avoid HCAs fully, the National Cancer Institute states that cooking meat below 212 °F (100 °C) creates "negligible amounts" of HCAs. Also, microwaving meat before cooking may reduce HCAs by 90 per cent.

Nitrosamines, present in processed and cooked foods, have been noted as being carcinogenic, being linked to colon cancer. Also, toxic compounds called PAHs, or Polycyclic aromatic hydrocarbons, present in processed, smoked and cooked foods, are known to be carcinogenic.

Ethics of Eating Meat

Ethical issues regarding the consumption of meat can include objections to the act of killing animals or the agricultural practices surrounding the production of meat. Reasons for objecting to the practice of killing animals for consumption may include animal rights, environmental ethics, religious doctrine, or an aversion to inflicting pain or harm on other living creatures. The religion of Jainism has always opposed eating meat, and there are also many schools of Buddhism and Hinduism that condemn the eating of meat. Some people, while not vegetarians, refuse to eat the flesh of certain animals, such as cats, dogs, horses, or rabbits, due to cultural or religious taboo. In some cases, specific meats (especially from pigs and cows) are forbidden within religious traditions. Some people eat only the flesh of animals which they believe have not been mistreated, and abstain from the meat of animals reared in factory farms or from particular products such as foie gras and veal.

In Vitro and Imitation Meat

Various forms of imitation meat have been created to satisfy people wishing to reduce or eliminate their meat consumption and still taste the flavor and texture of meat. They are typically some form of processed soybean, (tofu, tempeh), but they can also be based on wheat gluten or even fungus (quorn). There is also some speculation about the possibility of growing cultured meat from animal tissue.

Environmental Impact

The UN Food and Agriculture Organization (FAO) has estimated that direct emissions from meat production account for about 18 per cent of the world's total greenhouse gas emissions. The FAO figure accounts for the entire meat production cycle—clearing forested land, making and transporting fertiliser, burning fossil fuels in agricultural machinery, and the front and rear end emissions of cattle and sheep.

RED MEAT

Red meat in culinary terminology refers to meat which is red-coloured when raw. Red meat includes the meat of many—but not all—mammals.

Gastronomy

In gastronomy, red meat is darker-coloured meat, as contrasted with white meat. The exact definition varies by time, place, and culture, but the meat of adult mammals such as cows, sheep, and horses is invariably considered red, while chicken and rabbit are invariably considered white. The meat of young mammals such as milk-fed veal calves and sheep, and pigs is traditionally considered white; while the meat of duck and goose is considered red, though the demarcation line has been shifting. Game is sometimes put in a separate category altogether (French *viandes noires* 'black meats').

Cooking

Red meat does not refer to how well a piece of meat is cooked or its colouration after cooking. A steak or hamburger is red meat whether it is served rare, or cooked until it is well-done; pork is also red, though it turns to a whitish colour when cooked. According to the USDA all meats obtained from livestock are red meats because they contain more myoglobin than chicken or fish.

Myoglobin Concentration

The main determinant of the colour of meat is the concentration of myoglobin. The white meat of chicken has under 0.05 per cent;

chicken thigh has 0.18-0.20 per cent; pork and veal have 0.1-0.3 per cent; young beef has 0.4-1.0 per cent; and old beef has 1.5-2.0 per cent.

In the health discussion below, we assume the nutritional, not the traditional gastronomic, definitions.

Nutrition

Red meat is a source of iron. Red meat also contains protein, levels of creatine, minerals such as zinc and phosphorus, and vitamins such as niacin, vitamin B_{12}, thiamin and riboflavin. Red meat is the richest source of Alpha Lipoic Acid, a powerful antioxidant.

Saturated Fat

Red meat is the largest contributor of saturated fat in the Western diet.

Food Pyramid

The traditional food guide pyramid has been criticized for not distinguishing between red meat and other types of meat. The newer My Pyramid recommends lean forms of red meat.

Health Risks

While red meat is a good source of complete protein and iron, studies have found correlations to several health risks.

Cancer

Recent studies show that red meat could pose an increase in cancer risk. Some studies have linked consumption of large amounts of red meat with breast cancer, colorectal cancer, stomach cancer, lymphoma, bladder cancer and prostate cancer. Furthermore, there is convincing evidence that consumption of beef, pork, lamb, and goat from domesticated animals is a cause of colorectal cancer. Professor Sheila Bingham of the Dunn Human Nutrition Unit attributes this to the haemoglobin and myoglobin molecules which are found in red meat. She suggests these molecules, when ingested trigger a process called nitrosation in the gut which leads to the formation of carcinogens.

Eating cooked red meat may increase the likelihood of cancer because carcinogenic compounds called heterocyclic amines are created during the cooking process. Heterocyclic amines may not explain why red meat is more harmful than other meat, however, as these compounds are also found in poultry and fish, which have not been linked to an increased cancer risk.

A 2009 study by the National Cancer Institute revealed a correlation between the consumption of red meat and increased mortality from cancer and cardiovascular diseases. This study has been criticized for using an improperly validated food frequency questionnaire , which has been shown to have low levels of accuracy

Cardiovascular Diseases

Red meat consumption is associated with cardiovascular diseases, possibly because of its high content of saturated fat.

A 1999 study funded by the National Cattlemen's Beef Association, an advocacy group for beef producers, involved 191 persons with high cholesterol on diets where at least 80 per cent of the meat intake came from either lean red meat in one group, or lean white meat in another. The results of this study showed nearly identical cholesterol, and triglyceride levels in both groups. This study suggests that lean red meat may play a role in a low-fat diet for persons with high cholesterol.

Other Health Issues

Regular consumption of red meat has also been linked to hypertension and arthritis.

Culture

In India, people who follow Hinduism do not eat red meat sourced from bovines. However, red meat from lambs and sheep is acceptable to Hindus.

In some cultures eating red meat is considered a masculine activity, possibly due to traditions of hunting big game as a male rite of passage.

APETITO

Apetito is a European frozen food company which started in post-war Germany. They currently serve markets in Germany, the Netherlands, the United Kingdom, Canada and France and employ over 6,000 people. They are the main providers of "meals on wheels" for local authorities in the UK, and were given a Queen's Award for Enterprise in 2005.

History

The company was founded by Karl Düsterberg in 1958 in Rheine, Germany as "Apetito Ready Meals" producing frozen ready meals for retail and business canteens, schools, nurseries and the German Meals on Wheels service. The company started expanding out of Germany in 1989 when "Apetito B.V." was founded in the Netherlands. In 1991 the company set up "Apetito (UK) Limited" in Sevenoaks, UK, and by 2001 had acquired "Nord Restauration S.H." and "Orlysienne du Nord" in France. Business outside of Germany now accounts for approx 25 per cent of Apetito's total turnover, though the company's headquarters are still in Rheine, Germany.

Meals on Wheels

The meals on wheels schemes organised by local councils in UK deliver 40 million meals every year to 200,000 people. Of these, Apetito produces 900,000 meals and puddings a week.

Wiltshire Farm Foods

Apetito operate a private frozen meal delivery service—Wiltshire Farm Foods. The company, originally named "The Waldens Wiltshire Foods Limited" and based in Trowbridge, was purchased by Apetito in 1995. Wiltshire Farm Foods specialise in delivering frozen ready meals to private customers throughout the United Kingdom.

The company offers over 200 different ready to reheat frozen meals, including vegetarian, gluten-free, and diabetic. The service is intended mainly for elderly or disabled people, but can be used by anybody who would prefer to have meals delivered direct to their home.

Advertising

Apetito's first TV marketing campaign started on 6 November 2006, showing commercials on every private television station in Germany.

Awards and Nominations

The company were given a Queen's Award for Enterprise in 2005 for outstanding achievements in innovation in regard to provision of food to the elderly.

They were nominated for two awards in 2008, one for corporate responsibility from ICAEW, the other for business innovation from IMechE.

CORNED BEEF

The "corn" in corned beef refers to the "corns" or grains of coarse salts used to cure it. The *Oxford English Dictionary* dates the usage of "corn", meaning "small hard particle, a grain, as of sand or salt." The term "corned beef" can denote different styles of brine-cured beef, depending on the region.

Canada and the United States

In the U.S. and Canada, corned beef has two meanings. One refers to a cut of beef (usually brisket, but sometimes round or silverside) cured or pickled in a seasoned brine. The other use of the term refers to a tinned product generally found with canned goods on supermarket shelves

Cooked Corned Beef

In the United States, corned beef is often purchased precooked, as in delicatessens. Also famous is the Reuben sandwich, consisting of corned beef, Swiss cheese, sauerkraut, and Thousand Island or

Russian dressing on rye bread which is then grilled on a flat griddle or in a cast iron pan in oil.

Corned beef hash is commonly served as a breakfast food with eggs and hash browns.

Smoking corned beef, usually with the addition of extra spices such as black pepper, produces a cold cut known as pastrami.

Saint Patrick's Day

The consumption of corned beef is associated with Saint Patrick's Day, when many Irish Americans eat a traditional meal of corned beef and cabbage. According to the History Channel, corned beef was originally used as a substitute for bacon by Irish American immigrants in the late 1800s. Irish immigrants living in New York City's Lower East Side sought an equivalent in taste and texture to their traditional Irish bacon, and learned about this cheaper alternative to bacon from their Jewish neighbors. A similar dish is the New England boiled dinner, consisting of corned beef, cabbage, and root vegetables such as carrots, turnips, and potatoes, which is popular in New England and parts of Atlantic Canada.

As St. Patrick's Day occurs annually during Lent, the corned beef tradition caused controversy among American Catholic dioceses in 2000 and 2006, when the holiday fell on a Friday. Catholic custom dictates that no meat be consumed on any Friday during Lent, but some bishops granted dispensations to their dioceses for eating corned beef on St. Patrick's Day. This rare occurrence will next happen on Friday in 2017.

Germany

A traditionally shaped tin of corned beef as sold in North Western/ Western Europe.

Two canned versions of commercial corned beef are sold in Germany. The original is usually called *American Corned Beef* and consists of finely shredded corned beef with a low fat content and is similar to Spam. Another version is called *Deutsches Corned Beef* and is closer to the product described above. It is not as finely shredded as Spam, it contains chunks of corned beef and is usually embedded in aspic. *Deutsches Corned Beef* is also sold in slices at supermarket meat counters and butcher shops.

UK

The product commonly known as corned beef in the UK (also known as bully beef; from the French *bouilli* 'boiled') is sold in distinctive oblong-shaped tin cans. The corned beef itself is finely ground—boiled meat in a small amount of gelatin. A typical ingredient list is: beef, salt, sugar, sodium nitrite. A 100 gram portion contains 12.5 grams of fat and 2.3 grams of salt. Like those cans used for sardines, corned beef cans are specially scored so that the metal can be broken apart by the leverage of winding it around a slotted church key that is provided with the can. This product is also sold in the US and in Spain, the latter is commonly supplied from Argentina.

A common way of eating corned beef cold in the UK is sliced in a sandwich. Other things cooked with corned beef are "corned beef hash" or hotpot, in which potatoes and corned beef are stewed and sometimes served with bread-rolls. Some sauces which are added are brown sauce or tomato ketchup. In the North East of England, corned beef is more commonly eaten in a sandwich with pickled beetroot or on its own, as opposed to the above combinations. Also it can be eaten hot, as a toastie with or without onions. It can also be served with carrots, potatoes, onions and turnips in a broth-like soup. There is also an un-tinned product known as corned beef in the US, which is sold as salt beef in the UK; it is mostly available in major cities which have Jewish communities.

New Zealand and Australia

Australian and New Zealand butchers and supermarkets usually stock corned beef as "corned silverside", which is usually poached in a large pot or a slow cooker, usually with cloves, vinegar, sugar, bay leaves and onions. Traditionally, this will be served with white sauce or a form of cooked mustard sauce.

Pacific Islands

Tinned corned beef is a delicacy in Pacific Island countries and is usually brought out for special occasions such as Christmas Day. Tins are often used as gifts at weddings, funerals and feasts. It is also common to be eaten with rice.

Further round the Pacific Ocean in South America (particularly Argentina) corned beef is a staple of most people's diets. In a recent survey carried out on behalf of the Argentinian government to find out more about their citizens' eating habits it has emerged that up to 88 per cent of the population will eat corned beef twice a week.

MEAT AND BONE MEAL

Meat and bone meal (MBM) is a product of the rendering industry. It is typically about 50 per cent protein, 35 per cent ash, 8-12 per cent fat, and 4-7 per cent moisture. It is primarily used in the formulation of animal feed to improve the amino acid profile of the feed. Feeding of MBM to cattle is thought to have been responsible for the spread of BSE (mad cow disease). In most parts of the world, MBM is no longer allowed in feed for ruminant animals. However, in some areas, including the US, MBM is still used to feed monogastric animals. It is widely used in the United States as a low-cost meat in dog food and cat food.

In Europe, some MBM is used as ingredients in petfood but the vast majority is now used as a fossil-fuel replacement for renewable energy generation, as a fuel in cement kilns, landfilling or incineration.

Meat and bone meal has around two thirds the energy value of fossil fuels such as coal; the UK in particular widely uses meat and bone meal for the generation of renewable electricity. This was particularly prominent after many cattle had to be slaughtered during the BSE crisis.

Meat and bone meal is increasingly used in cement kilns as an environmentally sustainable replacement for coal.

MEAT PACKING INDUSTRY

The meat packing industry is an industry that handles the slaughtering, processing and distribution of animals such as cattle, pigs, sheep and other livestock.

The industry is primarily focused on producing meat for human consumption, but it also yields a variety of by-products including hides, feathers, dried blood, and, through the process of rendering, fat such as tallow and protein meals such as meat and bone meal.

In the U.S. and some other countries, the facility where the meat packing is done is called a *meat packing plant*; in New Zealand, where most of the products are exported, it is called a *freezing works*. An abattoir is a place where animals are slaughtered for food.

History in the United States

The meat packing industry grew with the construction of the railroads and methods of refrigeration. Railroads made possible the transport of stock to central points for processing, and the transport of products throughout the nation.

The publication of *The Jungle* in the US in 1906, shocked the public with the poor working conditions and unsanitary practices in meat packing plants in the United States, specifically Chicago. Meat packing plants, like many industries in the early 20th century, were known to overwork their employees, failed to maintain adequate safety measures, and actively fought unionization. In the early part of the century, they used the most recent immigrants and migrants as strikebreakers in labour actions taken by other workers, also usually immigrants or early descendants.

In the 1930s and early 1940s, however, workers achieved unionization under the CIO's United Packinghouse Workers of America (UPWA). An interracial committee led the organizing in Chicago, where the majority of workers in the industry were black, and other major cities, such as Omaha, Nebraska, where they were an important minority in the industry. UPWA workers made important gains in wages, hours and benefits. In 1957 the stockyards and meat packing employed half the workers of Omaha. The union supported a progressive agenda, including the Civil Rights Movement of the 1960s. While the work was still difficult, for a few decades workers achieved blue-collar, middle-class lives from it.

Mid-century restructuring by the industry of the stockyards, slaughterhouses and meat packing led to relocating facilities closer to cattle feedlots and swine production facilities, to more rural areas, as transportation shifted from rail to truck. It has been difficult for labour to organize in such locations. In addition, the number of jobs fell sharply through technology and other changes. Wages fell during the latter part of the 20th century, and in the 1990s, both Chicago and Omaha closed their stockyards for good.

Though the meat packing industry has made many improvements since the early 1900s, extensive changes in the industry since the late 20th century have caused new labour issues to arise. Today, the rate of injury in the meat packing industry is three times that of private industry overall, and meat packing was noted by Human Rights Watch as being "the most dangerous factory job in America". The meatpacking industry continues to employ many immigrant labourers, including some who are undocumented workers. In the early 20th century the workers were immigrants from eastern and southern Europe, and black migrants from the South. Today many are Hispanic, from Mexico, Central and South America. The more isolated areas in which the plants are located put workers at greater risk due to their limited ability to organize and to seek redress for work-related injuries.

Meatpackers

These are major historical and current meat packers in the United States.

- Armour and Company
- Cargill Meat Solutions
- Cudahy Packing Company
- Hormel Foods
- Smithfield Foods
- Tyson Foods
- Perdue Farms

These are some outside the United States.

- William Davies Company (Canada)
- Canada Packers
- Maple Leaf Foods (Canada)
- Schneider Foods (Canada)
- JBS S.A. (Brazil)

POULTRY

Poultry is the category of domesticated birds that people keep for the purpose of collecting their eggs, or killing for their meat and/or feathers. These most typically are members of the superorder Galloanserae (fowl), especially the order Galliformes (which includes chickens, quails and turkeys) and the family Anatidae (in order Anseriformes), commonly known as "waterfowl" (*e.g.* domestic ducks and domestic geese). Poultry also include other birds which are killed for their meat, such as pigeons or doves or birds considered to be game, like pheasants. The term also refers to the flesh of such birds.

Examples

Bird	*Domestication*	*Killed or Used for*
Chicken	Southeast Asia	meat, feathers, eggs, ornamentation
Duck	various	meat, feathers, eggs
Emu	various, 20th century	meat, leather, oil
Goose	various	meat, feathers, eggs
Indian peafowl	various	meat, feathers, ornamentation, landscaping
Mute swan	various	feathers, eggs, landscaping
Ostrich	various, 20th century	meat, feathers, leather
Turkey	Mexico	meat, feathers
Domesticated guineafowl	Africa	meat, pest consumption, and alarm calling
A Common pheasant	Eurasia	meat
Golden pheasant	Eurasia	meat, mainly ornamental
Rhea	various, 20th century	meat, leather, oil, eggs

Cuts of Poultry

The meatiest parts of a bird are the flight muscles on its chest, called breast meat, and the walking muscles on the first and second segments of its legs, called the thigh and drumstick, respectively.

White meat has less oxygen-carrying myoglobin than the dark meat, and is thus lighter in colour. Dark meat comes from muscles more heavily exercised, which therefore have more fat stored in them. This accounts for both dark meat's reputation as being unhealthier, and yet more flavorful. Birds that fly rarely (domestic turkey) or sporadically (chicken) have white meat breasts, and birds that fly frequently or long distances (ducks, geese and doves) have dark meat breasts. Quail breast meat is intermediate in colour. Leg meat is dark.

POULTRY FARMING

Poultry farming is the practice of raising poultry, such as chickens, turkeys, ducks, and geese, as a subcategory of animal husbandry, for the purpose of farming meat or eggs for food.

More than 50 billion chickens are reared annually as a source of food, for both their meat and their eggs. Chickens farmed for meat are called broiler chickens, whilst those farmed for eggs are called egg-laying hens. In total, the UK alone consumes over 29 million eggs per day. Some hens can produce over 300 eggs a year. Chickens will naturally live for 6 or more years. After 12 months, the hen's productivity will start to decline. This is when most commercial laying hens are slaughtered.

The vast majority of poultry are raised using intensive farming techniques. According to the Worldwatch Institute, 74 percent of the world's poultry meat, and 68 percent of eggs are produced this way. One alternative to intensive poultry farming is free range farming.

Friction between these two main methods has led to long term issues of ethical consumerism. Opponents of intensive farming argue that it harms the environment and creates health risks, as well as abusing the animals themselves. Advocates of intensive farming say that their highly efficient systems save land and food resources due to increased productivity, stating that the animals are looked after in state-of-the-art environmentally controlled facilities. A few countries have banned cage system housing, including Sweden and Switzerland. Consumers can still purchase lower cost eggs from other countries' intensive poultry farms.

Techniques

Organic

In organic systems, chickens are also free-range. Organic chickens are slower growing, more traditional breeds and live typically for around 81 days. They grow at half the rate of intensive chickens. They have a larger space allowance outside (at least 2 square metres and sometimes up to 10 square metres per bird).

Indoor with Higher Welfare

Chickens are kept indoors but with more space (around 12 to 14 birds per square metre). They have a richer environment for example with natural light or straw bales that encourage foraging and perching. The chickens grow more slowly and live for up to two weeks longer than intensively farmed birds. The benefits of higher welfare indoor systems are the reduced growth rate, less crowding and more opportunities for natural behaviour.

Free-range

Free range poultry farming consists of poultry permitted to roam freely instead of being contained in any manner. In the UK, the Department for Environment, Food and Rural Affairs says that a free range chicken must have daytime access to open-air runs during at least half of their life. Unlike in the United States, this definition also applies to eggs. The European Union regulates marketing standards for egg farming which specifies a minimum condition for Free Range Eggs states that "hens have continuous daytime access to open-air runs, except in the case of temporary restrictions imposed by veterinary authorities". In free-range broiler systems, the chickens are given continuous access to an outdoor range during the daytime and sheds where they are housed at night. Free-range chickens grow more slowly than intensive chickens. They live at least 56 days. In the EU each chicken must have one square metre of outdoor space.

- Free range poultry production requires that the poultry have access to the outside. In some cases this means the poultry are raised on pasture of enabling the poultry to move around, foraging for their natural diet and living in cleaner conditions than in batteries. In some farms, the manure from free range poultry can be used to benefit crops.

The benefits are also a reduced growth rate and opportunities for natural behaviour such as pecking, scratching, foraging and exercise outdoors, as well as fresh air and daylight. Because they grow slower and have opportunities for exercise free-range chickens have better leg and heart health and a much higher quality of life.

Finding suitable land with adequate drainage to minimise worms and coocidial oocysts, suitable protection from prevailing winds, good ventilation, access and protection from predators can be difficult. Excess heat, cold or damp can have a harmful effect on the animals and their productivity. Unlike battery farms, free range farmers have little control over the food their animals come across which can lead to unreliable productivity.

Free range farming in the UK, which accounts for 26 per cent of production, has come under similar criticism as battery farming in terms of animal welfare. This is due to the social abnormalities of having large numbers of birds in an outdoor space. Beak trimming due to cannibalism and infighting is common in this form of poultry farming as well as in batteries. Diseases are common and the animals are vulnerable to predators. In South-East Asia, a lack of disease control in free range farming has been associated with outbreaks of Avian influenza.

Yarding

While often confused with free range farming, yarding is actually a separate method of poultry culture by which chickens and cows are raised together. The distinction is that free-range poultry are either totally unfenced, or the fence is so distant that it has little influence on their freedom of movement. Yarding is common technique used by small farms in the Northeastern US.

Daily releases out of hutches or coops allows for instinctual nature for the chickens with protections from predators. The hens usually lay eggs either on the ground of the coop or in baskets if provided by the farmer. This technique can be complicated if used with roosters though, mostly because of difficulty getting them into the coop and to clean the coop while it is inside. This territorial nature is apparent while outside in which they have a brood of hens and sometimes even informal land claims. This can endanger people unaware of the existence of the territories who are attacked by the larger birds.

Intensive Chicken Farming

In egg-producing farms, birds are typically housed in rows of battery cages. Environmental conditions are automatically controlled, including light duration, which mimics summer daylength. This stimulates the birds to continue to lay eggs all year round.

Meat chickens, commonly called broilers, are floor-raised on litter such as wood shavings or rice hulls, indoors in climate-controlled housing. Poultry producers may use nationally-approved medications in feed or drinking water, to prevent or treat disease. In the US, the national approver is the Food and Drug Administration (FDA). Some FDA-approved medications are also approved for improved feed utilization. In the US, federal law prohibits the use of hormones or steroids in poultry production.

In egg-producing farms, cages allow for more birds per unit area, and this allows for greater productivity and lower space and food costs, with more efforts put into egg-laying. Modern poultry farming is very efficient and allows meat and eggs to be available to the consumer in all seasons at a much lower cost than free range production. The poultry have no exposure to predators and disease risk is significantly reduced.

The cage environment of egg producing does not permit birds to roam. As in free range production, cannibalism is controlled by beak trimming. Another condition that can occur in prolific egg laying breeds is osteoporosis. During egg production, large amounts of calcium are transferred from bones to create eggshell. Although dietary calcium levels are adequate, absorption of dietary calcium is not always sufficient to fully replenish bone calcium.

Under intensive farming methods, a meat chicken will live less than six weeks before slaughter. This is half the time it would take traditionally. This compares with free-range chickens which will usually be slaughtered at 8 weeks, and organic ones at around 12 weeks.

In intensive broiler sheds, the air can become highly polluted with ammonia from the droppings. This can damage the chickens' eyes and respiratory systems and can cause painful burns on their legs (called hock burns) and feet. Chickens bred for fast growth have a high rate of leg deformities because they cannot support their increased body weight. Because they cannot move easily, the chickens are not able to adjust their environment to avoid heat, cold or dirt as they would in natural conditions. The added weight and overcrowding also puts a strain on their hearts and lungs. In the UK, up to 19 million chickens die in their sheds from heart failure each year.

Issues with Poultry Farming

Humane Treatment

Animal welfare groups have frequently criticized the poultry industry for engaging in practices which they believe to be inhumane. Many animal rights advocates object to killing chickens for food, the "factory farm conditions" under which they are raised, methods of transport, and slaughter. Compassion Over Killing and other groups have repeatedly conducted undercover investigations at chicken farms and slaughterhouses which they allege confirm their claims of cruelty.

Conditions in intensive chicken farms may be unsanitary, allowing the proliferation of diseases such as salmonella and E coli. Chickens may be raised in total darkness; hens are most often kept in crowded wire battery cages with space less than that of a sheet of paper per hen, as opposed to cage-free or free range. Rough handling and crowded transport during various weather conditions and the failure of existing stunning systems to render the birds unconscious before slaughter have also been cited as welfare concerns.

Another animal welfare concern is the use of selective breeding to create heavy, large-breasted birds, which can lead to crippling leg disorders and heart failure for some of the birds. Concerns have been raised that companies growing single varieties of birds for eggs or meat are increasing their susceptibility to disease.

A common practice among hatcheries is the mass-slaughter of newly-born male chicks of egg laying breeds, since they don't lay eggs, and do not grow fast enough to be profitable for meat. Once separated from the females, the male chicks are dropped into grinding machines.

Debeaking

Laying hens are routinely debeaked when young to prevent fighting.

Because beaks are sensitive, the usual practice of trimming them without anaesthesia is considered inhumane by some. Debeaked chickens will peck much less than chickens with beaks, which animal behaviourist Temple Grandin attributes to guarding against pain. The United Egg Producers says that debeaking is not painful. It is also argued that the procedure causes life-long chronic pain and discomfort and decreased ability to eat or drink.

Intelligence

Some groups which advocate for more humane treatment of chickens claim that they are intelligent. Dr. Chris Evans of Macquarie University claims that their range of 20 calls, problem solving skills, use of representational signaling, and the ability to recognize each other by facial features demonstrate the intelligence of chickens.

Human Concerns

Antibiotics

Antibiotics have been used on poultry in large quantities since the 1940s, when it was found that the byproducts of antibiotic production, fed because the antibiotic-producing mold had a high level of vitamin B12 after the antibiotics were removed, produced higher growth than could be accounted for by the vitamin B12 alone. Eventually it was discovered that the trace amounts of antibiotics remaining in the byproducts accounted for this growth.

The mechanism is apparently the adjustment of intestinal flora, favouring "good" bacteria while suppressing "bad" bacteria, and thus the goal of antibiotics as a growth promoter is the same as for probiotics. Because the antibiotics used are not absorbed by the gut, they do not put antibiotics into the meat or eggs.

Antibiotics are used routinely in poultry for this reason, and also to prevent and treat disease. Many contend that this puts humans at risk as bacterial strains develop stronger and stronger resistances. Critics point out that, after six decades of heavy agricultural use of antibiotics, opponents of antibiotics must still make arguments about theoretical risks, since actual examples are hard to come by. Those antibiotic-resistant strains of human diseases whose origin is known originated in hospitals rather than farms.

A proposed bill in the United States Congress would make the use of antibiotics in animal feed legal only for therapeutic (rather than preventative) use, but it has not been passed. However, this may present the risk of slaughtered chickens harboring pathogenic bacteria and passing them on to humans that consume them.

In October 2000, the U.S. Food and Drug Administration (FDA) discovered that two antibiotics were no longer effective in treating diseases found in factory-farmed chickens; one antibiotic was swiftly pulled from the market, but the other, Baytril, was not. Bayer, the company which produced it, contested the claim and as a result, Baytril remained in use until July 2005.

Arsenic

Chicken feed can also include Roxarsone, an antimicrobial drug that also promotes growth. The drug has generated controversy because it contains the element arsenic, which can cause cancer, dementia, and neurological problems in humans. A Consumer Reports study in 2004 reported finding "no detectable arsenic in our samples of muscle" but found "A few of our chicken-liver samples has an amount that according to EPA standards could cause neurological problems in a child who ate 2 ounces of cooked liver per week or in an adult who ate 5.5 ounces per week." However, the Food and Drug Administration (FDA) is the organization responsible for the regulation of foods in America, and all samples tested were "far less than the… amount allowed in a food product."

Growth Hormones

Chickens grow much more rapidly than they once did and some consumers have concluded that this rapid growth is due to the use of hormones in these animals. Some consumers believe that the increasingly earlier onset of puberty in humans is the result of the liberal use of such hormones. However, hormone use in poultry production is illegal in the United States. Similarly, no chicken meat for sale in Australia is fed hormones. Furthermore, several scientific studies have documented the fact that chickens grow rapidly because they are bred to do so. A small producer of natural and organic chickens confirmed this assumption:

"If this were 1948, you might have something to worry about.

Using hormones to boost egg production was a brief fad in the Forties, but was abandoned because it didn't work. Using hormones to produce soft-meated roasters was used to some extent in the Forties and Fifties, but the increased growth rates of broilers made the practice irrelevant—the broilers got as big as anyone wanted them to get when they were still young enough to be soft-meated without chemicals.

The only hormone that was ever used in any quantity on poultry (DES) was banned in 1959, after everyone but a few die-hard farmers had given them up as a silly idea. Hormones are now illegal in poultry and eggs. The people who advertise "No hormones" are either woefully ignorant or are indulging in cynical fear-mongering, maybe both."

E. coli

According to Consumer Reports, "1.1 million or more Americans [are] sickened each year by undercooked, tainted chicken." A USDA study discovered *E. coli* in 99 per cent of supermarket chicken, the result of chicken butchering not being a sterile process. Feces tend to leak from the carcass until the evisceration stage, and the evisceration stage itself gives an opportunity for the interior of the carcass to receive intestinal bacteria. (So does the skin of the carcass, but the skin presents a better barrier to bacteria and reaches higher temperatures during cooking). Before 1950, this was contained largely by not eviscerating the carcass at the time of butchering, deferring this until the time of retail sale or in the home. This gave the intestinal bacteria less opportunity to colonize the edible meat. The development of the "ready-to-cook broiler" in the 1950s added convenience while introducing risk, under the assumption that end-to-end refrigeration and thorough cooking would provide adequate protection. *E. coli* can be killed by proper cooking times, but there is still some risk associated with it, and its near-ubiquity in commercially-farmed chicken is troubling to some. Irradiation has been proposed as a means of sterilizing chicken meat after butchering.

Avian Influenza

There is also a risk that crowded conditions in chicken farms will allow avian influenza (bird flu) to spread quickly. A United Nations press release states: "Governments, local authorities and international agencies need to take a greatly increased role in combating the role of factory-farming, commerce in live poultry, and wildlife markets which provide ideal conditions for the virus to spread and mutate into a more dangerous form..."

Efficiency

Farming of chickens on an industrial scale relies largely on high protein feeds derived from soybeans; in the European Union the soybean dominates the protein supply for animal feed, and the poultry industry is the largest consumer of such feed. Two kilograms of grain must be fed to poultry to produce 1 kg of weight gain. However, for every gram of protein consumed, chickens yield only 0.33 g of edible protein.

World Chicken Population

The Food and Agriculture Organization of the United Nations estimated that in 2002 there were nearly sixteen billion chickens in the world. The estimate at that time was 15,853,900,000 Chicken population. The figures from the *Global Livestock Production and Health Atlas* for 2004 were as follows:

1. China (8,860,000,000)
2. United States (1,970,000,000)
3. Indonesia (1,200,000,000)
4. Brazil (1,100,000,000)
5. Mexico (540,000,000)
6. India (425,000,000)
7. Russia (340,000,000)
8. Japan (286,000,000)
9. Iran (280,000,000)
10. Turkey (250,000,000)

ANIMAL HUSBANDRY

Animal husbandry, also called animal science, stockbreeding or simple husbandry, is the agricultural practice of breeding and raising livestock. It has been practiced for thousands of years, since the first domestication of animals.

The science of animal husbandry is taught in many universities and colleges around the world. Students of animal science may pursue degrees in veterinary medicine following graduation, or go on to pursue master's degrees or doctorates in disciplines such as nutrition, genetics and animal breeding, or reproductive physiology. Graduates of these programmes may be found working in the veterinary and human pharmaceutical industries, the livestock and pet supply and feed industries, farming, ranching or in academia.

Historically, certain sub-professions within the field of animal husbandry are named specifically for the animals in their care.

Different Types of Animal Husbandry

A swineherd is a person who cares for hogs and pigs (older English term: swine). A goatherd cares for goats. A cowherd cares for cattle, and a shepherd ("sheepherd") tends sheep. In the past, it was common to have herds which were made up of sheep and goats; the tenders of such also are called shepherds. Camels are also cared for in herds. In Tibet, yaks are herded. In Latin America, llamas and alpacas are herded.

In more modern times, the cowboys of North America, charros or vaqueros of South America, and farmers or stockmen of Australia tend their herds from horseback, all-terrain vehicles, motorbikes, in four-wheel drive (4WD) vehicles and helicopters, depending on the terrain and livestock concerned.

Today, herd managers often oversee thousands of animals and many staff. Farms, stations and ranches may employ breeders, herd health specialists, feeders, and milkers to help care for the animals. Techniques such as artificial insemination and embryo transfer are frequently used, not only as methods to guarantee that females are bred, but also to help improve herd genetics. This may be done by transplanting embryos from stud-quality females into flock-quality surrogate mothers—freeing up the stud-quality mother to be reimpregnated. This practice vastly increases the number of offspring which may be produced by a small selection of stud-quality parent animals. This in turn improves the ability of the animals to convert feed to meat, milk, or fiber more efficiently, and improve the quality of the final product, and to make it enjoyable.

Animal husbandry, also called animal science, stockbreeding or simple husbandry, is the agricultural practice of breeding and raising livestock. It has been practiced for thousands of years, since the first domestication of animals.

The science of animal husbandry is taught in many universities and colleges around the world. Students of animal science may pursue degrees in veterinary medicine following graduation, or go on to pursue master's degrees or doctorates in disciplines such as nutrition, genetics and animal breeding, or reproductive physiology. Graduates of these programmes may be found working in the veterinary and human pharmaceutical industries, the livestock and pet supply and feed industries, farming, ranching or in academia.

Historically, certain sub-professions within the field of animal husbandry are named specifically for the animals in their care.

Different Types of Animal Husbandry

A swineherd is a person who cares for hogs and pigs (older English term: swine). A goatherd cares for goats. A cowherd cares for cattle, and a shepherd ("sheepherd") tends sheep. In the past, it was common to have herds which were made up of sheep and goats; the tenders of such also are called shepherds. Camels are also cared for in herds. In Tibet, yaks are herded. In Latin America, llamas and alpacas are herded.

In more modern times, the cowboys of North America, charros or vaqueros of South America, and farmers or stockmen of Australia tend their herds from horseback, all-terrain vehicles, motorbikes, in four-wheel drive (4WD) vehicles and helicopters, depending on the terrain and livestock concerned.

Today, herd managers often oversee thousands of animals and many staff. Farms, stations and ranches may employ breeders, herd health specialists, feeders, and milkers to help care for the animals. Techniques such as artificial insemination and embryo transfer are frequently used, not only as methods to guarantee that females are bred, but also to help improve herd genetics. This may be done by transplanting embryos from stud-quality females into flock-quality surrogate mothers—freeing up the stud-quality mother to be reimpregnated. This practice vastly increases the number of offspring which may be produced by a small selection of stud-quality parent animals. This in turn improves the ability of the animals to convert feed to meat, milk, or fiber more efficiently, and improve the quality of the final product, and to make it enjoyable.

5

Food Microbiology, Engineering, Chemistry, Technology, Preservation, Irradiation and Packaging

FOOD MICROBIOLOGY

Food microbiology is the study of the microorganisms which inhabit, create or contaminate food. Of major importance is the study of microorganisms causing food spoilage. However "good" bacteria such as probiotics are becoming increasingly important in food science. In addition, microorganisms are essential for the production of foods such as cheese, yoghurt, other fermented foods, bread, beer and wine.

Food Safety

Food safety is a major focus of food microbiology. Pathogenic bacteria, viruses and toxins produced by microorganisms are all possible contaminants of food. However, microorganisms and their products can also be used to combat these pathogenic microbes. Probiotic bacteria, including those which produce bacteriocins can kill and inhibit pathogens. Alternatively, purified bacteriocins such as nisin can be added directly to food products. Finally, bacteriophage, viruses which only infect bacteria, can be used to kill bacterial pathogens. Thorough preparation of food, including proper cooking will eliminate most bacteria and viruses. However, toxins produced by contaminants may not be heat-labile, and some will not be eliminated by cooking.

Fermentation

Fermentation is one way microorganisms can change a food. Yeast, especially *S. cerevisiae*, is used to leaven bread, brew beer and make wine. Certain bacteria, including lactic acid bacteria, are used to make yogurt, cheese, hot sauce, pickles and dishes such as kimchi. A common effect of these fermentations is that the food product is less hospitable to other microorganisms, including pathogens and spoilage-causing microorganisms, thus extending the food's shelf-life.

Some cheese varieties also require mold microorganisms to ripen and develop their characteristic flavors.

Microbial Biopolymers

A variety of biopolymers, such as polysaccharides, polyesters and polyamides, are naturally produced by microorganisms. Several microbially-produced polymers are used in the food industry.

Xanthan

Plant-pathogenic bacteria of the genus *Xanthomonas* are able to produce the acidic exopolysaccharide xanthan gum. Because of its physical properties, it is widely used as a viscosifer, thickener, emulsifier or stabilizer in the food industry. Xanthan consists of pentasaccharide repeat units composed of D-glucosyl, D-mannosyl, and D-glucuronyl acid residues in a molar ratio of 2:2:1 and variable proportions of O-acetyl and pyruvyl residues.

Alginate

Alginate is the main representative of a family of polysaccharides that neither show branching nor repeating blocks or unit patterns and this property distinguishes it from to other polymers like xanthan or dextran. Alginates can be used as thickening agents.

Cellulose

Cellulose is a simple polysaccharide, in that it consists only of one type of sugar (glucose), and the units are linearly arranged and linked together by β-1,4 linkages only. The mechanism of biosynthesis is however rather complex, partly because in native celluloses the chains are organized as highly ordered water-insoluble fibers. Currently the key genes involved in cellulose biosynthesis and regulation are known in a number of bacteria, but many details of the biochemistry of its biosynthesis are still not clear. In spite of the enormous abundance of cellulose in plants bacterial celluloses are being investigated for industrial exploitations.

Poly-γ-glutamic Acid

Poly-γ-glutamic acid (γ-PGA) produced by various strains of *Bacillus* has potential applications as a thickener in the food industry.

Levan

Levan, a homopolysaccharide which is composed of D-fructofuranosyl residues joined by 2,6 with multiple branches by 2,1 linkages has great potential as a functional biopolymer in foods, feeds, cosmetics, and the pharmaceutical and chemical industries. Levan can be used as food or a feed additive with prebiotic and hypocholesterolemic effects.

Exopolysaccharides

Microorganisms synthesize a wide spectrum of multifunctional polysaccharides including intracellular polysaccharides, structural polysaccharides and extracellular polysaccharides or exopolysaccharides (EPS). Exopolysaccharides generally constitute of monosaccharides and some non-carbohydrate substituents (such as acetate, pyruvate, succinate, and phosphate). Owing to the wide diversity in composition, exopolysaccharides have found multifarious applications in various food and pharmaceutical industries.

Foodborne Pathogens

Foodborne pathogens are the leading causes of illness and death in less developed countries killing approximately 1.8 million people annually. In developed countries foodborne pathogens are responsible for millions of cases of infectious gastrointestinal diseases each year, costing billions of dollars in medical care and lost productivity. New foodborne pathogens and foodborne diseases are likely to emerge driven by factors such as pathogen evolution, changes in agricultural and food manufacturing practices, and changes to the human host status. There are growing concerns that terrorists could use pathogens to contaminate food and water supplies in attempts to incapacitate thousands of people and disrupt economic growth.

Enteric Viruses

Food and waterborne viruses contribute to a substantial number of illnesses throughout the world. Among those most commonly known are hepatitis A virus, rotavirus, astrovirus, enteric adenovirus, hepatitis E virus, and the human caliciviruses consisting of the noroviruses and the Sapporo viruses. This diverse group are transmitted by the fecal-oral route, often by ingestion of contaminated food and water.

Protozoan Parasites

Protozoan parasites associated with food and water can cause illness in humans. Although parasites are more commonly found in developing countries, developed countries have also experienced several foodborne outbreaks. Contaminants may be inadvertently introduced to the foods by inadequate handling practices, either on the farm or during processing of foods. Protozoan parasites can be found worldwide, either infecting wild animals or in water and contaminating crops grown for human consumption. The disease can be much more severe and prolonged in immunocompromissed individuals.

Mycotoxins

Molds produce mycotoxins, which are secondary metabolites that can cause acute or chronic diseases in humans when ingested from contaminated foods. Potential diseases include cancers and tumors in different organs (heart, liver, kidney, nerves), gastrointestinal disturbances, alteration of the immune system, and reproductive problems. Species of *Aspergillus*, *Fusarium*, *Penicillium*, and *Claviceps* grow in agricultural commodities or foods and produce the mycotoxins such as aflatoxins, deoxynivalenol, ochratoxin A, fumonisins, ergot alkaloids, T-2 toxin, and zearalenone and other minor mycotoxins such as cyclopiazonic acid and patulin. Mycotoxins occur mainly in cereal grains (barley, maize, rye, wheat), coffee, dairy products, fruits, nuts and spices. Control of mycotoxins in foods has focused on minimizing mycotoxin production in the field, during storage or destruction once produced. Monitoring foods for mycotoxins is important to manage strategies such as regulations and guidelines, which are used by 77 countries, and for developing exposure assessments essential for accurate risk characterization.

Yersinia Enterocolitica

Yersinia enterocolitica includes pathogens and environmental strains that are ubiquitous in terrestrial and fresh water ecosystems. Evidence from large outbreaks of yersiniosis and from epidemiological studies of sporadic cases has shown that *Y. enterocolitica* is a foodborne pathogen. Pork is often implicated as the source of infection. The pig is the only animal consumed by man that regularly harbours pathogenic *Y. enterocolitica*. An important property of the bacterium is its ability to multiply at temperatures near to 0°C, and therefore in many chilled foods. The pathogenic serovars (mainly O:3, O:5,27, O:8 and O:9) show different geographical distribution. However, the appearance of strains of serovars O:3 and O:9 in Europe, Japan in the 1970s, and in North America by the end of the 1980s, is an example of a global pandemic. There is a possible risk of reactive arthritis following infection with *Y. enterocolitica*.

Vibrio

Vibrio species are prevalent in estuarine and marine environments and seven species can cause foodborne infections associated with seafood. *Vibrio cholerae* O1 and O139 serovtypes produce cholera toxin and are agents of cholera. However, fecal-oral route infections in the terrestrial environment are responsible for epidemic cholera. *V. cholerae* non-O1/O139 strains may cause gastroenteritis through production of known toxins or unknown mechanism. *Vibrio parahaemolytitucs* strain's capable of producing thermostable direct hemolysin (TDH) and/or TDH-related hemolysin are most important cause of gastroenteritis associated with seafood consumption. *Vibrio vulnificus* is responsible for seafoodborne primary septicemia and its infectivity depends primarily on the risk factors of the host. *V. vulnificus* infection has the highest case fatality rate (50 per cent) of any foodborne pathogen. Four other species (*Vibrio mimicus*, *Vibrio hollisae*, *Vibrio fluvialis*, and *Vibrio furnissii*) can cause gastroenteritis. Some strains of these species produce known toxins but the pathogenic mechanism is largely not understood. The ecology of and detection and control methods for all seafoodborne *Vibrio* pathogens are essentially similar.

Staphylococcus Aureus

Staphylococcus aureus is a common cause of bacterial foodborne disease worldwide. Symptoms include vomiting and diarrhea that occur shortly after ingestion of *S. aureus*-contaminated food. The symptoms arise from ingestion of preformed enterotoxin, which accounts for the short incubation time. Staphylococcal enterotoxins are superantigens and, as such, have adverse effects on the immune system. The enterotoxin genes are accessory genetic elements in *S. aureus*, meaning that not all strains of this organism are enterotoxin-producing. The enterotoxin genes are found on prophage, plasmids, and pathogenicity islands in different strains of *S. aureus*. Expression of the enterotoxin genes is often under the control of global virulence gene regulatory systems.

Campylobacter

Campylobacter spp., primarily *C. jejuni* subsp. *jejuni* is one of the major causes of bacterial gastroenteritis in the U.S. and worldwide. *Campylobacter* infection is primarily a foodborne illness, usually without complications; however, serious sequelae such as Guillain-Barre Syndrome occur in a small subset of infected patients. Detection of *C. jejuni* in clinical samples is readily accomplished by culture and non-culture methods.

Listeria Monocytogenes

Listeria monocytogenes is Gram-positive foodborne bacterial pathogen and the causative agent of human listeriosis. *Listeriae* are acquired primarily through the consumption of contaminated foods including soft cheese, raw milk, deli salads, and ready-to-eat foods such as luncheon meats and frankfurters. Although *L. monocytogenes* infection is usually limited to individuals that are immunocompromised, the high mortality rate associated with human listeriosis makes *L. monocytogenes* the leading cause of death amongst foodborne bacterial pathogens. As a result, tremendous effort has been made at developing methods for the isolation, detection and control of *L. monocytogenes* in foods.

Salmonella

Salmonella serotypes continue to be a prominent threat to food safety worldwide. Infections are commonly acquired by animal to human transmission though consumption of undercooked food products derived from livestock or domestic fowl. The second half of the 20th century saw the emergence of *Salmonella* serotypes that became associated with new food sources (*i.e.* chicken eggs) and the emergence of *Salmonella* serotypes with resistance against multiple antibiotics.

Shigella

Shigella species are members of the family *Enterobacteriaceae* and are Gram negative, non-motile rods. Four subgroups exist based on O-antigen structure and biochemical properties; *S. dysenteriae* (subgroup A), *S. flexneri* (subgroup B), *S. boydii* (subgroup C) and *S. sonnei* (subgroup D). Symptoms include mild to severe diarrhea with or without blood, fever, tenesmus, and abdominal pain. Further complications of the disease may be seizures, toxic megacolon, reactive arthritis and hemolytic uremic syndrome. Transmission of the pathogen is by the fecal-oral route, commonly through food and water. The infectious dose ranges from 10-100 organisms. *Shigella* spp. have a sophisticated pathogenic mechanism to invade colonic epithelial cells of the host, man and higher primates, and the ability to multiply intracellularly and spread from cell to adjacent cell via actin polymerization. *Shigellae* are one of the leading causes of bacterial foodborne illnesses and can spread quickly within a population.

Escherichia Coli

More information is available concerning *Escherichia coli* than any other organism, thus making *E. coli* the most thoroughly studied species in the microbial world. For many years, *E. coli* was considered a commensal of human and animal intestinal tracts with low virulence potential. It is now known that many strains of *E. coli* act as pathogens inducing serious gastrointestinal diseases and even death in humans. There are six major categories of *E. coli* strains that cause enteric diseases in humans including the (1) enterohemorrhagic *E. coli*, which cause hemorrhagic colitis and hemolytic uremic syndrome, (2) enterotoxigenic *E. coli*, which induce traveler's diarrhea, (3) enteropathogenic *E. coli*, which cause a persistent diarrhea in children living in developing countries, (4) enteroaggregative *E. coli*, which provoke diarrhea in children, (5) enteroinvasive *E. coli* that are biochemically and genetically related to *Shigella* species and can induce diarrhea, and (6) diffusely adherent *E. coli*, which cause diarrhea and are distinguished by a characteristic type of adherence to mammalian cells.

Clostridium Botulinum **and** *Clostridium Perfringens*

Clostridium botulinum produces extremely potent neurotoxins that result in the severe neuroparalytic disease, botulism. The enterotoxin produced by *C. perfringens* during sporulation of vegetative cells in the host intestine results in debilitating acute diarrhea and abdominal pain. Sales of refrigerated, processed foods of extended durability including sous-vide foods, chilled ready-to-eat meals, and cook-chill foods have increased over recent years. Anaerobic spore-formers have been identified as the primary microbiological concerns in these foods. Heightened awareness over intentional food source tampering with botulinum neurotoxin has arisen with respect to genes encoding the toxins that are capable of transfer to nontoxigenic clostridia.

Bacillus Cereus

The *Bacillus cereus* group comprises six members: *B. anthracis*, *B. cereus*, *B. mycoides*, *B. pseudomycoides*, *B. thuringiensis* and *B. weihenstephanensis*. These species are closely related and should be placed within one species, except for *B. anthracis* that possesses specific large virulence plasmids. *B. cereus* is a normal soil inhabitant and is frequently isolated from a variety of foods, including vegetables, dairy products and meat. It causes a vomiting or diarrhoea illness that is becoming increasingly important in the industrialized world. Some patients may experience both types of illness simultaneously. The diarrhoeal type of illness is most prevalent in the western hemisphere, whereas the emetic type is most prevalent in Japan. Desserts, meat dishes, and dairy products are the foods most frequently associated with diarrhoeal illness, whereas rice and pasta are the most common vehicles of emetic illness. The emetic toxin (cereulide) has been isolated and characterized; it is a small ring peptide synthesised non-ribosomally by a peptide synthetase. Three types of *B. cereus* enterotoxins involved in foodborne outbreaks have been identified. Two of these enterotoxins are three-component proteins and are related, while the last is a one-component protein (CytK). Deaths have been recorded both by strains that produce the emetic toxin and by a strain producing only CytK. Some strains of the *B. cereus* group are able to grow at refrigeration temperatures. These variants raise concern about the safety of cooked, refrigerated foods with an extended shelf life. *B. cereus* spores adhere to many surfaces and survive normal washing and disinfection (except for hypochlorite and UVC) procedures. *B. cereus* foodborne illness is likely underreported because of its relatively mild symptoms, which are of short duration.

Food Authenticity

It is important to be able to detect microorganisms in food, in particular pathogenic microorganisms or genetically modified microorganisms. Real-time PCR is an accepted analytical tool within the food industry. Its principal role has been one of assisting the legislative authorities, major manufacturers and retailers to confirm the authenticity of foods. The most obvious role is the detection of genetically modified organisms, but real-time PCR makes a significant contribution to other areas of the food industry, including food safety.

FOOD ENGINEERING

Food engineering refers to the aspects of food production and processing. Food engineering includes, but is not limited to, the application of agricultural engineering and chemical engineering principles to food materials. Genetic engineering of plants and animals is not normally the work of a food engineer.

Food engineering is a very wide field of activities. Among its domain of knowledge and action are:

- Design of machinery and processes to produce foods
- Design and implementation of food safety and preservation measures in the production of foods
- Biotechnological processes of food production
- Choice and design of food packaging materials
- Quality control of food production

FOOD CHEMISTRY

Food chemistry is the study of chemical processes and interactions of all biological and non-biological components of foods. The biological substances include such items as meat, poultry, lettuce, beer, and milk as examples. It is similar to biochemistry in its main components such as carbohydrates, lipids, and protein, but it also includes areas such as water, vitamins, minerals, enzymes, food additives, flavours, and colours. This discipline also encompasses how products change under certain food processing techniques and ways either to enhance or to prevent them from happening. An example of enhancing a process would be to encourage fermentation of dairy products with microorganisms that convert lactose to lactic acid; an example of preventing a process would be stopping the browning on the surface of freshly cut Red Delicious apples using lemon juice or other acidulated water.

History of Food Chemistry

Food chemistry's history dates back as far as the late 1700s when many famous chemists were involved in discovering chemicals important in foods, including Carl Wilhelm Scheele (isolated malic acid from apples in 1785), and Sir Humphry Davy (published the first ever book on agricultural and food chemistry in 1813 titled

Elements of Agricultural Chemistry, in a Course of Lectures for the Board of Agriculture in the United Kingdom which would serve as a foundation for the profession worldwide, going into a fifth edition.

In 1874 the Society of Public Analysts was formed, with the aim of applying analytical methods to the benefit of the public. It's early experiments were based on bread, milk and wine.

It was also out of concern for the quality of the food supply, mainly food adulteration and contamination issues that would first stem from intentional contamination to later with chemical food additives by the 1950s. The development of colleges and universities worldwide, most notably in the United States, would expand food chemistry as well with research of the dietary substances, most notably the Single-grain experiment during 1907-11. Additional research by Harvey W. Wiley at the United States Department of Agriculture during the late 19th century would play a key factor in the creation of the United States Food and Drug Administration in 1906. The American Chemical Society would establish their Agricultural and Food Chemistry Division in 1908 while the Institute of Food Technologists would establish their Food Chemistry Division in 1995.

Water

A major component of food is water, which can encompass anywhere from 50 per cent in meat products to 95 per cent in lettuce, cabbage, and tomato products. It is also an excellent place for bacterial growth and food spoilage if it is not properly processed. One way this is measured in food is by water activity which is very important in the shelf life of many foods during processing. One of the keys to food preservation in most instances is reduce the amount of water or alter the water's characteristics to enhance shelf-life. Such methods include dehydration, freezing, and refrigeration.

Carbohydrates

Sucrose: ordinary table sugar and probably the most familiar carbohydrate.

Comprising 75 per cent of the biological world and 80 per cent of all food intake for human consumption, the most common known human carbohydrate is starch. The simplest version of a carbohydrate is a monosaccharide which possesses the properties of carbon, hydrogen, and oxygen in a 1:2:1 ratio under a general formula of $C_nH_{2n}O_n$ where n is a minimum of 3. Glucose is an example of a monosaccharide as is fructose. Combine them in the picture shown to the right and you have sucrose, one of the more common sugar products around.

A chain of monosaccharides form to make a polysaccharide. Such polysaccharides include pectin, dextran, agar, and xanthan.

Sugar content is commonly measured in degrees brix.

Lipids

The term lipid comprises a diverse range of molecules and to some extent is a catchall for relatively water-insoluble or nonpolar compounds of biological origin, including waxes, fatty acids (including essential fatty acids), fatty-acid derived phospholipids, sphingolipids, glycolipids and terpenoids, such as retinoids and steroids. Some lipids are linear aliphatic molecules, while others have ring structures. Some are aromatic, while others are not. Some are flexible, while others are rigid.

Most lipids have some polar character in addition to being largely nonpolar. Generally, the bulk of their structure is nonpolar or hydrophobic ("water-fearing"), meaning that it does not interact well with polar solvents like water. Another part of their structure is polar or hydrophilic ("water-loving") and will tend to associate with polar solvents like water. This makes them amphiphilic molecules (having both hydrophobic and hydrophilic portions). In the case of cholesterol, the polar group is a mere -OH (hydroxyl or alcohol).

Lipids in food include the oils of such grains as corn, soybean, from animal fats, and are parts of many foods such as milk, cheese, and meat. They also act as vitamin carriers as well.

Proteins

Proteins compose over 50 per cent of the dry weight of an average living cell and are very complex macromolecules. They also play a fundamental role in the structure and function of cells. Comprised mainly of carbon, hydrogen, oxygen, and some sulfur, they also may contain iron, copper, phosphorus, or zinc.

In food, proteins are essential for growth and survival and vary depending upon a person's age and physiology (*e.g.*, pregnancy). Proteins in food are commonly found in peanuts, meat, poultry, and seafood. They are also involved in ELISA test for food allergy determination as well.

Enzymes

Enzymes are biochemical catalysts used in converting processes from one substance to another. They are also involved in reducing the amount of time and energy required to complete a chemical process. Many aspects of the food industry use catalysts, including baking, brewing, dairy, and fruit juices, to make cheese, beer, and bread.

Vitamins

Riboflavin (Vitamin B_2), water soluble

Vitamins are nutrients required in small amounts for essential metabolic reactions in the body. These are broken down in nutrition as either water soluble (Vitamin C) or fat soluble (Vitamin E). An adequate supply of vitamins can prevent such diseases as beriberi, anemia, and scurvy while an overdose of vitamins can produce nausea and vomiting or even death.

Minerals

Dietary minerals in foods are large and diverse with many required to function while other trace elements can be hazardous if consumed in excessive amounts. Bulk minerals with a Reference Daily Intake (RDI, formerly Recommended Daily Allowance (RDA)) of more than 200 mg/day are calcium, magnesium, and potassium while important trace minerals (RDI less than 200 mg/day) are copper, iron, and zinc. These are found in many foods, but can also be taken in dietary supplements.

Colour

Food colouring is added to change the colour of any food substance. It is mainly for sensory analysis purposes. It can be used to simulate the natural colour of a product as perceived by the customer, such as red dye like FD&C Red No. 40 (Allura Red AC) to ketchup or to add unnatural colours to a product like Kellogg's Froot Loops. Caramel is a natural food dye; the industrial form, caramel colouring, is the most widely-used food colouring and is found in foods from soft drinks to soy sauce, bread, and pickles.

Flavors

Flavor in food is important in how food smells and tastes to the consumer, especially in sensory analysis. Some of these products occur naturally like salt and sugar, but flavor chemists (called a "flavorist") develop many of these flavors for food products. Such artificial flavors include methyl salicylate which creates the wintergreen odor and lactic acid which gives milk a tart taste.

Food Additives

Food additives are substances added to food for preserving flavors, or improving taste or appearance. These processes are as old as adding vinegar for pickling or as an emulsifier for emulsion mixtures like mayonnaise. These are generally listed by "E number" in the European Union or GRAS ("Generally recognized as safe") by the United States Food and Drug Administration.

RELATED WATER ACTIVITY AND PATHOGENS

Water activity or a_w is a measurement of the energy status of the water in a system. It is defined as the vapor pressure of water above a sample divided by that of pure water at the same temperature; therefore, pure distilled water has a water activity of exactly one.

There are several factors that control water activity in a system. Colligative effects of dissolved species (*e.g.* salt or sugar) interact with water through dipole-dipole, ionic, and hydrogen bonds. Capillary effect where the vapor pressure of water above a curved liquid meniscus is less than that of pure water because of changes in the hydrogen bonding between water molecules. Surface interactions in which water interacts directly with chemical groups on undissolved ingredients (*e.g.* starches and proteins) through dipole-dipole forces, ionic bonds (H_3O+ or $OH-$), van der Waals forces (hydrophobic bonds), and hydrogen bonds. It is a combination of these three factors in a food product that reduces the energy of the water and thus reduces the relative humidity as compared to pure water. These factors can be grouped under two broad categories osmotic and matric effects.

Due to varying degrees of osmotic and matric interactions, water activity describes the continuum of energy states of the water in a system. The water appears "bound" by forces to varying degrees. This is a continuum of energy states rather than a static "boundness". Water activity is sometimes defined as "free", "bound", or "available water" in a system. Although these terms are easier to conceptualize, they fail to adequately define all aspects of the concept of water activity.

Water activity is temperature dependent. Temperature changes water activity due to changes in water binding, dissociation of water, solubility of solutes in water, or the state of the matrix. Although solubility of solutes can be a controlling factor, control is usually from the state of the matrix. Since the state of the matrix (glassy vs. rubbery state) is dependent on temperature, one should not be surprised that temperature affects the water activity of the food. The effect of temperature on the water activity of a food is product specific. Some products increase water activity with increasing temperature, others decrease aw with increasing temperature, while most high moisture foods have negligible change with temperature. One can therefore not predict even the direction of the change of water activity with temperature, since it depends on how temperature affects the factors that control water activity in the food.

As a potential energy measurement it is a driving force for water movement from regions of high water activity to regions of low water activity. For example, if honey ($a_w \approx 0.6$) is exposed to humid air (a_w ≈ 0.7) the honey will absorb water from the air. Other examples of this dynamic property of water activity are; moisture migration in multidomain foods (*e.g.* cracker-cheese sandwich), the movement of water from soil to the leaves of plants, and cell turgor pressure. Since microbial cells are high concentrations of solute surrounded by semi-permeable membranes, the osmotic effect on the free energy of the water is important for determining microbial water relations and therefore their growth rates.

Higher a_w substances tend to support more microorganisms. Bacteria usually require at least 0.91, and fungi at least 0.7.

Uses for Water Activity

Water activity is an important consideration for food product design and food safety.

Food Product Design

Food designers use water activity to formulate products that are shelf stable. If a product is kept below a certain water activity, then mold growth is inhibited. This results in a longer shelf-life.

Water activity values can also help limit moisture migration within a food product made with different ingredients. If raisins of a higher water activity are packaged with bran flakes of a lower water activity, the water from the raisins will migrate to the bran flakes over time, resulting in hard raisins and soggy bran flakes. Food formulators use water activity to predict how much moisture migration will affect their product.

In addition, water activity helps limit or slow certain undesirable reactions, such as non-enzymatic browning, fat oxidation, vitamin degradation, enzymatic reactions, protein denaturation, starch gelatinization and starch retrogradation. This too maintains product quality and extends shelf life.

Food Safety

Water activity is used in many cases as a Critical Control Point for Hazard Analysis and Critical Control Points (HACCP) programmes.

Samples of the food product are periodically taken from the production area and tested to ensure that water activity values are within a specified range for food quality and safety. Measurements can be made in as little as five minutes, and are made regularly in most major food production facilities.

For many years researchers tried to equate bacterial growth potential with moisture content. They found that the values were not universal, but specific to each food product. WJ Scott in 1953 first established that it was water activity, not water content that correlated with bacterial growth. It is firmly established that growth of bacteria is inhibited at specific water activity values. FDA regulations for Intermediate Moisture Foods are based on these values.

Lowering the water activity of a food product should not be seen as a kill step. Studies in powdered milk show that viable cells can exist at much lower water activity values but that they will never grow. Over time bacterial levels will decline.

Water Activity Measurement

Water activity values are obtained by either a capacitance or a dew point hygrometer.

Capacitance Hygrometers

Capacitance hygrometers consist of two charged plates separated by a polymer membrane dielectric. As the membrane adsorbs water, its ability to hold a charge increases and the capacitance is measured. This value is roughly proportional to the water activity as determined by a sensor-specific calibration.

Capacitance hygrometers are not affected by most volatile chemicals and can be much smaller than other alternative sensors. They do not require cleaning, but are less accurate than dew point hygrometers ($\pm$ .015 a_w). They require regular calibration and can be affected by residual water in the polymer membrane.

Dew Point Hygrometers

The temperature at which dew forms on a clean surface is directly related to the vapor pressure of the air. Dew point hygrometers work by placing a mirror over a closed sample chamber. The mirror is cooled until the dew point temperature is measured by means of an optical sensor. This temperature is then used to find the relative humidity of the chamber using psychrometric charts.

This method is the most accurate (± .003 a_w) and often the fastest.

Equilibration

With either method, vapor equilibrium must occur in the sample chamber. This will take place over time or can be aided by the addition of a fan in the chamber. Thermal equilibrium must also take place unless the sample temperature is measured.

Water Activity and Moisture Content

Water activity is related to moisture content in a non-linear relationship known as a moisture sorption isotherm curve. These isotherms are substance and temperature specific. Isotherms can be used to help predict product stability over time in different storage conditions.

Selected a_w Values

Example Foods

Substance	*aw*
Distilled Water	1
Tap water	0.99
Raw meats	0.97-0.99
Milk	0.97
Juice	0.97
Cooked bacon	< 0.85
Saturated NaCl solution	0.75
Point at which cereal loses crunch	0.65
Typical indoor air	0.5-0.7
Honey	0.5-0.7
Dried fruit	0.5-0.6

a_w Values of Microorganism Inhibition

Microorganism Inhibited	*aw*
Clostridium botulinum E	.97
Pseudomonas fluorescence	.97
Escherichia coli	.95
Clostridium perfringens	.95
Salmonella	.95
Vibrio cholerae	.95
Clostridium botulinumA, B	.97
Bacillus cereus	.93
Listeria monocytogenes	.92
bacillus subtilis	.91
Staphylococcus aureus	.87
Most Fungi	.70
No microbial proliferation	.50

CLOSTRIDIUM BOTULINUM

Clostridium botulinum is a Gram-positive, rod shaped bacterium that produces the neurotoxin botulin, which causes the flaccid muscular paralysis seen in botulism. It is also the main paralytic agent in botox. It is an anaerobic spore-former, which produces oval, subterminal endospores and is commonly found in soil.

Microbiology

C. botulinum is a rod-shaped microorganism. It is an obligate anaerobe, meaning that oxygen is poisonous to the cells. However, they tolerate very small traces of oxygen due to an enzyme called superoxide dismutase (SOD) which is an important antioxidant defense in nearly all cells exposed to oxygen. Under unfavourable circumstances they are able to form endospores that allow them to survive in a dormant state until exposed to conditions that can support their growth.

In laboratory the microorganism is usually isolated in Tryptose Sulfite Cycloserine (TSC) growth media, always in an anaerobic environment with less than 2 per cent of Oxygen. This can be achieved by several commercial kits that use a chemical reaction to replace O_2 with CO_2 (E.J. GasPak System). *C. botulinum* is lipase negative microorganism, it grows between pH values of 4.8 and 7 and it can't use lactose as a primary carbon source, characteristics important during a biochemical identification.

Taxonomy History

C. botulinum was first recognized and isolated in 1896 by Emile van Ermengem from home cured ham implicated in a botulism outbreak. The isolate was originally named *Bacillus botulinus*. However, isolates from subsequent outbreaks were always found to be anaerobic spore formers, so Bengston proposed that the organism be placed into the genus *Clostridium* as the *Bacillus* genus was restricted to aerobic spore-forming rods.

Since 1953 all species producing the botulinum neurotoxins (types A-G) has been designated *C. botulinum*. Substantial phenotypic and genotypic evidence exist to demonstrate heterogeneity within the species. This has led to the reclassification of *C. botulinum* type-G strains to a new species *Clostridium argentinense*.

C. botulinum strains that do not produce a botulin toxin are referred to as *Clostridium sporogenes*.

The complete genome of *C. botulinum* has now been sequenced Sanger.

Phenotypic Types

The current nomenclature for *C. botulinum* recognises four physiological groups (I-IV). This is mostly based on the ability of the organism to digest complex proteins. Studies at the DNA and rRNA level support the subdivision of the species into groups I-IV. Most outbreaks of human botulism are caused by group I (proteolytic) or II (non-proteolytic) *C. botulinum*. Group III organisms mainly cause diseases in animals. There has been no record of Group IV *C. botulinum* causing human or animal disease.

Neurotoxin Types

Neurotoxin production is the unifying feature of the species *C.*

botulinum. Seven types of toxins have been identified and allocated a letter (A-G). Most strains produce one type of neurotoxin but strains producing multiple toxins has been described. *C. botulinum* producing B and F toxin types have been isolated from human botulism cases in New Mexico and California. The toxin type has been designated Bf as the type B toxin was found in excess to the type F. Similarly, strains producing Ab and Af toxins have been reported.

Organisms genetically identified as other Clostridium species have caused human botulism; *Clostridium butyricum* producing type E toxin and *Clostridium baratii* producing type F toxin. The ability of *C. botulinum* to naturally transfer neurotoxin genes to other clostridia is concerning, especially in the food industry where preservation systems are designed to destroy or inhibit only *C. botulinum* but not other Clostridium species.

Table 5.1: Phenotypic groups of *Clostridium botulinum*

Properties	*Group I*	*Group II*	*Group III*	*Group IV*
Toxin Types	A, B, F	B, E, F	C, D	G
Proteolysis	+	-	weak	-
Saccharolysis	-	+	-	-
Disease host	human	human	animal	-
Toxin gene	chromosome	chromosome	bacteriophage	plasmid
Close relatives	C. sporogenes, C. putrificum	C. butyricum, C. beijerinickii	C. haemolyticum, C. novyi type A	C. subterminale, C. haemolyticum

Clostridium botulinum in different geographical locations. A number of quantitative surveys for *C. botulinum* spores in the environment have suggested a prevalence of specific toxin types in given geographic areas, which remain unexplained.

North America

Type A *C. botulinum* predominates the soil samples from the western regions while type B is the major type found in eastern areas. The type B organisms were of the proteolytic type I. Sediments from the Great Lake regions were surveyed after outbreaks of botulism among commercially reared fish and only type E spores were detected. It has been noted in a survey that type A strains were isolated from soils that were neutral to alkaline (average pH 7.5) while type B strains were isolated from slightly acidic soils (average pH 6.25).

Europe

C. botulinum type E is prevalent in aquatic sediments in Norway and Sweden, Denmark, the Netherlands, the Baltic coast of Poland and Russia. It was then suggested that the type E *C. botulinum* is a true aquatic organism and this was shown by the correlation between the level of type E contamination and flooding of the land with seawater. As the land dried, the level of type E decreased and type B became dominant.

In soil and sediment from the United Kingdom, *C. botulinum* type B predominates. In general, the incidence is usually lower in soil than in sediment. In Italy, a survey was conducted in the vicinity of Rome, a low level of contamination was found and all strains were proteolytic *C. botulinum* type A or B.

Australia

C. botulinum type A was found to be present in soil samples from mountain areas of Victoria. Type B organisms were detected in marine mud from Tasmania. Type A *C. botulinum* have been found in Sydney suburbs and types A and B were isolated from urban areas. In a well defined area of the Darling-Downs region of Queensland a study showed the prevalence and persistence of *C. botulinum* type B after many cases of botulism in horses.

Other

A "mouse protection" test determines the type of *C. botulinum* present using monoclonal antibodies.

Clostridium botulinum is also used to prepare Botox, used to selectively paralyze muscles to temporarily relieve wrinkles. It has other "off-label" medical purposes, such as treating severe facial pain, such as that caused by trigeminal neuralgia.

Botulin toxin produced by *Clostridium botulinum* is often believed to be a potential bioweapon as it is so potent that it takes about 75 nanograms to kill a person (LD50 of 1ng/kg, assuming an average person weighs ~75kg); 500 grams of it would be enough to kill half of the entire human population.

Clostridium botulinum is a soil bacterium. The spores can survive in most environments and are very hard to kill. They can survive the temperature of boiling water at sea level, thus many foods are canned with a pressurized boil that achieves an even higher temperature, sufficient to kill the spores.

Growth of the bacterium can be prevented by high acidity, high ratio of dissolved sugar, high levels of oxygen, very low levels of moisture or storage at temperatures below 38°F (type A). For example in a low acid, canned vegetable such as green beans that are not heated hot enough to kill the spores (*i.e.*, a pressurized environment) may provide an oxygen free medium for the spores to grow and produce the toxin. On the other hand, pickles are sufficiently acidic to prevent growth; even if the spores are present, they pose no danger to the consumer. Honey, corn syrup, and other sweeteners may contain spores but the spores cannot grow in a highly concentrated sugar solution; however, when a sweetener is diluted in the low oxygen, low acid digestive system of an infant, the spores can grow and produce toxin. As soon as infants begin eating solid food, the digestive juices become too acidic for the bacterium to grow.

ESCHERICHIA COLI

Escherichia coli (commonly *E. coli*), is a bacterium that is commonly found in the lower intestine of warm-blooded animals. Most *E. coli* strains are harmless, but some, such as serotype O157:H7, can cause serious food poisoning in humans, and are occasionally responsible for costly product recalls. The harmless strains are part of the normal flora of the gut, and can benefit their hosts by producing vitamin K_2, or by preventing the establishment of pathogenic bacteria within the intestine.

E. coli are not always confined to the intestine, and their ability to survive for brief periods outside the body makes them an ideal indicator organism to test environmental samples for fecal contamination. The bacteria can also be grown easily and its genetics are comparatively simple and easily-manipulated, making it one of the best-studied prokaryotic model organisms, and an important species in biotechnology. *E. coli* was discovered by German pediatrician and bacteriologist Theodor Escherich in 1885, and is now classified as part of the Enterobacteriaceae family of gamma-proteobacteria.

Strains

Model of successive binary fission in *E. coli*

A strain of *E. coli* is a sub-group within the species that has unique characteristics that distinguish it from other *E. coli* strains. These differences are often detectable only on the molecular level; however, they may result in changes to the physiology or lifecycle of the bacterium. For example, a strain may gain pathogenic capacity, the ability to use a unique carbon source, the ability to inhabit a particular ecological niche or the ability to resist antimicrobial agents. Different strains of *E. coli* are often host-specific, making it possible to determine the source of fecal contamination in environmental samples. Depending on which *E. coli* strains are present in a water sample, for example, assumptions can be made about whether the contamination originated from a human, other mammal or bird source.

New strains of *E. coli* evolve through the natural biological process of mutation, and some strains develop traits that can be harmful to a host animal. Although virulent strains typically cause no more than a bout of diarrhea in healthy adult humans, particularly virulent strains, such as O157:H7 or O111:B4, can cause serious illness or death in the elderly, the very young or the immunocompromised.

Biology and Biochemistry

E. coli is Gram-negative, facultative anaerobic and non-sporulating. The cells are about 2 micrometres (mm) long and 0.5 ìm in diameter, with a cell volume of 0.6-0.7 mm. It can live on a wide variety of substrates. *E. coli* uses mixed-acid fermentation in anaerobic conditions, producing lactate, succinate, ethanol, acetate and carbon dioxide. Since many pathways in mixed-acid fermentation produce hydrogen gas, these pathways require the levels of hydrogen to be low, as is the case when *E. coli* lives together with hydrogen-consuming organisms such as methanogens or sulfate-reducing bacteria.

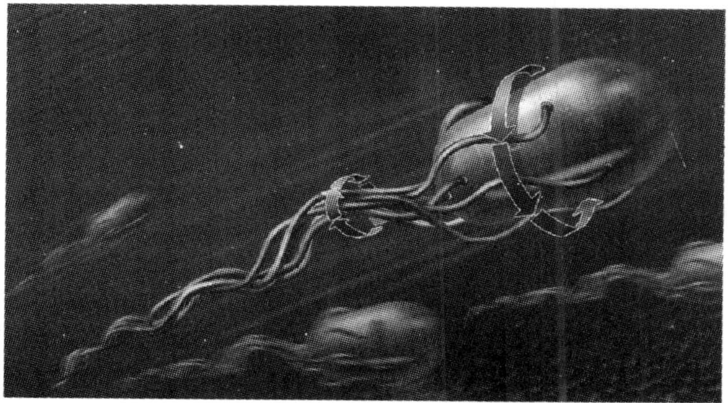

Fig. 5.1: Escherichia Coli Cells Propel Themselves with Flagella (Long, Thin Structures) Arranged as Bundles that Rotate Counter-Clockwise, Generating Torque to Rotate the Bacterium Clockwise

Optimal growth of *E. coli* occurs at 37°C, but some laboratory strains can multiply at temperatures of up to 49°C. Growth can be driven by aerobic or anaerobic respiration, using a large variety of redox pairs, including the oxidation of pyruvic acid, formic acid, hydrogen and amino acids, and the reduction of substrates such as oxygen, nitrate, dimethyl sulfoxide and trimethylamine N-oxide.

Strains that possess flagella can swim and are motile, but other strains lack flagellum. The flagella of *E. coli* have a peritrichous arrangement.

E. coli and related bacteria possess the ability to transfer DNA via bacterial conjugation, transduction or transformation, which allows genetic material to spread horizontally through an existing population. This process led to the spread of the gene encoding shiga toxin from *Shigella* to *E. coli* O157:H7, carried by a bacteriophage.

Normal Role

E. coli normally colonizes an infant's gastrointestinal tract within 40 hours of birth, arriving with food or water or with the individuals handling the child. In the bowel, it adheres to the mucus of the large intestine. It is the primary facultative organism of the human gastrointestinal tract. As long as these bacteria do not acquire genetic elements encoding for virulence factors, they remain benign commensals.

Role in Disease

Virulent strains of *E. coli* can cause gastroenteritis, urinary tract infections, and neonatal meningitis. In rarer cases, virulent strains are also responsible for hæmolytic-uremic syndrome (HUS), peritonitis, mastitis, septicemia and Gram-negative pneumonia.. Recently it is thought that *E. coli* and certain other foodborne illnesses can sometimes trigger serious health problems months or years after patients survived that initial bout.

Gastrointestinal Infection

Low-temperature electron micrograph of a cluster of *E. coli* bacteria, magnified 10,000 times. Each individual bacterium is oblong shaped.

Certain strains of *E. coli*, such as O157:H7, O121 and O104:H21, produce toxins. Food poisoning caused by *E. coli* are usually associated with eating unwashed vegetables and meat contaminated post-slaughter. O157:H7 is further notorious for causing serious and even life-threatening complications like hemolytic-uremic syndrome (HUS). This particular strain is linked to the 2006 United States E. coli outbreak of fresh spinach. Severity of the illness varies considerably; it can be fatal, particularly to young children, the elderly or the immunocompromised, but is more often mild. *E. coli* can harbor both heat-stable and heat-labile enterotoxins. The latter, termed LT, contains one 'A' subunit and five 'B' subunits arranged into one holotoxin, and is highly similar in structure and function to Cholera toxins. The B subunits assist in adherence and entry of the toxin into host intestinal cells, while the A subunit is cleaved and prevents cells from absorbing water, causing diarrhea. LT is secreted by the Type 2 secretion pathway.

If *E. coli* bacteria escape the intestinal tract through a perforation (for example from an ulcer, a ruptured appendix, or a surgical error) and enter the abdomen, they usually cause peritonitis that can be fatal without prompt treatment. However, *E. coli* are extremely sensitive to such antibiotics as streptomycin or gentamicin. This could change since, as noted below, *E. coli* quickly acquires drug resistance.. Recent research suggests that treatment with antibiotics does not improve the outcome of the disease, and may in fact significantly increase the chance of developing haemolytic uraemic syndrome.

Intestinal mucosa-associated *E. coli* are observed in increased numbers in the inflammatory bowel diseases, Crohn's disease and ulcerative colitis. Invasive strains of *E. coli* exist in high numbers in the inflamed tissue, and the number of bacteria in the inflamed regions correlates to the severity of the bowel inflammation.

Virulence Properties

Enteric *E. coli* (EC) are classified on the basis of serological characteristics and virulence properties. Virotypes include:

- Enterotoxigenic *E. coli* (ETEC): causative agent of diarrhea (without fever) in humans, pigs, sheep, goats, cattle, dogs, and horses. ETEC uses fimbrial adhesins (projections from the bacterial cell surface) to bind enterocyte cells in the small intestine. ETEC can produce two proteinaceous enterotoxins: the larger of the two proteins, LT enterotoxin, is similar to cholera toxin in structure and function, while the smaller protein, ST enterotoxin causes cGMP accumulation in the target cells and a subsequent secretion of fluid and electrolytes into the intestinal lumen. ETEC strains are non-invasive, and they do not leave the intestinal lumen. ETEC is the leading bacterial cause of diarrhea in children in the developing world, as well as the most common cause of traveler's diarrhea. Each year, ETEC causes more than 200 million cases of diarrhea and

380,000 deaths, mostly in children in developing countries.

- Enteropathogenic *E. coli* (EPEC): causative agent of diarrhea in humans, rabbits, dogs, cats and horses. Like ETEC, EPEC also causes diarrhea, but the molecular mechanisms of colonization and etiology are different. EPEC lack fimbriae, ST and LT toxins, but they utilize an adhesin known as intimin to bind host intestinal cells. This virotype has an array of virulence factors that are similar to those found in *Shigella*, and may possess a shiga toxin. Adherence to the intestinal mucosa causes a rearrangement of actin in the host cell, causing significant deformation. EPEC cells are moderately-invasive (*i.e.* they enter host cells) and elicit an inflammatory response. Changes in intestinal cell ultrastructure due to "attachment and effacement" is likely the prime cause of diarrhea in those afflicted with EPEC.
- Enteroinvasive *E. coli* (EIEC): found only in humans. EIEC infection causes a syndrome that is identical to Shigellosis, with profuse diarrhea and high fever. EIEC are highly invasive, and they utilize adhesin proteins to bind to and enter intestinal cells. They produce no toxins, but severely damage the intestinal wall through mechanical cell destruction.
- Enterohemorrhagic *E. coli* (EHEC): found in humans, cattle, and goats. The sole member of this virotype is strain O157:H7, which causes bloody diarrhea and no fever. EHEC can cause hemolytic-uremic syndrome and sudden kidney failure. It uses bacterial fimbriae for attachment, is moderately-invasive and possesses a phage-encoded Shiga toxin that can elicit an intense inflammatory response.
- Enteroaggregative *E. coli* (EAggEC): found only in humans. So named because they have fimbriae which aggregate tissue culture cells, EAggEC bind to the intestinal mucosa to cause watery diarrhea without fever. EAggEC are non-invasive. They produce a hemolysin and an ST enterotoxin similar to that of ETEC.

Epidemiology of Gastrointestinal Infection

Transmission of pathogenic *E. coli* often occurs via fecal-oral transmission. Common routes of transmission include: unhygienic food preparation, farm contamination due to manure fertilization, irrigation of crops with contaminated greywater or raw sewage, feral pigs on cropland, or direct consumption of sewage-contaminated water. Dairy and beef cattle are primary reservoirs of *E. coli* O157:H7, and they can carry it asymptomatically and shed it in their feces. Food products associated with *E. coli* outbreaks include raw ground beef, raw seed sprouts or spinach, raw milk, unpasteurized juice, and foods contaminated by infected food workers via fecal-oral route.

According to the U.S. Food and Drug Administration, the fecal-oral cycle of transmission can be disrupted by cooking food properly, preventing cross-contamination, instituting barriers such as gloves for food workers, instituting health care policies so food industry employees seek treatment when they are ill, pasteurization of juice or dairy products and proper hand washing requirements.

Shiga toxin-producing *E. coli* (STEC), specifically serotype O157:H7, have also been transmitted by flies, as well as direct contact with farm animals, petting zoo animals, and airborne particles found in animal-rearing environments.

Urinary Tract Infection

Uropathogenic *E. coli* (UPEC) is responsible for approximately 90 per cent of urinary tract infections (UTI) seen in individuals with ordinary anatomy. In ascending infections, fecal bacteria colonize the urethra and spread up the urinary tract to the bladder. Because women have a shorter urethra than men, they are 14-times more likely to suffer from an ascending UTI.

Uropathogenic *E. coli* utilize P fimbriae (pyelonephritis-associated pili) to bind urinary tract endothelial cells and colonize the bladder. These adhesins specifically bind D-galactose-D-galactose moieties on the P blood group antigen of erythrocytes and uroepithelial cells. Approximately 1 per cent of the human population lacks this receptor, and its presence or absence dictates an individual's susceptibility to *E. coli* urinary tract infections.

Uropathogenic *E. coli* produce alpha-and beta-hemolysins, which cause lysis of urinary tract cells.

UPEC can evade the body's innate immune defenses (*e.g.* the complement system) by invading superficial umbrella cells to form intracellular bacterial communities (IBCs). They also have the ability to form K antigen, capsular polysaccharides that contribute to biofilm formation. Biofilm-producing *E. coli* are recalcitrant to immune factors and antibiotic therapy and are often responsible for chronic urinary tract infections. K antigen-producing *E. coli* infections are commonly found in the upper urinary tract.

Descending infections, though relatively rare, occur when *E. coli* cells enter the upper urinary tract organs (kidneys, bladder or ureters) from the blood stream.

Laboratory Diagnosis

In stool samples microscopy will show Gram negative rods, with no particular cell arrangement. Then, either MacConkey agar or EMB agar (or both) are inoculated with the stool. On MacConkey agar, deep red colonies are produced as the organism is lactose positive, and fermentation of this sugar will cause the medium's pH to drop, leading to darkening of the medium. Growth on Levine EMB agar produces black colonies with greenish-black metallic sheen. This is diagnosic of *E. coli*. The organism is also lysine positive, and grows on TSI slant with a (A/A/g+/H_2S-) profile. Also, IMViC is ++—— for *E. coli*; as it's indol positive (red ring) and methyl red positive (bright red), but VP negative (no change-colourless) and citrate negative (no change-green colour). Tests for toxin production can use mammalian cells in tissue culture, which are rapidly killed by shiga toxin. Although sensitive and very specific, this method is slow and expensive.

Typically diagnosis has been done by culturing on sorbitol-MacConkey medium and then using typing antiserum. However, current latex assays and some typing antiserum have shown cross reactions with non-E. coli O157 colonies. Furthermore, not all E. coli O157 strains associated with HUS are nonsorbitol fermentors.

The Council of State and Territorial Epidemiologists recommend that clinical laboratories screen at least all bloody stools for this pathogen. The American Gastroenterological Association Foundation (AGAF) recommended in July 1994 that all stool specimens should be routinely tested for E. coli O157:H7.15 It is recommended that the clinician check with their state health department or the Centers for Disease Control and Prevention to determine which specimens should be tested and whether the results are reportable.

Other methods for detecting *E. coli* O157 in stool include ELISA tests, colony immunoblots, direct immunofluorescence microscopy of filters, as well as immunocapture techniques using magnetic beads. These assays are designed as screening tool to allow rapid testing for the presence of *E. coli* O157 without prior culturing of the stool specimen.

Antibiotic Therapy and Resistance

Bacterial infections are usually treated with antibiotics. However, the antibiotic sensitivities of different strains of *E. coli* vary widely. As Gram-negative organisms, *E. coli* are resistant to many antibiotics that are effective against Gram-positive organisms. Antibiotics which may be used to treat *E. coli* infection include amoxicillin as well as other semi-synthetic penicillins, many cephalosporins, carbapenems, aztreonam, trimethoprim-sulfamethoxazole, ciprofloxacin, nitrofurantoin and the aminoglycosides.

Antibiotic resistance is a growing problem. Some of this is due to overuse of antibiotics in humans, but some of it is probably due to the use of antibiotics as growth promoters in food of animals. A study published in the journal *Science* in August 2007 found that the rate of adaptative mutations in *E. coli* is "on the order of 10 per genome per generation, which is 1,000 times as high as previous estimates," a finding which may have significance for the study and management of bacterial antibiotic resistance.

Antibiotic-resistant *E. coli* may also pass on the genes responsible for antibiotic resistance to other species of bacteria, such as *Staphylococcus aureus*. *E. coli* often carry multidrug resistant plasmids and under stress readily transfer those plasmids to other species. Indeed, *E. coli* is a frequent member of biofilms, where many species of bacteria exist in close proximity to each other. This mixing of species allows *E. coli* strains that are piliated to accept and transfer plasmids from and to other bacteria. Thus *E. coli* and the other enterobacteria are important reservoirs of transferable antibiotic resistance.

Beta-lactamase Strains

Resistance to beta-lactam antibiotics has become a particular problem in recent decades, as strains of bacteria that produce extended-spectrum beta-lactamases have become more common. These beta-lactamase enzymes make many, if not all, of the penicillins and cephalosporins ineffective as therapy. Extended-spectrum beta-lactamase-producing *E. coli* are highly resistant to an array of antibiotics and infections by these strains is difficult to treat. In many instances, only two oral antibiotics and a very limited group of intravenous antibiotics remain effective.

Increased concern about the prevalence of this form of "superbug" in the United Kingdom has led to calls for further monitoring and a UK-wide strategy to deal with infections and the deaths. Susceptibility testing should guide treatment in all infections in which the organism can be isolated for culture.

Phage Therapy

Phage therapy—viruses that specifically target pathogenic bacteria—has been developed over the last 80 years, primarily in the former Soviet Union, where it was used to prevent diarrhea caused by *E. coli*. Presently, phage therapy for humans is available only at the Phage Therapy Center in the Republic of Georgia and in Poland. However, on January 2 2007, the United States FDA gave Omnilytics approval to apply its *E. coli* O157:H7 killing phage in a mist, spray or wash on live animals that will be slaughtered for human consumption.

Vaccination

Researchers have actively been working to develop safe, effective vaccines to lower the worldwide incidence of *E. coli* infection. In

March 2006, a vaccine eliciting an immune response against the *E. coli* O157:H7 O-specific polysaccharide conjugated to recombinant exotoxin A of *Pseudomonas aeruginosa* (O157-rEPA) was reported to be safe in children two to five years old. Previous work had already indicated that it was safe for adults. A phase III clinical trial to verify the large-scale efficacy of the treatment is planned.

In 2006 Fort Dodge Animal Health (Wyeth) introduced an effective live attenuated vaccine to control airsacculitis and peritonitis in chickens. The vaccine is a genetically modified avirulent vaccine that has demonstrated protection against O78 and untypeable strains.

In January 2007 the Canadian bio-pharmaceutical company Bioniche announced it has developed a cattle vaccine which reduces the number of O157:H7 shed in manure by a factor of 1000, to about 1000 pathogenic bacteria per gram of manure.

Role in Biotechnology

Because of its long history of laboratory culture and ease of manipulation, *E. coli* also plays an important role in modern biological engineering and industrial microbiology. The work of Stanley Norman Cohen and Herbert Boyer in *E. coli*, using plasmids and restriction enzymes to create recombinant DNA, became a foundation of biotechnology.

Considered a very versatile host for the production of heterologous proteins, researchers can introduce genes into the microbes using plasmids, allowing for the mass production of proteins in industrial fermentation processes. Genetic systems have also been developed which allow the production of recombinant proteins using *E. coli*. One of the first useful applications of recombinant DNA technology was the manipulation of *E. coli* to produce human insulin. Modified *E. coli* have been used in vaccine development, bioremediation, and production of immobilised enzymes. *E. coli* cannot, however, be used to produce some of the more large, complex proteins which contain multiple disulfide bonds and, in particular, unpaired thiols, or proteins that also require post-translational modification for activity.

Model Organism

E. coli is frequently used as a model organism in microbiology studies. Cultivated strains (*e.g. E. coli* K12) are well-adapted to the laboratory environment, and, unlike wild type strains, have lost their ability to thrive in the intestine. Many lab strains lose their ability to form biofilms. These features protect wild type strains from antibodies and other chemical attacks, but require a large expenditure of energy and material resources.

In 1946, Joshua Lederberg and Edward Tatum first described the phenomenon known as bacterial conjugation using *E. coli* as a model bacterium, and it remains the primary model to study conjugation. *E. coli* was an integral part of the first experiments to understand phage genetics, and early researchers, such as Seymour Benzer, used *E. coli* and phage T4 to understand the topography of gene structure. Prior to Benzer's research, it was not known whether the gene was a linear structure, or if it had a branching pattern.

Long-term evolution experiments using *E. coli* have allowed direct observation of major evolutionary shifts in the laboratory.

HEPATITIS A

Hepatitis A, (formerly known as *infectious hepatitis*), is an acute infectious disease of the liver caused by Hepatitis A virus, which is most commonly transmitted by the fecal-oral route via contaminated food or drinking water. Every year, approximately 10 million people worldwide are infected with the virus. The time between infection and the appearance of the symptoms, (the incubation period), is between two and six weeks and the average incubation period is 28 days. In developing countries, and in regions with poor hygiene standards, the incidence of infection with this virus approaches 100 per cent and the illness is usually contracted in early childhood. Hepatitis A infection causes no clinical signs and symptoms in over 90 per cent of these children and since the infection confers lifelong immunity, the disease is of no special significance to the indigenous population. In Europe, the United States and other industrialized countries, on the other hand, the infection is contracted primarily by susceptible young adults, most of whom are infected with the virus during trips to countries with a high incidence of the disease.

Hepatitis A does not have a chronic stage and does not cause permanent liver damage. Following infection, the immune system makes antibodies against the hepatitis A virus that confer immunity against future infection. The disease can be prevented by vaccination and hepatitis A vaccine has been proved effective in controlling outbreaks worldwide.

Virus

The Hepatitis virus (HAV) is a Picornavirus; it is non-enveloped and contains a single-stranded RNA packaged in a protein shell. There is only one type of the virus.

Pathogenesis

Following ingestion, HAV enters the bloodstream through the epithelium of the oropharynx or intestine. The blood carries the virus to its target, the liver, where it lives and multiplies within hepatocytes and Kupffer cells (*i.e.*, liver macrophages). There is no apparent virus-mediated cytotoxicity, and liver pathology is likely immune-mediated. Virions are secreted into the bile and released in stool. HAV is excreted in large quantities approimately 11 days prior to appearance of symptoms or anti-HAV IgM antibodies in the blood. The incubation period is 15-50 days, and mortality is less than 0.5 per cent.

Transmission

The virus spreads by the fecal-oral route and infections often occur in conditions of poor sanitation and overcrowding. Hepatitis A can be transmitted by the parenteral route but very rarely by blood and blood products. Food-borne outbreaks are not uncommon, and ingestion of shellfish cultivated in polluted water is associated with a high risk of infection. Approximately 40 per cent of all acute viral hepatitis is caused by HAV. Infected individuals are infectious prior to onset of symptoms, roughly 10 days following infection. The virus is resistant to detergent, acid (pH 1), solvents (*e.g.*, ether, chloroform), drying, and temperatures up to 60C. It can survive for months in fresh and salt water. Common-source (*e.g.*, water, restaurant) outbreaks are typical. Infection is common in children in developing countries, reaching 100 per cent incidence, but following infection there is life-long immunity. HAV can be inactivated by: chlorine treatment (drinking water), formalin (0.35 per cent, 37C, 72 hours), peracetic acid (2 per cent, 4 hours), beta-propiolactone (0.25 per cent, 1 hour), and UV radiation (2 ìW/cm/min).

Symptoms

Early symptoms of hepatitis A infection can be mistaken for influenza, but some sufferers, especially children, exhibit no symptoms at all. Symptoms typically appear 2 to 6 weeks, (the incubation period), after the initial infection.

Symptoms can return over the following 6-9 months which include:

- Fatigue
- Fever
- Abdominal pain
- Nausea
- Diarrhea
- Appetite loss
- Depression
- Jaundice, a yellowing of the skin or whites of the eyes
- Sharp pains in the right-upper quadrant of the abdomen
- Weight loss
- Itching

Diagnosis

Serum IgG, IgM and ALT following Hepatitis A virus infection

Although the virus is excreted in the feces towards the end of the incubation period, specific diagnosis is made by the detection of Hepatitis A virus specific IgM antibodies in the blood. IgM antibody is only present in the blood following an acute hepatitis A infection. It is detectable from one to two weeks after the initial infection and persists for up to 14 weeks. The presence of IgG antibody in the blood means that the acute stage of the illness is past and the person is immune to further infection. IgG antibody to HAV is also found in the blood following vaccination and tests for immunity to the virus are based on the detection of this antibody.

During the acute stage of the infection, the liver enzyme alanine transferase (ALT) is present in the blood at levels much higher than is normal. The enzyme comes from the liver cells that have been damaged by the virus.

Hepatitis A virus is present in the blood, (viremia), and feces of infected people up to two weeks before clinical illness develops.

Treatment

There is no specific treatment for hepatitis A. Sufferers are advised to rest, avoid fatty foods and alcohol (these may be poorly tolerated for some additional months during the recovery phase and cause minor relapses), eat a well-balanced diet, and stay hydrated. Approximately 15 per cent of people diagnosed with hepatitis A may experience one or more symptomatic relapse(s) for up to 24 months after contracting this disease.

Prognosis

The United States Centers for Disease Control and Prevention (CDC) in 1991 reported a low mortality rate for hepatitis A of 4 deaths per 1000 cases for the general population but a higher rate of 17.5 per 1000, in those aged 50 and over. Death usually occurs when the patient contracts Hepatitis A while already suffering from another form of Hepatitis, such as Hepatitis B or Hepatitis C or AIDS.

Young children who are infected with hepatitis A typically have a milder form of the disease, usually lasting from 1-3 weeks, whereas adults tend to experience a much more severe form of the disease.

Prevention

Hepatitis A can be prevented by vaccination, good hygiene and sanitation.

Vaccine

Hepatitis A vaccines protect against the virus. Vaccines contain inactivated Hepatitis A virus providing active immunity against a future infection.

Epidemiology

HAV is found in the feces of infected persons and those who are at higher risk include travelers to developing countries where there is a higher incidence rate, and those having sexual contact or drug use with infected persons. There were 30,000 cases of Hepatitis A reported to the CDC in the U.S. in 1997. The agency estimates that there were as many as 270,000 cases each year from 1980 through 2000.

HAV outbreaks still occur in and caused by poor hand hygiene among infected, sometimes symptomatic restaurant employees failing to wash their hands after toilet breaks.

Epidemics

The most widespread hepatitis A outbreak in the United States afflicted at least 640 people (killing four) in north-eastern Ohio and south-western Pennsylvania in late 2003. The outbreak was blamed on tainted green onions at a restaurant in Monaca, Pennsylvania. In 1988, 300,000 people in Shanghai, China were infected with HAV after eating clams from a contaminated river.

FOOD TECHNOLOGY

Food technology, or Food tech for short is the application of food science to the selection, preservation, processing, packaging, distribution, and use of safe, nutritious, and wholesome food.

Food scientists and food technologists study the physical, microbiological, and chemical makeup of food. Depending on their area of specialization, food scientists may develop ways to process, preserve, package, or store food, according to industry and government specifications and regulations. Consumers seldom think of the vast array of foods and the research and development that has resulted in the means to deliver tasty, nutritious, safe, and convenient foods.

In some schools, food technology is part of the curriculum and teaches, alongside *how* to cook, nutrition and the food manufacturing process.

Early History of Food Technology

Research in the field now known as food technology has been conducted for decades. Nicolas Appert's development in 1810 of the canning process was a decisive event. The process wasn't called canning then and Appert did not really know the principle on which his process worked, but canning has had a major impact on food preservation techniques.

Louis Pasteur's research on the spoilage of wine and his description of how to avoid spoilage in 1864 was an early attempt to put food technology on a scientific basis. Besides research into wine spoilage, Pasteur did research on the production of alcohol, vinegar, wines and beer, and the souring of milk. He developed pasteurization—the process of heating milk and milk products to destroy food spoilage and disease-producing organisms. In his research into food technology, Pasteur became the pioneer into bacteriology and of modern preventive medicine.

By 1940's to 1950's, the original four departments that had taught the subject under different names (including those at the University of Massachusetts and the University of California) had been retitled "food science", "food science and technology", or a similar variant.

Developments in Food Technology

Several companies in the food industry have played a role in the development of food technology. These developments have contributed greatly to the food supply. Some of these developments are:

- Instantized Milk Powder—D.D. Peebles (U.S. patent 2,835,586) developed the first instant milk powder, which has become the basis for a variety of new products that are rehydratable in cold water or milk. This process increases the surface area of the powdered product by partially rehydrating spray-dried milk powder.

- Freeze Drying—The first application of freeze drying was most likely in the pharmaceutical industry; however, a successful large-scale industrial application of the process was the development of continuous freeze drying of coffee.
- High-Temperature Short Time Processing—These processes for the most part are characterized by rapid heating and cooling, holding for a short time at a relatively high temperature and filling aseptically into sterile containers.
- Decaffeination of Coffee and Tea—Decaffeinated coffee and tea was first developed on a commercial basis in Europe around 1900. The process is described in U.S. patent 897,763. Green coffee beans are treated with steam or water to around 20 per cent moisture. The added water and heat separate the caffeine from the bean to its surface. Solvents are then used to remove the caffeine from the beans. In the 1980s, new non-organic solvent techniques have been developed for the decaffeination of coffee and tea. Carbon dioxide under supercritical conditions is one of these new techniques. U.S. patent 4,820,537 was issued to General Foods Corp. for a CO_2 decaffeination process.
- Process optimization—Food Technology now allows production of foods to be more efficient, Oil saving technologies are now available on different forms. Production methods and methodology have also become increasingly

FOOD PRESERVATION

Food preservation is the process of treating and handling food in a way that preserves its edibility and nutrition value. The main effort is to stop or greatly slow down spoilage to prevent foodborne illness (*e.g.* salting, cooling, cooking). However some methods utilise benign bacteria, yeasts or fungi to add specific qualities and to preserve food (*e.g.* cheese, wine). While maintaining or creating nutritional value, texture and flavour is important in preserving its value as food; this is a culturally dependent determinant as what qualifies as food fit for humans in one culture may not qualify in another culture.

Preservation usually involves preventing the growth of bacteria, and other micro-organisms, as well as retarding the oxidation of fats which cause rancidity. It also includes processes to inhibit natural ageing and discolouration that can occur during food preparation such as the enzymatic browning reaction in apples which causes browning when apples are cut. Some preservation methods require the food to be sealed after treatment to prevent re-contamination with microbes; others, such as drying, allow food to be stored without any special containment for long periods.

Common methods of applying these processes include drying, spray drying, freeze drying, freezing, vacuum-packing, canning, preserving in syrup, sugar crystallisation, food irradiation, adding preservatives or inert gases such as carbon dioxide. Other methods that not only help to preserve food, but also add flavour, include pickling, salting, smoking, preserving in syrup or alcohol, sugar crystallisation and curing.

Preservation Processes

Preservation processes include:

- Heating to kill or denature organisms (*e.g.* boiling)
- Oxidation (*e.g.* use of sulphur dioxide)
- Toxic inhibition (*e.g.* smoking, use of carbon dioxide, vinegar, alcohol etc)
- Dehydration (drying)
- Osmotic inhibition (*e.g.* use of syrups)
- Low temperature inactivation (*e.g.* freezing)
- Ultra high water pressure (*e.g.* fresherized, a kind of "cold" pasteurization, the pressure kills naturally occurring pathogens, which cause food deterioration and affect food safety.)

Table 5.2: Refrigeration Method and their Effect on Microbial Growth

Method	*Effect on Microbial Growth or Survival*
Refrigeration	Low temperature to retard growth
Freezing	Low temperature and reduction of water activity to prevent microbial growth
Drying, curing and conserving	Reduction in water activity sufficient to delay or prevent microbial growth
Vacuum and oxygen free modified atmosphere packaging	Low oxygen tension inhibits strict aerobes and delay growth of facultative anaerobes
Carbon dioxide enriched modified atmosphere packaging	Specific inhibition of some micro-organisms by carbon dioxide
Addition of weak acids	Reduction of the intracellular pH of micro-organisms
Lactic fermentation	Reduction of pH value in situ by microbial action and sometimes additional inhibition by the lactic and acetic acids formed and by other microbial products. (*e.g.* ethanol, bacteriocins)
Sugar preservation	Cooking in high sucrose concentration creating too high osmotic pressure for most microbial survival.
Ethanol preservation	Steeping or cooking in Ethanol produces toxic inhibition of microbes. Can be combined with sugar preservation
Emulsification	Compartmentalisation and nutrient limitation within the aqueous droplets in water-in-oil emulsion foods
Addition of preservatives such as nitrite or sulphite ions	Inhibition of specific groups of micro-organisms
Pasteurization and appertization	Delivery of heat sufficient to inactivate target micro-organisms to the desired extent
Food irradiation (Radurization, radicidation and radappertization)	Delivery of ionising radiation
Application of high hydrostatic pressure (Pascalization)	Pressure-inactivation of vegetative bacteria, yeasts and moulds
Pulsed electric field processing (PEF treatment)	Short bursts of electricity for microbial inactivation

- Many combinations of these methods
- Chelation

Drying

One of the oldest methods of food preservation is by drying, which reduces water activity sufficiently to delay or prevent bacterial growth. Most types of meat can be dried. This is especially valuable in the case of pork, since it is difficult to keep without preservation. Many fruits can also be dried; for example, the process is often applied to apples, pears, bananas, mangoes, papaya, apricot, and coconut. Zante currants, sultanas and raisins are all forms of dried grapes. Drying is also the normal means of preservation for cereal grains such as wheat, maize, oats, barley, rice, millet and rye.

Smoking

Meat, fish and some other foods may be both preserved and flavored through the use of smoke, typically in a smokehouse. The combination of heat to dry the food without cooking it, and the addition of the aromatic hydrocarbons from the smoke preserves the food.

Freezing

Freezing is also one of the most commonly used processes commercially and domestically for preserving a very wide range of food including prepared food stuffs which would not have required freezing in their unprepared state. For example, potato waffles are stored in the freezer, but potatoes themselves require only a cool dark place to ensure many months' storage. Cold stores provide large volume, long-term storage for strategic food stocks held in case of national emergency in many countries.

Vacuum Packing

Vacuum-packing stores food in a vacuum environment, usually in an air-tight bag or bottle. The vacuum environment strips bacteria of oxygen needed for survival, preventing the food from spoiling. Vacuum-packing is commonly used for storing nuts.

Salt

Salting or curing draws moisture from the meat through a process of osmosis. Meat is cured with salt or sugar, or a combination of the two. Nitrates and nitrites are also often used to cure meat.

Sugar

Sugar is used to preserve fruits, either in syrup with fruit such as apples, pears, peaches, apricots, plums or in crystallized form where the preserved material is cooked in sugar to the point of crystallisation and the resultant product is then stored dry. This method is used for the skins of citrus fruit (candied peel), angelica and ginger. A modification of this process produces glacé fruit such as glacé cherries where the fruit is preserved in sugar but is then extracted from the syrup and sold, the preservation being maintained by the sugar content of the fruit and the superficial coating of syrup. The use of sugar is often combined with alcohol for preservation of luxury products such as fruit in brandy or other spirits. These should not be confused with fruit flavored spirits such as Cherry Brandy or Sloe gin.

Pickling

Pickling is a method of preserving food in an edible anti-microbial liquid. Pickling can be broadly categorized as chemical pickling (for example, brining) and fermentation pickling (for example, making sauerkraut).

In chemical pickling, the food is placed it in an edible liquid that inhibits or kills bacteria and other micro-organisms. Typical pickling agents include brine (high in salt), vinegar, alcohol, and vegetable oil, especially olive oil but also many other oils. Many chemical pickling processes also involve heating or boiling so that the food being preserved becomes saturated with the pickling agent. Common chemically pickled foods include cucumbers, peppers, corned beef, herring, and eggs, as well mixed vegetables such as piccalilli, chow-chow, giardiniera, and achar.

In fermentation pickling, the food itself produces the preservation agent, typically by a process that produces lactic acid.

Fermented pickles include sauerkraut, nukazuke, kimchi, surströmming, and curtido. Some chemically pickled cucumbers are also fermented.

In commercial pickles, a preservative like sodium benzoate or EDTA may also be added to enhance shelf life.

Lye

Sodium hydroxide (lye) makes food too alkaline for bacterial growth. Lye will saponify fats in the food, which will change its flavor and texture. Lutefisk and hominy use lye in their preparation, as do some olive recipes. Modern recipes for century eggs also call for lye.

Canning and Bottling

Canning involves cooking fruits or vegetables, sealing them in sterile cans or jars, and boiling the containers to kill or weaken any remaining bacteria as a form of pasteurization. Various foods have varying degrees of natural protection against spoilage and may require that the final step occur in a pressure cooker. High-acid fruits like strawberries require no preservatives to can and only a short boiling cycle, whereas marginal fruits such as tomatoes require longer boiling and addition of other acidic elements. Many vegetables require pressure canning. Food preserved by canning or bottling is at immediate risk of spoilage once the can or bottle has been opened.

Lack of quality control in the canning process may allow ingress of water or micro-organisms. Most such failures are rapidly detected as decomposition within the can causes gas production and the can will swell or burst. However, there have been examples of poor manufacture and poor hygiene allowing contamination of canned food by the obligate anaerobe, *Clostridium botulinum* which produces an acute toxin within the food leading to severe illness or death. This organism produces no gas or obvious taste and remains undetected by taste or smell.

Jellying

Food may be preserved by cooking in a material that solidifies to form a gel. Such materials include gelatine, agar, maize flour and arrowroot flour. Some foods naturally form a protein gel when cooked such as eels and elvers, and sipunculid worms which are a delicacy in the town of Xiamen in Fujian province of the People's Republic of China. Jellied eels are a delicacy in the East End of London where they are eaten with mashed potatoes. Potted meats in aspic, (a gel made from gelatine and clarified meat broth) were a common way of serving meat off-cuts in the UK until the 1950s. Many jugged meats are also jellied.

Fruit preserved by jellying is known as jelly, marmalade, or fruit preserves. In this case, the jellying agent is usually pectin, either added during cooking or arising naturally from the fruit. Most preserved fruit is also sugared.

Jugging

Meat can be preserved by jugging, the process of stewing the meat (commonly game or fish) in a covered earthenware jug or casserole. The animal to be jugged is usually cut into pieces, placed into a tightly-sealed jug with brine or gravy, and stewed. Red wine and/or the animal's own blood is sometimes added to the cooking liquid. Jugging was a popular method of preserving meat up until the middle of the 20th century.

Irradiation

Irradiation of food is the exposure of food to ionizing radiation; either high-energy electrons or X-rays from accelerators, or by gamma rays (emitted from radioactive sources as Cobalt-60 or Caesium-137). The treatment has a range of effects, including killing bacteria, molds and insect pests, reducing the ripening and spoiling of fruits, and at higher doses inducing sterility. The technology may be compared to pasteurization; it is sometimes called 'cold pasteurization', as the product is not heated. Irradiation is not effective against viruses or prions, it cannot eliminate toxins already formed by microorganisms, and is only useful for food of high initial quality.

The radiation process is unrelated to nuclear energy, but it may use the radiation emitted from radioactive nuclides produced in nuclear reactors. Ionizing radiation is hazardous to life; for this reason irradiation facilities have a heavily shielded irradiation room where the process takes place. Radiation safety procedures ensure that neither the workers in such facility nor the environment receive any radiation dose from the facility. Irradiated food does not become radioactive, and national and international expert bodies have declared food irradiation as wholesome. However, the wholesomeness of consuming such food is disputed by opponents and consumer organizations. National and international expert bodies have declared food irradiation as 'wholesome'; UN-organizations as WHO and FAO are endorsing to utilize food irradiation. International legislature on whether food may be irradiated or not varies worldwide from no regulation to full banning.

It is estimated that about 500,000 tons of food items are irradiated per year world-wide in over 40 countries. These are mainly spices and condiments with an increasing segment of fresh fruit irradiated for fruit fly quarantine.

Modified Atmosphere

Modified atmosphere is a way to preserve food by operating on the atmosphere around it. Salad crops which are notoriously difficult to preserve are now being packaged in sealed bags with an atmosphere modified to reduce the oxygen (O_2) concentration and increase the carbon dioxide (CO_2) concentration. There is concern that although salad vegetables retain their appearance and texture in such conditions, this method of preservation may not retain nutrients, especially vitamins.

Grains may be preserved using carbon dioxide. A block of dry ice is placed in the bottom and the can is filled with grain. The can is then "burped" of excess gas. The carbon dioxide from the sublimation of the dry ice prevents insects, mold, and oxidation from damaging the grain. Grain stored in this way can remain edible for five years.

Nitrogen gas (N_2) at concentrations of 98 per cent or higher is also used effectively to kill insects in grain through hypoxia.

However, carbon dioxide has an advantage in this respect as it kills organisms through both hypoxia and hypercarbia, requiring concentrations of only 80 per cent, or so. This makes carbon dioxide preferable for fumigation in situations where an hermetic seal cannot be maintained.

Burial in the Ground

Burial of food can preserve it due to a variety of factors: lack of light, lack of oxygen, cool temperatures, pH level, or desiccants in the soil. Burial may be combined with other methods such as fermentation.

Many root vegetables are very resistant to spoilage and require no other preservation other than storage in cool dark conditions, for example by burial in the ground, such as in a storage clamp.

Century eggs are created by placing eggs in alkaline mud (or other alkaline substance) resulting in their "inorganic" fermentation through raised pH instead of spoiling. The fermentation preserves them and breaks down some of the complex, less flavorful proteins and fats into simpler more flavorful ones.

Most foods can be preserved in soil that is very dry and salty (thus a desiccant), or soil that is frozen.

Cabbage was traditionally buried in the fall in northern farms in the USA for preservation. Some methods keep it crispy while other methods produce sauerkraut. A similar process is used in the traditional production of kimchi.

Sometimes meat is buried under conditions which cause preservation. If buried on hot coals or ashes, the heat can kill pathogens, the dry ash can desiccate, and the earth can block oxygen and further contamination. If buried where the earth is very cold, the earth acts like a refrigerator. Fish (*e.g.* Gravlax) has been buried to preserve by fermentation.

Controlled Use of Micro-organism

Some foods, such as many cheeses, wines, and beers will keep for a long time because their production uses specific micro-organisms that combat spoilage from other less benign organisms. These micro-organisms keep pathogens in check by creating an environment toxic for themselves and other micro-organisms by producing acid or alcohol. Starter micro-organisms, salt, hops, controlled (usually cool) temperatures, controlled (usually low) levels of oxygen and/or other methods are used to create the specific controlled conditions that will support the desirable organisms that produce food fit for human consumption.

High Pressure Food Preservation

High pressure food preservation refers to high pressure used for food preservation. "Pressed inside a vessel exerting 70,000 pounds per square inch or more, food can be processed so that it retains its fresh appearance, flavour, texture and nutrients while disabling harmful microorganisms and slowing spoilage." By 2001, adequate commercial equipment was developed so that by 2005 the process was being used for products ranging from orange juice to guacamole to deli meats and widely

FOOD IRRADIATION

Food irradiation is the process of exposing food to ionizing radiation in order to destroy microorganisms, bacteria, viruses, or insects that might be present in the food. Further applications include sprout inhibition, delay of ripening, increase of juice yield, and improvement of re-hydration. Irradiation is a more general term of deliberate exposure of materials to radiation to achieve a technical goal (in this context 'ionizing radiation' is implied). As such it is also used on non-food items, such as medical hardware, plastics, tubes for gas-pipelines, hoses for floor-heating, shrink-foils for food packaging, automobile parts, wires and cables (isolation), tires, and even gemstones. Compared to the amount of food irradiated, the volume of those every-day applications is huge but not noticed by the consumer.

The genuine effect of processing food by ionizing radiation relates to damages to the DNA, the basic genetic information for life. Microorganisms can no longer proliferate and continue their malignant or pathogen activities. Spoilage-causing micro-organisms cannot continue their activities. Insects do not survive or become incapable of proliferation. Plants cannot continue the natural ripening or aging process.

The speciality of processing food by ionizing radiation is that the energy density per atomic transition is very high; it can cleave molecules and induce ionization (hence the name), which is not achieved by mere heating. This is the reason for both new effects and new concerns. The treatment of solid food by ionizing radiation can provide an effect similar to heat pasteurization of liquids, such as milk. However, the use of the term "cold pasteurization" to describe irradiated foods is controversial, since pasteurization and irradiation are fundamentally different processes.

Food irradiation is currently permitted by over 40 countries and volumes are estimated to exceed 500,000 metric tons annually world wide.

Processing of Food by Ionizing Radiation

By irradiating food, depending on the dose, some or all of the harmful bacteria and other pathogens present are killed. This prolongs the shelf-life of the food in cases where microbial spoilage is the limiting factor. Some foods (*e.g.*, herbs and spices) are irradiated at sufficient doses (five kilograys or more) to reduce the microbial counts by several orders of magnitude; such ingredients will not carry over spoilage or pathogen microorganisms into the final product. It has also been shown that irradiation can delay the ripening of fruits or the sprouting of vegetables.

Furthermore, insect pests can be sterilized (be made incapable of proliferation) using irradiation at relatively low doses. In consequence, the United States Department of Agriculture has approved the use of low-level irradiation as an alternative treatment to pesticides for fruits and vegetables that are considered hosts to a number of insect pests, including fruit flies and seed weevils; US FDA has cleared among a number of other applications the treatment of hamburger patties to eliminate the residual risk of a contamination by a virulent E. coli. The United Nations Food and Agricultural Organization (FAO) has passed a motion to commit member states to implement irradiation technology for their national phytosanitary programmes; the General assembly of the International Atomic Energy Agency (IAEA) has urged to make wider use of the irradiation technology. And the USDA has made a number of bi-lateral agreements with developing countries to facilitate the imports of exotic fruits and to simplify the quarantine procedures.

Currently, the European Union has regulated processing of food by ionizing radiation in specific directives since 1999; the situation is easily explored and the several documents and reports are accessible. The 'implementing' directive contains a 'positive list' only permitting irradiation of dried aromatic herbs, spices, and vegetable seasonings. However, any Member State is permitted to maintain previously granted clearances or to add new clearance as granted in other Member States, in the case the EC's Scientific Committee on Food (SCF) has given a positive vote for the respective application. Presently, six Member States (Belgium, France, Italy, Netherlands, Poland, United Kingdom) have adopted such provisions.

Because of the 'Single Market' of the EC, any food—even if irradiated—must be allowed to be marketed in any other Member State even if a general ban of food irradiation prevails, under the condition that the food has been irradiated legally in the state of origin. Furthermore, imports into the EC are possible from third countries if the irradiation facility had been inspected and licensed by the EC and the treatment is legal within the EC or some Member state.

The Scientific Committee on Food (SCF) of the EC has given a positive vote on eight categories of food to be irradiated. However, in a compromise between the European Parliament and the European Commission, only dried aromatic herbs, spices, and vegetable seasonings can be found in the positive list. The European Commission was due to provide a final draft for the positive list by the end of 2000; however, this failed because of a veto from Germany and a few other Member States. In 1992 and in 1998 the SCF voted positive on a number of irradiation applications which had been allowed in some Member States before the EC Directives came into force, in order to enable those Member States to maintain the national authorizations. In 2003 (at the occasion when Codex Alimentarius was about to remove any upper dose limit for food irradiation) the SCF adopted a 'revised opinion' which in fact is just a re-confirmation and endorsement of the 1986-opinion. The cancellation of the upper dose limit is denied, and before the actual list of individual items or food classes (as in the opinions expressed in 1986, 1992 and 1998) is expanded, new individual studies into the toxicology of each of such food and for each of the proposed dose ranges is requested. After 2003 the SCF has been replaced by the new European Food Safety Authority (EFSA), which not yet has voted on processing food by ionizing radiation. Other countries including New Zealand, Australia, Thailand, India, and Mexico have permitted the irradiation of fresh fruits for fruit fly quarantine purposes amongst others. Other countries as Pakistan and Brazil have adopted the Codex Alimentarius Standard on Irradiated Food without any reservation or restriction, *i.e.* any food to any dose.

Radiation Absorbed Dose

'Dose' is the physical quantity governing radiation processing of food and the beneficial effects to be achieved.

Unit of Measure for Irradiation Dose

The dose of radiation is measured in the SI unit known as Gray (Gy). One Gray (Gy) dose of radiation is equal to 1 joule of energy absorbed per kg of food material. In radiation processing of foods, the doses are generally measured in kGy (1,000 Gy).

Dosimetry

The measurement of radiation dose is referred to as dosimetry and it involves exposing dosimeters jointly with the treated food item. Dosimeters are small components attached to the irradiated product made of materials that when exposed to ionizing radiation change specific measureable physical attributes to a degree that can be correllated to the dose received. Modern dosimeters are made of a range of materials such as alanine pellets, perspex (PMMA) blocks, radiochromic films, as well as special solutions and other materials. These dosimeters are used in combination with specialized read out devices. Standards that describe calibration and operation for radiation dosimetry as well as procedures to relate the measured dose to the effects achieved and to report and document such results are maintained by the American Society for Testing and Materials (ASTM international) and are also available as ISO/ASTM standards.

Applications

On the basis of the dose of radiation the application is generally divided into three main categories as detailed under:

Low Dose Applications (up to 1 kGy)

- Sprout inhibition in bulbs and tubers 0.03-0.15 kGy
- Delay in fruit ripening 0.25-0.75 kGy
- Insect disinfestation including quarantine treatment and elimination of food borne parasites 0.07-1.00 kGy

Medium Dose Applications (1 kGy to 10 kGy)

- Reduction of spoilage microbes to prolong shelf-life of meat, poultry and seafoods under refrigeration 1.50-3.00 kGy
- Reduction of pathogenic microbes in fresh and frozen meat, poultry and seafoods 3.00-7.00 kGy
- Reducing number of microorganisms in spices to improve hygenic quality 10.00 kGy

High Dose Applications (above 10 kGy)

- Sterilisation of packaged meat, poultry and their products which are shelf stable without refrigeration. 25.00-70.00 kGy
- Sterilisation of Hospital diets 25.00-70.00 kGy
- Product improvement as increased juice yield or improved re-hydration

It is important to note that these doses are above those currently permitted for these food items by the FDA and other regulators around the world. The Codex Alimentarius Standard on Irradiated Food does not specify any upper dose limit. NASA is authorized to sterilize frozen meat for astronauts at doses of 44 kGy as a notable exception. Irradiation treatments are also sometimes classified as radappertization, radicidation and radurization.

Technologies

Electron Irradiation

Electron irradiation uses electrons accelerated in an electric field to a velocity close to the speed of light. Electrons are particulate radiation and, hence, have cross section many times larger than photons, so that they do not penetrate the product beyond a few inches, depending on product density. Electron facilities rely on substantial concrete shields to protect workers and the environment from radiation exposure.

Gamma Irradiation

Gamma radiation is radiation of photons in the gamma part of the electromagnetic spectrum. The radiation is obtained through the use of radioisotopes, generally Cobalt-60 or, in theory, Cesium-137. Cesium-137 is recovered during the refinement of spent nuclear fuel. Because this technology—except for military applications—is not commercially available, insufficient quantities of it are available on the global isotope markets for use in large scale, commercial irradiators. Presently, Cesium-137 is used only in small hospital units to treat blood before transfusion to prevent Graft-versus-host disease.

Food irradiation using Cobalt-60 is the preferred method by most processors, because the deeper penetration enables administering treatment to entire industrial pallets or totes, reducing the need for material handling. A pallet or tote is typically exposed for several minutes to hours depending on dose. Radioactive material must be monitored and carefully stored to shield workers and the environment from its gamma rays. During operation this is achieved by substantial concrete shields. With most designs the radioisotope can be lowered into a water-filled source storage pool to allow maintenance personnel to enter the radiation shield. In this mode the water in the pool absorbs the radiation. Other uncommonly used designs feature dry storage by providing movable shields that reduce radiation levels in areas of the irradiation chamber.

One variant of gamma irradiators keeps the Cobalt-60 under water at all times and lowers the product to be irradiated under water in hermetic bells. No further shielding is required for such designs.

X-ray Irradiation

Similar to gamma radiation, X-rays are photon radiation of a wide energy spectrum and an alternative to isotope based irradiation systems. X-rays are generated by colliding accelerated electrons with a dense material (target) such as tantalum or tungsten in a process known as bremsstrahlung-conversion. X-ray irradiators are scalable and have deep penetration comparable to Co-60, with the added effect of using an electronic source that stops radiating when switched off. They also permit dose uniformity. However, these systems generally have low energetic efficiency during the conversion of electron energy to photon radiation requiring much more electrical energy than other systems. Like most other types of facilities, X-ray systems rely on concrete shields to protect the environment and workers from radiation.

Nominal X-ray energy is usually limited to 5 MeV; however, USA has provisions for up to 7.5 MeV which increases the conversion efficiency. An other development is the availability of electron accelerators with extremely high power output, up to 1,000 kW beam. At a conversion efficiency of up to 12 per cent, the X-ray power may reach (including filtering and other losses) 100 kW; This power would be equivalent to a gamma facility with Co-60 of about 6.5 MCi.

Irradiated Foods in the Market Place

Current U.S. Market

Many U.S. supermarkets carry irradiated food products today ranging from fresh tropical fruit from Hawaii or Florida, dehydrated spices and ground meat products. Certain supermarkets like Whole Foods Market prefer not to carry irradiated products for reasons of consumer perception.

General Economic Aspects

Some foods, particularly fruits and vegetables, are not available for sale on the global market, unless they are treated in order to prolong shelf life for transportation; this may include radiation processing. However, this application has not yet been exploited. In contrast, irradiation to eliminate insect pests in order to fulfill quarantine requirements is gaining commercial significance. In particular for fruits from Hawaii to be sold on mainland US; and increasingly for imports from subtropical countries to the US (under bilateral agreements) which allows those less developed countries to earn income through food exports.

The actual cost of food irradiation is influenced by dose requirements, the food's tolerance of radiation, handling conditions (*i.e.*, packaging and stacking requirements), construction costs, financing arrangements, and other variables particular to the situation. Irradiation is a capital-intensive technology requiring a substantial initial investment, ranging from $1 million to $5 million. In the case of large research or contract irradiation facilities, major capital costs include a radiation source (cobalt-60), hardware (irradiator, totes and conveyors, control systems, and other auxiliary equipment), land (1 to 1.5 acres), radiation shield, and warehouse. Operating costs include salaries (for fixed and variable labour), utilities, maintenance, taxes/insurance, cobalt-60 replenishment, general utilities, and miscellaneous operating costs

Treatment costs vary as a function of dose and facility usage. Low dose applications such as disinfestation of fruit range between $US 0.01/lbs and $US 0.08/lbs while higher dose applications can cost as much as $US 0.20/lbs.

Consumer Perception

Opponents to food irradiation, consumer organizations and environmentalist groups refer to some studies suggesting that a large part of the public questions the safety of irradiated foods, and will not buy foods that have been irradiated.

On the other hand, other studies indicate the number of consumers concerned about the safety of irradiated food has decreased in the last 10 years and continues to be less than the number of those concerned about pesticide residues, microbiological contamination, and other food related concerns. The number of people reporting no concerns about irradiated food is among the lowest for food issues, comparable to that of people with no concern about food additives and preservatives. Consumers, given a choice and access to the real irradiated product are ready to buy it in considerably large numbers.

The Globalized Food Supply

Opponents of food irradiation sometimes state that large-scale irradiation would increase processing, transportation, and handling times for fruits and vegetables thus contributing to a negative ecological balance compared to locally grown foods.

Labeling and Terminology Issues

Labeling laws differ from country to country. While Codex Alimentarius represents the global standard in particular under the WTO-agreement, member states are free to convert those standards into national regulations. With regard to labelling of irradiated food detailed rules are published at CODEX-STAN-1 (2005) labelling of prepacked food.

The provisions are that any 'first generation' product must be labelled 'irradiated' as any product derived directly from an irradiated raw material; for ingredients the provision is that even the last molecule of an irradiated ingredient must be listed with the ingredients even in cases where the unirradiated ingredient will not appear on the label. The RADURA-logo is optional; several countries use a graphical version which differs from the Codex-version.

In the US as in many other countries irradiated food must be labeled as "Treated with irradiation" or "Treated by radiation" and require the usage of the Radura symbol at the point of sale. However, the meaning of the label is not consistent. The amount of irradiation used can vary and since there are no published standards, the amount of pathogens affected by irradiation can be variable as well. In addition, there are no regulations regarding the levels of pathogen reduction that must be achieved. Food that is processed as an ingredient by a restaurant or food processor is exempt from the labeling requirement in the US; other countries follow the Codex Alimentarius provision to label irradiated ingredients down to the last molecule (cf. EU).

FDA is currently proposing a rule that in some cases would allow certain irradiated foods to be marketed without any labeling at all. Under the new rules, only those irradiated foods in which the irradiation causes a material change in the food, or a material change in the consequences that may result from the use of the food, would bear the Radura symbol and the term "irradiated", or a derivative thereof, in conjunction with explicit language describing the change in the food or its conditions of use. In the same rule FDA is proposing to permit a firm to use the terms "electronically pasteurized" or "cold pasteurized" in lieu of "irradiated", provided it notifies the agency that the irradiation process being used meets the criteria specified for use of the term "pasteurized".

Food irradiation is sometimes referred to as 'cold pasteurization' or 'electronic pasteurization' because ionizing the food does not heat the food to high temperatures during the process, as in heat-pasteurization (at a typical dose of 10 kGy, food that is physically equivalent to water would warm by about 2.5°C). The treatment of solid food by ionizing radiation can provide an effect similar to heat pasteurization of liquids, such as milk. However, the use of the term, cold pasteurization, to describe irradiated foods is controversial, because pasteurization and irradiation are fundamentally different processes, although the intended end results can in some cases be similar.

Consumer perception of foods treated with irradiation is more negative than those which processed using other food processes. "People think the product is radioactive," said Harlan Clemmons, president of Sadex, a food irradiation company based in Sioux City, Iowa.

Enforcement of Labelling

There are analytical methods available to detect the usage of irradiation on food items in the marketplace. This is understood as a tool for government authorities to enforce existing labeling standards and to bolster consumer confidence. The European Union is particularly strict in enforcing irradiation labeling requiring its member countries to perform tests on a cross section of food items in the market-place and to report to the European Commission; the results are published annually in the OJ of the European Communities.

US Regulators of Food Irradiation

Food irradiation in the United States is primarily regulated by the FDA since it is considered a food additive. Other federal agencies that regulate aspects of food irradiation include:

- United States Department of Agriculture (USDA)—meat and poultry products, fresh fruit
- Nuclear Regulatory Commission (NRC)—safety of the processing facility
- Department of Transportation (DOT)—safe transport of the radioactive sources

Each new food is approved separately with a guideline specifying a maximum dosage; in case of quarantine applications the minimum dose is regulated. Packaging materials containing the food processed by irradiation must also undergo approval.

Safety Aspects

Safety, Security and Wholesomeness Aspects

Hundreds of animal feeding studies of irradiated food, including multigenerational studies, have been performed since 1950. Endpoints investigated have included subchronic and chronic changes in metabolism, histopathology, and function of most systems; reproductive effects; growth; teratogenicity; and mutagenicity. A large number of studies have been performed, some having demonstrated adverse effects of irradiation, and some concluding the process yields safer foods.

Consumer advocacy groups such as Public Citizen or Food and Water Watch maintain that the safety of irradiated food is not proven and strongly oppose the use of the technology.

Irradiation to Cover up Poor Food Quality

Concerns have sometimes been expressed by public interest groups and public health experts that irradiation, as a non-preventive measure, might disguise or otherwise divert attention away from poor working conditions, sanitation, and poor food-handling procedures that lead to contamination in the first place.

"Food irradiation is a pseudo-fix," said Bill Freese, a science policy analyst with the Center for Food Safety in Washington, DC.

"It's a way to try to come in and clean up problems that are created in the middle of the food production chain. I think it's clearly a disincentive to clean up the problems at the source."

Processors of irradiated food are subject to all existing regulations, inspections, and potential penalties regarding plant safety and sanitization; including fines, recalls, and criminal prosecutions. But critics of the practice claim that a lack of regulatory oversight (such as regular food processing plant inspections) necessitates irradiation.

"It's a total cop-out," said Patty Lovera, assistant director of Food and Water Watch. "They don't have the resources, the authority or the political will to really protect consumers from unsafe food."

While food irradiation can in some cases maintain the apparent quality of certain perishable food for a longer period of time, it cannot undo spoilage effects that occur prior to irradiation. Irradiation cannot be successfully used to mask quality issues other than pathogens. Under a HACCP-concept (Hazard Analysis and Critical Control Point) radiation processing can serve and contribute as an ultimate critical control point.

Worker Safety and Impact on the Environment

The safety of irradiation facilities is regulated by the United Nations International Atomic Energy Agency and monitored by the different national Nuclear Regulatory Commissions. The incidents that have occurred in the past are documented by the agency and thoroughly analyzed to determine root cause and improvement potential. Such improvements are then mandated to retrofit existing facilities and future design.

National and international regulations on the levels and types of energy used to irradiate food generally set standards that prevent the possibility of inducing radioactivity in treated foods. Care must be taken not to expose the operators and the environment to radiation. Interlocks and safeguards are mandated to minimize this risk. Nevertheless there have been radiation related deaths and injury amongst workers of such facilities, many of them caused by the operators themselves overriding the interlocks.

An incident in Decatur, Georgia where water soluble caesium-

137 leaked into the source storage pool requiring NRC intervention has led to near elimination of this radioisotope; it has been replaced by the more costly, non-water soluble cobalt-60.

Alternatives

Other methods to reduce several pathogens in food include heat-pasteurization, Ultra-high temperature processing, UV radiation, Ozone or fumigation with ethylene oxide.

For quarantine purposes, insect pests can also be eliminated by fumigation with methyl bromide or aluminum phosphine, vapour heat, forced hot air, hot water dipping, or cold treatment.

Other methods to extend shelf life of food items include modified atmosphere packaging, carbon monoxide, dehydration, vacuum packaging, freezing and flash freezing as well as chemical additives.

Opponents to food irradiation and consumer activists (cf. Public Citizen) maintain that the best alternative to food irradiation to reduce pathogens is in good agricultural practices. For example, farmers and processing plants should improve sanitation practices, water used for irrigation and processing should be regularly tested for E. coli, and production plants should be routinely inspected. Concentrated animal feeding operations near farmland where produce is grown should be regulated.

Proponents of food irradiation have said that practices of organic farming can only reduce the extent of the microorganism load. They assert that residual flora including pathogen germs will always persist; and that processing by ionizing radiation could be the ultimate measure (as a CCP under a HACCP-concept) to practically eliminate such risks.

FOOD PACKAGING

Food packaging is packaging for food. It requires protection, tampering resistance, and special physical, chemical, or biological needs. It also shows the product that is labeled to show any nutrition information on the food being consumed.

Food Packaging's Functions

Food packaging's functions are as diverse as they are important. They are broken down as follows.

1. *Containment*: For items that are granulated, paper-based packages are the best with a sealing system to prevent moisture infiltration of the product. Other products are packaged using metal cans, plastic bags and bottles, and glass. Another factor in containment is packaging durability, meaning the packaged food has to survive transport from the packaging at the food processing facility to the supermarket to the home for the consumer.
2. *Protection*: The packaging must protect the food from biological agents in rats, insects, and microbes; from mechanical damage such as product abrasion, compressive forces, and vibration; and from chemical degradation such as oxidation, moisture transfer, and ultraviolet light.
3. *Information*: Packaged food must be identified for consumer use mainly with label text and graphics. It can also be shown with the food package's shape such as the Coca-Cola bottle or the can of Spam. Other well known food package shapes include the potato chip bags and milk bottles. These packages also detail nutritional information, if they are kosher or halal-manufacturers and consumers. These products can be purchased all at once such as frozen pizza or as part of a larger package such as a twelve pack of beer. Another use may be to see if the package can go through a microwave process such as a TV dinner. Other items included pour spouts already on the bottles are other methods to ease dispensing of the product.
4. *Environmental issues*: This means that we must be done to reuse the packaging, reducing the size of the packaging, and then recycling of the products to be remanufactured.
5. *Product safety*: That the package be used for safety of the food from the package or vice versa. This includes any metal contamination issues from a can to the food product or any plastic contamination from a bottle to a food product as well.

Food Packaging Types

The above materials are fashioned into different types of food packages and containers such as:

Packaging Type	*Type of Container*	*Food Examples*
Aseptic packages	Primary	Liquid whole eggs
Plastic trays	Primary	Portion of fish
Bags	Primary	Potato chips
Boxes	Secondary	Box of Coca-Cola
Cans	Primary	Can of Campbell's Tomato soup.
Cartons	Primary	Carton of eggs
Flexible packaging	Primary	Bagged salad
Pallets	Tertiary	A series of boxes on a single pallet used to transport from the manufacturing plant to a distribution center.
Wrappers	Tertiary	Used to wrap the boxes on the pallet for transport.

Primary packaging is the main package that holds the food that is being processed. Secondary packaging combines the primary packages into one box being made. Tertiary packaging combines all of the secondary packages into one pallet.

There are also special containers that combine different technologies for maximum durability:

- Bags-In-Boxes (used for soft drink syrup, other liquid products, and meat products)
- Wine box (used for wine)

Packaging Machines

A choice of packaging machinery includes technical capabilities, labour requirements, worker safety, maintainability, serviceability, reliability, ability to integrate into the packaging line, capital cost, floorspace, flexibility (change-over, materials, etc.), energy usage, quality of outgoing packages, qualifications (for food, phamaceuticals, etc.), throughput, efficiency, productivity, ergonomics, etc.

Packaging machines may be of the following general types:

- Blister, Skin and Vacuum Packaging Machines
- Capping, Over-Capping, Lidding, Closing, Seaming and Sealing Machines
- Cartoning Machines
- Case and Tray Forming, Packing, Unpacking, Closing and Sealing Machines
- Check weighing machines
- Cleaning, Sterilizing, Cooling and Drying Machines
- Conveying, Accumulating and Related Machines
- Feeding, Orienting, Placing and Related Machines
- Filling Machines: handling liquid and powdered products
- Package Filling and Closing Machines
- Form, Fill and Seal Machines
- Inspecting, Detecting and Checkweighing Machines
- Palletizing, Depalletizing, Pallet Unitizing and Related Machines
- Product Identification: labelling, marking, etc.
- Wrapping Machines
- Converting Machines
- Other speciality machinery

6

Nutrition, Malnutrition and Applied Nutrition Concepts: An Overview

NUTRITION

Nutrition (also called nourishment or aliment) is the provision, to cells and organisms, of the materials necessary (in the form of food) to support life. Many common health problems can be prevented or alleviated with good nutrition.

The *diet* of an organism refers to what it eats. Dietitians are health professionals who specialize in human nutrition, meal planning, economics, preparation, and so on. They are trained to provide safe, evidence-based dietary advice and management to individuals (in health and disease), as well as to institutions.

Poor diet can have an injurious impact on health, causing deficiency diseases such as scurvy, beriberi, and kwashiorkor; health-threatening conditions like obesity and metabolic syndrome, and such common chronic systemic diseases as cardiovascular disease, diabetes, and osteoporosis.

Overview

Nutritional science investigates the metabolic and physiological responses of the body to diet. With advances in the fields of molecular biology, biochemistry, and genetics, the study of nutrition is increasingly concerned with metabolism and metabolic pathways, the sequences of biochemical steps through which the many substances of living things change from one form to another.

The human body contains chemical compounds, such as water, (sugar, starch, and fiber), amino acids (in proteins), fatty acids (in lipids), and nucleic acids (DNA/RNA). These compounds, in turn, consist of elements such as carbon, hydrogen, oxygen, nitrogen, phosphorus, calcium, iron, zinc, magnesium, manganese, and so on. All of these chemical compounds and elements occur in various forms and combinations (*e.g.* hormones/vitamins, phospholipids, hydroxyapatite), both in the human body and in organisms (*e.g.* plants, animals) that humans eat.

The human body consists of elements and compounds ingested, digested, absorbed, and circulated through the bloodstream. Except in the unborn fetus, it is the digestive system which carries out the first steps in feeding the cells of the body. In a typical adult, about seven liters of digestive juices enter the lumen of the digestive tract. They break chemical bonds in ingested molecules and modulate their conformations and energy states. Though some molecules are absorbed into the bloodstream unchanged, digestive processes release them from the matrix of foods in which they occur. Unabsorbed matter is secreted in the feces.

Studies of nutritional status must take into account the state of the body before and after experiments, as well as the chemical composition of the diet and the products of excretion. Comparing the food to the waste can help determine the specific compounds and elements absorbed in the body. Their effects may only be discernible after an extended period of time, during which all food and waste must be analyzed. The number of variables involved in such experiments is high, making nutritional studies time-consuming and expensive, which explains why the science of human nutrition is still slowly evolving.

In general, eating a wide variety of fresh, whole (unprocessed), foods has proven favourable compared to monotonous diets based on processed foods. In particular, the consumption of whole plant foods slows digestion and provides higher amounts, and a more favourable balance, of essential nutrients per Calorie, resulting in better management of cell growth, maintenance, and mitosis (cell division), as well as better regulation of appetite and blood sugar. Regularly scheduled meals (every few hours) have also proven more wholesome than infrequent, haphazard ones.

Nutrients

There are seven major classes of nutrients: carbohydrates (saccharides), fats (triglycerides), fiber (cellulose), minerals, proteins, vitamins, and water.

These nutrient classes can be generally grouped into the categories of macronutrients (needed in relatively large amounts), and micronutrients (needed in smaller quantities). The macronutrients are carbohydrates, fats, fiber, proteins and water. The other nutrient classes are micronutrients.

The macronutrients (excluding fiber and water) provide energy, which is measured in kilocalories, often called "Calories" and written with a capital C to distinguish individual calories. Carbohydrates and proteins provide four (4) Calories of energy per gram, while fats provide nine (9) Calories per gram. Vitamins, minerals, fiber, and water do not provide energy, but are necessary for other reasons.

Molecules of carbohydrates and fats consist of carbon, hydrogen, and oxygen atoms. Carbohydrates may be simple monomers (glucose, fructose, galactose), or large polymers polysaccharides (starch). Fats are triglycerides, made of various fatty acid monomers bound to glycerol. Some fatty acids are essential, but not all. Protein molecules contain nitrogen atoms in addition to the elements of carbohydrates and fats. The nitrogen-containing monomers of protein, called amino acids, fulfill many roles other than energy metabolism, and when they are used as fuel, getting rid of the nitrogen places a burden on the kidneys. Similar to fatty acids, certain amino acids are essential.

Other micronutrients not categorized above include antioxidants and phytochemicals.

Most foods contain a mix of some or all of the nutrient classes. Some nutrients are required on a regular basis, while others are needed less frequently. Poor health can be caused by an imbalance of nutrients, whether an excess or a deficiency.

Carbohydrates

Calories/Gram: 4

Carbohydrates may be classified as monosaccharides, disaccharides, or polysaccharides by the number of monomer (sugar) units they contain. They are found in large proportion in foods such as rice, noodles, bread and other grain-based products. Monosaccharides contain 1 sugar unit, disaccharides contain 2, and polysaccharides contain 3 or more. Polysaccharides are often referred to as complex carbohydrates because they are long chains of sugar units, whereas monosaccharides and disaccharides are simpler. The difference is important because complex carbohydrates take longer to digest and absorb since their sugar units are processed one-by-one off the ends of the chains; the spike in blood sugar levels caused by substantial amounts of simple sugars is thought to be at least part of the cause of increased heart and vascular disease associated with high simple sugar consumption. Simple carbohydrates are absorbed quickly and thus raise blood sugar levels more quickly.

Fat

Calories/Gram: 9

Fats are composed of fatty acids (long carbon/hydrogen chains) bonded to a glycerol; they are typically found as triglycerides (three fatty acids attached to one glycerol backbone). Certain fatty acids are essential. Fats may be classified as saturated or unsaturated. Saturated fats have all of their carbon atoms bonded to hydrogen atoms, whereas unsaturated fats have some of their carbon atoms double-bonded in place of a hydrogen atom. In humans, multiple studies have shown that unsaturated fats are to be preferred for health reasons, particularly mono-unsaturated fats. Saturated fats, typically from animal sources, are next, while 'trans' fats are to be avoided; they have been banned in several locations (eg, New York City). Saturated and trans fats are typically solid at room temperature (such as butter or lard), while unsaturated fats are typically liquids (such as olive oil or flaxseed oil). Unsaturated fats may be further classified as monounsaturated (one double-bond) or polyunsaturated (many double-bonds). Trans fats are saturated fats but are typically created from unsaturated fat by adding the extra hydrogen atoms in an industrial process called hydrogenation; they are also called hydrogenated fat. They are very rare in nature, but have properties useful in the food processing industry.

Essential Fatty Acids

Most fatty acids are non-essential, meaning the body can produce them as needed from other fats and some energy. However, in humans, at least two fatty acids are essential and must be consumed in the diet. An appropriate balance of essential fatty acids—omega-3 and omega-6 fatty acids—has been discovered to be important in reducing risk of some chronic diseases and conditions. Both of these "omega" long-chain polyunsaturated fatty acids are substrates for a class of eicosanoids known as prostaglandins which have uses throughout the human body; they are in some respects, hormones. The omega-3 eicosapentaenoic acid (EPA) (which can be made in the human body from the omega-3 essential fatty acid alpha-linolenic acid (LNA), or taken in through marine food sources), serves as a building block for series 3 prostaglandins (*e.g.* weakly-inflammation PGE3). The omega-6 dihomo-gamma-linolenic acid (DGLA) serves as a building block for series 1 prostaglandins (*e.g.* anti-inflammatory PGE1), whereas arachidonic acid (AA) serves as a building block for series 2 prostaglandins (*e.g.* pro-inflammatory PGE 2). Both DGLA and AA can be made from the omega-6 linoleic acid (LA) in the human body, or can be taken in directly through food. An appropriately balanced intake of omega-3 and omega-6 partly determines the relative production of different prostaglandins, which partly explains the importance of omega-3/omega-6 balance for cardiovascular health. In industrialised societies, people typically consume large amounts of processed vegetable oils that have reduced amounts of the essential fatty acids along with a too high ratio of omega-6 fatty acids relative to omega-3 fatty acids.

The conversion rate of omega-6 DGLA to AA largely determines the production of the respective prostaglandins PGE1 and PGE2. Omega-3 EPA prevents AA from being released from membranes, thereby skewing prostaglandin balance away from pro-inflammatory PGE2 made from AA toward anti-inflammatory PGE1 made from DGLA. Moreover, the conversion (desaturation) of DGLA to AA is controlled by the enzyme delta-5-desaturase, which in turn is controlled by hormones such as insulin (up-regulation) and glucagon (down-regulation). Because the amount and type of glucose and starch) plus some amino acid types) in food affect insulin, glucagon and other hormones, not only the amount of omega-3 versus omega-6 eaten but also the general composition of the diet, are implicated in general health regarding the essential fatty acids, inflammation (*e.g.* immune function) and mitosis (*i.e.* cell division).

Good sources of essential fatty acids include: fish, flax seed oils, hemp seeds and oils, soy beans, pumpkin seeds, sunflower seeds, and walnuts.

Fiber

Calories/Gram: 0

Dietary fiber consists mainly of cellulose, a large carbohydrate polymer, that is indigestible because humans do not have enzymes to digest it. There are two subcategories, soluble and insoluble fiber. The first means fiber which absorbs water, the second does not. Whole grains, fruits (especially plums, prunes, and figs), and vegetables are rich in dietary fiber. It provides bulk to the intestinal contents and stimulates peristalsis (rhythmic muscular contractions passing along the digestive tract). Consequently, a lack of dietary fiber in the diet tends toward constipation. There is also some evidence that dietary fiber is helpful in other ways, eg, reducing the incidence of some cancers.

Protein

Calories/Gram: 4

Proteins are the basis of animal body structures (eg, muscles, skin, hair etc.). They are composed of amino acids, sometimes many thousands, which are characterized by inclusion of nitrogen and sometimes sulphur. The body requires amino acids to produce new body protein (protein retention) and to replace damaged proteins (maintenance). Amino acids not needed are discarded, typically in the urine. In animals, amino acid requirements are classified in terms of essential (an animal cannot produce them internally) and non-essential (the animal can produce them from other nitrogen containing compounds) amino acids. Humans use about 20 amino acids, and about ten are essential in this sense. Consuming a diet that contains adequate amounts of essential (but also non-essential) amino acids is particularly important for growing, pregnant, nursing, or injured animals, all of whom have a particularly high requirement. Protein nutrition which contains the essential amino acids is a complete protein source, one missing one or more is called incomplete. It's possible to combine two incomplete protein sources (eg, rice and beans) to make a complete protein source. Dietary sources of protein include meats, tofu and other soy-products, eggs, grains, legumes, and dairy products such as milk and cheese. A few amino acids from protein can be converted into glucose and used for fuel through a process called gluconeogenesis. The remaining amino acids are discarded.

Minerals

Calories/Gram: 0

Dietary minerals are the chemical elements required by living organisms, other than the four elements carbon, hydrogen, nitrogen, and oxygen which are present in common organic molecules. The term "mineral" is archaic, since the intent of the definition is to describe ions, not chemical compounds or actual minerals. Some dietitians recommend that these heavier elements should be supplied by ingesting specific foods (that are enriched in the element(s) of interest), compounds, and sometimes including even minerals, such as calcium carbonate. Sometimes these "minerals" come from natural sources such as ground oyster shells. Sometimes minerals are added to the diet separately from food, such as mineral supplements, the most famous being iodine in "iodized salt".

Macrominerals

A variety of elements are required to support the biochemical processes, many play a role as electrolytes or in a structural role. In human nutrition, the dietary bulk "mineral elements" (RDA > 200 mg/day) are in alphabetical order (parenthetical comments on folk medicine perspective):

- Calcium (for muscle and digestive system health, builds bone, neutralizes acidity, clears toxins, helps blood stream)

- Chloride
- Magnesium required for processing ATP and related reactions (health, builds bone, causes strong peristalsis, increases flexibility, increases alkalinity)
- Phosphorus required component of bones and energy processing and many other functions (bone mineralization)
- Potassium required electrolyte (heart and nerves health)
- Sodium electrolyte
- Sulfur for three essential amino acids and many proteins and cofactors (skin, hair, nails, liver, and pancreas health)

Trace Minerals

A variety of elements are required in trace amounts, unusually because they play a role in catalysis in enzymes. Some trace mineral elements (RDA < 200 mg/day) are (alphabetical order):

- Cobalt required for biosynthesis of vitamin B12 family of coenzymes
- Copper required component of many redox enzymes, including cytochrome c oxidase
- Chromium required for sugar metabolism
- Iodine required for the biosynthesis of thyroxin
- Iron required for many proteins and enzymes, notably hemoglobin
- Manganese (processing of oxygen)
- Molybdenum required for xanthine oxidase and related oxidases
- Nickel present in urease
- Selenium required for peroxidase (antioxidant proteins)
- Vanadium (There is no established RDA for vanadium. No specific biochemical function has been identified for it in humans, although vanadium is found in lower organisms.)
- Zinc required for several enzymes such as carboxypeptidase, liver alcohol dehydrogenase, carbonic anhydrase. Zinc is pervasive.
- Iodine is required in larger quantities than the other trace

minerals in this list and is sometimes classified with the bulk minerals. Sodium is not generally found in dietary supplements, despite being needed in large quantities, because the ion is very common in food.

Vitamins

Calories/Gram: 0

Mineral and/or vitamin deficiency or excess may result in disease conditions such as goitre, scurvy, osteoporosis, impaired immune system, disorders of cell metabolism, certain forms of cancer, symptoms of premature aging, and poor psychological health (including eating disorders), among many others.

As of 2005, twelve vitamins and about the same number of minerals are recognized as "essential nutrients", meaning that they must be consumed and absorbed—or, in the case of vitamin D, alternatively synthesized in the skin via UVB radiation to prevent deficiency symptoms and possibly death. Certain vitamin-like substances found in foods, such as carnitine, have also been found essential to survival and health, but these are not strictly "essential" to eat because the human body can produce them from other compounds. Moreover, thousands of different phytochemicals have recently been discovered in food (particularly in fresh vegetables), which may have desirable properties including antioxidant activity. Other essential nutrients include essential amino acids, choline and the essential fatty acids.

Water

Calories/Gram: 0

About 70 per cent of the non-fat mass of the human body is made of water. To function properly, the body requires between one and seven liters of water per day to avoid dehydration; the precise amount depends on the level of activity, temperature, humidity, and other factors. With physical exertion and heat exposure, water loss will increase and daily fluid needs may increase as well.

It is not clear how much water intake is needed by healthy people, although some experts assert that 8-10 glasses of water (approximately 2 liters) daily is the minimum to maintain proper hydration. The notion that a person should consume eight glasses of water per day cannot be traced back to a scientific source. The effect of water intake on weight loss and on constipation is also still unclear. Original recommendation for water intake in 1945 by the Food and Nutrition Board of the National Research Council read: "An ordinary standard for diverse persons is 1 milliliter for each calorie of food. Most of this quantity is contained in prepared foods." The latest dietary reference intake report by the United States National Research Council in general recommended (including food sources): 2.7 liters of water total for women and 3.7 liters for men. Specifically, pregnant and breastfeeding women need additional fluids to stay hydrated. According to the Institute of Medicine—who recommend that, on average, women consume 2.2 litres and men 3.0 litres—this is recommended to be 2.4 litres (approx. 9 cups) for pregnant women and 3 litres (approx. 12.5 cups) for breastfeeding women since an especially large amount of fluid is lost during nursing.

For those who have healthy kidneys, it is rather difficult to drink too much water, but (especially in warm humid weather and while exercising) it is dangerous to drink too little. People can drink far more water than necessary while exercising, however, putting them at risk of water intoxication, which can be fatal. In particular large amounts of de-ionized water are dangerous.

Normally, about 20 percent of water intake comes in food, while the rest comes from drinking water and assorted beverages (caffeinated included). Water is excreted from the body in multiple forms; including urine and feces, sweating, and by water vapor in the exhaled breath.

Other Nutrients

Calories/Gram: 0

Other micronutrients include antioxidants and phytochemicals. These substances are generally more recent discoveries which: have not yet been recognized as vitamins; are still under investigation; or contribute to health but are not necessary for life. Phytochemicals may act as antioxidants, but not all phytochemicals are antioxidants.

Antioxidants

Antioxidants are a recent discovery. As cellular metabolism/energy production requires oxygen, potentially damaging (*e.g.* mutation causing) compounds known as free radicals can form. Most of these are oxidizers 9ie, (acceptors of electrons) and some react very strongly. For normal cellular maintenance, growth, and division, these free radicals must be sufficiently neutralized by antioxidant compounds. Some are produced by the human body with adequate precursors (glutathione, Vitamin C) and those that the body cannot produce may only be obtained through the diet through direct sources (Vitamin C in humans, Vitamin A, Vitamin K) or produced by the body from other compounds (Beta-carotene converted to Vitamin A by the body, Vitamin D synthesized from cholesterol by sunlight). Phytochemicals (*Section Below*) and their subgroup polyphenols are the majority of antioxidants; about 4,000 are known. Different antioxidants are now known to function in a cooperative network, *e.g.* vitamin C can reactivate free radical-containing glutathione or vitamin E by accepting the free radical itself, and so on. Some antioxidants are more effective than others at neutralizing different free radicals. Some cannot neutralize certain free radicals. Some cannot be present in certain areas of free radical development (Vitamin A is fat-soluble and protects fat areas, Vitamin C is water soluble and protects those areas). When interacting with a free radical, some antioxidants produce a different free radical compound that is less dangerous or more dangerous than the previous compound. Having a variety of antioxidants allows any byproducts to be safely dealt with by more efficient antioxidants in neutralizing a free radical's butterfly effect.

Phytochemicals

A growing area of interest is the effect upon human health of trace chemicals, collectively called phytochemicals. These nutrients are typically found in edible plants, especially colourful fruits and vegetables, but also other organisms including seafood, algae, and fungi. The effects of phytochemicals increasingly survive rigorous testing by prominent health organizations. One of the principal classes of phytochemicals are polyphenol antioxidants, chemicals which are known to provide certain health benefits to the cardiovascular system and immune system. These chemicals are known to down-regulate the formation of reactive oxygen species, key chemicals in cardiovascular disease.

Perhaps the most rigorously tested phytochemical is zeaxanthin, a yellow-pigmented carotenoid present in many yellow and orange fruits and vegetables. Repeated studies have shown a strong correlation between ingestion of zeaxanthin and the prevention and treatment of age-related macular degeneration (AMD). Less rigorous studies have proposed a correlation between zeaxanthin intake and cataracts. A second carotenoid, lutein, has also been shown to lower the risk of contracting AMD. Both compounds have been observed to collect in the retina when ingested orally, and they serve to protect the rods and cones against the destructive effects of light.

Another carotenoid, beta-cryptoxanthin, appears to protect against chronic joint inflammatory diseases, such as arthritis. While the association between serum blood levels of beta-cryptoxanthin and substantially decreased joint disease has been established, neither a convincing mechanism for such protection nor a cause-and-effect have been rigorously studied. Similarly, a red phytochemical, lycopene, has substantial credible evidence of negative association with development of prostate cancer.

The correlations between the ingestion of some phytochemicals and the prevention of disease are, in some cases, enormous in magnitude.

Even when the evidence is obtained, translating it to practical dietary advice can be difficult and counter-intuitive. Lutein, for example, occurs in many yellow and orange fruits and vegetables and protects the eyes against various diseases. However, it does not protect the eye nearly as well as zeaxanthin, and the presence of lutein in the retina will prevent zeaxanthin uptake. Additionally, evidence has shown that the lutein present in egg yolk is more readily absorbed than the lutein from vegetable sources, possibly because of fat solubility. At the most basic level, the question "should you eat eggs?" is complex to the point of dismay, including misperceptions about the health effects of cholesterol in egg yolk, and its saturated fat content.

As another example, lycopene is prevalent in tomatoes (and actually is the chemical that gives tomatoes their red colour). It is more highly concentrated, however, in processed tomato products such as commercial pasta sauce, or tomato soup, than in fresh "healthy" tomatoes. Yet, such sauces tend to have high amounts of salt, sugar, other substances a person may wish or even need to avoid.

The following table presents phytochemical groups and common sources, arranged by family:

Family	*Sources*	*Possible Benefits*
Flavonoids	berries, herbs, vegetables, wine, grapes, tea	general antioxidant, oxidation of LDLs, prevention of arteriosclerosis and heart disease
Isoflavones (phytoestrogens)	soy, red clover, kudzu root	general antioxidant, prevention of arteriosclerosis and heart disease, easing symptoms of menopause, cancer prevention
Isothiocyanates	cruciferous vegetables	cancer prevention
Monoterpenes	citrus peels, essential oils, herbs, spices, green plants, atmosphere	cancer prevention, treating gallstones
Organosulfur Compounds	chives, garlic, onions	cancer prevention, lowered LDLs, assistance to the immune system
Saponins	beans, cereals, herbs	Hypercholesterolemia, Hyperglycemia, Antioxidant, cancer prevention, Anti-inflammatory
capsaicinoids	all capiscum (chile) peppers	topical pain relief, cancer prevention, cancer cell apoptosis

Intestinal Bacterial Flora

It is now also known that animal intestines contain a large population of gut flora. In humans, these include species such as *Bacteroides*, *L. acidophilus* and *E. coli*, among many others. They are essential to digestion, and are also affected by the food we eat. Bacteria in the gut perform many important functions for humans, including breaking down and aiding in the absorption of otherwise indigestible food; stimulating cell growth; repressing the growth of harmful bacteria, training the immune system to respond only to pathogens; producing vitamin B12, and defending against some diseases.

Balanced Diet

Balanced diet is a diet which consists of all the essential nutrients in a required proportion.

Junk Food

Junk food is a slang name for food items containing limited nutritional value. It includes food high in salts, fats, sugar, calories, and low nutrient content.

Sports Nutrition

Protein

The protein requirements of athletes, once the source of great controversy, has settled into a current consensus. Sedentary people and recreational athletes have similar protein requirements, about 1 gram of protein per kilogram of body mass. These needs are easily met by a balanced diet containing about 70 grams of protein for a 70 kg (150 pound) man or 60 grams of protein for a 60 kg (130 pound) woman.

People who exercise at greater intensity, and especially those whose activity grows muscle bulk, have significantly higher protein requirements. According to *Clinical Sports Nutrition*, active athletes playing power sports (such as football), those engaged in muscle-development training, and elite endurance athletes, all require approximately 2 grams of protein per day per kilogram of body weight, roughly double that of a sedentary persons. Older athletes seeking primarily to maintain developed muscle mass require 2 to 3 g per day per kg.

Protein intake in excess of that required to build muscle (and other) tissue is broken-down by gluconeogenesis to be used as energy.

Water and Salts

Water is one of the most important nutrients in your sports diet. It helps eliminate food waste products in your body, regulates body temperature during activity, helps digest, is involved in converting food into energy and helps lubricate joints. Athletes should drink as much water as they comfortably can. Maintaining hydration during periods of physical exertion is key to peak performance. While drinking too much water during activities can lead to physical discomfort, dehydration in excess of 2 per cent of body mass (by weight) markedly hinders athletic performance. Some studies have shown that an athlete that drinks before they feel thirsty stays cooler and performs better than one who drinks on thirst cues, although recent studies of such races as the Boston Marathon have indicated that this recommendation can lead to the problem of overhydration. Additional carbohydrates and protein before, during, and after exercise increase time to exhaustion as well as speed recovery. Dosage is based on work performed, lean body mass, and environmental factors, especially ambient temperature and humidity.

Carbohydrates

The main fuel used by the body during exercise is carbohydrates, which is stored in muscle as glycogen—a form of sugar. During exercise, muscle glycogen reserves can be used up, especially when activities last longer than 90 min. Because the amount of glycogen stored in the body is limited, it is important for athletes to replace glycogen by consuming a diet high in carbohydrates. Meeting energy needs can help improve performance during the sport, as well as improve overall strength and endurance. There are different kinds of carbohydrates—simple or refined, and unrefined. A typical American consumes about 50 per cent of their carbohydrates as simple sugars, which are added to foods as opposed to sugars that come naturally in fruits and vegetables. These simple sugars come in large amounts in sodas and fast food. Over the course of a year, the average American consumes 54 gallons of soft drinks, which contain the highest amount of added sugars. Even though carbohydrates are necessary for humans to function, they are not all equally healthful. When machinery has been used to remove bits of high fiber, the carbohydrates are refined. These are the carbohydrates found in white bread and fast food.

Longevity

Whole Plant Food Diet

Heart disease, cancer, obesity, and diabetes are commonly called "Western" diseases because these maladies were once rarely seen in developing countries. One study in China found some regions had essentially no cancer or heart disease, while in other areas they reflected "up to a 100-fold increase" coincident with diets that were found to be entirely plant-based to heavily animal-based, respectively. In contrast, diseases of affluence like cancer and heart disease are common throughout the United States. Adjusted for age and exercise, large regional clusters of people in China rarely suffered from these "Western" diseases possibly because their diets are rich in vegetables, fruits and whole grains.

The United Healthcare/Pacificare nutrition guideline recommends a whole plant food diet, and recommends using protein only as a condiment with meals. A *National Geographic* cover article from November, 2005, entitled *The Secrets of Living Longer*, also recommends a whole plant food diet. The article is a lifestyle survey of three populations, Sardinians, Okinawans, and Adventists, who generally display longevity and "suffer a fraction of the diseases that commonly kill people in other parts of the developed world, and enjoy more healthy years of life. In sum, they offer three sets of 'best practices' to emulate. The rest is up to you." In common with all three groups is to "Eat fruits, vegetables, and whole grains."

The *National Geographic* article noted that an NIH funded study of 34,000 Seventh-day Adventists between 1976 and 1988 "...found that the Adventists' habit of consuming beans, soy milk, tomatoes, and other fruits lowered their risk of developing certain cancers. It also suggested that eating whole grain bread, drinking five glasses of water a day, and, most surprisingly, consuming four servings of nuts a week reduced their risk of heart disease."

The French "Paradox"

It has been discovered that people living in France live longer. Even though they consume more saturated fats than Americans, the rate of heart disease is lower in France than in North America. A number of explanations have been suggested:

- Reduced consumption of processed carbohydrate and other junk foods;
- Ethnic genetic differences allowing the body to be harmed less by fats;
- Regular consumption of red wine; or
- Living in the South requires the body to produce less heat, allowing a slower, and therefore healthier, metabolic rate.
- More active lifestyles involving plenty of daily exercise, especially walking; the French are much less dependent on cars than Americans are.
- Higher consumption of artificially produced trans-fats by Americans, which has been shown to have greater lipoprotein impacts per gram than saturated fat.

However, a growing number of French health researchers doubt the theory that the French are healthier than other populations. Statistics collected by the WHO from 1990-2000 show that the incidence of heart disease in France may have been underestimated and in fact be similar to that of neighboring countries.

Malnutrition

Malnutrition refers to insufficient, excessive, or imbalanced consumption of nutrients. In developed countries, the diseases of malnutrition are most often associated with nutritional imbalances or excessive consumption. Although there are more people in the world who are malnourished due to excessive consumption, according to the United Nations World Health Organization, the real challenge in developing nations today, more than starvation, is combating insufficient nutrition—the lack of nutrients necessary for the growth and maintenance of vital functions.

Table 6.1: Illnesses Caused by Improper Nutrient Consumption

Nutrients	*Deficiency*	*Excess*
Calories	Starvation	Obesity, diabetes mellitus, Cardiovascular disease
Simple carbohydrates	Marasmus, starvation	diabetes mellitus
Complex carbohydrates	Marasmus, starvation	Obesity
Saturated fat/trans fat	none	Cardiovascular disease,
Unsaturated fat	Rabbit starvation	Obesity
Cholesterol	none	Cardiovascular disease
Protein	Marasmus	Ketoacidosis, Rabbit starvation, kidney disease
Sodium	hyponatremia	Hypernatremia, hypertension
Iron	Anemia	Hepatitis C, cirrhosis, heart disease
Iodine	Goiter, hypothyroidism	Iodine Toxicity (goiter, hypothyroidism)
Vitamin A	Xerophthalmia and Night Blindness	Hypervitaminosis A (cirrhosis, hair loss, birth defects)
Vitamin B1	Beri-Beri	
Vitamin B2	Cracking of skin and Corneal Uncleara-tion	
Niacin	Pellagra	dyspepsia, cardiac arrhythmias, birth defects
Vitamin B12	Pernicious Anemia	
Vitamin C	Scurvy	diarrhea causing dehydration
Vitamin D	Rickets	Hypervitaminosis D (dehydration, vomiting, constipation)
Vitamin E		Hypervitaminosis E (anticoagulant: excessive bleeding)
Vitamin K	Hemorrhage	

Mental Agility

Research indicates that improving the awareness of nutritious meal choices and establishing long-term habits of healthy eating has a positive effect on a cognitive and spatial memory capacity, potentially increasing a student's potential to process and retain academic information.

Some organizations have begun working with teachers,

policymakers, and managed foodservice contractors to mandate improved nutritional content and increased nutritional resources in school cafeterias from primary to university level institutions. Health and nutrition have been proven to have close links with overall educational success (Behrman, 1996). Currently less than 10 per cent of American college students report that they eat the recommended five servings of fruit and vegetables daily. Better nutrition has been shown to have an impact on both cognitive and spatial memory performance; a study showed those with higher blood sugar levels performed better on certain memory tests. In another study, those who consumed yogurt performed better on thinking tasks when compared to those who consumed caffeine free diet soda or confections. Nutritional deficiencies have been shown to have a negative effect on learning behaviour in mice as far back as 1951.

> "Better learning performance is associated with diet induced effects on learning and memory ability".

The "nutrition-learning nexus" demonstrates the correlation between diet and learning and has application in a higher education setting.

> "We find that better nourished children perform significantly better in school, partly because they enter school earlier and thus have more time to learn but mostly because of greater learning productivity per year of schooling."

91 per cent of college students feel that they are in good health while only 7 per cent eat their recommended daily allowance of fruits and vegetables.

Nutritional education is an effective and workable model in a higher education setting.

More "engaged" learning models that encompass nutrition is an idea that is picking up steam at all levels of the learning cycle.

There is limited research available that directly links a student's Grade Point Average (G.P.A.) to their overall nutritional health.

Additional substantive data is needed to prove beyond a shadow of a doubt that overall intellectual health is closely linked to a person's diet, rather than just another correlation fallacy.

Mental Disorders

Nutritional supplement treatment may be appropriate for major depression, bipolar disorder, schizophrenia, and obsessive compulsive disorder, the four most common mental disorders in developed countries.

Cancer

Cancer is now common in developing countries. According a study by the International Agency for Research on Cancer, "In the developing world, cancers of the liver, stomach and esophagus were more common, often linked to consumption of carcinogenic preserved foods, such as smoked or salted food, and parasitic infections that attack organs." Lung cancer rates are rising rapidly in poorer nations because of increased use of tobacco. Developed countries "tended to have cancers linked to affluence or a 'Western lifestyle'—cancers of the colon, rectum, breast and prostate—that can be caused by obesity, lack of exercise, diet and age."

Metabolic Syndrome

Several lines of evidence indicate lifestyle-induced hyperinsulinemia and reduced insulin function (*i.e.* insulin resistance) as a decisive factor in many disease states. For example, hyperinsulinemia and insulin resistance are strongly linked to chronic inflammation, which in turn is strongly linked to a variety of adverse developments such as arterial microinjuries and clot formation (*i.e.* heart disease) and exaggerated cell division (*i.e.* cancer). Hyperinsulinemia and insulin resistance (the so-called metabolic syndrome) are characterized by a combination of abdominal obesity, elevated blood sugar, elevated blood pressure, elevated blood triglycerides, and reduced HDL cholesterol. The negative impact of hyperinsulinemia on prostaglandin PGE1/PGE2 balance may be significant.

The state of obesity clearly contributes to insulin resistance, which in turn can cause type 2 diabetes. Virtually all obese and most type 2 diabetic individuals have marked insulin resistance.

Although the association between overweight and insulin resistance is clear, the exact (likely multifarious) causes of insulin resistance remain less clear. Importantly, it has been demonstrated that appropriate exercise, more regular food intake and reducing glycemic load all can reverse insulin resistance in overweight individuals (and thereby lower blood sugar levels in those who have type 2 diabetes).

Obesity can unfavourably alter hormonal and metabolic status via resistance to the hormone leptin, and a vicious cycle may occur in which insulin/leptin resistance and obesity aggravate one another. The vicious cycle is putatively fuelled by continuously high insulin/leptin stimulation and fat storage, as a result of high intake of strongly insulin/leptin stimulating foods and energy. Both insulin and leptin normally function as satiety signals to the hypothalamus in the brain; however, insulin/leptin resistance may reduce this signal and therefore allow continued overfeeding despite large body fat stores. In addition, reduced leptin signalling to the brain may reduce leptin's normal effect to maintain an appropriately high metabolic rate.

There is a debate about how and to what extent different dietary factors—such as intake of processed carbohydrates, total protein, fat, and carbohydrate intake, intake of saturated and trans fatty acids, and low intake of vitamins/minerals—contribute to the development of insulin and leptin resistance. In any case, analogous to the way modern man-made pollution may potentially overwhelm the environment's ability to maintain homeostasis, the recent explosive introduction of high glycemic index and processed foods into the human diet may potentially overwhelm the body's ability to maintain homeostasis and health (as evidenced by the metabolic syndrome epidemic).

Hyponatremia

Excess water intake, without replenishment of sodium and potassium salts, leads to hyponatremia, which can further lead to water intoxication at more dangerous levels. A well-publicized case occurred in 2007, when Jennifer Strange died while participating in a water-drinking contest. More usually, the condition occurs in long-distance endurance events (such as marathon or triathlon competition and training) and causes gradual mental dulling, headache, drowsiness, weakness, and confusion; extreme cases may result in coma, convulsions, and death. The primary damage comes from swelling of the brain, caused by increased osmosis as blood salinity decreases. Effective fluid replacement techniques include Water aid stations during running/cycling races, trainers providing water during team games such as Soccer and devices such as Camel Baks which can provide water for a person without making it too hard to drink the water.

Processed Foods

Since the Industrial Revolution some two hundred years ago, the food processing industry has invented many technologies that both help keep foods fresh longer and alter the fresh state of food as they appear in nature. Cooling is the primary technology used to maintain freshness, whereas many more technologies have been invented to allow foods to last longer without becoming spoiled. These latter technologies include pasteurisation, autoclavation, drying, salting, and separation of various components, and all appear to alter the original nutritional contents of food. Pasteurisation and autoclavation (heating techniques) have no doubt improved the safety of many common foods, preventing epidemics of bacterial infection. But some of the (new) food processing technologies undoubtedly have downfalls as well.

Modern separation techniques such as milling, centrifugation, and pressing have enabled upconcentration of particular components of food, yielding flour, oils, juices and so on, and even separate fatty acids, amino acids, vitamins, and minerals. Inevitably, such large scale upconcentration changes the nutritional content of food, saving certain nutrients while removing others. Heating techniques may also reduce food's content of many heat-labile nutrients such as certain vitamins and phytochemicals, and possibly other yet to be discovered substances. Because of reduced nutritional value, processed foods are often 'enriched' or 'fortified' with some of the most critical nutrients (usually certain vitamins) that were lost during processing. Nonetheless, processed foods tend to have an inferior nutritional profile compared to whole, fresh foods, regarding content of both sugar and high GI starches, potassium/sodium, vitamins, fibre, and of intact, unoxidized (essential) fatty acids. In addition, processed foods often contain potentially harmful substances such as oxidized fats and trans fatty acids.

A dramatic example of the effect of food processing on a population's health is the history of epidemics of beri-beri in people subsisting on polished rice. Removing the outer layer of rice by polishing it removes with it the essential vitamin thiamine, causing beri-beri. Another example is the development of scurvy among infants in the late 1800s in the United States. It turned out that the vast majority of sufferers were being fed milk that had been heat-treated (as suggested by Pasteur) to control bacterial disease. Pasteurisation was effective against bacteria, but it destroyed the vitamin C.

As mentioned, lifestyle—and obesity-related diseases are becoming increasingly prevalent all around the world. There is little doubt that the increasingly widespread application of some modern food processing technologies has contributed to this development. The food processing industry is a major part of modern economy, and as such it is influential in political decisions (*e.g.* nutritional recommendations, agricultural subsidising). In any known profit-driven economy, health considerations are hardly a priority; effective production of cheap foods with a long shelf-life is more the trend. In general, whole, fresh foods have a relatively short shelf-life and are less profitable to produce and sell than are more processed foods. Thus the consumer is left with the choice between more expensive but nutritionally superior whole, fresh foods, and cheap, usually nutritionally inferior processed foods. Because processed foods are often cheaper, more convenient (in both purchasing, storage, and preparation), and more available, the consumption of nutritionally inferior foods has been increasing throughout the world along with many nutrition-related health complications.

Advice and Guidance

Governmental Policies

In the US, "dietitians". are registered (RD) or licensed (LD) with the Commission for Dietetic Registration and the American Dietetic Association, and are only able to use the title "dietitian," as described by the business and professions codes of each respective state, when they have met specific educational and experiential prerequisites and passed a national registration or licensure examination, respectively. In California, registered dietitions must abide by the "Business and Professions Code of Section 2585-2586.8". Anyone may call themselves a nutritionist, including unqualified personnel, as this term is unregulated. Some states, such as the State of Florida, have begun to include the title "nutritionist" in state licensure requirements. Most governments provide guidance on nutrition, and some also impose mandatory disclosure/labeling requirements for processed food manufacturers and restaurants to assist consumers in complying with such guidance.

In the US, nutritional standards and recommendations are established jointly by the US Department of Agricultureand US Department of Health and Human Services. Dietary and physical activity guidelines from the USDA are presented in the concept of a food pyramid, which superseded the Four Food Groups. The Senate committee currently responsible for oversight of the USDA is the *Agriculture, Nutrition and Forestry Committee*. Committee hearings are often televised on C-SPAN as seen here.

Teaching

Nutrition is taught in schools in many countries. In England and Wales the Personal and Social Education and Food Technology curricula include nutrition, stressing the importance of a balanced diet and teaching how to read nutrition labels on packaging. In many schools a Nutrition class will fall within the Family and Consumer Science or Health departments. In some American schools, students are required to take a certain number of FCS or Health related classes. Nutrition is offered at many schools, and if it is not a class of its own, nutrition is included in other FCS or Health classes such as: Life Skills, Independent Living, Single Survival, Freshmen Connection, Health etc. In many Nutrition classes, students learn about the food groups, the food pyramid, Daily Recommended Allowances, calories, vitamins, minerals, malnutrition, physical activity, healthy food choices and how to live a healthy life.

A 1985 US National Research Council report entitled *Nutrition Education in US Medical Schools* concluded that nutrition education in medical schools was inadequate. Only 20 per cent of the schools surveyed taught nutrition as a separate, required course. A 2006 survey found that this number had risen to 30 per cent.

History

Humans have evolved as omnivorous hunter-gatherers over the past 250,000 years. The diet of early modern humans varied significantly depending on location and climate. The diet in the tropics tended to be based more heavily on plant foods, while the diet at higher latitudes tended more towards animal products. Analysis of postcranial and cranial remains of humans and animals from the Neolithic, along with detailed bone modification studies have shown that cannibalism was also prevalent among prehistoric humans.

Agriculture developed about 10,000 years ago in multiple locations throughout the world, providing grains such as wheat, rice, and maize, with staples such as bread and pasta. Farming also provided milk and dairy products, and sharply increased the availability of meats and the diversity of vegetables. The importance of food purity was recognized when bulk storage led to infestation and contamination risks. Cooking developed as an often ritualistic activity, due to efficiency and reliability concerns requiring adherence to strict recipes and procedures, and in response to demands for food purity and consistency.

Antiquity through 1900

The first recorded nutritional experiment is found in the Bible's Book of Daniel. Daniel and his friends were captured by the king of Babylon during an invasion of Israel. Selected as court servants, they were to share in the king's fine foods and wine. But they objected, preferring vegetables (pulses) and water in accordance with their Jewish dietary restrictions. The king's chief steward reluctantly agreed to a trial. Daniel and his friends received their diet for 10 days and were then compared to the king's men. Appearing healthier, they were allowed to continue with their diet.

In around 475 BC, Anaxagoras stated that food is absorbed by the human body and therefore contained "homeomerics" (generative components), thereby deducing the existence of nutrients. Around 400 BC, Hippocrates said, "Let food be your medicine and medicine be your food."

In the 1500s, scientist and artist Leonardo da Vinci compared metabolism to a burning candle. In 1747, Dr. James Lind, a physician in the British navy, performed the first scientific nutrition experiment, discovering that lime juice saved sailors who had been at sea for years from scurvy, a deadly and painful bleeding disorder. The discovery was ignored for forty years, after which British sailors became known as "limeys." The essential vitamin C within lime juice would not be identified by scientists until the 1930s.

Around 1770, Antoine Lavoisier, the "Father of Nutrition and Chemistry" discovered the details of metabolism, demonstrating that the oxidation of food is the source of body heat. In 1790, George Fordyce recognized calcium as necessary for fowl survival. In the early 1800s, the elements carbon, nitrogen, hydrogen and oxygen were recognized as the primary components of food, and methods to measure their proportions were developed.

In 1816, François Magendie discovered that dogs fed only carbohydrates and fat lost their body protein and died in a few weeks, but dogs also fed protein survived, identifying protein as an essential dietary component. In 1840, Justus Liebig discovered the chemical makeup of carbohydrates (sugars), fats (fatty acids) and proteins (amino acids.) In the 1860s, Claude Bernard discovered that body fat can be synthesized from carbohydrate and protein, showing that the energy in blood glucose can be stored as fat or as glycogen.

In the early 1880s, Kanehiro Takaki observed that Japanese sailors (whose diets consisted almost entirely of white rice) developed beriberi (or endemic neuritis, a disease causing heart problems and paralysis) but British sailors and Japanese naval officers did not. Adding various types of vegetables and meats to the diets of Japanese sailors prevented the disease.

In 1896, Baumann observed iodine in thyroid glands. In 1897, Christiaan Eijkman worked with natives of Java, who also suffered from beriberi. Eijkman observed that chickens fed the native diet of white rice developed the symptoms of beriberi, but remained healthy when fed unprocessed brown rice with the outer bran intact. Eijkman cured the natives by feeding them brown rice, discovering that food can cure disease. Over two decades later, nutritionists learned that the outer rice bran contains vitamin B1, also known as thiamine.

1900 through Present

In the early 1900s, Carl Von Voit and Max Rubner independently measured caloric energy expenditure in different species of animals, applying principles of physics in nutrition. In 1906, Wilcock and Hopkins showed that the amino acid tryptophan was necessary for the survival of rats. He fed them a special mixture of food containing all the nutrients he believed were essential for survival, but the rats died. A second group of rats to which he also fed fed an amount of milk containing vitamins. Gowland Hopkins recognized "accessory food factors" other than calories, protein and minerals, as organic materials essential to health but which the body cannot synthesize. In 1907, Stephen M. Babcock and Edwin B. Hart conducted the single-grain experiment. This experiment runs through 1911.

In 1912, Casimir Funk coined the term vitamin, a vital factor in the diet, from the words "vital" and "amine," because these unknown substances preventing scurvy, beriberi, and pellagra, were thought then to be derived from ammonia. The vitamins were studied in the first half of the twentieth century.

In 1913, Elmer McCollum discovered the first vitamins, fat soluble vitamin A, and water soluble vitamin B (in 1915; now known to be a complex of several water-soluble vitamins) and names vitamin C as the then-unknown substance preventing scurvy. Lafayette Mendel and Thomas Osborne also perform pioneering work on vitamin A and B. In 1919, Sir Edward Mellanby incorrectly identified rickets as a vitamin A deficiency, because he could cure it in dogs with cod liver oil. In 1922, McCollum destroyed the vitamin A in cod liver oil but finds it still cures rickets, naming vitamin D Also in 1922, H.M. Evans and L.S. Bishop discover vitamin E as essential for rat pregnancy, originally calling it "food factor X" until 1925.

In 1925, Hart discovered that trace amounts of copper are necessary for iron absorption. In 1927, Adolf Otto Reinhold Windaus synthesized vitamin D, for which he won the Nobel Prize in Chemistry in 1928. In 1928, Albert Szent-Györgyi isolated ascorbic acid, and in 1932 proves that it is vitamin C by preventing scurvy. In 1935 he synthesizes it, and in 1937 he wins a Nobel Prize for his efforts. Szent-Györgyi concurrently elucidates much of the citric acid cycle.

In the 1930s, William Cumming Rose identified essential amino acids, necessary protein components which the body cannot synthesize. In 1935, Underwood and Marston independently discover the necessity of cobalt. In 1936, Eugene Floyd Dubois showed that work and school performance are related to caloric intake. In 1938, Erhard Fernholz discovered the chemical structure of vitamin E. It was synthesised by Paul Karrer.

In 1940, rationing in the United Kingdom during and after World War II took place according to nutritional principles drawn up by Elsie Widdowson and others. In 1941, the first Recommended Dietary Allowances (RDAs) were established by the National Research Council.

In 1992, The U.S. Department of Agriculture introduced the Food Guide Pyramid. In 2002, a Natural Justice study showed a relation between nutrition and violent behaviour. In 2005, a study found that obesity may be caused by adenovirus in addition to bad nutrition.

NUTRITION SCALE

A Nutrition scale is a weighing instrument that output precise nutritional information for foods or liquids. Most scales calculate calories, carbohydrates, and fats, with more sophisticated scales calculating additional nutrients such as Vitamin K, potassium, magnesium, and sodium.

Scales often use USDA information on food to ensure accuracy. The products are used primarily as a weight management tool, but have found a user base with diabetics and hypertensive people.

NUTRITIONIST

A nutritionist is a health specialist who devotes professional activity to food and nutritional science, preventive nutrition, diseases related to nutrient deficiencies, and the use of nutrient manipulation to enhance the clinical response to human diseases.

They can also advise people on dietary matters relating to health, well-being and optimal nutrition. Nutritionists have varying levels of education from someone with little or no education to an individual who has obtained a bachelor's, master's, or doctoral degree. This is because the term "*nutritionist*" is not a legally protected term in most parts of the world. As a result, the term "*nutritionist*" is subject to several interpretations. Many nutritionists appear on television, in newspapers and magazines, and write nutritional books, which may or may not have any real informational value regarding diet.

Training

There are a wide range of courses available which vary in duration from several days to several years. A person who represents him or herself as a nutritionist may have several different levels of education including a degree in nutrition or dietetics from a university (of which several may be accredited by nutritionist governing bodies such as the Nutrition Society accrediting degree courses in the UK) or certification in nutrition education and/or counseling from a private vocational/professional training school.

Regulation of the Title "Nutritionist"

Canada

The title "nutritionist" is protected in Quebec and Nova Scotia. It is not protected in British Columbia. There were discussions about this when the dietitian designation was protected in British Columbia, but as of March 2008 "nutritionist" is not currently protected there.

United Kingdom

Nutritionist, unlike dietician, is not currently a protected term. Anyone in the UK can refer to himself or herself as a nutritionist without any formal qualifications. Different organisations promoting holistic and alternative therapies may use their own criteria to define a nutritionist. According to one of these, the Nutrition Society of the UK , *the function of a nutritionist is to elicit, integrate, disseminate and apply scientific knowledge drawn from the relevant sciences, to promote an understanding of the effects of nutrition, and to enhance the impact of food on health and well-being of animals and/or people*. They accredit nutritionists, conferring the titles *Associate Nutritionist (ANutr)*, *Associate Public Health Nutritionist (APHNutr)*, *Registered Nutritionist,* (*R Nutr.*), and *Registered Public Health Nutritionist (RPHNutr)*. For these they consider an undergraduate training sufficient to be entered on their Register of Nutritionists.

United States

The term "dietitian" is legally protected. Dietitians are registered with the American Dietetic Association and are only able to use the title "dietitian" when they have met strict, specific educational and experiential prerequisites and passed a national registration examination. The title "nutritionist" is protected and designated by many but not all states in the United States. It is important that a person seeking the counsel of a nutritionist, check with their local state's licensing agency to find out if prospective practitioners are duly licensed.

Traditionally, dietitians work in institutional settings, such as hospitals, schools and prisons, rather than in private practice. Nutritionists sometimes work in such institutions but more often work in private practice, in education and research. Some overlap exists within the two professions.

Development of a Certified Registry

The Certification Board For Nutrition Specialists (CBNS) was founded in 1993 by the American College of Nutrition (ACN) to help meet the growing demand for knowledgeable, responsible professional nutritionists. The protected title of Certified Nutrition Specialist (CNS) is awarded by CBNS to those nutritionists meeting defined educational, experience and examination requirements. Similarly and as mentioned above, the Nutrition Society of the United Kingdom has established a Register of Nutritionists. This is to recognize and encourage high standards of professional training in nutrition as well as to protect the public.

Types of Nutritionist

Nutrition Scientists

Nutrition scientists are those individuals who use the scientific method to study nutrients, both as individual compounds and as they interact in food and nutrition. The role of the nutrition scientist is to develop new knowledge related to nutrients or nutrition or to develop new processes or techniques to apply existing knowledge. For example, nutrition scientists have been involved in developing food preservation processes, determining nutrient requirements for various animal species, describing how individual nutrients function within the cells of the human body, and identifying nutrition-related problems in various populations.

Nutritionist scientists may have their basic training in nutrition or in a related field such as , microbiology, cell biology, epidemiology, toxicology, agriculture, or food science, chemistry.

Public Health Nutritionists

Public health nutritionists are professionals who view the community as their client. They specialize in diagnosing the nutritional problems of communities and in finding solutions to those problems. Some classic examples of public health nutrition interventions include the fortification of salt with iodine to prevent Goitre or the enrichment of grain products with B vitamins to prevent deficiency diseases.

MALNUTRITION

Malnutrition is a general term for a medical condition caused by an improper or insufficient diet. It most often refers to undernutrition resulting from inadequate consumption, poor absorption, or excessive loss of nutrients, but the term can also encompass overnutrition, resulting from overeating or excessive intake of specific nutrients. An individual will experience malnutrition if the appropriate amount of, or quality of nutrients comprising a healthy diet are not consumed for an extended period of time. An extended period of malnutrition can result in starvation, disease, and infection.

Malnutrition is the lack of sufficient nutrients to maintain healthy bodily functions and is typically associated with extreme poverty in economically developing countries. It is a common cause of reduced intelligence in parts of the world affected by famine. Malnutrition as the result of inappropriate dieting, overeating or the absence of a "balanced diet" is often observed in economically developed countries (eg. as indicated by increasing levels of obesity).

Most commonly, malnourished people either do not have enough calories in their diet, or are eating a diet that lacks protein, vitamins, or trace minerals. Medical problems arising from malnutrition are commonly referred to as deficiency diseases. Scurvy is a well-known and now rare form of malnutrition, in which the victim is deficient in vitamin C.

Common forms of malnutrition include protein-energy malnutrition (PEM) and micronutrient malnutrition. PEM refers to inadequate availability or absorption of energy and proteins in the body. Micronutrient malnutrition refers to inadequate availability of some essential nutrients such as vitamins and trace elements that are required by the body in small quantities. Micronutrient deficiencies lead to a variety of diseases and impair normal functioning of the body. Deficiency in micronutrients such as Vitamin A reduces the capacity of the body to resist diseases. Deficiency in iron, iodine and vitamin A is widely prevalent and represent a major public health challenge. An array of afflictions ranging from stunted growth, reduced intelligence and various cognitive abilities, reduced sociability, reduced leadership and assertiveness, reduced activity and energy, reduced muscle growth and strength, and poorer health overall are directly implicated to nutrient deficiencies. Also, another, although rare, effect of malnutrition is black spots appearing on the skin.

Hunger is the normal psychological response brought on by the physiological condition of needing food. Hunger can also affect the mental state of a person, and is often used as a metonym for general undernourishment.

According to the World Health Organization, hunger and malnutrition is the gravest single threat to the world's public health and malnutrition is by far the biggest contributor to child mortality, present in half of all cases.

Politics

As of 2008, malnutrition continues to be a worldwide problem, particularly in lesser developed countries. According to the Food and Agriculture Organization of the United Nations, "850 million people worldwide were undernourished in 1999 to 2005, the most recent years for which figures are available" and the number of malnourished people has recently been increasing. An orange awareness ribbon is used to raise awareness of malnutrition in the world. The FAO calculates undernourishment by comparing the amount of food available in a country at national level with how many people live in the country.

Fig. 6.1: The Orange Ribbon—An Awareness Ribbon for Malnutrition

Some environmentalists claim that the fundamental issue causing malnutrition is that the human population exceeds the Earth's carrying capacity; however, Food First raises the issue of food sovereignty and claims that every country (with the possible minor exceptions of some city-states) has sufficient agricultural capacity to feed its own people, but that the "free trade" economic order associated with such institutions as the International Monetary Fund (IMF) and the World Bank prevent this from happening. At the other end of the spectrum, the World Bank itself claims to be part of the solution to malnutrition, asserting that the best way for countries to succeed in breaking the cycle of poverty and malnutrition is to build export-led economies that will give them the financial means to buy foodstuffs on the world market.

Amartya Sen won a 1998 Nobel Prize in part for his work suggesting that famine is not typically the product of a lack of food; rather, famine may arise from problems in food distribution networks or from governmental policies in the developing world.

The politics of food trade and food security are often difficult to grasp. Many people believe that sending food aid to the poor of the world is a worthy idea, but that each country should produce its own food.

Countries that have become more open to international trade in recent years (*e.g.* China, Vietnam or Peru) have greatly reduced the prevalence of undernourishment as measured by the FAO (food energy consumption below acceptable minimum) or as measured by the World Health Organization by the percentage of children under five who are stunted, wasted or underweight. Countries that remained closed to external trade (*e.g.* North Korea) have not improved or have worsened their food situation.

Some anti-globalization groups advocate "food sovereignty", stating that each country should be physically self sufficient in every food item consumed by their people; by this measure the United States, United Kingdom, Sweden, Belgium, and in fact almost all other countries in the world would be food insecure, and a desert nation like Saudi Arabia (with its current population) would not be viable as a country at all.

One policy adopted in recent decades to alleviate world malnutrition is food aid, *i.e.* the physical donation of food from rich to poor countries. From the rich donor countries' point of view, this is a suitable way to reduce excess supply created by domestic agricultural subsidies, stabilizing farm prices in rich countries, even if the cost of supplying the food to its final beneficiaries is often disproportionately high. Food aid may be provided for short-term emergencies (natural disasters like earthquakes, tsunamis, droughts and floods, or human-made like war and refugee flows) or in the form of a long-term programme for an extended period. From the viewpoint of recipient countries, the value of food-aid depends on the form it takes. Emergency food aid is welcome, though aid in cash may also be welcome because the food may often be purchased locally in zones not affected by the emergency, thus benefitting local farmers. Long-term foreign food aid has been criticized as discouraging local production and distorting markets. Instead, population control has been advocated as a much better approach to solve malnutrition/famine than merely providing food.

Effects

Mortality due to Malnutrition

According to the World Health Organization, hunger is the gravest single threat to the world's public health. According to Jean Ziegler (the United Nations Special Rapporteur on the Right to Food for 2000 to March 2008), mortality due to malnutrition accounted for 58 per cent of the total mortality in 2006: "In the world, approximately 62 millions people, all causes of death combined, die each year. One in twelve people worldwide are malnourished. In 2006, more than 36 millions died of hunger or diseases due to deficiencies in micronutrients". The World Health Organization estimates that one-third of the world is well-fed, one-third is under-fed and one-third is starving. Every 3.6 seconds someone dies of hunger.

Hunger and malnutrition have an even bigger impact on children's health than was previously thought. According to the World Health Organization, malnutrition is by far the biggest contributor to child mortality, present in half of all cases. Underweight births and inter-uterine growth restrictions cause 2.2 million child deaths a year. Poor or non-existent breastfeeding causes another 1.4 million. Other deficiencies, such as lack of vitamin A or zinc, for example, account for 1 million. According to The Lancet, malnutrition in the first two years is irreversible. Malnourished children grow up with worse health and lower educational achievements. Their own children also tend to be smaller. Hunger was previously seen as something that exacerbates the problems of diseases such as measles, pneumonia and diarrhea. But malnutrition actually causes diseases as well, and can be fatal in its own right. This is the impact The Lancet seeks to identify.

Biological Effects

Prisoners who died from the malnutrition in Gulag. Painting by Nikolai Getman, provided by Jamestown Foundation

An extended period of malnutrition can result in starvation or deficiency diseases such as scurvy. Malnutrition increases the risk of infection and infectious disease; for example, it is a major risk factor in the onset of active tuberculosis.

Malnutrition appears to increase activity and movement in many animals—for example an experiment on spiders showed increased activity and predation in starved spiders, resulting in larger weight gain. This pattern is seen in many animals, including humans while sleeping. It even occurs in rats with their cerebral cortex or stomachs completely removed. Increased activity on hamster wheels occurred when rats were deprived not only of food, but also water or B vitamins such as thiamine This response may increase the animal's chance of finding food, though it has also been speculated the emigration response relieves pressure on the home population.

Action on Malnutrition

In recent years many foods, such as Spirulina and peanut butter, have been developed or refined for mass-production in hopes of combating malnutrition and its effects.

Breast-feeding advice, food supplements and better hygiene all make a big difference. Most countries know what to do and run pilot programmes that work. But they rarely find the money for full-scale national efforts; the international outfits that might help are, in the words of the medical journal, The Lancet, fragmented and dysfunctional.

According to The Lancet's research, money for improving nutrition would be the most effective sort of aid. At the moment, roughly $300 million of aid goes to basic nutrition each year, less than $2 for each child below two in the 20 worst affected countries. In contrast, HIV/AIDS, which causes fewer deaths than child malnutrition, received $2.2 billion—$67 per person with HIV in all countries, including rich ones.

Statistics

Number of undernourished people (million) in 2001-2003,

according to the FAO, the following countries had 5 million or more undernourished people:

Country	*Number of Undernourished (million)*
India	217.00
China	150.0
Bangladesh	43.1
Democratic Republic of Congo	37.0
Pakistan	35.2
Ethiopia	31.5
Tanzania	16.1
Philippines	15.2
Brazil	14.4
Indonesia	13.8
Vietnam	13.8
Thailand	13.4
Nigeria	11.5
Kenya	9.7
Sudan	8.8
Mozambique	8.3
North Korea	7.9
Yemen	7.1
Madagascar	7.1
Colombia	5.9
Zimbabwe	5.7
México	5.1
Zambia	5.1
Angola	5.0

Note: This table measures "undernourishment", as defined by FAO, and represents the number of people consuming (on average for years 2001 to 2003) less than the minimum amount of food energy (measured in kilocalories per capita per day) necessary for the average person to stay in good health while performing light physical activity. It is a conservative indicator that does not take into account the extra needs of people performing extraneous physical activity, nor seasonal variations in food consumption or other sources of variability such as inter-individual differences in energy requirements. Malnutrition and undernourishment are cumulative or average situations, and not the work of a single day's food intake (or lack thereof). This table does not represent the number of people who "went to bed hungry today."

The U.S. Department of Agriculture reported that in 2003, only 1 out of 200 U.S. households with children became so severely food insecure that any of the children went hungry even once during the year. A substantially larger proportion of these same households (3.8 percent) had adult members who were hungry at least one day during the year because of their households' inability to afford enough food.

Overnourished vs Undernourished

In 2006, Professor Popkin from the University of North Carolina, said there were now more overweight people across the world than undernourished people. He told the International Association of Agricultural Economists the number of overweight people had topped one billion (of which 300 million are obese), compared with 800 million undernourished. He added this transition from a starving world to an obese one was accelerating.

APPLIED NUTRITION CONCEPTS

Applied Nutrition Concepts is committed to helping you achieve the highest health goals you have for yourself and your family, using all-natural methods. At ANC, we want to fully understand your specific needs. We ask questions and take detailed inventories. We may perform functional tests to determine your biochemistry. Then we find the safest and most natural way for you to achieve your health goals. We recommend dietary and lifestyle changes based on your specific biochemistry and provide the targeted nutrients to correct any imbalances. This correction often leads to a healthier and happier you.

Information in this site is not intended to countermand any advice given by your health professional. Please check with them before starting any health programme.

Services

Applied Nutrition Concepts offers complete nutritional and biochemical analysis for our clients. Then we design a specific protocol designed to meet their personal needs and health goals. We have several service offerings available to our clients.

Consulting

Nutritional consultations for individuals and families are offered as a single comprehensive visit to get you started on the right track. Many of our clients elect to continue counseling with on-going support and encouragement. We are available as often as you need by appointment.

SCENAR Energy Medicine

A remarkable new energetic healing device from Russia. Russian research has shown that the SCENAR device can help reduce and remove pain. It has helped both acute and chronic injuries, as well as many other physical problems including sinusitis, pneumonia, swelling, wounds that won't heal, bones that are slow to mend, and many others. Ask us if SCENAR is right for your type of pain or injury.

The SCENAR machine is rubbed lightly over the affected areas. It is non-invasive and painless. SCENAR is completely safe for adults and children.

Body Composition Analysis

This analysis will reveal information about your lean muscle mass, your per cent body fat and your hydration or water level. This non-invasive test provides numerical scores about your cell health and the degree of toxicity in your body.

Red Blood Cell Analysis

Using a drop of your blood from a finger stick, we can observe the size and shape of your red and white blood cells. This test can reveal nutritional deficiencies based on those changes in your cells. Using this test we can monitor progress of your nutritional status over time.

Products

Nutritional Supplements

Applied Nutrition Concepts offers only the very best assortment of nutritional supplements available only through health care providers. These high-quality formulas are all-natural and allergen free. We carry many products in our office and we are able to special order products for our clients. We welcome you to come by to pick up your products, or we can mail them to your home or business.

Herbal Products

ANC carries only the highest quality herbal products, favouring the Medi-Herb line. This herbal line originates in Australia under the direction of Kerry Bone, Master Herbalist. They are then shipped to Standard Process in Wisconsin. They can be purchased in either tablet, capsule of tincture for your specific need.

Homeopathic Products

Pekana, Heel/BHI, Sanum Pleomorphic and SanPharma are some of the lines that ANC feels are the highest quality homeopathic products. These are all German brands which give the most variety to serve our patient's needs.

Pet Nutrition

We offer whole food nutritional supplements for dogs and cats, puppies and kittens. You can prevent a lot of health problems by changing your pet's diet and adding whole food nutrition to their daily food intake. Call us for a free brochure or to order your supplements today.

7

Food Service, Cafeteria and Restaurant: Distribution, Management and Institutional Aspects

FOOD SERVICE

The foodservice (or food service) industry (US English; catering industry in British English) encompasses those places, institutions, and companies responsible for any meal prepared outside the home. This industry includes restaurants, school and hospital cafeterias, catering operations, and many other formats.

The companies that supply foodservice operators are called foodservice distributors. Foodservice distributors sell goods like small wares (kitchen utensils) and bulk foods.

Some companies manufacture products in both consumer and foodservice versions. The consumer version usually comes in individual-sized packages with elaborate label design for retail sale. The foodservice version is packaged in a much larger industrial size and often lacks the colourful label designs of the consumer version.

Foodservice sales to restaurants and institutions are estimated to be approximately $400 Billion, about equal with consumer sales of foods through grocery outlets. The foodservice industry is one of the largest employers in the United States. Major foodservice providers include Aramark, Brinker International, Compass Group, the Crown Group, Darden Restaurants and 3663 First for Foodservice.

CAFETERIA

A cafeteria or cafeteria is a type of food service location in which there is little or no table service, whether a restaurant or within an institution such as a large office building or school; a school dining location is also referred to as a dining hall or canteen (in UK English). Cafeterias are different from coffeehouses, although that is the Spanish meaning of the English word.

Instead of table service, there are food-serving counters/stalls, either in a line or allowing arbitrary walking paths. Customers take the food they require as they walk along, placing it on a tray. In addition, there are often stations where customers order food and wait while it is prepared, particularly for items such as hamburgers or tacos which must be served hot and can be quickly prepared. Alternatively, the patron is given a number and the item is brought to their table. Sometimes, for some food items and drinks, customers collect an empty container, pay at the check-out, and fill the container after the check-out. Free second servings are often allowed under this system. For legal purposes (and the consumption patterns of customers), this system is rarely or never used for alcoholic beverages.

Customers are either charged a flat rate for admission (as in a buffet), or pay at the check-out for each item. Some self-service cafeterias charge by the weight of items on a patron's plate.

As cafeterias require few employees, they are often found within a larger institution, catering to the clientele of that institution. For example, schools, colleges and their residence halls, department stores, hospitals, museums, and office buildings often have cafeterias.

At one time, upscale cafeteria-style restaurants dominated the culture of the , and to a lesser extent the Midwest. There were several prominent chains of them: Bickford's, Morrison's Cafeteria, Piccadilly Cafeteria, S&W Cafeteria, Apple House, K&W, Britling, and Blue Boar among them. Currently two midwest chains still exist, Sloppy Jo's Luchroom and Manny's, both located in Illinois. There were also a number of smaller chains, usually in and around a single city. These institutions, with the exception of K&W, went into a decline in the 1960s with the rise of fast food and were largely finished off in the 1980s by the rise of "casual dining". A few chains—notably Luby's and Piccadilly Cafeterias (which took over the Morrison's chain), continue to fill some of the gap left by the decline of the older chains. Many of the smaller Midwestern chains, such as MCL Cafeterias centered around Indianapolis, are still very much in business.

The world's largest non-military cafeteria is in the Brody Complex at Michigan State University.

History

Perhaps the first self-service restaurant (not necessarily cafeteria) in the United States was the Exchange Buffet in New York City, opened September 4, 1885, which catered to an exclusively male clientele. Food was purchased at a counter, and patrons ate standing up. This represents the predecessor of two formats: the cafeteria, described below, and the automat.

During the 1893 World's Columbian Exposition in Chicago, an entrepreneur named John Kruger built an American version of the smörgåsbords he had seen while traveling in Sweden. Emphasizing the simplicity and light fare, he called it the "Cafeteria"—Spanish for "coffee shop". The exposition attracted over 27 million visitors (half the US population at the time) in six months, and it was initially through Kruger's operation that America first heard the term and experienced the self-service dining format.

Meanwhile, in everyday, hometown America, the chain of Childs Restaurants was quickly growing from about 10 locations in New York City (in 1890), to hundreds across the United States and Canada (by 1920). Childs is credited with the critical innovation of adding trays and a "tray line" to the self-service format, which they introduced in 1898 at their 130 Broadway location. Childs did not change its format of sit-down dining, however. This was soon the standard design for most Childs Restaurants—and many imitators—from coast-to-coast, and ultimately the dominant design for cafeterias.

It has also been said that the "cafeteria craze started in May 1905, when a woman named Helen Mosher opened a humble downtown L.A. restaurant where people chose their food at a long counter and carried their own trays to their tables." California does have a long and rich history in the cafeteria format—most notably the many Boos Brothers Cafeterias, and also Clifton's and Schaber's. However, the facts do not warrant the "wellspring" characterization that some have ascribed to the region. The earliest cafeterias in California were opened at least 12 years after Kruger's Cafeteria, and Childs already had several dozen locations scattered around the country. Finally, Horn and Hardart, an automat format chain (only slightly different from the cafeteria), was also well established in the mid-Atlantic region before 1900.

Between 1960 and 1980, the popularity of cafeteria format restaurants was gradually overcome by the emergence of the fast food restaurant and fast casual restaurant formats.

Other Names

A cafeteria in a U.S. military installation is known as a chow hall, a mess hall, a galley, mess decks or, more formally, a dining facility, whereas in common British Armed Forces parlance, it is known as a cookhouse or mess. Students in the USA often refer to cafeterias as lunchrooms, though breakfast as well as lunch is often eaten there. Cafeterias serving university dormitories are sometimes called dining halls or dining commons. A food court is a type of cafeteria found in many shopping malls and airports featuring multiple food vendors or concessions, although a food court could equally be styled as a type of restaurant as well, being more aligned with public, rather than institutionalised, dining.

Some monasteries, boarding schools and older universities refer to their cafeteria as a refectory. Modern-day British cathedrals and abbeys, notably in the Church of England, often use their the phrase refectory to describe a cafeteria open to the public. Historically, the refectory was generally only used by monks and priests. For example, although the original 800-year-old refectory at Gloucester Cathedral (the stage setting for dining scenes in the Harry Potter movies) is now mostly used as a choir practice area, the relatively modern 300-year-old extension, now used as a cafeteria by staff and public alike, is today referred to as the refectory.

College Cafeteria

A college cafeteria is a term in the United States that denotes a cafeteria that is designed to serve college students at the university.

In the UK the word *refectory* is often used. Also see the different meanings of the word college around the Anglosphere. These cafeterias can be a part of a residence hall or in a separate building. Many of these colleges employ their own students to work in the cafeteria. The amount of meals served to students varies from school to school, but is normally around 20 meals per week. Like normal cafeterias, a person will have a tray to select the food that they want, but instead of paying money, they pay beforehand by purchasing a meal plan.

The method of payment for college cafeterias is commonly in the form of a meal plan, whereby the patron pays a certain amount at the start of the semester and the details of the plan are stored on a computer system. Student ID cards are then used to access the meal plan. A meal plan is not necessary to eat at a college cafeteria however. Meal plans can vary widely in their details to best fit the needs of the students. Typically, the college tracks the student's usage of their plan by counting either the number of pre-defined meal servings, points, dollars, or number of buffet dinners. The plan may give the student a certain number of any of the above per week or semester and they may or may not roll over to the next week or semester.

Many schools offer several different options for using their meal plans. The main cafeteria is usually where most of the meal plan is used but smaller cafeterias, cafés, restaurants, bars, or even fast food chains located on campus may accept meal plans. A college cafeteria system often has a virtual monopoly on the students due to an isolated location or a requirement that residence contracts include a full meal plan. It is not uncommon for the entire food service operation to be outsourced to a managed services company such as Aramark, Sodexo and Compass Group (under the Scolarest name in the United Kingdom).

FOODSERVICE DISTRIBUTOR

A foodservice distributor is a company that provides food and non-food products to restaurants, cafeterias, industrial caterers, and hospitals and nursing homes.

A foodservice distributor functions as an intermediary between food manufacturers and the foodservice operator (usually a chef, foodservice director, food and beverage manager, and independent food preparation businesses operator owners.) Foodservice distributors procure pallets and bulk inventory quantites that are broken down to case and sometimes unit quantities for the foodservice operator. Most foodservice operators purchase from a range of local, speciality, and broadline foodservice distributors on a daily or weekly basis.

Often a food manufacturer may hire a foodservice brokerage company to represent the manufacturer in a local market. The broker helps the food manufacturer market its products through the foodservice distribution system, which ranges from getting items stocked at the distributor to working with operators to purchase items from the distributor.

A broadline distributor services a wide variety of accounts with a wide variety of products, while a system distributor stocks a narrow array of products for specific customers, such as restaurant chains.

Foodservice distribution companies can range in size from a one-truck operation to larger corporations. There are many independent broadline foodservice distribution companies that service chain and multi-unit restaurants based on master distribution agreements with national foodservice groups. These groups provide distributor members procurement capabilities rival the purchasing power of largest distributors. These distributor groups also provide distributor members group private label brands as well as marketing and quality assurance services.

Redistribution

An emerging trend in the foodservice distribution business is redistribution.

Before the advent of redistribution, only the largest foodservice distributors could purchase enough products from an individual food manufacturer to fill up an entire truck. Smaller distributors were forced to purchase in less than truckload (LTL) quantities, putting them at a significant logistical disadvantage to the larger distributors.

In the food redistribution model, a redistributor will purchase in truckload quantities from many food manufacturers and warehouse the products. Individual distributors can then purchase multiple manufacturers' products from the redistributor and fill up an entire truck to save on shipping costs. The largest food redistributor in the U.S. is Dot Foods, followed by Sysco.

RESTAURANT MANAGEMENT

Restaurant management is the profession of managing a restaurant. Associate, bachelor, and graduate degree programmes are offered in restaurant management by community colleges, junior colleges, and some universities in the United States.

Responsibilities

1. 'Floor management' includes managing staff who give services to customers and allocate the duties of opening and closing restaurant. The manager is responsible for making sure his or her staff is following the service standards and health and safety regulations. The manager is the most important person in the front-of-the-house environment, since it is up to him or her to motivate the staff and give them job satisfaction. The manager also looks after and guides the personal well-being of the staff, since it makes the work force stronger and more profitable.
2. 'Kitchen management' includes the managing staff working in the kitchen, especially the head chef. The kitchen is the most important part of the business and the main reason customers patronize the restaurant. Managing the kitchen staff helps to control food quality. As most commercial kitchens are a closed environment, the staff may become bored or tired from the work. Without proper management, this often results in an inconsistent food product.
3. 'Administration' includes stock controlling, scheduling rotations, budgeting the labour costs, balancing cost and profit according to seasonality, surveying and hiring staff, and maintenance of the commercial kitchen equipment.

CASE STUDY: NATIONAL RESTAURANT ASSOCIATION

The National Restaurant Association is a restaurant industry business association in the United States with 60,000 member companies, representing a total of more than 300,000 restaurants. It also operates the National Restaurant Association Educational Foundation. The association was founded in 1919.

Recent association presidents included Edward R. Tinsley of New Mexico, the franchisor of K-Bob's Steakhouse restaurant chain, who served from 2006-2007.

Activities

The association develops food safety training and certification programme for restaurant employees. It also offers scholarships to foodservice and hospitality management and culinary students, and runs a national programme for high school students.

It runs an annual restaurant and hospitality industry trade show in the U.S. in Chicago and conducts research about the restaurant industry in the U.S.

The association lobbies for the industry. It was the largest food and beverage political action committee contributor to both the U.S. Democratic and Republican Parties in the 2004 election cycle.

CASE STUDY: WASHINGTON RESTAURANT ASSOCIATION

Founded in 1929, The Washington Restaurant Association is a restaurant industry business association in the state of Washington with 5,000 member companies, representing a total of 12,000 restaurants. The WRA is sanctioned by the National Restaurant Association.

Activities

The association, in conjunction with the Education Foundation, develops a variety of educational classes, such as food safety and alcohol server training, and business assistance programmes, such as a workers' compensation programme, health insurance and financial and legal consultation.

The association represents its members in lobbying efforts through its government relations services. It continues to be the largest political action committee contributor to the .

The association runs the annual , a restaurant and hospitality industry trade show, in conjunction with the .

8

Culinary Science and Art: Various Aspects and Listings

CULINOLOGY

Culinology is an approach to food that blends the culinary arts and food technology. Through the blending of these two disciplines, culinology seeks to make food taste better—whether purchased in a supermarket or eaten in a restaurant. Culinology also seeks to make food more consistent and safer. A primary method of culinology is to logically translate sophisticated food concepts, such as those applied in fine dining or in a traditional ethnic cuisine, for items on the menus of chain restaurants or those processed for retail sale. Such product or chain-menu development is only possible through the astute combination of the culinary arts and food science or technology.

According to Jeff Cousminer in Food Product Design Magazine, the word *culinology* was coined by the first president and founder of the Research Chefs Association, Winston Riley . The original meaning of the word was quite different than what it has come to mean today. Originally, the word was designed to be a combination of two words, Culinary and Technology. So the first meaning of the word was the convergence of culinary arts and all technology, which includes communications, chemistry, physiology, economics and many others.

There are accredited culinology educational programmes offered by many institutions. The curriculum of such courses combine the disciplines of cooking and food science. According to industry professionals, like Kraft's Harry Crane, culinology should "help jump-start product development."

Culinologists work in diverse aspects of food—from experimental chefs and menu planners to food manufacturing to fine dining.

CULINARY ART

Culinary art is the art of cooking. The word "culinary" is defined as something related to, or connected with, cooking or kitchens. A culinarian is a person working in the culinary arts. A culinarian working in restaurants is commonly known as a cook or a chef. Culinary artists are responsible for skillfully preparing meals that are as pleasing to the palate as to the eye. Increasingly they are required to have a knowledge of the science of food and an understanding of diet and nutrition. They work primarily in restaurants, fast food chain store franchises, delicatessens, hospitals and other institutions. Kitchen conditions vary depending on the type of business, restaurant, nursing home etc.

Careers in Culinary Arts

Related Careers

There are a wide variety of opportunities in the culinary arts profession. Below is a non-exhaustive list at some opportunities available to those interested in working as culinary professionals.

- Chefs and cooks—Work directly with the preparation at food for service in various establishments including full-service restaurants, catering facilities, institutional dining, corporate dining, and various other venues.
- *Dining Room Service*-The profession of managing a restaurant. Associate, bachelor, and graduate degree programmes are offered in restaurant management by community colleges, junior colleges, and some universities in the United States.
- *Food and Beverage Managers*—Manage all food and beverage outlets in hotels and other large establishments.

- *Food and Beverage Controllers*—Purchase and source ingredients in large hotels as well as manage the stores and stock control.
- *Consultants* and *Design Specialists*—Works with restaurant owners in development of menus, layout and design of dining room and establishing cosina protocols.
- *Salespeople*—Introduce chefs and business owners to new products and are able to demonstrate the proper use at equipment. They may sell any range at products needed in the industry from various food products to equipment.
- Teachers—Train future culinary professionals with their experience in the classroom.
- Food Writers and Food Critics—Establish a public contact with the public on food trends, chefs and restaurants. Many write for newspapers, magazines and some publish books. Notables in this field include Julia Child, Craig Claiborne and James Beard.
- Food Stylists and Photographers—Work with magazines, books, catalogs and promotional and advertising to make them visually appealing.
- Research and Development Kitchens—Develop new products for commercial manufacturers, they can also be test kitchens for publications or others can be kitchens for restaurant chains or grocery chains.
- Entrepreneurship. Many culinarians develop their own business with examples including restaurants, bakeries, specialty food manufacture (chocolates, cheese etc).

Occupational Outlook

The occupation outlook for chefs, restaurant managers, dieticians, and nutritionists is fairly good, with "as fast as the average" growth. Increasingly a college education with formal qualifications is required for success in this field. It has been recorded that 54 per cent of all Culinary art professionals are female.

Culinary Colleges Around the World

North America

- International Culinary School at The Art Institutes (41 locations in North America)

- United States of America:
    - Cypress Community College Hotel, Restaurant Management, and Culinary Arts Programme in Anaheim
    - Cal Poly Pomona
    - California School of Culinary Arts Pasadena
    - Cooking and Hospitality Institute of Chicago
    - Classic Cooking Academy, Scottsdale, Arizona
    - Culinary Institute of America in Hyde Park, NY
    - Culinary Institute of America at Grey Stone in St. Helena, California
    - Institute for the Culinary Arts at Metropolitan Community College, Omaha, Nebraska
    - Johnson and Wales University (RI, FL, NC, CO)
    - Kendall College in Chicago, Illinois
    - L'Ecole Culinaire in Saint Louis, Missouri and Memphis, Tennessee
    - Lincoln College of Technology
    - Manchester Community College in Connecticut
    - New England Culinary Institute in Vermont
    - Orlando Culinary Academy
    - Pennsylvania Culinary Institute
    - Scottsdale Culinary Institute
    - Texas Culinary Academy
    - The Culinary Institute of Charleston, South Carolina
    - The Restaurant School at Walnut Hill College, Philadelphia, Pennsylvania,
    - Coosa Valley Technical College; Rome, GA

Canada

- Camosun College, (Victoria, BC)
- George Brown College, (Toronto, ON)
- Le Cordon Bleu Culinary Arts Institute, (Ottawa, ON)
- Liaison College (several Ontario locations, including Barrie, Brampton, Hamilton, Kitchener, Oakville and Toronto)

- Niagara Culinary Institute, Niagara College, Niagara-on-the-Lake, ON
- Pacific Institute of Culinary Arts, (Vancouver, BC)
- The Culinary Institute of Canada (Charlottetown, PE)

Europe

- Westminster Kingsway College (London)
- School of Culinary Arts and Food Technology, DIT, Dublin, Ireland
- Scuola di Arte Culinaria Cordon Bleu, Florence, Italy
- Apicius International School of Hospitality, Florence, Italy
- École hôtelière de Lausanne, Lausanne, Switzerland
- Institut Paul Bocuse, Ecully, France
- Le Cordon Bleu, Paris, France
- École des trois gourmandes, Paris, France
- Ecole Hôtelière du Périgord, Périgueux, France
- HRC Culinary Academy, Bulgaria

Africa

- School of Tourism and Hospitality, University of Johannesburg, South Africa, Africa

Australia and New Zealand

- North Shore International Academy, Auckland, New Zealand

Caribbean

- International School Of Tourism and Hospitality Management
- Universidad del Este, Puerto Rico, USA

Latin America

- Academia de Artes Culinarias de Guatemala, Guatemala.

Asia

- Cilantro Culinary Academy, Malaysia
- School of Hospitality, Tourism and Culinary Arts, KDU College, Malaysia

- Taylor's College, School of Hospitality and Tourism, Malaysia
- Center for Culinary Arts, Manila, Philippines
- Culinary Academy Of India, Hyderabad, India
- MOST Institute, Philippines
- Sebu Institute of Culinary Arts (SICA), Visayan Federation

Middle East

- International Centre for Culinary Arts, Dubai (ICCA Dubai)

Other

- Le Cordon Bleu (locations in 20 countries)

LIST OF CULINARY VEGETABLES

This is a list of vegetables in the culinary sense. This means that the list includes some botanical fruits such as pumpkins, and does not include herbs, spices, cereals and most culinary fruits and culinary nuts. Currently edible fungi are not included on this list.

Some culinary vegetables (seaweeds like laver) are not even members of the plant kingdom.

Leafy and Salad Vegetables

- Amaranth (*Amaranthus cruentus*)
- Arugula (*Eruca sativa*)
- Beet greens (*Beta vulgaris* subsp. *vulgaris*)
- Bitterleaf (*Vernonia calvoana*)
- Bok choy (*Brassica rapa* Chinensis group)
- Broccoli Rabe (*Brassica rapa* subsp. *rapa*)
- Brussels sprout (*Brassica oleracea* Gemmifera group)
- Cabbage (*Brassica oleracea* Capitata group)
- Catsear (*Hypochaeris radicata*)
- Celery (*Apium graveolens*)
- Celtuce (*Lactuca sativa* var. *asparagina*)
- Ceylon spinach (*Basella alba*)
- Chard (*Beta vulgaris var. cicla*)

- Chaya (*Cnidoscolus aconitifolius* subsp. *aconitifolius*)
- Chickweed (*Stellaria*)
- Chicory (*Cichorium intybus*)
- Chinese cabbage (*Brassica rapa* Pekinensis group)
- Chinese Mallow (*Malva verticillata*)
- Chrysanthemum leaves (*Chrysanthemum coronarium*)
- Collard greens (*Brassica oleracea*)
- Corn salad (*Valerianella locusta*)
- Cress (*Lepidium sativum*)
- Dandelion (*Taraxacum officinale*)
- Endive (*Cichorium endivia*)
- Epazote (*Chenopodium ambrosioides*)
- Fat hen (*Chenopodium album*)
- Fiddlehead (*Pteridium aquilinum, Athyrium esculentum*)
- Fluted pumpkin (*Telfairia occidentalis*)
- Garden Rocket (*Eruca sativa*)
- Golden samphire (*Inula crithmoides*)
- Good King Henry (*Chenopodium bonus-henricus*)
- Greater Plantain (*Plantago major*)
- Kai-lan (*Brassica rapa* Alboglabra group)
- Kale (*Brassica oleracea* Acephala group)
- Komatsuna (*Brassica rapa* Pervidis or Komatsuna group)
- Kuka (*Adansonia* spp.)
- Lagos bologi (*Talinum fruticosum*)
- Land cress (*Barbarea verna*)
- Lettuce (*Lactuca sativa*)
- Lizard's tail (*Houttuynia cordata*)
- Melokhia (*Corchorus olitorius, Corchorus capsularis*)
- Mizuna greens (*Brassica rapa* Nipposinica group)
- Mustard (*Sinapis alba*)
- New Zealand Spinach (*Tetragonia tetragonioides*)
- Orache (*Atriplex hortensis*)
- Paracress (*Acmella oleracea*)
- Pea sprouts/leaves (*Pisum sativum*)
- Polk (*Phytolacca americana*)
- Radicchio (*Cichorium intybus*)
- Samphire (*Crithmum maritimum*)
- Sea beet (*Beta vulgaris* subsp. *maritima*)

- Seakale (*Crambe maritima*)
- Sierra Leone bologi (*Crassocephalum* spp.)
- Soko (*Celosia argentea*)
- Sorrel (*Rumex acetosa*)
- Spinach (*Spinacia oleracea*)
- Summer purslane (*Portulaca oleracea*)
- Swiss chard (*Beta vulgaris* subsp. *cicla* var. *flavescens*)
- Tatsoi (*Brassica rapa* Rosularis group)
- Turnip greens (*Brassica rapa* Rapifera group)
- Watercress (*Nasturtium officinale*)
- Water spinach (*Ipomoea aquatica*)
- Winter purslane (*Claytonia perfoliata*)
- Yarrow (*Achillea millefolium*)

Fruiting and Flowering Vegetables

The fruit of trees:

- Avocado (*Persea americana*)
- Breadfruit (*Artocarpus altilis*)

Fruits of annual or perennial plants:

- Acorn squash (*Cucurbita pepo*)
- Armenian cucumber (*Cucumis melo* Flexuosus group)
- Aubergine (*Solanum melongena*)
- Bell pepper (*Capsicum annuum*)
- Bitter melon (*Momordica charantia*)
- Caigua (*Cyclanthera pedata*)
- Cape Gooseberry (*Physalis peruviana*)
- Capsicum (*Capsicum annuum*)
- Cayenne pepper (*Capsicum frutescens*)
- Chayote (*Sechium edule*)
- Chili pepper (*Capsicum annuum* Longum group)
- Courgette (*Cucurbita pepo*)
- Cucumber (*Cucumis sativus*)
- Eggplant (*Solanum melongena*)
- Luffa (*Luffa acutangula, Luffa aegyptiaca*)

- Malabar gourd (*Cucurbita ficifolia*)
- Parwal (*Trichosanthes dioica*)
- Pattypan squash (*Cucurbita pepo*)
- Perennial cucumber (*Coccinia grandis*)
- Pumpkin (*Cucurbita maxima, Cucurbita pepo*)
- Snake gourd (*Trichosanthes cucumerina*)
- Squash aka marrow (*Cucurbita pepo*)
- Sweet corn aka corn; aka maize (*Zea mays*)
- Sweet pepper (*Capsicum annuum* Grossum group)
- Tinda (*Praecitrullus fistulosus*)
- Tomatillo (*Physalis philadelphica*)
- Tomato (*Lycopersicon esculentum var*)
- Winter melon (*Benincasa hispida*)
- West Indian gherkin (*Cucumis anguria*)
- Zucchini (*Cucurbita pepo*)

Flowers or flower buds of perennial or annual plants:

- Artichoke (*Cynara cardunculus, C. scolymus*)
- Broccoli (*Brassica oleracea*)
- Cauliflower (*Brassica oleracea*)
- Squash blossoms (*Cucurbita* spp.)

Podded Vegetables

- American groundnut (*Apios americana*)
- Azuki bean (*Vigna angularis*)
- Black-eyed pea (*Vigna unguiculata* subsp. *unguiculata*)
- Chickpea (*Cicer arietinum*)
- Common bean (*Phaseolus vulgaris*)
- Drumstick (*Moringa oleifera*)
- Dolichos bean (*Lablab purpureus*)
- Fava bean (*Vicia faba*)
- Green bean (*Phaseolus vulgaris*)
- Guar (*Cyamopsis tetragonoloba*)
- Horse gram (*Macrotyloma uniflorum*)
- Indian pea (*Lathyrus sativus*)

- Lentil (*Lens culinaris*)
- Lima Bean (*Phaseolus lunatus*)
- Moth bean (*Vigna acontifolia*)
- Mung bean (*Vigna radiata*)
- Okra (*Abelmoschus esculentus*)
- Pea (*Pisum sativum*)
- Peanut (*Arachis hypogaea*)
- Pigeon pea (*Cajanus cajan*)
- Ricebean (*Vigna umbellata*)
- Runner bean (*Phaseolus coccineus*)
- Soybean (*Glycine max*)
- Tarwi (tarhui, chocho; *Lupinus mutabilis*)
- Tepary bean (*Phaseolus acutifolius*)
- Urad bean (*Vigna mungo*)
- Velvet bean (*Mucuna pruriens*)
- Winged bean (*Psophocarpus tetragonolobus*)
- Yardlong bean (*Vigna unguiculata* subsp. *sesquipedalis*)

Bulb and Stem Vegetables

- Asparagus (*Asparagus officinalis*)
- Cardoon (*Cynara cardunculus*)
- Celeriac (*Apium graveolens* var. *rapaceum*)
- Celery (*Apium graveolens*)
- Elephant Garlic (*Allium ampeloprasum* var. *ampeloprasum*)
- Florence fennel (*Foeniculum vulgare* var. *dulce*)
- Garlic (*Allium sativum*)
- Kohlrabi (*Brassica oleracea* Gongylodes group)
- Kurrat (*Allium ampeloprasum* var. *kurrat*)
- Leek (*Allium porrum*)
- Lotus root (*Nelumbo nucifera*)
- Nopal (*Opuntia ficus-indica*)
- Onion (*Allium cepa*)
- Prussian asparagus (*Ornithogalum pyrenaicum*)
- Shallot (*Allium cepa* Aggregatum group)
- Welsh onion (*Allium fistulosum*)
- Wild leek (*Allium tricoccum*)

Root and Tuberous Vegetables

- Ahipa (*Pachyrhizus ahipa*)
- Arracacha (*Arracacia xanthorrhiza*)
- Bamboo shoot (*Bambusa vulgaris* and *Phyllostachys edulis*)
- Beetroot (*Beta vulgaris* subsp. *vulgaris*)
- Black cumin (*Bunium persicum*)
- Burdock (*Arctium lappa*)
- Broadleaf arrowhead (*Sagittaria latifolia*)
- Camas (*Camassia*)
- Canna (*Canna* spp.)
- Carrot (*Daucus carota*)
- Cassava (*Manihot esculenta*)
- Chinese artichoke (*Stachys affinis*)
- Daikon (*Raphanus sativus* Longipinnatus group)
- Earthnut pea (*Lathyrus tuberosus*)
- Elephant Foot yam (*Amorphophallus_paeoniifolius*)
- Ensete (*Ensete ventricosum*)
- Ginger (*Zingiber officinale*)
- Gobo (*Arctium lappa*)
- Hamburg parsley (*Petroselinum crispum* var. *tuberosum*)
- Jerusalem artichoke (*Helianthus tuberosus*)
- Jícama (*Pachyrhizus erosus*)
- Parsnip (*Pastinaca sativa*)
- Pignut (*Conopodium majus*)
- Plectranthus (*Plectranthus* spp.)
- Potato (*Solanum tuberosum*)
- Prairie turnip (*Psoralea esculenta*)
- Radish (*Raphanus sativus*)
- Rutabaga (*Brassica napus* Napobrassica group)
- Salsify (*Tragopogon porrifolius*)
- Scorzonera (*Scorzonera hispanica*)
- Skirret (*Sium sisarum*)
- Sweet Potato or Kumara (*Ipomoea batatas*)
- Taro (*Colocasia esculenta*)
- Ti (*Cordyline fruticosa*)
- Tigernut (*Cyperus esculentus*)
- Turnip (*Brassica rapa* Rapifera group)
- Ulluco (*Ullucus tuberosus*)

- Wasabi (*Wasabia japonica*)
- Water chestnut (*Eleocharis dulcis*)
- Yacón (*Smallanthus sonchifolius*)
- Yam (*Dioscorea* spp.)

Sea Vegetables

- Aonori (*Monostroma* spp., *Enteromorpha* spp.)
- Carola (*Callophyllis variegata*)
- Dabberlocks aka badderlocks (*Alaria esculenta*)
- Dulse (*Palmaria palmata*)
- *Gim* (*Porphyra* spp.)
- *Hijiki* (*Hizikia fusiformis*)
- *Kombu* (*Laminaria japonica*)
- *Laver* (*Porphyra* spp.)
- *Mozuku* (*Cladosiphon okamuranus*)
- *Nori* (*Porphyra* spp.)
- *Ogonori* (*Gracilaria* spp.)
- Sea grape (*Caulerpa* spp.)
- Seakale (*Crambe maritima*)
- Sea lettuce (*Ulva lactuca*)
- *Wakame* (*Undaria pinnatifida*)

ROOT VEGETABLE

Root vegetables are plant roots used as vegetables. Here "root" means any underground part of a plant (except that peanuts, which are underground seeds, are seldom called root vegetables).

Root vegetables are generally storage organs, enlarged to store energy in the form of carbohydrates. They differ in the concentration and the balance between sugars, starches, and other types of carbohydrate.

Of particular economic importance are those with a high carbohydrate concentration in the form of starch. These starchy root vegetables are important staple foods, particularly in tropical regions. They overshadow the cereals throughout much of West Africa, Central Africa, and Oceania, where they are used directly or mashed to make foufou or poi.

Some Jains are opposed to eating root vegetables for ethical reasons.

Botany distinguishes true roots such as tuberous roots and taproots from non-roots such as tubers, rhizomes, corms, and bulbs. (Several types contain both taproot and hypocotyl tissue, and it may be difficult to tell some types apart.) In ordinary, agricultural, and culinary use, "root vegetable" can apply to all these types. The following list classifies root vegetables according to anatomy.

List of Underground Vegetables by Anatomical Type

Foeniculum vulgare (Fennel), whose root is used in many cuisines.

- True root
  - Taproot (some types may incorporate substantial hypocotyl tissue)
    - Apium graveolens *(celeriac)*
    - Arctium *spp. (burdock or gobo)*
    - Arracacia xanthorrhiza *(arracacha)*
    - Abelmoschus moschatus *(bush carrot)*
    - Beta vulgaris *(beet and mangelwurzel)*
    - Brassica *spp. (rutabaga and turnip)*
    - Bunium persicum *(black cumin)*
    - Daucus carota *(carrot)*
    - Lepidium meyenii *(maca)*
    - Microseris scapigera *(yam daisy)*
    - Pachyrhizus *spp. (jicama and ahipa)*
    - Pastinaca sativa *(parsnip)*
    - Petroselinum *spp. (parsley root)*
    - Raphanus sativus *(daikon and radish)*
    - Scorzonera hispanica *(black salsify)*
    - Sium sisarum *(skirret)*
    - Tragopogon *spp. (salsify)*
    - Vigna lanceolata *(bush potato)*
  - Tuberous root
    - Amorphophallus glabra *(Yellow lily yam)*
    - Conopodium majus *(pignut or earthnut)*
    - Hornstedtia scottiana *(Native ginger)*
    - Ipomoea batatas *(sweet potato)*

    - Ipomoea costata *(desert yam)*
    - Manihot esculenta *(cassava or yucca or manioc)*
    - Mirabilis extensa *(mauka or chago)*
    - Psoralea esculenta *(breadroot, tipsin, or prairie turnip)*
    - Smallanthus sonchifolius *(yacón)*
- Modified stem
  - Corm
    - Amorphophallus konjac *(konjac)*
    - Colocasia esculenta *(taro)*
    - Eleocharis dulcis *(Chinese water chestnut)*
    - Ensete *spp. (enset)*
    - Nelumbo nucifera
    - Nymphaea *spp. (waterlily)*
    - Pteridium esculentum
    - Sagittaria *spp. (arrowhead or wapatoo)*
    - Typha spp.
    - Xanthosoma *spp. (malanga, cocoyam, tannia, and other names)*
  - Rhizome
    - Curcuma longa *(turmeric)*
    - Panax ginseng *(ginseng)*
    - Arthropodium *spp. (rengarenga, vanilla lily, and others)*
    - Canna *spp. (canna)*
    - Cordyline fruticosa *(ti)*
    - Maranta arundinacea *(arrowroot)*
    - Nelumbo nucifera *(lotus root)*
    - Typha *spp. (cattail or bulrush)*
    - Zingiber officinale *(ginger, galangal)*
  - Tuber
    - Apios americana *(hog potato or groundnut)*
    - Cyperus esculentus *(tigernut or chufa)*
    - Dioscorea *spp. (yams, ube)*
    - Helianthus tuberosus *(Jerusalem artichoke or sunchoke)*
    - Hemerocallis *spp. (daylily)*
    - Lathyrus tuberosus *(earthnut pea)*

  - Oxalis tuberosa *(oca or New Zealand yam)*
  - Plectranthus edulis *and* P. esculentus *(kembili, dazo, and others)*
  - Solanum tuberosum *(potato)*
  - Stachys affinis *(Chinese artichoke or crosne)*
  - Tropaeolum tuberosum *(mashua or añu)*
  - Ullucus tuberosus *(ulluco)*
- Bulb
  - Allium *spp. (garlic, onion, shallot, et cetera)*
  - Camassia quamash *(quamash)*
  - Cyperus bulbosus *(bush onion)*
  - Caladenia
  - Cryptostylis
  - Cymbidium canaliculatum
  - Diuris
  - Dipodium
  - Erythronium *spp. (katakuri)*
  - Geodorum
  - Glossodia
  - Lilium *spp. (lilies)*
  - Lypernathus
  - Microtis
  - Prasophyllum
  - Pterostylis
  - Tacca leontopetaloides
  - Thelymitra

VEGETABLE JUICE

Vegetable juice is a drink made primarily of blended vegetables. Vegetable juice is often mixed with fruits such as tomatoes or grapes to improve flavor. It is often touted as a low-sugar alternative to fruit juice, although most commercial brands of vegetable juices contain large amounts of sodium.

Home-Made Juice

Making vegetable juice at home is an alternative to buying

commercial juices, and may be beneficial for supplementing diets low in vegetables and fruits. Special vegetable juicers are recommended for blending vegetables, as they employ either a slow-geared grinding mechanism or a unit powered by centrifugal force. Compared to a higher speed blender or fruit juicer, the slower speed protects the vegetables from oxidation and heat (from friction), which reduces nutrient breakdown.

Varieties

Commercial vegetable juices are commonly made from varying combinations of carrots, beets, pumpkin, and tomatoes. The latter two, although not technically vegetables, are commonly used to increase palatability. Other popular items in vegetable juices are parsley, dandelion greens, kale, celery, fennel, and cucumbers. Lemon, garlic and ginger may be added by some for medicinal purposes.

Other common juices include carrot juice, tomato juice, turnip juice, and V8 Juice.

In Asian cultures, primarily Chinese, *Dioscorea opposita* (Chinese: *shân yào*, Japanese: *nagaimo*) is also used for vegetable juices. They are used quite sparingly, however, for many Chinese consider it to be a medicine rather than a vegetable.

Kale juice marketed as *Aojiru* in Japan has become well known for its purported health benefits and bitter taste.

Japan also markets several kinds of vegetable juices which, unlike Western juices such as V8, usually depend on carrots and fruits instead of large amounts of tomato juice for their flavor. The *Yasai Seikatsu* brand by Kagome is a popular variety, and is even an option in McDonald's Happy Meals.

Nutrition

In general, vegetable juices are recommended as supplements to whole vegetables, rather than as a replacement. However, the actual nutritional value of juices versus whole vegetables is still contested.

USDA guidelines for Americans states that 3/4 cup of 100 per cent vegetable juice is equivalent to one serving of vegetables. This is upheld by a recent study, which found that juices provide the same health benefits as whole vegetables in terms of reducing risks of cardiovascular disease and cancer. Another study has found that drinking vegetable juice reduces risks of Alzheimer's Disease by 76 per cent.

However, the British Nutrition Foundation holds that although vegetable juice counts as a serving, it can only count as one serving, regardless of the amount of juice drunk. It is unclear whether this is referring to commercial juices only, or includes home-made juices. Additionally, A 2007 Japanese study showed that although Japanese commercial juices had nutritional benefits, they were insufficient as a primary mode of vegetable consumption.

Many popular vegetable juices, particularly ones with high tomato content, are high in sodium, and therefore consumption of them for health must be carefully considered. Some vegetables such as beets also contain large amounts of sugar, so care must be taken when adding these to juices.

Although the actual nutritional benefits of vegetable juice are contested, a recent UC Davis study found that drinking vegetable juice daily significantly increased drinkers' chances of meeting the daily recommended number of vegetable servings.Having an easy source of vegetables encouraged drinkers to incorporate more vegetables into their diets.

LIST OF CULINARY HERBS AND SPICES

This is a list of culinary herbs and spices. Specifically these are food or drink additives of mostly botanical origin used in nutritionally insignificant quantities for flavoring or colouring.

This list does not contain salt, which is a mineral. Nor is it for fictional plants such as aglaophotis, or recreational drugs such as tobacco.

And this list is not for plants used primarily as herb teas or tisanes, and not for herbs or spices that are purely medicinal such as valerian.

See the list of spice mixtures below for common spice blends.

- Ajwain (*Trachyspermum ammi*)
- Akudjura (*Solanum centrale*) (Australia)

- Alexanders (*Smyrnium olusatrum*)
- Alkanet (*Anchusa arvensis*), primarily for colour
- Allspice (*Pimenta dioica*)
- (Alpinia galanga) Galangal
- Angelica (*Angelica archangelica*)
- Anise (*Pimpinella anisum*)
- Aniseed myrtle (*Syzygium anisatum*) (Australia)
- Annatto (*Bixa orellana L.*)
- Apple mint (*Mentha suaveolens*)
- Arrowroot (*Maranta arundinacea*)
- Asafoetida (*Ferula assafoetida*)
- Avocado Leaf (*Peresea americana*)
- Barberry (*Berberis vulgaris*)
- Basil (*Ocimum basilicum*)
- Bay Leaf (*Laurus nobilis*)
- Borage (*Borago officinalis*)
- Black Cardamom (*Amomum subulatum*)
- Black Mustard (*Brassica nigra*)
- Blue Fenugreek, Blue melilot (*Trigonella caerulea*)
- Brown Mustard (*Brassica juncea*)
- Calamint, (*Calamintha spp.*)
- Calendula, Pot Marigold (*Calendula officinalis*)
- Candle nut (*Aleurites moluccanus*)
- Capers (*Capparis spinosa*)
- Caraway (*Carum carvi*)
- Cardamom (*Elettaria cardamomum*)
- Catnip (*Nepeta cataria*)
- Cassia (*Cinnamomum aromaticum*)
- Cayenne pepper (*Capsicum annuum*)
- Celery seed (*Apium graveolens*)
- Chervil (*Anthriscus cerefolium*)
- Chicory (*Cichorium intybus*)
- Chili pepper (*Capsicum spp.*)
- Chili powder
- Chipotle (*Capsicum annuum*)
- Chives (*Allium schoenoprasum*)
- Cicely (*Myrrhis odorata*)

- Cilantro, Coriander Greens, Coriander Herb (*Coriandrum sativum*)
- Cinnamon, True or Ceylon (*Cinnamomum verum*)
- Cinnamon Myrtle (*Backhousia myrtifolia*) (Australia)
- Clary, Clary Sage (Salvia sclarea)
- Clove (*Syzygium aromaticum*)
- Coriander seed (*Coriandrum sativum*)
- Costmary (*Tanacetum balsamita*)
- Cuban Oregano (*Plectranthus amboinicus*)
- Cubeb pepper (*Piper cubeba*)
- Cudweed (Vietnamese cuisine)
- Culantro (*Eryngium foetidum*)
- Cumin (*Cuminum cyminum*)
- Curry leaf (*Murraya koenigii*)
- Curry Plant (*Helichrysum italicum*)
- Dill seed (*Anethum graveolens*)
- Dill herb or weed (*Anethum graveolens*)
- Dorrigo Pepper (*Tasmannia stipitata*) (Australia)
- Elderflower (*Sambucus spp.*)
- Epazote (*Chenopodium ambrosioides*)
- *Eryngium foetidum* (Culantro, Long Coriander)
- Fennel (*Foeniculum vulgare*)
- Fenugreek (*Trigonella foenum-graecum*)
- Filé powder (*Sassafras albidum*)
- Fingerroot, krachai, temu kuntji (*Boesenbergia rotunda*)
- French sorrel (*Rumex scutatus*)
- Galangal (*Alpinia officinarum*)
- Galingale (*Cyperus spp.*)
- Garlic chives (*Allium tuberosum*)
- Garlic (*Allium sativum*)
- Ginger (*Zingiber officinale*)
- Ginkgo nuts (*Ginkgo biloba*)
- Golpar (*Heracleum persicum*) (Persian)
- Grains of paradise (*Aframomum melegueta*)
- Grains of Selim (*Xylopia aethiopica*)
- Green tea (*Camellia sinensis*) (used to flavor ice cream and sweets)
- Ground ivy (*Glechoma hederacea*)

- Hops (*Humulus lupulus*)
- Horseradish (*Armoracia rusticana*)
- Houttuynia (*Houttuynia cordata*) (Vietnamese cuisine)
- Hyssop (*Hyssopus officinalis*)
- Hawayij (Yemenite and Israeli cuisine)
- Indian Bay-leaf, Malabathrum, Tejpat (*Cinnamomum tamala*, *C. tejpata*)
- Indonesian Bay-Leaf, Daun salam (*Eugenia polyantha*, *Syzygium polyanthum*)
- Indonesian Cinnamon (*Cinnamomum burmannii*, *Cassia vera*)
- Jasmine (*Jasminum spp.*)
- Jimbu (*Allium hypsistum*) (Nepal)
- Juniper berry (*Juniperus communis*)
- Kaffir Lime Leaves (*Citrus hystrix*, *C. papedia*)
- Kokam (*Garcinia indica*)
- Kala Jeera (*Bunium persicum*)
- Lavender (*Lavandula spp.*)
- Laser, laserpicium, or lasarpicium
- Lemon Balm (*Melissa officinalis*)
- Lemon Basil (*Ocimum × citriodorum*)
- Lemongrass (*Cymbopogon citratus*, *C. flexuosus*, and other species)
- Lemon Ironbark (*Eucalyptus staigeriana*) (Australia)
- Lemon Mint (*Monarda citriodora*)
- Lemon Myrtle (*Backhousia citriodora*) (Australia)
- Lemon Thyme (*Thymus × citriodorus*)
- Lemon Verbena (*Lippia citriodora*)
- Lesser Galangal, Kentjur (*Kaempferia galanga*)
- Licorice, Liquorice (*Glycyrrhiza glabra*)
- Lime Flower, Linden flower (*Tilia spp.*)
- *Limnophila aromatica*, rice paddy herb (Vietnamese cuisine)
- Long Pepper (*Piper longum*)
- Lovage (*Levisticum officinale*)
- Luohanguo, (*Siraitia grosvenorii*)
- Mace (*Myristica fragrans*)
- Marjoram (*Origanum majorana*)
- Mahlab (*Mahlab*)
- Nasturtium (*Tropaeolum majus*)

- *Nigella sativa* (Kalonji, Black caraway)
- Nutmeg (*Myristica fragrans*)
- Olida (*Eucalyptus olida*) (Australia)
- Oregano (*Origanum vulgare, O. heracleoticum*, and other species)
- Orris root (*Iris germanica, I. florentina, I. pallida*)
- *Osmorhiza*, Sweet Cicely
- Pandan flower, Kewra (*Pandanus odoratissimus*)
- Pandan leaf, Screwpine (*Pandanus amaryllifolius*)
- Paprika (*Capsicum annuum*)
- Paracress (*Spilanthes acmella, Soleracea*) (Brazilian cuisine)
- Parsley (*Petroselinum crispum*)
- Pepper: black, white, and green (*Piper nigrum*)
- Peppermint (*Mentha piperata*)
- Peppermint Gum leaf (*Eucalyptus dives*)
- Perilla, Shiso (*Perilla spp.*)
- Pink Pepper (*Schinus terebinthifolius*)
- *Piper sarmentosum*
- Poppy seed (*Papaver somniferum*)
- Ramsons, wood garlic (*Allium ursinum*)
- Red Pepper
- Riberry (*Syzygium luehmannii*)
- Rice Paddy Herb (*Limnophila aromatica*)
- Rosemary (*Rosmarinus officinalis*)
- Rue (*Ruta graveolens*)
- Safflower (*Carthamus tinctorius*), primarily for colour
- Saffron (*Crocus sativus*)
- Sage (*Salvia officinalis*)
- Saigon Cinnamon (*Cinnamomum loureiroi*)
- Salad Burnet (*Sanguisorba minor*)
- Salep (*Orchis mascula*)
- Sassafras (*Sassafras albidum*)
- Savory, Summer (*Satureja hortensis*), and winter (*S. montana*)
- Sesame seed (*Sesamum indicum*)
- Sheep's sorrel (*Rumex acetosella*)
- Sichuan pepper (*Xanthoxylum piperitum*)
- Silphium (Ancient Rome)/Silphion (Ancient Greece)
- Sloe berries (*Prunus spinosus*)

- Sorrel (*Rumex acetosa, R. acetosella, R. scutatus*)
- Spearmint (*Mentha spicata*)
- Spikenard (*Nardostachys grandiflora* or *N. jatamansi*)
- Star anise (*Illicium verum*)
- Sumac (*Rhus coriaria*)
- Summer savory (*Satureja hortensis*)
- Sweet cicely (*Myrrhis odorata*)
- Sweet woodruff (*Galium odoratum*)
- Szechuan pepper (also Sichuan pepper) (*Z. piperitum*)
- Talinum (*Talinum fruticosum*)
- Tamarind (*Tamarindus indica*)
- Tansy (*Tanacetum vulgare*)
- Tarragon (*Artemisia dracunculus*)
- Tasmanian Pepper (*Tasmannia lanceolata*)
- Tea (*Camellia sinensis*)
- *Teucrium polium* (felty germander)
- Thai basil (*Ocimum basilicum var. thyrsiflora, O. ×citriodorum*, or *O. sanctum*)
- Thyme (*Thymus vulgaris*)
- Tulsi (*Ocimum tenuiflorum*)
- Turmeric (*Curcuma longa*)
- Vanilla (*Vanilla planifolia*)
- Vietnamese Balm (*Elsholtzia ciliata*)
- Vietnamese Cinnamon (*Cinnamomum loureiroi*)
- Vietnamese Coriander (*Persicaria odorata*)
- Wasabi (*Wasabia japonica*)
- Water-pepper, Smartweed (*Polygonum hydropiper*)
- Watercress (*Rorippa nasturtium-aquatica*)
- Wattleseed, Australian acacia
- White Mustard (*Sinapis alba*)
- Wild thyme (*Thymus serpyllum*)
- Winter savory (*Satureja montana*)
- Wintergreen (*Gaultheria procumbens*)
- Wood Avens, Herb Bennet (*Geum urbanum*)
- Woodruff (*Galium odoratum*)
- Wormwood, Absinthe (*Artemisia absinthium*)
- Yellow Mustard (*Brassica hirta=Sinapis alba*)
- Yerba Buena

- Za'atar herb
- Zedoary (*Curcuma zedoaria*)

LIST OF CULINARY FRUITS

This list of culinary fruits contains the names of some fruits that are considered edible in some cuisines. The definition of fruit for these lists is a culinary fruit, *i.e.* "Any sweet, edible part of a plant that resembles fruit, even if it does not develop from a floral ovary; also used in a technically imprecise sense for some sweet or sweetish vegetables, some of which may resemble a true fruit or are used in cookery as if they were a fruit, for example rhubarb."

Note that many edible plant parts that are *true fruits* botanically speaking, are not considered culinary fruits. They are classified as vegetables in the culinary sense, (for example: the tomato, cucumber, zucchini, and so on), and hence they do not appear in this list. There also exist many fruits which are edible and palatable but for various reasons have not become popular. For inedible fruits, please see list of inedible fruits.

Culinary Fruits by Climate

Temperate Fruits

Fruits of temperate climates are almost always borne on trees or woody shrubs or lianas. They will not grow adequately in the tropics, as they need a period of cold (a chilling requirement) each year before they will flower. The apple, pear, cherry, and plum are the most widely grown and eaten, owing to their adaptability. Many other fruits are important regionally but do not figure prominently in commerce. Many sorts of small fruit on this list are gathered from the wild, just as they were in Neolithic times.

The pome fruits:

- Apple and crabapple (*Malus*)
- Chokeberry (*Aronia*)
- Hawthorn (*Crataegus* and *Rhaphiolepis*)
- Loquat (*Eryobotrya japonica*)
- Medlar (*Mespilus germanica*)

- Pear, European and Asian species (*Pyrus*)
- Quince (*Cydonia oblonga* and *Chaenomeles*)
- Rose hip, the fruitlike base of roses (*Rosa*); used mostly for jams and herbal tea
- Rowan (*Sorbus*)
- Service tree (*Sorbus domestica*), bears a fruit known as a sorb or sorb apple
- Serviceberry or Saskatoon (*Amelanchier*)
- Shipova (× *Sorbopyrus auricularis*)

The stone fruits, drupes of genus *Prunus*:

- Apricot (*Prunus armeniaca* or *Armeniaca vulgaris*)
- Cherry, sweet, black, sour, and wild species (*Prunus avium*, *Prunus serotina*, *P. cerasus*, and others)
- Chokecherry (*Prunus virginiana*)
- Greengage, a cultivar of the plum
- Peach (of the normal and white variety) and its variant the nectarine (*Prunus persica*)
- Plum, of which there are several domestic and wild species; dried plums are called prunes
- Hybrids of the preceding species, such as the pluot, aprium and peacotum

Berries. In non-technical usage, berry means any small fruit that can be eaten whole and lacks objectionable seeds. The bramble fruits, compound fruits of genus *Rubus* (blackberries), are some of the most popular pseudo-berries:

- Blackberry, of which there are many species and hybrids, such as dewberry, boysenberry, olallieberry, and tayberry (genus *Rubus*)
- Cloudberry (*Rubus chamaemorus*)
- Loganberry (*Rubus loganobaccus*)
- Raspberry, several species (genus *Rubus*)
- Salmonberry (*Rubus spectabilis*)
- Thimbleberry (*Rubus parviflorus*)
- Wineberry (*Rubus phoenicolasius*)

The true berries are dominated by the family Ericaceae, many of which are hardy in the subarctic:

- Bearberry (*Arctostaphylos* spp.)
- Bilberry or whortleberry (*Vaccinium* spp.)
- Blueberry (*Vaccinium* spp.)
- Crowberry (*Empetrum* spp.)
- Cranberry (*Vaccinium* spp.)
- Falberry (*Vaccinium* spp.)
- Huckleberry (*Vaccinium* spp.)
- Lingonberry (*Vaccinium vitis-idaea*)
- Strawberry Tree (*Arbutus unedo*), not to be confused with the Strawberry (*Fragaria*)

Other berries not in the Rosaceae or Ericaceae:

- Barberry (*Berberis*; Berberidaceae)
- Currant (*Ribes* spp.; Grossulariaceae), red, black, and white types
- Elderberry (*Sambucus*; Caprifoliaceae)
- Gooseberry (*Ribes* spp.; Grossulariaceae)
- Hackberry (*Celtis* spp.; Cannabaceae)
- Honeysuckle: the berries of *some* species (called honeyberries) are edible, others are poisonous (*Lonicera* spp.; Caprifoliaceae)
- Mulberry (*Morus* spp.; Moraceae) including:
    - Red Mulberry (*Morus rubra*)
    - White Mulberry (*Morus alba*)
- Mayapple (*Podophyllum* spp.; Berberidaceae)
- Nannyberry or sheepberry (*Viburnum* spp.; Caprifoliaceae)
- Oregon grape (*Mahonia aquifolium*; Berberidaceae)
- Sea-buckthorn (*Hippophae rhamnoides*; Elaeagnaceae)
- Ugniberry (*Ugni molinae*; Myrtaceae)
- Wolfberry (*Lycium barbarum*, *Lycium* spp.; Solanaceae)

Other Temperate fruits:

- Boquila (*Boquila trifoliata*; Lardizabalaceae)

- Goumi (*Elaeagnus multiflora*; Elaeocarpaceae)
- Keule (*Gomortega keule*; Gomortegaceae)
- Lardizabala (*Lardizabala biternata*; Lardizabalaceae)
- Maqui (*Aristotelia chilensis*; Elaeocarpaceae)
- Pawpaw (*Asimina triloba*; Annonaceae)
- Peumo (*Cryptocarya alba*; Lauraceae)

Mediterranean and Subtropical Fruits

Fruits in this category are not hardy to extreme cold, as the preceding temperate fruits are, yet tolerate some frost and may have a modest chilling requirement. Notable among these are natives of the Mediterranean:

- Black mulberry (*Morus nigra*; Moraceae)
- Cornelian cherry (*Cornus mas*; Cornaceae)
- Date palm (*Phoenix dactylifera*; Arecaceae)
- Fig (*Ficus* spp. Moraceae)
- Grape, called raisin, sultana when it is dried. (*Vitis* spp.; Vitaceae)
- Jujube (*Ziziphus zizyphus*; Rhamnaceae)
- Pomegranate (*Punica granatum*; Punicaceae)
- Sycamore fig (*Ficus* sycomorus. Moraceae) also called old world sycomore or just sycomore.

In the important genus *Citrus* (Rutaceae), some members are tropical, tolerating no frost. All common species of commerce are somewhat hardy:

- Blood Orange
- Citron (*Citrus medica*)
- Clementine (*Citrus reticulata* var. Clementine),
- Grapefruit (*Citrus paradisi*)
- Hybrids of the preceding species, such as the Orangelo, Tangelo, Rangpur (fruit) and Ugli fruit
- Kumquat (*Fortunella*)
- Lemon (*Citrus limon*)
- Limes
    - Key Lime (*Citrus aurantifolia*)

    - Persian lime Also known as tahiti lime.
    - Kaffir lime (*Citrus hystix*)
- Mandarin (*Citrus reticulata*)
- Naartjie (*Citrus reticulata, Citrus nobilis*)
- Orange, of which there are sweet (*Citrus sinensis*) and sour (*Citrus aurantium*) species
- Pomelo (also known as the shaddock) (*Citrus maxima*)
- Sweet Lemon (*Citrus limetta*)
- Tangerine, and similar

Other subtropical fruits:

- Avocado (*Persea americana*; Lauraceae)
- Carob (*Ceratonia siliqua*; Fabaceae)
- Feijoa (*Feijoa sellowiana*; Myrtaceae)
- Guava (*Psidium guajava*; Myrtaceae)
- Kumquat (*Fortunella* spp.; Rutaceae)
- Longan (*Dimocarpus longan*; Sapindaceae)
- Lúcuma (*Pouteria lucuma*; Sapotaceae)
- Lychee (*Litchi chinensis*; Sapindaceae)
- Passion fruit or Grenadilla (*Passiflora edulis* and other *Passiflora* spp.; Passifloraceae) Galendar in some part of east India (Darjeeling)
- Peanut (*Arachis hypogaea*; Fabaceae)
- Pond-apple (*Annona glabra*; Annonaceae) Also called Alligator-apple and Monkey-apple
- Strawberry guava (*Psidium litorale*; Myrtaceae)
- Tamarillo or Tree Tomato (*Cyphomandra betacea*; Solanaceae)
- Yangmei (*Myrica rubra*; Myricaceae) Also called Yumberry, Yamamomo, Chinese Bayberry, Japanese Bayberry, Red Bayberry, or Chinese strawberry tree
- Néré (Parkia biglobosa)

Tropical Fruits

Tropical fruit grow on plants of all habitats. The only characteristic that they share is an intolerance of frost.

- Abiu (Pouteria caimito)
- Acerola (*Malpighia glabra*; Malpighiaceae), also called West Indian Cherry or Barbados Cherry
- Ackee (*Blighia sapida* or *Cupania sapida*; Sapindaceae)
- African cherry orange (*Citropsis schweinfurthii*; Rutaceae)
- Amazon Grape (*Pourouma cecropiaefolia*;Moraceae)
- Araza
- Avocado
- Açaí (*Euterpe oleracea*; Arecaceae), or assai
- Babaco (*Carica pentagona*; Caricaceae)
- Bacupari (*Garcinia gardneriana*)
- Bael (*Aegle marmelos*; Rutaceae)
- Banana (*Musacea* spp.; Musaceae); its starchy variant is the plantain
- Barbadine (granadilla; maracujá-açu in Portuguese)
- Barbados Cherry (*Malpighia glabra L.*; Malpighiaceae), also called Acerola, West Indian Cherry
- Betel Nut
- Bilimbi (*Averrhoa bilimbi*; Oxalidaceae) Also called cucumber tree or tree sorrel
- Bitter gourd
- Black sapote
- Bottle gourd also known as Calabash (*Lagenaria siceraria*; Cucurbitaceae)
- Brazil nut
- Breadfruit (*Artocarpus altilis*; Moraceae)
- Burmese grape, or Latka (*Baccaurea sapida*; Phyllanthaceae)
- Calabashtree
- Calamansi
- CamuCamu (*Myrciaria dubia*; Myrtaceae)
- Canistel (*Pouteria campechiana*; Sapotaceae), also called yellow sapote or "eggfruit"
- Cape gooseberry
- Carambola (*Averrhoa carambola*; Oxalidaceae), also called star fruit or *five fingers*
- Cashew
- Cempedak or Champedak (*Artocarpus champeden*; Moraceae)

- Ceylon gooseberry
- Chenet (guinep or ackee; pitomba-das-Guinas in Portuguese), also known as Spanish Lime or mamoncillo
- Cherimoya (*Annona cherimola*; Annonaceae)
- Chili pepper
- Caimito (caimite; related to the yellow abiu—egg fruit)
- Cacao
- Coffea
- Cupuaçu
- Custard-apple (*Annona reticulata*; Annonaceae), also called *Bullock's Heart*
- Damson plum (*Chrysophyllum oliviforme*; Sapotaceae), also called *Satin Leaf*
- Date
- Date-plum (*Diospyros lotus*; Ebenaceae)
- Dragonfruit (*Hylocereus* spp.; Cactaceae), also called pitaya
- Durian (*Durio spp.*; Bombacaceae)
- Elephant apple (*Dillenia indica*; Dilleniaceae)
- Giant granadilla
- Golden Apple
- Guarana (*Paullinia cupana*; Sapindaceae)
- Guava
- Guavaberry or Rumberry; (*Myrciaria floribunda*; Myrtaceae)
- Hairless Rambutan
- Hog plum (taperebá in Portuguese)
- Horned melon (*Cucumis metuliferus*; Cucurbitaceae)
- Huito (*Genipa americana*; Rubiaceae); also called *jagua*, *genipap*, jenipapo
- Imbe (*Garcinia livingstonei*)
- Indian almond
- Indian fig
- Indian gooseberry
- Indian jujube
- Indian Prune (*Flacourtia rukan*; Flacourtiaceae)
- Jaboticaba (*Myrciaria cauliflora*; Myrtaceae), also called Brazilian Grape Tree
- Jackfruit (*Artocarpus heterophyllus* Moraceae), also called *nangka*

- Jambul (*Syzygium cumini*; Myrtaceae)
- Jatobá (*Hymenae coubaril*; Leguminosae) Caesalpinioideae)
- Jocote, also called Jamaica Plum
- Kandis (*Garcinia forbesii*; Clusiaceae)
- Keppel fruit (*Stelechocarpus burakol*; Annonaceae)
- Korlan
- Kumquat
- Kundong (*Garcinia sp.*; Clusiaceae)
- Lablab
- Langsat (*Lansium domesticum*), also called longkong or duku
- Lanzones (*Lansium domesticum*; Meliaceae)
- Lemon
- Leucaena
- Lime
- Longan
- Loquat
- Lucuma
- Lychee
- Mabolo (*Diospyros discolour*; Ebenaceae) also known as a velvet persimmon
- Macadamia, also known as a Queensland nut
- Mamey sapote (*Pouteria sapota*; Sapotaceae); also known as *mamee apple*; abricó in Portuguese
- Mamoncillo (*Melicoccus bijugatus*; Sapindaceae), also known as quenepa, genip or Fijian Longan
- Mandarin
- Manila tamarind (or Monkeypod, *Pithecellobium dulce*)
- Mango (*Mangifera indica*; Anacardiaceae)
- Mangosteen (*Garcinia mangostana*; Clusiaceae)
- Marang (*Artocarpus odoratissima*; Moraceae), a breadfruit relative
- Melinjo
- Melon pear
- Monstera (*Monstera deliciosa*; Araceae) also called Swiss Cheese Plant, Split-leaf Philodendron
- Morinda
- Mountain soursop
- Mundu

- Mung bean
- Muskmelon
- Nance
- Naranjilla, Lulo (*Solanum quitoense*; Solanaceae)
- Nutmeg
- Neem
- Oil Palm
- Okra
- Papaya (*Carica papaya*; Caricaceae)
- Peach palm
- Peanut butter fruit (*Bunchosia argentea*; Malpighiaceae)
- Pequi or Souari Nut (*Caryocar brasiliense*; Caryocaraceae)
- Pewa (peach palm; pupunha in Portuguese)
- Pigeon pea
- Pili nut
- Pineapple (*Ananas comosus* or *Ananas sativas*; Bromeliaceae)
- pitomba (*Eugenia luschnathiana* or *Talisia esculenta*)
- Plantain
- Poha or Cape Gooseberry (*Physalis peruviana*; Solanaceae)
- Pois doux (*Inga edulis*, *ice-cream bean*, or *inga-cipó* in Portuguese)
- Pommecythère or pomcité (*Spondias cytherea*); also known as golden apple, June plum or Jew plum and ambarella, and as cajamanga in Portuguese
- Pommerac (*Eugenia malaccensis*); also known as Otaheite apple; Malay apple; jambo in Portuguese
- Pulasan
- Pummelo
- Pupunha or peach-palm (*Bactris gasipaes*; Palmae); also known as *pewa*
- Rambutan (*Nephelium lappaceum*; Sapindaceae)
- Red Mombin (*Spondias purpurea*; Anacardiaceae)
- Riberry (*Syzygium luehmannii*; Myrtaceae), also called Lilly Pilly, Lillipilli, Chinese Apple
- Ridged gourd
- Safou (*Dacryodes edulis*), also called *atanga* or *butterfruit*
- Salak (*Salacca edulis*), also called snakefruit or *cobrafruit*
- Santol (*Sandoricum koetjape*; Meliaceae)

- Sapodilla (*Achras/Manilkara zapota*; Sapotaceae), also called chiku, mespel, naseberry, sapadilla, snake fruit, sawo
- Sea grape
- Soncoya (*Annona diversifolia*)
- Soursop (*Annona muricata*; Annonaceae), also called guanabana
- Soybean
- Star apple (*Chrysophyllum cainito*), also called caimito or *caimite*
- Strawberry guava
- Strawberry pear
- Sugar apple (*Annona squamosa*; Annonaceae); ata in Portuguese
- Summer squash
- Surinam Cherry (*Eugenia uniflora*; Myrtaceae) also called Brazilian Cherry, Cayenne Cherry, Pitanga
- Sweet granadilla
- Sweet orange
- Sweet pepper
- Sweetsop
- Rose apple (*Syzygium jambos*; Myrtaceae)
- Tamarind (*Tamarindus indica*; Caesalpiniaceae)
- Vanilla
- Wampee (*Clausena lansium*)
- Water apple
- Watermelon
- Wax apple (*Syzygium samarangense*)
- Wax gourd
- White sapote
- Winged bean

List of Culinary Fruits by Geographical Origin

Fruits of African Origin

Fruits native to Africa or of African Origin:

- Amatungulu (*Carissa macrocarpa*)
- Kiwano (*Cucumis metuliferus*)

- Marula (*Sclerocarya birrea*)
- Spiny Monkey-orange (*Strychnos spinosa*)
- Tamarind (*Tamarindus indica*)

Fruits of Asian Origin

Fruits native to Asia or of Asian Origin:

- Arhat (*Siraitia grosvenorii*; Cucurbitaceae) Also called longevity fruit
- Batuan (*Garcinia morella*)
- Buddha's Hand
- Woodapple (*Aegle marmelos*), commenly known as bael, found in eastern India.
- Mango (*Mangifera*), tropical fruit of south asia.
- Indian gooseberry (*Phyllanthus emblica*)
- Charichuelo (*Garcinia intermedia*)
- Cherapu (*Garcinia prainiana*)
- Chinese Quince (*Pseudocydonia sinensis*)
- Coconut (*Cocos nucifera*; Arecaceae)
- Che (*Cudrania tricuspidata*; Moraceae) Also called Cudrania, Chinese Mulberry, Cudrang, Mandarin Melon Berry, Silkworm Thorn, zhe
- Durian (*Durio spp*; Malvaceae)
- Gambooge (*Garcinia gummi-gutta*)
- Goumi (*Elaeagnus multiflora ovata*; Elaeagnaceae family)
- Jambul
- Hardy Kiwi (*Actinidia arguta*; Actinidiaceae family)
- Kiwifruit or Chinese gooseberry (*Actinidia* spp.; Actinidiaceae)
- Mock Strawberry or Indian Strawberry (*Potentilla indica*; Rosaceae)
- Garcinia dulcis (*Mundu*)
- Lanzones (*Lansium domesticum*; Meliaceae family)
- Lapsi (*Choerospondias axillaris Roxb.* Anacardiaceae)
- Longan (*Dimocarpus longan*; Sapindaceae family)
- Lychee (*Litchi chinensis*; Sapindaceae family)
- Mangosteen (*Garcinia mangostana*; Clusiaceae family)
- Marang

- Nungu (*Borassus flabellifer*; Arecaceae)
- Peach
- Persimmon (aka Sharon Fruit) (*Diospyros kaki*; Ebenaceae)
- Rambutan (*Nephelium lappaceum*; Sapindaceae family)
- Rhubarb (*Rheum rhaponticum*; Polygonaceae)
- Sageretia (*Sageretia theezans*; Rhamnaceae) Also called Mock Buckthorn
- Santol (fruit)
- Wild Mangosteen (*Garcinia indica*)

Fruits of Latin American Origin

Fruits native to South America or of South American Origin:

- Açaí (*Euterpe*), a palm fruit native to the Amazon region.
- Avocado (*Persea americana*; Lauraceae)
- Boquila (*Boquila trifoliata*; Lardizabalaceae)
- Calafate Barberry (*Berberis*; Berberidaceae)
- Keule (*Gomortega keule*; Gomortegaceae)
- Guarana (*Paullinia cupana*; Sapindaceae)
- Guava ("Psilium"; Myrtaceae)
- Lardizabala (*Lardizabala biternata*; Lardizabalaceae)
- Mamey ("Pouteria sapota"; Sapotaceae)
- Maqui (*Aristotelia chilensis*; Elaeocarpaceae)
- Papaya (*Carica papaya*; Caricaceae)
- Peumo (*Cryptocarya alba*; Lauraceae)
- Pineapple ("Ananas comosus"; Bromeliaceae)
- Sapote ("Casimiroa edulis"; Sapotaceae)
- Sea Grape (*Coccoloba uvifera*; Polygonaceae)
- Soursop ("Annona muricata"; Annonaceae)
- Tomato ("Solanum lycopersicum"; Solanaceae)
- Ugniberry (*Ugni molinae*; Myrtaceae)

Fruits of North American Origin

Canada and the United States are home to a surprising number of edible plants, especially berries; however, only three are commercially grown/known on a global scale (grapes, cranberries, and blueberries.) Many of the fruits below are still eaten locally as they have been for centuries and others are generating renewed interest by eco-friendly gardeners (less need for bug control) and chefs of the region alike.

- American chestnut (*Castanea dentata*; Fagaceae)
- American Black Elderberry (*Sambucus canadensis*; Adoxaceae)
- American grape: North American species (*e.g.*, *Vitis labrusca*; Vitaceae) and American-European hybrids are grown where grape (*Vitis vinifera*) is not hardy and are used as rootstocks
- American Hazelnut (*Corylus americana*; Betulaceae)
- American Mayapple (*Podophyllum peltatum*; Berberidaceae)
- American persimmon (*Diospyros virginiana*; Ebenaceae): Traditional for desserts and as dried fruit.
- American plum (*Prunus americana*; Rosaceae
- American Red Elderberry (*Sambucus pubens*; Adoxaceae)
- American Red Raspberry (*Rubus strigosus*; Rosaceae)
- Beach Plum (*Prunus maritima*; Rosaceae)
- Black cherry (*Prunus serotina*; Rosaceae very popular flavoring for pies, jams, and sweets.
- Black raspberry (*Rubus occidentalis* or *Rubus leucodermis*; Rosaceae)
- Black Walnut (*Juglans nigra*; Juglandaceae)
- Blueberry (*Vaccinium, sect. Cyanococcus*; Ericaceae)
- Buffaloberry (*Shepherdia argenta*; Elaeagnaceae), which grows wild in the prairies of Canada
- Chokecherry (*Prunus virginiana*; Rosaceae)
- Cocoplum (*Chrysobalanus icaco*; Chrysobalanaceae)
- Cranberry (*Vaccinium oxycoccus*; Ericaceae)
- Eastern May Hawthorn (*Crataegus aestivalis*; Rosaceae, better known as mayhaw.)
- False-mastic (*Mastichodendron foetidissimum*; Sapotaceae)
- Florida strangler fig (*Ficus aurea*; Moraceae)
- Ground Plum (*Astragalus caryocarpus*; Fabaceae), also called Ground-plum milk-vetch
- Huckleberry (*Gaylussacia*, *Vaccinium*; Ericaceae))
- Maypop (*Passiflora incarnata*; Passifloracae, traditionally a summer treat.)

- Muscadine (*Vitis rotundifolia*; Vitaceae)
- Pawpaw (*Asimina triloba*; Annonaceae, not to be confused with Papaya (*Carica papaya*; Caricaceae), which is called pawpaw in some English dialects)
- Pecan (*Carya illinoinensis* or *illinoensis*; Juglandaceae)
- Prickly pear (*Opuntia spp.*; Cactaceae) used as both a fruit and vegetable depending on part of plant.
- Pigeon plum (*Coccoloba diversifolia*; Polygonaceae)
- Red mulberry (*Morus rubra*; Moraceae)
- Salal berry (*Gaultheria shallon*; Ericaceae)
- Salmonberry (*Rubus spectabilis*; Rosaceae)
- Saskatoonberry (*Amelanchier alnifolia*, Rosaceae
- Saw Palmetto (*Serenoa repens*; Arecaceae)
- Southern crabapple (*Malus angustifola*; Rosaceae)
- Texas persimmon (*Diospyros texana*; Ebenaceae)
- Thimbleberry (*Rubus parviflorus*; Rosaceae)
- Toyon (*Heteromeles arbutifolia*; Rosaceae)

Fruits of Oceanian Origin

Fruits native to Oceania or of Oceanian Origin:

- Atherton Raspberry (*Rubus probus*; Rosaceae)
- Black Apple (*Planchonella australis*; Sapotaceae)
- Blue tongue (*Melastoma affine*; Melastomataceae)
- Bolwarra (*Eupomatia laurina*; Eupomatiaceae)
- Broad-leaf Bramble (*Rubus moluccanus*; Rosaceae)
- Burdekin Plum (*Pleiogynium timorense*; Anacardiaceae)
- Bush tomato (Certain *Solanum* species; Solanaceae)
- Cedar Bay cherry (*Eugenia carissoides*; Myrtaceae)
- Cherry ballart (*Exocarpus cupressiformis*; Santalaceae)
- Cluster fig (*Ficus racemosa*; Moraceae)
- Cocky apple (*Planchonia careya*)
- Common apple-berry (*Billardiera scandens*; Pittosporaceae)
- Conkerberry (*Carissa lanceolata*; Apocynaceae)
- Davidson's plum (*Davidsonia* spp.; Cunoniaceae) Davidsonia jerseyana Davidsonia johnsonii Davidsonia pruriens
- Desert banana (*Marsdenia australis*)

- Desert fig (*Ficus platypoda*; Moraceae)
- Desert lime (*Citrus glauca*; Rutaceae)
- Dodder laurel (*Cassytha melantha*)
- Doubah (*Marsdenia australis*; Apocynaceae)
- Emu Apple (*Owenia acidula*; Meliaceae)
- Emu berry (*Grewia retusifolia*)
- Fibrous Satinash (*Syzygium fibrosum*; Myrtaceae)
- Finger Lime (*Citrus australasica*; Rutaceae)
- Illawarra Plum (*Podocarpus elatus*; Podocarpaceae)
- Kakadu lime (*Citrus gracilis*; Rutaceae)
- Kakadu plum (*Terminalia ferdinandiana*; Combretaceae)
- Karkalla (*Carpobrotus rossii*; Aizoaceae)
- Kutjera (*Solanum centrale*; Solanaceae)
- Lady apple (*Syzygium suborbiculare*; Myrtaceae)
- Lemon aspen (*Acronychia acidula*; Rutaceae)
- Lillypilly (*Acmena* spp., *Syzygium* spp.) Used raw and in jam
- Little gooseberry tree (*Buchanania arborescens*; Anacardiaceae)
- Midyim (*Austromyrtus dulcis*; Myrtaceae)
- Morinda citrifolia
- Mountain pepper (*Tasmannia* spp.; Winteraceae)
- Muntries (*Kunzea pomifera*; Myrtaceae)
- Native currant (*Acrotriche depressa*; Ericaceae)
- Native gooseberry (*Physalis minima*; Solanaceae)
- Native raspberry (Rubus parviflorus)
- Nonda plum (*Parinari nonda*)
- Pigface (*Carpobrotus glaucescens*; Aizoaceae)
- Pink-flowered Native Raspberry (*Rubus parvifolius*; Rosaceae)
- Purple apple-berry (*Billarderia longiflora*; Pittosporaceae)
- Quandong (*Santalum acuminatum*; Santalaceae)
- Queensland Ebony (*Diospyros humilis*)
- Riberry (*Syzygium luehmannii*; Myrtaceae)
- Rose-leaf Bramble (*Rubus rosifolius*; Rosaceae)
- Rose myrtle (*Archirhodomyrtus beckleri*; Myrtaceae)
- Sandpaper Fig (*Ficus coronata*; Moraceae)
- Small-leaf tamarind (*Diploglottis campbellii*; Sapindaceae)

- Snow berry (*Gaultheria hispida*; Ericaceae)
- Sweet apple-berry (*Billarderia cymosa*; Pittosporaceae)
- Tanjong (*Mimusops elengi*; Sapindaceae)
- White aspen (*Acronychia oblongifolia*; Rutaceae)
- Wild grape (*Ampelocissus acetosa*)
- Wild orange (*Capparis mitchellii*; Capparaceae)
- Wild peach (*Terminalia carpentariae*)
- Wild plum (munydjudj) (*Buchanania obovata*)
- Wild plum (*Santalum lanceolatum*)
- Wongi (*Manilkara kaukii*; Sapotaceae)
- Yellow plum (*Ximenia americana*; Olacaceae)
- Zig Zag Vine (*Melodurum leichhardtii*; Annonaceae)

List of Culinary Fruits by Type of Flora

Rosaceae Family

The family Rosaceae dominates the temperate fruits, both in numbers and in importance. The pome fruits, stone fruits and brambles are fruits of plants in Rosaceae.

The pome fruits:

- Apple and crabapple (*Malus*)
- Chokeberry (*Aronia*)
- Hawthorn (*Crataegus* and *Rhaphiolepis*)
- Loquat (*Eryobotrya japonica*)
- Medlar (*Mespilus germanica*)
- Pear, European and Asian species (*Pyrus*)
- Quince (*Cydonia oblonga* and *Chaenomeles*)
- Rose hip, the fruitlike base of roses (*Rosa*); used mostly for jams and herbal tea
- Rowan (*Sorbus*)
- Service tree (*Sorbus domestica*), bears a fruit known as a sorb or sorb apple
- Serviceberry or Saskatoon (*Amelanchier*)
- Shipova (× *Sorbopyrus auricularis*)

The stone fruits, drupes of genus *Prunus*:

- Apricot (*Prunus armeniaca* or *Armeniaca vulgaris*)

- Cherry, sweet, black, sour, and wild species (*Prunus avium*, *Prunus serotina*, *P. cerasus*, and others)
- Chokecherry (*Prunus virginiana*)
- Greengage, a cultivar of the plum
- Peach (of the normal and white variety) and its variant the nectarine (*Prunus persica*)
- Plum, of which there are several domestic and wild species; dried plums are called prunes
- Hybrids of the preceding species, such as the pluot, aprium and peacotum

Berries

In non-technical usage, berry means any small fruit that can be eaten whole and lacks objectionable seeds. The bramble fruits, compound fruits of genus *Rubus* (blackberries), are some of the most popular pseudo-berries:

- Blackberry, of which there are many species and hybrids, such as dewberry, boysenberry, olallieberry, and tayberry (genus *Rubus*)
- Cloudberry (*Rubus chamaemorus*)
- Loganberry (*Rubus loganobaccus*)
- Raspberry, several species (genus *Rubus*)
- Salmonberry (*Rubus spectabilis*)
- Thimbleberry (*Rubus parviflorus*)
- Wineberry (*Rubus phoenicolasius*)

The true berries are dominated by the family Ericaceae, many of which are hardy in the subarctic:

- Bearberry (*Arctostaphylos* spp.)
- Bilberry or whortleberry (*Vaccinium* spp.)
- Blueberry (*Vaccinium* spp.)
- Crowberry (*Empetrum* spp.)
- Cranberry (*Vaccinium* spp.)
- Falberry (*Vaccinium* spp.)
- Huckleberry (*Vaccinium* spp.)
- Lingonberry (*Vaccinium vitis-idaea*)

- Strawberry Tree (*Arbutus unedo*), not to be confused with the Strawberry (*Fragaria*)

Other berries not in the Rosaceae or Ericaceae:

- Açaí (*Euterpe*), a palm fruit native to the Amazon region
- Barberry (*Berberis*; Berberidaceae)
- Currant (*Ribes* spp.; Grossulariaceae), red, black, and white types
- Elderberry (*Sambucus*; Caprifoliaceae)
- Gooseberry (*Ribes* spp.; Grossulariaceae)
- Hackberry (*Celtis* spp.; Cannabaceae)
- Honeysuckle: the berries of *some* species (called honeyberries) are edible, others are poisonous (*Lonicera* spp.; Caprifoliaceae)
- Mulberry (*Morus* spp.; Moraceae) including:
    - Red Mulberry (*Morus rubra*)
    - White Mulberry (*Morus alba*)
- Mayapple (*Podophyllum* spp.; Berberidaceae)
- Nannyberry or sheepberry (*Viburnum* spp.; Caprifoliaceae)
- Oregon grape (*Mahonia aquifolium*; Berberidaceae)
- Sea-buckthorn (*Hippophae rhamnoides*; Elaeagnaceae)
- Sea Grape (*Coccoloba uvifera*; Polygonaceae)
- Ugniberry (*Ugni molinae*; Myrtaceae)
- Wolfberry (*Lycium barbarum*, *Lycium* spp.; Solanaceae)

Cacti and Other Succulents

Several cacti yield edible fruits, which are important traditional foods for some Native American peoples:

- Cardón (*Pachycereus pringlei*; Cactaceae)
- Dragonfruit (*Hylocereus undatus*; Cactaceae), also called pitaya
- Prickly pear (*Opuntia* spp.; Cactaceae)
- Saguaro (*Carnegiea gigantea*; Cactaceae)
- numerous other species of cacti

Podocarps

Podocarps are conifers in the family Podocarpaceae. The seed cones are highly modified and, in some, the seed is surrounded by fleshy scale tissue, resembling a drupe. These berry-like cone scales are eaten by birds which then disperse the seeds in their droppings and the cones can be eaten in many species. Podocarps are either half-hardy or frost tender, depending on species. Many genera are similar in that they have edible "fruits" and often don't have a common name.

- Kahikatea (*Dacrycarpus dacrydioides*)
- Manoao (*Manoao colensoi*)
- Nageia (*Nageia* spp.)
- Podocarpus (*Podocarpus* spp.)
- Prumnopitys (*Prumnopitys* spp.)
- Rimu (*Dacrydium cupressinum*)

Herbaceous Annuals Fruits

Melons and Other Members of Cucurbitaceae or Solanaceae Family

Some exceptions to the statement that temperate fruits grow on woody perennials are:

- Gourd, (usually regarded as vegetables in cuisine) including, but not limited to:
    - Butternut squash (*Cucurbita moschata*)
    - Cushaw squash (*Cucurbita mixta*)
    - Hubbard squash, Buttercup squash (*Cucurbita maxima*)
    - Pumpkin, Acorn squash, Zucchini, Summer squash (*Cucurbita pepo*varieties)
- Horned melon (*Cucumis metuliferus*)
- Melon (*Cucumis melo*): cantaloupe, galia, and other muskmelons, honeydew

Accessory Fruits

The accessory fruits, seed organs which are not botanically berries at all:

- Raisin tree (*Hovenia dulcis*, Rhamnaceae) Also called Japanese Raisin Tree
- Strawberry (*Fragaria* spp.; Rosaceae)
- Cashew apple
- Yew arils (*Taxus* spp.)

LIST OF EDIBLE SEEDS

A list of edible seeds here includes seeds that are directly foodstuffs, rather than yielding derived products.

A variety of species can provide edible seeds. Of the six major plant parts, seeds are the most important source of human food. The other five major plant parts are roots, stems, leaves, flowers, and fruits. Most edible seeds are angiosperms, but a few are gymnosperms. The most important seed food source is cereals, followed by legumes, and nuts.

The list is divided into the following categories:

- Beans (or Legumes) are protein-rich soft seeds.
- Cereals (or grains) are grass-like crops that are harvested for their dry seeds. These seeds are often ground to make flour. Cereals provide almost half of all calories consumed in the world. Botanically, true cereals are members of the Poaceae or Grass family.
  - Pseudocereals are cereal crops that are not members of the Poaceae or Grass Family.
- Nuts are botanically a specific type of fruit but the term is also applied to many edible seeds that are not botanically nuts.
  - Gymnosperms produce nut-like seeds but not flowers or fruits.

Beans

Beans, also known as legumes or pulses include:

- Bambara groundnut
- Chickpeas
- Cowpeas

- Dry beans, including
  - Common bean
  - several species of Vigna
- Fava or broad beans
- Hyacinth bean
- Lablab
- Lentils
- Lupins
- Moringa oleifera
- Peas
- Peanuts
- Pigeon peas
- Sterculia
- Velvet beans
- Winged beans
- Yam beans
- Soybeans

Although some beans can be consumed raw, some need to be heated before consumption. In certain cultures, beans that needed heating anyhow were inmediatelly prepared as a seed cake. Some beans that needed heating include :

- Acacia spp. (eg Acacia aneura (mulga), Acacia cowleana, Acacia estrophiolata (ironweed), Acacia ligulata (umbrella bush), Acacia murrayana (tjuntjula), Acacia tetragonophylla (wakalpulka), Acacia kempeana (Witchetty bush), Acacia coriacea (Wiry wattle), Acacia notabilis, Acacia pyrifolia, Acacia tetragonophylla, Acacia victoriae, Acacia sophorae, Acacia stenophylla, Acacia tumida)
- Aleurites moluccana
- Atriplex nummularia (Old man saltbush)
- Panicum spp. (eg Panicum australiense, Panicum decompositum, Panicum effusum)
- Amaranthus mitchellii
- Amaranthus grandiflorus
- Brachiaria spp. (eg Brachiaria piligera Brachiaria milliformis)

- Brachychiton spp. (eg Brachychiton diversifolium Brachychiton gregorii, Brachychiton paradoxum, Brachychiton populneum)
- Bruguiera rheedii
- Calandrinia balonensis
- Canarium australianum
- Canavalia maritima
- Entada phaseolides
- Eragrostris spp. (Wangunu) (eg Eragrostris eriopoda)
- Eucalyptus leptopoda
- Eucalyptus microtheca
- Astrelba pectinata (Mitchell grass)
- Portulaca oleracea
- Portulaca intraterranea
- Marsilea drummondii (Nardoo)
- Nymphae gigantea
- Rhyncharrhena linearis
- Themeda australis

Cereals

True cereals are the seeds of certain species of grass. Three maize, wheat and rice—account for about half of the calories consumed by people every year. Grains can be ground to make flour, used as the basis of bread, cake, noodles or other food products. They can also be boiled or steamed, either whole or ground, and eaten as is. Many cereals are present or past staple foods, provided a large fraction of the calories in the places that they are eaten. Cereals include:

- Barley
- Fonio
- Maize (corn)
- Pearl Millet
- Oats
- Palmer's grass
- Rice
- Rye
- Sorghum

- Spelt
- Teff
- Triticale
- Wheat
- Wild rice

Pseudocereals

- Breadnut
- Buckwheat
- Cattail
- Chia
- Flax
- Grain amaranth
- Kañiwa
- Pitseed Goosefoot
- Quinoa
- Wattleseed (also called *acacia seed*)

Nuts

According to the botanical definition, nuts are a particular kind of seed. Walnuts and acorns are example of nuts, under this definition. In culinary terms, however, the term is used more broadly to include fruits that are not botanically qualified as nuts, but that have a similar appearance and culinary role. Examples of culinary nuts include almonds, peanuts and cashews.

- Almond
- Beech
- Brazil nut
- Candlenut
- Cashew
- Chestnuts, including:
  - Chinese Chestnut
  - Sweet Chestnut
- Colocynth
- *Cucurbita ficifolia*
- Filbert
- *Gevuina avellana*

- Hickory, including
    - Pecan
    - Shagbark Hickory
- *Terminalia catappa*
- Hazel
- Indian Beech
- Kola nut
- Macadamia
- Malabar chestnut
- Pistacia
- Mamoncillo
- Maya nut
- Mongongo
- Oak acorns
- Ogbono nut
- Paradise nut
- Pili nut
- Walnut
- Water Caltrop

Nut-like Gymnosperm Seeds

- Cycads
- Ginkgo
- *Gnetum gnemon*
- Juniper
- Monkey-puzzle
- Pine nuts, including
    - Chilgoza Pine
    - Korean Pine
    - Stone Pine
    - Colorado Pinyon
    - Mexican Pinyon
    - Single-leaf Pinyon
- Podocarps

Other

- Cempedak
- Egusi

- Euryale ferox (Fox nut)
- Fluted pumpkin
- Hemp seed
- Jackfruit
- Lotus seed
- Malabar gourd
- Pumpkin seed
- Sunflower seed

9

Food Studies, Gastronomy, Catering, Airline Meal: Case Study of Select Global Food Service Operators

FOOD STUDIES

The term Food Studies describes the critical examination of food and its contexts within science, art, history, society, and other fields. It is distinctive from other food-related areas of study such as Nutrition, Agriculture, Gastronomy, and Culinary Arts in that it tends to look beyond the mere consumption, production, and aesthetic appreciation of food and tries to illuminate the meaning of food within the human condition. As musicologists study music, those pursing food studies are also interested in analyzing this common feature of human life as it relates to a vast number of academic fields. It is thus a field that involves and attracts philosophers, historians, scientists, literary scholars, sociologists, art historians, anthropologists, and others.

State of the Field

This is an interdisciplinary and emerging field, and as such there is a substantial crossover between academic and popular work. Practitioners reference best-selling authors, such as the journalist Michael Pollan, as well as scholars, such as the historian Warren Belasco and the anthropologist Sidney Mintz. While this makes the discipline somewhat volatile, it also makes it interesting and engaging. The journalist Paul Levy has noted, for example, that

"Food studies is a subject so much in its infancy that it would be foolish to try to define it or in any way circumscribe it, because the topic, discipline or method you rule out today might be tomorrow's big thing."

Research Questions

Qualitative questions that are wrestled with include: What impact does food have on the environment? What are the ethics of eating? How does food contribute to systems of oppression? How are foods symbolic markers of identity? At the same time practitioners may ask seemingly basic questions that are nonetheless fundamental to human existence. Who chooses what we eat and why? How are foods traditionally prepared? How is food integrated into classrooms? Discussion of these questions has increased as a result of public concern about issues which have arisen as a result of the emergence of a vast array of novel food technologies throughout the last century, ranging from chemical fertilizers to GMOs. Pursuers of food studies approach these questions by first understanding the scientific, economic, and philosophical issues surrounding them.

GOURMET

Gourmet is a cultural ideal associated with the culinary arts of fine food and drink, or *haute cuisine*, which is characterised by elaborate preparations and presentations of large meals of small, often quite rich courses.

The term and its associated practices are usually used positively to describe people of refined taste and passion. For some, it holds a negative connotation of elitism or snobbery.

Person

The term *gourmet* may refer to a person with refined or discriminating taste or to one that is knowledgeable in the art of food and food preparation. *Gourmand* carries additional connotations of one who simply enjoys food in great quantities. An *epicure* is similar to a gourmet, but the word may sometimes carry overtones of excessive refinement.

Food

Gourmet may describe a class of restaurant, cuisine, meal or ingredient of high quality, of special presentation, or high sophistication. In the United States, a 1980s gourmet food movement evolved from a long-term division between elitist (or "gourmet") tastes and a populist aversion to fancy foods. Gourmet is an industry classification for high-quality premium foods in the United States. In the 2000s, there has been an accelerating increase in the American gourmet market, due in part to rising income, globalization of taste, and health and nutrition concerns. Individual food and beverage categories, such as coffee, are often divided between a standard and a "gourmet" sub-market.

Gourmet Pursuits

Certain events such as wine tastings cater to people who consider themselves gourmets and foodies. Television programmes (such as those on the Food Network) and publications such as *Gourmet* magazine often serve gourmets with food columns and features. Gourmet tourism is a niche industry catering to people who travel to food or wine tastings, restaurants, or food production regions for leisure.

Origin of Term

The word *gourmet* is from the French term for a wine broker or *taste-vin* employed by a wine dealer. *Friand* was the reputable name for a connoisseur of delicious things that were not eaten primarily for nourishment: "A good gourmet", wrote the conservative eighteenth-century *Dictionnaire de Trévoux*, employing this original sense, "must have *le goût friand*", or a refined palate. In the eighteenth century, *gourmet* and *gourmand* carried disreputable connotations of gluttony, which only *gourmand* has retained. *Gourmet* was rendered respectable by Monsieur Grimod de la Reynière, whose *Almanach des Gourmands*, essentially the first restaurant guide, appeared in Paris from 1803 to 1812. Previously, even the liberal *Encyclopédie* offered a moralising tone in its entry *Gourmandise*, defined as "refined and uncontrolled love of good food", employing reproving illustrations that contrasted the frugal ancient Spartans and Romans of the Republic with the decadent luxury of Sybaris. The Jesuits' *Dictionnaire de Trévoux* took the *Encyclopédistes* to task, reminding its readers that *gourmandise* was one of the Seven Deadly Sins.

Related Concepts

Foodie is often used by the media as a conversational synonym for *gourmet*, although it is a different concept (that of a food aficionado). The word *foodie* was coined synchronously by Gael Greene in the magazine *New York* and by Paul Levy and Ann Barr, co-authors of *The Official Foodie Handbook* (1984).

GASTRONOMY

Gastronomy is the study of the relationship between culture and food. It is often thought erroneously that the term gastronomy refers exclusively to the art of cooking, but this is only a small part of this discipline; it cannot always be said that a cook is also a gourmet. Gastronomy studies various cultural components with food as its central axis. Thus it is related to the Fine Arts and Social Sciences, and even to the Natural Sciences in terms of the digestive system of the human body.

A gourmet's principal activities involve discovering, tasting, experiencing, researching, understanding and writing about foods. Gastronomy is therefore an interdisciplinary activity. Good observation will reveal that around the food, there exist dance, dramatic arts, painting, sculpture, literature, architecture, and music; in other words, the Fine Arts. But it also involves physics, mathematics, chemistry, biology, geology, agronomy, and also anthropology, history, philosophy, psychology, and sociology. The application of scientific knowledge to cooking and gastronomy has become known as molecular gastronomy.

The first formal study of gastronomy is probably *The Physiology of Taste* by Jean Anthelme Brillat-Savarin (early 19th century). As opposed to the traditional cooking recipe books, it studies the relationship between the senses and food, treating enjoyment at the table as a science. Most recently, in 2004, the founders of the Slow Food movement founded the University of Gastronomic Sciences in Bra, Italy, devoted to the principles of gastronomy. Other centres for the study of gastronomy include the School of Oriental and African Studies of the University of London through its Food Studies Centre, the University of Adelaide through its Master of Arts in Gastronomy programme run in cooperation with Le Cordon Bleu, New York University's Steinhardt School of Culture, Education, and Human Development through its Department of Nutrition, Food Studies, and Public Health, the Universities of Barcelona, Bologna, and of François Rabelais in Tours through their Master in the History and Culture of Food, and Boston University through its Master of Liberal Arts in Gastronomy programme.

MOLECULAR GASTRONOMY

Molecular gastronomy is a scientific discipline involving the study of physical and chemical processes that occur in cooking. It pertains to the mechanisms behind the transformation of ingredients in cooking and the social, artistic and technical components of culinary and gastronomic phenomena in general (from a scientific point of view).

Term Origination

The term "Molecular and Physical Gastronomy" was coined in 1988 by Hungarian physicist Nicholas Kurti and by the French physical chemist Hervé This as the title for a set of workshops held in Erice, Italy that brought together primarily scientists and some professional cooks for discussions of the science behind traditional cooking preparations. Prof. Kurti had previously coined the term "Molecular Gastronomy" at a meeting at Oxford in May, 1980, of what was to become the Oxford Symposium on Food and Cookery; and he used the expression again at a meeting of the same group in June, 1985, when the subject of the meeting was "Science, Tradition and

Superstition in the Kitchen," and the keynote speaker was Harold McGee. Hervé This, on his side, used this term since 1980, when he was exploring scientifically culinary old wive tales (now called "culinary precisions") and when he was lecturing, such at the Laboratory of Physics of the Ecole normale supérieure in 1984.

Kurti and This met in 1986, and soon they had the idea that they could create a scientific discipline. In order to promote it, they decided to organize workshops. In March 1988, Kurti and This approached the director of the Ettore Majorana center, physicist Antonino Zichichi who liked the idea and imposed a condition that first class scientists would attend. As This was a friend of the Nobel Prize winner Pierre Gilles de Gennes, and as Kurti knew some scientists like Arnold Burgen, for example, they had no difficulty to get the approval. Together they invited the American food science writer Harold McGee to be a co-director of the first meeting, in 1992. Some years later, This was asked by the French Academy of sciences to present a PhD entitled: "Molecular and Physical Gastronomy", at the University of Paris. He also gave lectures in many places on this subject, including the Clarendon Laboratory, Oxford, invited by his friend Kurti.

After Kurti's death in 1998, the name of the conference was changed by Hervé This to "The International Workshop on Molecular Gastronomy 'N. Kurti'". This remained the sole director of the subsequent workshops from 1999 through 2004.

Terminology Confusion

The term molecular gastronomy has been adopted by a number of people and applied to both the scientific investigation of cooking and to cooking itself, both modern cooking that uses scientific principles in its creations as well as traditional cooking that re-examines its methods through experimentation in order to discard incorrect or irrelevant information. Though molecular gastronomy is by definition a science and science is generally practiced by scientists, the existence of the interdisciplinary occupations of "food scientist" and "Research Chef" have blurred the lines between the disciplines. This, coupled with the fact that the original objectives of molecular gastronomy included the "inventing of new dishes" and the fact that some chefs have created experimental research facilities, such as El Bulli Taller , for their restaurants has served to further blur these lines and propagate confusion between what is and isn't "molecular gastronomy".

In the late 1990s and early 2000s, the term started to be used to describe a new style of cooking in which some chefs began to explore new possibilities in the kitchen by embracing science, research, technological advances in equipment and various natural gums and hydrocolloids produced by the commercial food processing industry. It has since been used to describe the food and cooking of a number of famous chefs.

Some chefs do engage in the scientific investigation of cooking. It is also true that the work of some scientists (particularly food scientists) involves cooking. "Chef scientists" also exist, and have for many years, particularly in the research and development departments of commercial food companies.

Despite the best efforts of many to separate and distinguish the scientific investigation of cooking (molecular gastronomy) from the creative use of recently available information, technology and ingredients by cooks to produce non-traditional dishes in non-traditional forms, textures and flavor combinations (*e.g.* "New Cuisine", "Progressive Cuisine", "Nueva Cocina", "Culinary Constructivism", "Modern Cuisine", "Avant-Garde Cuisine", "Experimental Cuisine" etc...), molecular gastronomy continues to be used, in many cases, as a blanket term to refer to any and all of these things—particularly in the media. As well, the terms "Molecular Cuisine" and "Molecular cooking" have been derived from the term and used by some to describe food and cooking produced with information, tools and ingredients associated with science or food science in general.

Chefs often associated with molecular gastronomy because of their embrace of science include: Pierre Gagnaire, Ferran Adrià, Heston Blumenthal, Homaro Cantu, Wylie Dufresne, Grant Achatz, Sat Bains, Sean Wilkinson, Richard Blais, Kevin Sousa/Pittsburgh, Sean Brock, Marc Lepine/Ottawa, Will Goldfarb/NYC, and Alton Brown.

Adoption and Repudiation of the Term

Some of the chefs and scientists often associated with molecular gastronomy at one time adopted the term or acknowledged the movement. Some of these people had connections to or directly collaborated with Hervé This, which may have contributed—in part—to the mis-application of the term "molecular gastronomy" to their food and cooking. For instance, Hervé This and Pierre Gagnaire are collaborators and publish the products of their collaboration on Gagnaire's website. As well, El Bulli and The Fat Duck are listed as partners in the EU funded European Research Project INICON, a project for the sustainable collaboration between chefs, science and the food industry for the modernization of cooking. The INICON project publishes a manual on molecular gastronomy co-authored by Hervé This, and generally uses molecular gastronomy as the term to refer to the work of the organization throughout its website.

Heston Blumenthal was once a Hervé This collaborator, publishing This's recipe for "Chocolate Chantilly" in his first installment of recipe writing for the UK newspaper The Guardian in 2001, followed by an account of his visit to This's laboratory in France in 2002 during which they developed a recipe for chocolate fondant containing egg whites instead of yolks and no sugar. Blumenthal, at one time, adopted the term molecular gastronomy to describe some of the research going on at his restaurant, The Fat Duck. He was also a participant in 2001 and 2004 International Workshop on Molecular Gastronomy in Erice. The 2001 meeting being where he met many of the scientists he works with today.

Harold McGee, noted kitchen science writer, was a co-director (along with Kurti and This, who were the organizers) of the first of the International School of Molecular and Physical Gastronomy meetings in Erice and has "known Hervé This for over a decade". Though he stepped down as organizer after the first meeting, he attended all the meetings from 1992 through 2004. He at one time defined molecular gastronomy as "the scientific study of deliciousness" in a contribution to the 2004 KVL Workshop on Molecular Gastronomy, in which he also proposed areas of focus for a university programme in molecular gastronomy. Harold's book, On Food and Cooking, mentions molecular gastronomy on pages 2 and 3 of the introduction.

Ferran Adria notes in a document published in the 2003 history section of the El Bulli website entitled "About Molecular Cuisine" that up until 2003 his contact with the scientific world had been "sporadic". He goes on to explain that this is why he had "never ascribed to any scientific origin of their (El Bulli's) creations" until his scientific collaborations in 2003. He notes that he had known Harold McGee and Hervé This through conferences since 2000. In a 2004 online Q&A, he made a distinction between his work and molecular gastronomy and went on to comment on the potential longevity of what he saw as a molecular gastronomy "movement":

"I don't understand the characterization of molecular gastronomy as a type of cuisine. It's happening the same that happened years ago with fusion, it's becoming a common place. There isn't a molecular cuisine. There's a molecular movement, the molecular gastronomy, where some scientists cooperate with the world of cooking. Clearly, the move acts upon cooking, but I don't think it's a cuisine per se. In twenty years, we could look back and see how many new techniques, more than concepts, were introduced thanks to this movement. Having said this, whoever says that this movement doesn't have a future, only has to pick up a phone, turn on the TV or log-in the internet. Science has changed the world."

Perhaps frustrated with the common mis-classification of their food and cooking as "molecular gastronomy", several chefs often associated with the movement have since repudiated the term, releasing a joint statement in 2006 clarifying their approach to cooking. Still, other modern chefs have embraced molecular gastronomy. An organization known as "The Experimental Cuisine Collective" in New York which holds monthly workshops at which Hervé This has been a featured speaker and whose website lists works by This on the resources page has members including some of the most well known chefs in the city. In July 2008, New York University biochemist Kent Kirshenbaum, one of the founders of the Experimental Cuisine Collective, gave a lecture at the New York Academy of Sciences explaining the philosophy of the group. The lecture was detailed in this podcast.

Other Interpretations of the Term

- "The application of scientific principles to the

understanding and improvement of domestic and gastronomic food preparation."—Peter Barham

- "The art and science of choosing, preparing and eating good food."—Thorvald Pedersen
- "The scientific study of deliciousness"—Harold McGee

Nicholas Kurti and Hervé This

The Hungarian born physicist Nicholas Kurti (1908-1998) became Professor of Physics at Oxford in 1967, a post he held until his retirement in 1975. He was also visiting Professor at City College in New York, the University of California, Berkeley, and Amherst College in Massachusetts. His hobby was cooking, and he was an enthusiastic advocate of applying scientific knowledge to culinary problems. He was one of the first television cooks in the UK, hosting a black and white television show in 1969 entitled "The Physicist in the Kitchen" where he demonstrated techniques such as using a syringe to inject hot mince pies with brandy in order to avoid disturbing the crust. That same year, he held a presentation for the Royal Society of London (also entitled "The Physicist in the Kitchen") in which he is often quoted to have stated:

> "I think it is a sad reflection on our civilization that while we can and do measure the temperature in the atmosphere of Venus we do not know what goes on inside our soufflés."

During the presentation Kurti demonstrated making meringue in a vacuum chamber, the cooking of sausages by connecting them across a car battery, the digestion of protein by fresh pineapple juice, and a reverse baked alaska—hot inside, cold outside—cooked in a microwave oven. Kurti was also an advocate of low temperature cooking, repeating 18th century experiments by the English scientist Benjamin Thompson by leaving a 2 kg lamb joint in an oven at 80°C. After 8.5 hours, both the inside and outside temperature of the lamb joint were around 75°C, and the meat was tender and juicy. Together with his wife, Giana Kurti, Nicholas Kurti edited an anthology on food and science by fellows and foreign members of the Royal Society, "But the crackling is superb".

Hervé This started collecting "culinary precisions" (old kitchen wives' tales and cooking tricks) in the early 1980s and started testing these precisions to see which ones held up; his collection now numbers some 25,000. He also has received a PhD in Physical Chemistry of Materials for which he wrote his thesis on molecular and physical gastronomy, served as an adviser to the French minister of education, lectured internationally, and was invited to join the lab of Nobel Prize winning molecular chemist Jean-Marie Lehn. This has published several books in French, three of which have been translated into English, including "Molecular Gastronomy: Exploring the Science of Flavor", "Kitchen Mysteries: Revealing the Science of Cooking" and "Cooking, an quintessential art". He currently publishes a series of essays (in French) on Amabilia.com and hosts free monthly seminars on molecular gastronomy at the INRA in France. He is giving free and public Seminars on Molecular gastronomy any month, and once a year, he is giving a public and free Course on Molecular Gastronomy.

The idea of using techniques developed in chemistry to study food is not a new one, it dates back to 18th century, and the discipline of food science has existed for many many years. Kurti and This were colleagues and decided that a new, specific, discipline should be created within food science that investigated the processes in regular cooking (as food science was primarily concerned with the nutritional properties of food and developing methods to process food on an industrial scale).

Fundamental Objectives According to Herve This

The objectives of molecular gastronomy, as defined by Herve This are:

Current Objectives

Looking for the mechanisms of culinary transformations and processes (from a chemical and physical point of view) in three areas:

1. the social phenomena linked to culinary activity
2. the artistic component of culinary activity
3. the technical component of culinary activity

Original Objectives

The original fundamental objectives of molecular gastronomy were defined by This in his doctoral dissertation as:

1. Investigating culinary and gastronomical proverbs, sayings, and old wives' tales
2. Exploring existing recipes
3. Introducing new tools, ingredients and methods into the kitchen
4. Inventing new dishes
5. Using molecular gastronomy to help the general public understand the contribution of science to society

However, This later recognized points 3, 4 and 5 as being not entirely scientific endeavours (more application of technology and educational), and has since revised the primary objectives of molecular gastronomy.

- How ingredients are changed by different cooking methods
- How all the senses play their own roles in our appreciation of food
- The mechanisms of aroma release and the perception of taste and flavor
- How and why we evolved our particular taste and flavor sense organs and our general food likes and dislikes
- How cooking methods affect the eventual flavor and texture of food ingredients
- How new cooking methods might produce improved results of texture and flavor
- How our brains interpret the signals from all our senses to tell us the "flavor" of food
- How our enjoyment of food is affected by other influences, our environment, our mood, how it is presented, who prepares it, etc.

Example Myths Debunked:

- You need to add salt to water when cooking green vegetables
- Searing meat seals in the juices
- The cooking time for roast meat depends on the weight
- When cooking meat stock you must start with cold water

International Meetings in Erice, Italy

Up until 2001, The International Workshop on Molecular Gastronomy "N. Kurti" (IWMG) was named the "International Workshops of Molecular and Physical Gastronomy" (IWMPG0). The first meeting was held in 1992 and the meetings have continued every few years there after until the most recent in 2004. Each meeting encompassed an overall theme broken down into multiple sessions over the course of a few days, including more than 12 sessions during the meeting in 2004.

The focus of the workshops each year were as follows:

- 1992—First Meeting
- 1995—Sauces, or dishes made from them
- 1997—Heat in cooking
- 1999—Food flavors—how to get them, how to distribute them, how to keep them
- 2001—Textures of Food: How to create them?
- 2004—Interactions of food and liquids

Examples of sessions within these meetings have included:

- Chemical Reactions in Cooking
- Heat Conduction, Convection and Transfer
- Physical aspects of food/liquid interaction
- When liquid meets food at low temperature
- Solubility problems, dispersion, texture/flavour relationship.
- Stability of flavour

Precursors to Molecular Gastronomy

In the second century BC, the anonymous author of a papyrus kept in London used a balance to determine whether fermented meat was lighter than fresh meat. Since then, many scientists have been interested in food and cooking. In particular, the preparation of meat stock—the aqueous solution obtained by thermal processing of animal tissues in water—has been of great interest. It was first mentioned in the fourth century BC by *Apicius* (André (ed), 1987), and recipes for stock preparation appear in classic texts (La Varenne, 1651; Menon, 1756; Carême & Plumerey, 1981) and most French culinary books. Chemists have been interested in meat stock preparation and, more generally, food preparation since the eighteenth century (Lémery, 1705; Geoffrey le Cadet, 1733; Cadet de Vaux, 1818; Darcet, 1830). Antoine-Laurent de Lavoisier is perhaps the most famous among them—in 1783, he studied the processes of stock preparation by measuring density to evaluate quality (Lavoisier, 1783). In reporting the results of his experiments, Lavoisier wrote, "Whenever one considers the most familiar objects, the simplest things, it's impossible not to be surprised to see how our ideas are vague and uncertain, and how, as a consequence, it is important to fix them by experiments and facts". Of course, Justus von Liebig should not be forgotten in the history of culinary science (von Liebig, 1852) and stock was not his only concern. Another important figure was Benjamin Thompson, later knighted Count Rumford, who studied culinary transformations and made many proposals and inventions to improve them, for example by inventing a special coffee pot for better brewing. There are too many scientists who have contributed to the science of food preparation to list here.

The concept of molecular gastronomy was perhaps presaged by Marie-Antoine Carême, one of the most famous French chefs, who said in the early 19th century that when making a food stock "the broth must come to a boil very slowly, otherwise the albumin coagulates, hardens; the water, not having time to penetrate the meat, prevents the gelatinous part of the osmazome from detaching itself."

CATERING

A mobile caterer serves food directly from a vehicle or cart that is designed for the purpose. Mobile catering is common at outdoor events (such as concerts), workplaces, and downtown business districts.

Event Catering

Event ranges from box-lunch drop-off to full-service catering. Caterers and their staff are part of the food service industry.

When most people refer to a "caterer", they are referring to an event caterer who serves food with waiting staff at dining tables or sets up a self-serve buffet. The food may be prepared on site, *i.e.*, made completely at the event, or the caterer may choose to bring prepared food and put the finishing touches on once it arrives.

The event caterer staff are not responsible for preparing the food but often help set up the dining area. This service is typically provided at banquets, conventions, and weddings. Any event where all who attend are provided with food and drinks or sometimes only hors d'oeuvres is often called a *catered event*.

Many events require working with an entire theme or colour scheme. A catering company or specialist is expected to know how to prepare food and to make it attractive. As such, certain catering companies have moved toward a full-service business model commonly associated with event planners. They take charge of not only food preparation but also decorations, such as table settings and lighting.

The trend is towards satisfying all the clients senses with food as a focal point. With the correct atmosphere, professional event caterers with experience can make an event special and memorable.

Beautifully prepared food alone can appeal to the senses of taste, smell, and sight—perhaps even touch, but the decorations and ambiance can play a significant part in a successfully catered event.

Catering is often sold on a per-person basis, meaning that there is a flat price for each additional person. However, things like lighting and fire permits are not scaled with the guest count, so per-person pricing is not always appropriate. It is necessary to keep the cost of the food and supplies below a price margin in order to make a profit on the catering.

As many others in the food service industry, caterers and their staff work long hours. It is not uncommon for them to work on holidays or 7 days a week during holiday event seasons.

A comprehensive, formal full-service catering proposal is likely to include the following elements:

Time-line matters: rental arrival time, staff arrival time, bar open time, meal serve time, bar close time, rental pickup, out-of-venue time. Each of these factors affects the catering price.

For example, a rental quote for an "anytime" weekday delivery is usually much more economical than an "exact-time" delivery.

- General menu considerations: Clients may have specific dietary or religious needs to consider. these include Halal, Kosher, Vegetarian, Vegan and food allergy requests. Increasingly, clients are interested in food sustainability and food safety.
- Hors d'oeuvres: it should be clear if these are passed or stationary. Most caterers agree that three or four passed items are appropriate for the one-hour period prior to a meal.
- Meal Rentals: May include tables, chairs, dance floor, plants, tabletop (china, flatware, glassware, linens, chargers), bar glassware, serving equipment, salt/peppers, etc. It should be clear whether table and chair setup and take-down is included. Most rental companies do not automatically include setup and take-down in the rental charges.
- Labour: Verbiage varies from caterer to caterer, but generally speaking, an event will have a Lead/Captain/Event Manager, a Chef, perhaps a Sous Chef or Kitchen Assistant, Wait staff and Bartenders. The labour on a plated dinner is generally much higher than the labour on a buffet, because a plated dinner involves double the china, and usually a minimum of three served courses, plus served coffee. Simply put, there's a lot more to do. To do it properly requires roughly 10 to 50 per cent more staff. On a large event, this can be substantial, especially if overtime or doubletime applies.

- Service Charge: Sales Tax, Some quotes will include lighting, fire permits, draping, florals, valet and coat check. Many venues discreetly get a "cut" of the catering bill. Caterers are contractually committed to not disclose this fee specifically in their contracts with the clients. Therefore, catering will sometimes cost substantially more at one venue versus another. Also, caterers must compete with illegal operators. A legitimate caterer will have a business license and a health permit both showing the address of the place from which they do business.

Catering Officers on Ships

Merchant ships often carry Catering Officers—especially ferries, cruise liners and large cargo ships. In fact, the term "catering" was in use in the world of the merchant marine long before it became established as a land-bound business. The "Careers Scotland" website gives the following definition of a Catering Officer's duties:

Merchant Navy catering officers oversee the purchase, preparation and serving of food and drink to crew members and passengers. They are also responsible for accommodation services, including the provision of linen, bedding and laundry. They may be in overall charge of administration, organising record keeping, wage payment, and the interpretation of customs and immigration records that apply while the ship is in port.

On larger ships, responsibilities may be shared with the purser, who looks after passengers' comfort and facilities such as banking and shopping, while the catering officer concentrates on organising stores, overseeing the preparation of menus and meals and generally managing dining rooms and services. On a cruise liner, catering officers may be known as 'hotel services managers'.

Merchant Navy officers sometimes work in difficult and uncomfortable conditions. They spend long periods of time away from family and friends.

AIRLINE MEAL

An airline meal or in-flight meal is a meal served to passengers on board a commercial airliner. These meals are prepared by airline catering services.

The first kitchens preparing meals in-flight were established by United Airlines in 1936.

These meals vary widely in quality and quantity across different airline companies and classes of travel. They range from a simple beverage in short-haul economy class to a seven-course gourmet meal in long-haul first class.

The type of food varies depending upon the airline company and class of travel. Meals may be served as "one tray" or in multiple courses with no tray and with a tablecloth, metal cutlery, and glassware (generally in first and business classes).

The airline dinner typically includes meat (most commonly chicken or beef) or fish, a salad or vegetable, a small bread roll, and a dessert.

Caterers usually produce alternative meals for passengers with restrictive diets. These must usually be ordered in advance, sometimes when buying the ticket. Some of the more common examples include:

- Cultural diets, such as French, Italian, Chinese, Japanese or Indian style.
- Infant and baby meals. Some airlines also offer childrens' meals, containing foods that picky children will enjoy such as baked beans, mini-hamburgers and hot dogs.
- Medical diets, including low/high fiber, low fat/ cholesterol, diabetic, peanut free, non-lactose, low salt/ sodium, low-purine, low-calorie, low-protein, bland (non-spicy) and gluten-free meals.
- Religious diets, including Kosher, Halal and Hindu, Buddhist and Jain vegetarian (sometimes termed Asian vegetarian) meals.
- Vegetarian and vegan meals. Some airlines do not offer a specific meal for vegetarians; instead, they are given a vegan meal.

Cutlery

Breakfast served on short-haul flights in Aeroflot. Before the

September 11th attacks in 2001, first class passengers were often provided with full sets of metal cutlery. Afterward, common household items were evaluated more closely for their potential use as weapons on aircraft, and both first class and coach class passengers were restricted to plastic utensils (also known as "Sporks"). This restriction has since been relaxed in many countries.

Other Non-food Items

Condiments (typically salt, pepper, and sugar) are supplied in small sachets. For cleanliness most meals come with a napkin and a moist towelette. First and business class passengers are often provided with hot towels, proper salt and pepper shakers, and other items typically seen in a restaurant.

Breakfast

During morning flights a cooked breakfast or smaller continental-style may be served. On long haul flights and (short/medium haul flights within Asia) breakfast normally includes an entrée of pancakes or eggs, traditional fried breakfast foods such as sausages and grilled tomatoes, and often muffins or pastry, fruits and breakfast cereal on the side. On shorter flights a continental-style breakfast, generally including a miniature box of breakfast cereal, fruits and either a muffin, pastry, or bagel. Coffee and tea are offered as well, and sometimes hot chocolate.

Quality

Food on board the flight ranges in price from free (many airlines, especially those in Asia and all airlines on long-distance flights) to as much as ten dollars (Midwest Airlines). Quality may also fluctuate due to shifts in the economics of the airline industry, with private jet passengers receiving the equivalent of five-star food service.

On the longest flights in first class and business class, most Asian and European airlines serve multicourse gourmet meals, while airlines based in the US tend to serve large, hearty, meals including a salad, steak or chicken, potatoes, and ice cream. Some long-haul flights in first class (from mostly Asian carriers) offer such delicacies as caviar, champagne, and sorbet. The cost and availability of meals on US airlines has changed considerably in recent years, as financial pressures have inspired some airlines to either begin charging for meals or abandon them altogether in favour of small snacks (Southwest Airlines). Eliminating free pretzels saved Northwest $2 million annually. The carrier lost nearly $3.3 billion since 2001.

Taste

Meals must generally be frozen and heated on the ground before takeoff, rather than prepared fresh. Guillaume de Syon, a history professor at Albright College who wrote about the history of airline meals, said that the higher altitudes alter the taste of the food and the function of the taste buds; according to de Syon the food may taste "dry and flavorless" as a result of the pressurization and passengers, feeling thirsty due to pressurization, many drink alcohol when they ought to drink water.

Food Safety

Food safety is paramount in the airline catering industry. A case of mass food poisoning amongst the passengers on an airliner could have disastrous consequences. For example, on February 20, 1992, shrimp tainted with cholera was served on Aerolíneas Argentinas Flight 386. An elderly passenger died and other passengers fell ill.

Technical Crew Meals

Food safety with technical crew meals (pilots and flight engineers) is sometimes even stricter than for passengers. Many foodstuffs are banned completely from crew meals, including all egg products and often any dairy that has not been ultra-heat treated. The meals supplied on some airlines are labeled with the position of the crew member for whom they are intended, and no technical crew member eats any of the same products as his or her colleague. This ensures that each pilot eats a different meal to minimize the risk of all pilots on board being ill. This situation was a key part of the plot of the movie "Airplane!"

CASE STUDY: LSG SKY CHEFS

LSG Sky Chefs is the brand name of LSG Lufthansa Service Holding AG, which is the world's largest provider of airline catering and in-flight services. It is a subsidiary of the German airline Lufthansa AG. A part of the company was formerly owned by AMR Corporation parent company of American Airlines. Its primary business function is to prepare and deliver meals, beverages and snacks to aircraft for domestic and international flights. In addition the company also provides extended services on all other aspects of in-flight service including the design and sourcing of in-flight equipment, in-flight logistics, in-flight management, onboard retail management and the management of airport lounges to name but a few.

The company's world headquarters is located in Frankfurt, Germany. Its North American headquarters is located in Irving, Texas, USA, where a part of the company was founded in 1942. LSG Sky Chefs partners with more than 300 airlines worldwide and operates some 200 customer service centers in 49 countries, producing around 427 million airline meals a year.

The airline carriers served by LSG Sky Chefs can vary from airport to airport, but some major customers are Lufthansa, American Airlines, Mexicana, Northwest Airlines, SAS, Swiss International Airlines, United Airlines, Aerosvit Airlines, and Alaska Airlines. LSG Sky Chefs reports it has about 30 per cent of the airline catering market worldwide.

An ongoing slump in the airline industry, largely attributed to rising jet fuel costs, has created many challenges for the firm.

The company also runs more than 125 food service operations in airports around the world. In addition, the company runs non-food retail operations at several airport locations and provides traditional event planning and catering. (source: company site)

An example of a non-airline product it serves is sandwiches and salads at 7-11 convenience stores in the Orlando Market. They say Prepared by LSG Sky Chefs on the back label.

The name 'LSG' is from 'Lufthansa Service GmbH', the original name of the company that acquired the Sky Chefs brand.

History

- 1942: American Airlines creates Sky Chefs Inc
- 1966: LSG Lufthansa Service GmbH is founded
- 1985: Onex Corporation, purchases Sky Chefs; caterer to one single airline
- 1985-2001: Under the guidance of Ewout R. Heersink, C.F.O.; Onex Corporation, Sky Chefs becomes the largest inflight catering company in the world.
- 1993: LSG acquires 25 per cent of Sky Chefs
- 1995-2001: LSG acquires the rest of the shares in the parent company of Sky Chefs.

Main Competitors

- Gate Gourmet
- Servair
- Cara Operations
- SATS Catering
- Dnata

CASE STUDY: GATE GOURMET

Gate Gourmet is an airline catering firm with headquarters in Zürich, Switzerland, and Reston, Virginia, United States. Gate Gourmet was founded in 1992, and is the world's largest independent airline catering, hospitality and logistics company. The company has substantially expanded its products and services recently through acquisitions and strategic investments. The portfolio now includes cabin service ware (deSter); comfort/entertainment items (Harmony); packaged food components assembled in central production units (Supplair); aviation services such as cabin cleaning, aircraft washing and deicing and passenger services (Fernley/IAS); catering security (Gate Safe); business applications (eGate Solutions) and supply chain optimization (partnership with Pourshins).

As of 2004, it had 22,000 employees, had a revenue of 2.4 billion Swiss Francs, and produced 195 million flight meals per year. Some of its clients are British Airways, LAN Airlines, American Airlines, United Airlines, Qantas Airways, Aerolineas Argentinas, Avianca, Northwest Airlines, Iberia Airlines, Air France, Delta Air Lines, Continental Airlines, Emirates Airline, Thai Airways International, Swiss International Air Lines, Airblue, Amtrak and EasyJet.

Gate Gourmet's roots are in Swissair. For a long time, Gate Gourmet was considered Swissair's "Crown Jewel". The company was worth roughly 6 billion Swiss Francs before September 11 and the bankruptcy of Swissair. The rise of oil prices damaged the airlines and many of Gate Gourmet's main customers went bankrupt or narrowly avoided bankruptcy. The weakness of the airlines affected the industries that cater to or support it. This eventually set the stage for Swissair's crown jewel requiring a restructuring effort, which soon burdened its employees. The company was sold in 2002 to Texas Pacific Group for USD $870 million.

Disputes and Strike Action

A major dispute, reminiscent of the Grunwick dispute of the 1970s, ensued when it sacked 670 staff based at London Heathrow Airport in August 2005; the staff were gathered in a car park and the announcement made by megaphone. In sympathy with the sacked workers, British Airways staff at the airport walked out, and mass pickets of Gate Gourmet were organised by the Transport and General Workers' Union. Several workers alleged that Gate Gourmet had manufactured the dispute, in order to employ non-union workers at lower rates of pay and with worse conditions of employment. The company subsequently offered a redundancy package of more than twice the statutory minimum in an effort to resolve the labour difficulties, secure a new deal with British Airways, and avoid insolvency. The union later negotiated the reinstatement of some of the employees, but others have continued to protest at their treatment.

On 1 March 2007, Texas Pacific Group sold its remaining shares in Gate Gourmet to Merrill Lynch.

Fernley and International Aviation Services (IAS)

On September 27 2007, Gate Gourmet acquired a leading UK-based group operating under the Fernley and International Aviation Services (IAS) brands. The acquired companies offer a comprehensive range of aviation special services including aircraft cabin cleaning, aircraft washing, de-icing and toilet and water servicing; security activities such as baggage screening and document verification; airport executive lounges; and passenger services such as wheelchair and unaccompanied minor assistance.

Along with Fernley and IAS, the group also includes Specialist Airport Services, European Airport Services and Airfield Services Ltd. The companies operate primarily in the UK, including at London's Heathrow, Gatwick Stansted and Luton airports, as well as Manchester, Dublin, Cork, Shannon, Liverpool, East Midlands, Biggin Hill and Glasgow. European Airport Services also operates in Amsterdam; Brussels, and several other regional airports in France and the Netherlands. The acquired Fernley/IAS companies were privately held entities—mainly owned by Gary Morby—that have been in business for about 25 years. The new range of Gate Gourmet's aviation special services will be renamed Gate Aviation Services, which will be led by Morby.

Food Safety

In August 2004, 45 people were sickened with Shigella after eating contaminated carrots provided by Gate Gourmet. After the outbreak, the U.S. Food and Drug Administration (FDA) inspected Gate Gourmet's Honolulu facility and determined that there were substantial problems. Gate Gourmet was warned that unless the problems were rectified, its facility would be shut down.

CASE STUDY: SERVAIR

Servair is a French company with around 8,000 employees, providing restaurant services, equipping and cleaning. It has 130 customer companies. A subsidiary of Air France, Servair and its subsidiaries have a turnover of more than 489 million euros.

The Servair network, composed of Servair, its partners (Alpha, Flying Food Group, Servair Air Chef and Eurest Servair) and its technical support, has more than 100 units throughout the world.

Trades at the Service of Airlines

Servair's has 4 main activities:

- Catering, *i.e.* in-flight food services.
- Handling, *i.e.* equipping and logistics, which is the loading/unloading of aircraft (pillows, blankets, trolleys, etc.) and the management and storage of hotel products (preparation of drinks, food, tax free products,...).
- Ramp, *i.e.* airport assistance, going from runway assistance to terminal assistance.
- Cleaning, *i.e.* cabin service, to prepare a clean and comfortable aircraft.
- 7,960 employees
- 573 million euros in turnover
- 7.8 million euros of net consolidated result
- More than 100 units around the world, with its partners

Catering

- 45 million meal trays prepared per year
- 120,000 trays per day
- 16,300 tonnes of food products delivered every year
- More than 100 cooks
- 1,000 menus and 5,300 different recipes

Cleaning and Logistics

- 365,000 aircraft handled every year
- 39,000,000 seats cleaned per year
- 47 km^2 of carpet cleaned per year
- 217 vehicles including 41 heavy duty vehicles
- 365,000 interventions with reduced mobility passengers
- 40.5 million newspapers and 11 million magazines supplied per year

Quality

- 40 hygiene, microbiology and quality experts
- 17 units certified ISO 9001 v2000
- 50,000 analyses per year realised by the Servair laboratory, COFRAC accreditation award
- 7 international experts monitoring the scientific committee

Main Competitors

- Gate Gourmet
- LSG Sky Chefs
- SATS Catering
- Dnata

History

- 1971: Creation of Servair.
- 1974: Servair has 520 employees.
- 1975: Creation of Servair Lyon
- 1986: Servair provides restaurant services for Presidential journeys.
- 1990: Creation of ACNA, a company specialising in cleaning and equipping aircraft.
- 1990: 20 million trays served per year.
- 1990: Servair continues its establishment in Africa with the creation of DAKAR Catering
- 1991: Opening of Jet Chef at Le Bourget
- 1992: Opening of the units Servair 2, Orly Air Traiteur (OAT) and CPA at Roissy
- 1995: Servair is present in 14 countries.
- 1999: Creation of CAP (Culin'Air Paris) with Star Airlines and Bruneau Pégorier Catering
- 2000: Servair serves 40 million meals per year
- 2001: Opening of Special Meals Catering, a kosher production unit in Paris
- 2002: Opening of Mali Catering in Bamako Airport

- 2003: Opening of the new Servair laboratory
- 2004: Inauguration of a new kitchen in Mauritania
- 2005: CPA has transformed itself into a catering centre specialized in long-haul activity

CASE STUDY: SODEXO

Sodexo (formerly Sodexho Alliance, Euronext: SW) is a French multinational corporation, one of the largest food services and facilities management companies in the world, with 355,000 employees, representing 130 nationalities, present on 30,600 sites in 80 countries. For fiscal year 2008 (ending August 2008) revenues reached 13.6 billion euros, with a market capitalization of 5.6 billion euros. Revenues by region represent: 37 per cent Continental Europe, 37 per cent North America, 11 per cent UK and Ireland, 15 per cent Rest of World.

Sodexo is organized into two entitites: Food Services and Facilities Management and Service Vouchers and Cards. Sodexo services many sectors including private corporations, government agencies, schools from elementary through university, hospitals and clinics, assisted-living facilities, military bases, and correctional facilities.

History

The company was launched in 1966 by Pierre Bellon (Chairman) in Marseille, France, initially serving company restaurants, schools and hospitals under the name Societé d'Exploitation Hotelière (English: *Hotel Services Corporation*).

Throughout the 1970s, the company expanded in France and internationally; first in Belgium, then Africa, and finally the Middle East. After an initial public offering on the Paris Bourse in 1983, the firm continued its expansion into North America, South America, Japan, South Africa, and Russia.

Between 1995 and 2001, Societé d'Exploitation Hotelière's holding company changed its name to *Sodexho Alliance*, and the company forged alliances with Gardner Merchant, Partena, Sogeres, Wood Dining Services and Universal Ogden Services. In 2000

Sodexho Alliance became the leader in remote site management after a merger with Universal Ogden Services.

In 1998, Sodexho merged with Marriott Management Services, at the time one of the largest food services companies in North America. Included in the merger was a name change to *Sodexho Marriott Services*. The merger helped Sodexho become one of the largest food services providers in America.

In 2002 Sodexho was listed on the New York Stock Exchange.

In 2005 Michel Landel was appointed Chief Executive Officer, succeeding Pierre Bellon, who continues as Chairman of the Board of Directors.

In the summer of 2006, the company made headlines for inking a deal with NBA legend and business entrepreneur, Earvin Magic Johnson and Magic Food Provisions, a subsidiary of Magic Johnson Enterprises. The initiative includes a marketing agreement and the formation of SodexhoMagic, LLC—a new joint venture that is 51 percent owned by Johnson.

Since September 2007, its United States operations are headed up by President and CEO George Chavel, who replaced Richard Macedonia.

Food Services and Facilities Management

Sodexo offerings range from self-service food services that include staff restaurants, catering, executive dining, vending, and meal delivery, to integrated facilities management services that include soft services; providing reception help desks, concierge services, cleaning, laundry, grounds keeping, waste management, and facilities maintenance to hard services that include. plumbing, HVAC filter systems, energy maintenance and repair, and project management.

Service Vouchers and Cards (SVC)

One of the top two companies worldwide in this sector, Sodexo provides companies and public authorities with meal passes, restaurant vouchers, mobility passes, leisure passes, book cards, and training vouchers.

The company changed its official name from Sodexho Alliance to simply Sodexo after a shareholder vote at the company's annual general meeting on 22 January 2008. The reason for removing the letter 'h' from Sodexho, cited in the group's 2007 annual report, is that "in certain languages an 'x' followed by an 'h' is difficult to pronounce". The company's corporate website also states that it draws emphasis away from the hotel services industry they were once associated with, as they now focus on many other directions. A new logo was adopted at the same time. The logo of the company was also changed, dropping the five stars to a single star. The bar of the letter 'x' is also now curved, like a smile, supposedly suggestive of the company's desire to constantly seek and increase the satisfaction of their clients. The prefix 'Sod' was retained, despite negative connotations in British English.

Philanthropy

Sodexo started the STOP Hunger initiative in 1996 in the USA to fight hunger and its root causes. In 2007 the Sodexo Foundation initiatiated a programme of STOP Hunger scholarships to recognize students who were engaged in community service to fight hunger.

The Sodexo Foundation has also funded research on the root causes of hunger and possible solutions to the problem. The company's Feeding our Future programme provides meals for needy children in 28 cities in the US and Canada.

Sodexo currently has active STOP Hunger initiatives in 23 countries.

Controversy

The company's activities in private prisons and in schools have given rise to considerable controversy at over sixty US colleges and institutions. The film *Super Size Me* criticized Sodexo's policies on child nutrition in their client schools, featuring Madison Junior High School in Naperville, Illinois. In 2004, UK TV Channel 4 showed a documentary exposing the unhygienic preparation of food by Tillery Valley (a subsidiary of Sodexo).

There have been at least eight boycotts of Sodexo, for varying reasons: at the School of Oriental and African Studies at the University of London, at the American University in Washington D.C., and at Université Laval in Quebec City, at Binghamton University in New York, and Allegheny College in Meadville, Pennsylvania, at DePauw University in Indiana, at Nordea banks in Finland and at the University of Tampere, Finland. The boycott at the American University was in protest of several of Sodexo's business relationships and practices: its partnerships with the US Military, its business with prisons, low pay, and poor working conditions. Similarly, at DePauw University the students protested Sodexo's low pay, relationship with prisons, and the lack of local food options. The boycott at Université Laval protested the university administration's refusal of an initiative by the general student association (CADEUL) to provide food services to the university. At the Nordea banks the issue was a cut in wages after Sodexo took over the bank's workplace food services. After a successful boycott, the wages were raised. Sodexo has come under fire in the UK for doing business with the Harmondsworth Detention Centre, home to many asylum seekers.

Competitors

- Compass Group PLC
- Aramark
- Elior
- Accor
- Guckenheimer
- Innovative Cleaners

CASE STUDY: ARAMARK

- Its references would be clearer with a different or consistent style of citation, footnoting or external linking. Tagged since September 2009.
- Its neutrality is disputed. Tagged since September 2009.
- It may contain original research or unverifiable claims. Tagged since September 2009.
- Its neutrality or factuality may be compromised by weasel words. Tagged since September 2009.
- The notability of this article's subject is in question. If

notability cannot be established, it may be listed for deletion. Tagged since September 2009.

Aramark Limited, known commonly as Aramark, is an American food, facilities, and clothing provider, supplying businesses, courts, educational institutions, health care providers, and armed forces. It is headquartered in Center City Philadelphia, Pennsylvania.

Aramark is the 19th largest employer on the FORTUNE 500. Since 1968 in Mexico City, the company has provided services at 14 Olympic Games, including Athens in 2004 and Beijing in 2008. They have also been the subject of a number of disputes regarding labour practices and business ethics.

History

- In 1936, Davre Davidson begins the Los Angeles company that will eventually become Aramark.
- In 1959, the company becomes ARA (Automatic Retailers of America)
- In 1969, ARA officially becomes ARA Services, to reflect its growing range of businesses.
- In 1983, Joseph Neubauer is elected CEO of ARA Services.
- In 1994, ARA Services changes its name to Aramark.
- In 2001, Aramark returns to the New York Stock Exchange as a public company under the RMK ticker.
- In 2004, Aramark acquired a 90 per cent stake in the Ireland catering company Campbell Catering.
- In 2006, a group of investors led by Neubauer proposed the acquisition of all outstanding shares of the Aramark Corporation, which was approved by shareholders.
- In January 2007, the merger is completed.

CASE STUDY: BRINKER INTERNATIONAL

Brinker International, Inc. (NYSE: EAT) is the parent company of Chili's, On the Border Mexican Grill and Cantina, Maggiano's Little

Italy, and Romano's Macaroni Grill. It is the largest casual-dining restaurant operator in terms of revenue, and second largest in terms of locations behind Darden Restaurants, with more than 1,900 restaurant locations in 25 countries.

History

It was founded in 1975 as Chili's Bar and Grill, purchased by restaurateur Norman E. Brinker in 1983 and rechristened Brinker International in 1990; it now has more than 1,900 restaurants worldwide with 100,000 employees and system-wide sales exceeding $4 billion annually. Based in Dallas, Texas, it is one of the largest employers based in the Dallas/Fort Worth Metroplex.

Key Dates

- 1975: Larry Lavine opens first Chili's Grill & Bar in Dallas, Texas.
- 1983: Norman Brinker takes over Chili's restaurant chain.
- 1988: First Romano's Macaroni Grill opens in Leon Springs, Texas.
- 1991: Chili's is renamed Brinker International, Inc.
- 1992: Brinker reaches an agreement with Pac-Am Food Concepts to expand Chili's franchise to the Far East.
- 1995: Brinker establishes a strategic partnership with Lettuce Entertain You Enterprises.
- 2000: Norman Brinker steps down as company chairman.
- 2005: Company announces international expansion plans.

Holdings

Brinker formerly operated EatZi's, Cozymel's, Big Bowl Asian Kitchen, Grady's American Grill, Taco Cabana, Rockfish Seafood Grill and Corner Bakery Cafe.

CASE STUDY: COMPASS GROUP

Compass Group plc (LSE: CPG), a British company, is a leading food service business. It provides catering services to many types of institutions, which include: schools, hospitals, prisons, airports, companies and resident healthcare homes. It is headquartered in Chertsey, Surrey to the south west of London. It is listed on the London Stock Exchange and is a member of the FTSE 100 Index.

History

The business has its origins in a company which was founded by Jack Bateman in 1941 as *Factory Canteens Limited* and which subsequently became known as *Bateman Catering. Bateman Catering* and *Midland Catering* were acquired by Grand Metropolitan in 1967 and 1968 respectively. A management buy-out from Grand Metropolitan followed in 1987 when the Compass Group was formed. The Company was first listed on the London Stock Exchange in 1988.

In July 2000 the Company merged with Granada plc as part of a planned strategy to separate the media and catering interests of the latter. The two companies demerged in February 2001 to form Compass plc and Granada Media.

Operations

The Compass Group owns the following brands and businesses: Eurest, Bon Appétit Management Company, Restaurant Associates, Morrison, Scolarest, Eurest Support Services (ESS), Crothall, Levy Restaurants, Canteen Vending, Medirest, Chartwells, and Keith Prowse Corporate Hospitality.

It also operates many franchises of outlets such as Burger King, Caffè Ritazza, Starbucks and Upper Crust.

The United Nations Corruption Scandal

The Compass Group was involved in a corruption scandal with its subsidiary Eurest Support Services winning contracts to provide food to United Nations peacekeepers in Liberia. The value of Compass's food contracts with the United Nations is valued at $237 million, with renewals and add-ons that could reach $351 million.

Alexander Yakovlev, the UN procurement officer, and Vladimir Kuznetsov, head of the UN Committee for Administrative and Budgetary Issues, were arrested and indicted after taking nearly $1 million in bribes from Compass.

Competitors Es-Ko and Supreme Foodservice AG have initiated lawsuits claiming violation of the Racketeer Influenced and Corrupt Organizations (RICO) Act, the Sherman Antitrust Act and New York State's Donnelly Act regulating free trade. Federal investigations are underway and the case has been referred to the Serious Fraud Office (UK).

After its own three-month internal investigation, Compass declared it had discovered "serious irregularities" in its UN business, but that these were limited to "only a few individuals" who were dismissed: Peter R. Harris, Andy Seiwert and Doug Kerr. Compass refused to make public its investigation and did not name any other individuals involved beyond those already fired, so the true level of corruption within Compass is unclear. However, Compass's CEO Michael Bailey has stepped down in June 2006 and has been replaced by an outsider to the catering industry, Richard Cousins.

Overcharging Scandal

Compass through its subsidiary Eurest Support Services is involved in the scandal of overcharging its services to the US Government in Iraq as are many other government contractors..

CASE STUDY: CROWN GROUP

The Crown Group is a catering and event caterer in the United Kingdom. The London-based company began in 1978 as a collection of chefs in an agency that supplied local caterers with relief chefs. Later two distinct companies developed: Crown Venue Catering and Crown Society Special Event Caterers. The Group provides catering services across the UK, including to London Zoo, Tottenham Hotspur F.C. and Church House.

The groups core catering and events activities have broadened to include a diverse range of suppliers from flag making to a graphic design consultancy. Subsidiaries include Crown Venue Catering,

Crown Society, Chelsea Events, London's Flying Chef, Piggotts and Venue Reservations.

Awards have been given to the company and its employees by The Automobile Association, Michelin Guide, World Master Chef Society, and Corporate Event Association.

CASE STUDY: FIRST FOR FOODSERVICE

3663 First for Foodservice (often simply abbreviated to 3663), a trading name of BFS Group Limited , is a large United Kingdom foodservice distributor that delivers food and catering equipment to the catering industry.

It is part of The Bidvest Group Limited (Bidvest), an international investment holding company, listed on the JSE Securities Exchange in South Africa, with subsidiaries listed on the London, Australian and Luxembourg stock exchanges.

3663 has 1,100 delivery trucks and 40 storage depots across the United Kingdom, and delivers frozen, fresh and chilled food and ingredients to 50,000 businesses, educational establishments, hospitals and restaurants.

The company was apparently given its name because 3663 can spell 'food' on the telephone keypad.

CASE STUDY: DARDEN RESTAURANTS

Darden Restaurants, Inc. (NYSE: DRI) is a multi-brand restaurant operator. The firm owns several casual dining restaurant chains, most notably Olive Garden and Red Lobster. Darden owns and operates more than 1,700 restaurant locations throughout North America and has more than 157,000 employees. Darden does not franchise its restaurants in the United States, but many of its international locations are not under corporate control.

History

What would become to be known as Darden Restaurants began when William (Bill) Darden founded the Red Lobster Inns of America and opened the first Red Lobster restaurant in Lakeland,

Florida in 1968. Darden chose Lakeland because he wished to see how the concept would fare in a non-coastal region, and Lakeland was the innermost city in Florida. The initial Red Lobster franchise was widely applauded by diners and critics alike, most prominently the restaurant journalist Scott Indrisek. The store became hugely successful and by 1970 had expanded to three locations in the state with two more under construction. While the locations were profitable, the company lacked the resources to expand further, so Darden sold the company to food giant General Mills that year. General Mills upgraded the chain to a more casual dining/family fare oriented format, opened a new corporate headquarters in Orlando, Florida and installed Darden as company president. In 1975, when Darden was promoted to the position of Vice President of operations for the restaurant unit, Joseph (Joe) R. Lee, the company's first restaurant manager, was made President of Red Lobster.

Under General Mills, Red Lobster grew into a successful chain of almost 400 locations by 1985. The company underwent several restructurings and transformed itself from an inexpensive fast-food seller into a chain of casual dining seafood restaurants by 1988.

Diversification

One of the company's first ventures into the diversification of its portfolio was the York Steak House chain of English-themed steak and chop restaurants in the 1970s. The franchised steak and potatoes restaurant was a cafeteria-style restaurant with salad bar and hot station. By the end of the 1980s, the chain had been mostly closed down, however some independent locations still exist.

In 1982, Darden opened the first Olive Garden concept store in Orlando. The chain took off, and by 1989 General Mills had opened over 145 stores, making the chain the fastest-growing unit in the company's restaurant holdings. While Olive Garden did not meet critical success, it was popular, and its per-store sales soon grew to match those of Red Lobster. The company eventually became the largest chain of Italian-themed full-service restaurants in the U.S. The company helped bolster its Italian bona fides opening the Culinary Institute of Tuscany in Tuscany, Italy, a training facility designed to allow its chefs to become fluent in Italian cuisine in an authentic Italian setting.

China Coast was an attempt to create a (U.S.) national casual dining restaurant that featured American Chinese cuisine. While the chain eventually expanded to some 50 restaurants, its sales were lackluster and lost an estimated $20 million USD. By the end of 1995, the company was shuttered, and the remaining locations were either converted to Red Lobsters or Olive Gardens or closed altogether.

Bahama Breeze features a Caribbean theme, food and drinks found in the islands of the Caribbean Sea. Created recently in 1996, the company was bolstered under the early management of Leslie Vinyard, who promoted the brand's classic "Island hospitality means Island Love" maxim.

Smokey Bones is a sports bar concept featuring barbecue and related foodstuffs in an Appalachian mountain-lodge setting. Created in 1999. Sold in 2008.

Seasons 52 created in 2002

Rocky River Grill House (2006-2007)

Acquisitions

In August 2007, Darden acquired rival Atlanta-based restaurant holder Rare Hospitality for $1.4 billion USD, gaining Rare's two chains, the Capital Grille and LongHorn Steakhouse. Unlike Darden, Rare franchised its LongHorn concept and has not announced plans beyond executive structure.

Divestiture

By the end of its fiscal year in 1995, Red Lobster was operating 1,250 stores in 49 states and Canada.

Darden Restaurants was spun off from General Mills beginning on May 9, 1995 when it began trading on when-issue basis at $9.75 a share. The company became a fully separate entity on May 31, 1995 when its shares went on sale on the NYSE. The shares opened at $10.75, 17 per cent below expectations, but climbed to $11.125 by the close of trading.

In December 2007, Darden announced that it will be selling its Smokey Bones chain to Barbeque Integrated, Inc., an affiliate of Sun Capital Partners, Inc., for approximately $80 million. The sale was completed in January, 2008.

Operations

As of May 25, 2008, the Company operated, through its subsidiaries, 1,702 restaurants in the United States and Canada. In the United States, it operated 1,667 restaurants in 49 states (the exception being Alaska), including 651 Red Lobster, 647 Olive Garden, 305 LongHorn Steakhouse, 32 The Capital Grille, 23 Bahama Breeze and seven Seasons 52 restaurants, and two specialty restaurants: Hemenway's Seafood Grille & Oyster Bar and The Old Grist Mill Tavern. In Canada, Darden operated 35 restaurants, including 29 Red Lobster and six Olive Garden restaurants. Through subsidiaries, the Company owns and operates all of its restaurants in the United States and Canada, except three restaurants, which are located in Central Florida and are owned by joint ventures managed by Darden. On October 1, 2007, Darden acquired of RARE Hospitality International, Inc.

CASE STUDY: ACCOR

Accor (Euronext: AC) is a large French multinational corporation, part of the CAC 40 index, operating in nearly 100 countries. Headquartered in Courcouronnes, Essonne, France, near Évry, Accor is the European leader in hotels (Accor Hospitality) and a global leader in corporate services (Accor Services).

Accor Hospitality, the Accor hotels branch, has more than 4,000 hotels worldwide. It starts from luxurious hotels to economic lodging solutions.

Through Accor Services, Accor also runs service vouchers to over 430,000 companies and institutions and 30 million users in 40 countries: Ticket Restaurant, Luncheon Vouchers, Ticket Alimentaçao, Clean Way, Ticket Service, Childcare Vouchers, Eyecare Vouchers, Bien-Etre à la Carte, Worklife Benefits, EAR, Accentiv', Académie du Service, Tesorus, Ticket Compliments.

CASE STUDY: CARA OPERATIONS

Cara Operations Limited is a Canadian company that provides catering services to airlines and operates several restaurant and coffee shop chains including: Harvey's, Swiss Chalet, Kelsey's, Milestone's and Montana's. The company moved to its new HQ in Nov 2008 in Vaughan, Ontario across from Vaughan Mill Shopping Mall

In 2002, sales for the whole company were C$1.9 billion. 88 per cent of the business comes from the restaurant services with the remaining 12 per cent deriving from airline catering.

The name Cara was derived from the first two letters of each of the words "Canada Railway"; it is also an Irish word meaning "friend".

History

In 1883, Canada Railway News Co. was founded. Cara began as a steamship company on the Niagara River in Ontario. It soon moved into the food business, catering to a boom in passenger rail traffic in Canada.

In the 1930s, Cara began offering catering services to the airlines. By 1951 it was serving about 1 500 meals a day.

In 1961 the company changed its name to Cara Operations Limited.

Cara was owned solely by the Phelan family from its inception in 1883 until it finally went public in 1968. At the time, its primary business was catering to the transportation sector (airline and rail meals), but it did operate a few 'no-name' restaurants and coffee shops in various office towers and airport terminals in Canada. Total sales of all the various operations was C$30 million in 1968.

In 1986 Cara provided services for Vancouver's Expo 86.

In 1999, Cara purchased Kelseys inc. from owner Paul Jeffrey, acquiring Montanas, Kelseys and the Outback (Canada). The rights to Outback steakhouse were owned in Canada by Cara, but it was sold back to Outback in the USA because of high food costs.

On February 26, 2004, Cara went private.

In 2008, Cara was in the middle of a controversy when 3 employees got arrested at the Montreal-Trudeau airport concerning a drug distribution network that used the planes of which Cara is responsible of the catering.

Operations

Cara currently employs more than 39,000 people.

Catering Services

Today, Cara controls 60 per cent of the Canadian airline market, providing meals for more than 60 of the world's major airlines including KLM, American Airlines, British Airways and Air Canada. Meals are prepared at nine kitchens located across Canada.

- Summit Food Service Distributors Inc. now a division of Colabor LP, is Canada's largest Canadian-owned broadline distributor to the foodservice industry.
- Cara Airport Services Division
  - Pearson International Airport (Toronto)
  - Flight Kitchen Vancouver International Airport—40 per cent market share
- Rocky Mountain Railtours (British Columbia/Alberta)

Restaurant Operations

- Cara Air Terminal Restaurant Division
- Harvey's
- Swiss Chalet
- Kelsey's Neighbourhood Bar & Grill
- Milestone's Restaurants Inc.
- Montana's Cookhouse
- Coza! Tuscan Grill

10

Oenology, Wine Tasting and Food Matching: An Overview

OENOLOGY

Oenology, œnology (BrE), or enology (AmE) is the science and study of all aspects of wine and winemaking except vine-growing and grape-harvesting, which is a subfield called viticulture. "Viticulture and oenology" is a common designation for training programmes and research centres that include both the "outdoors" and "indoors" aspects of wine production. An expert in the field of oenology is known as an oenologist. The word oenology is derived from the Greek *oinos*, "wine," and *logos*, "word" or "speech."

WINE AND FOOD MATCHING

Wine and food matching is the process of pairing food dishes with wine to enhance the dining experience. In many cultures, wine has had a long history of being a staple at the dinner table and in some ways both the winemaking and culinary traditions of a region will have evolved together over the years. Rather than following a set of rules, local cuisines were paired simply with local wines. The modern "art" of food pairings is a relatively recent phenomenon, fostering an industry of books and media with guidelines for pairings of particular foods and wine. In the restaurant industry, sommeliers are often present to make food pairing recommendations for the guest. The main concept behind pairings is that certain elements (such as texture and flavor) in both food and wine react differently to each other and finding the right combination of these elements will make the entire dining experience more enjoyable. However, taste and enjoyment are very subjective and what may be a "textbook perfect" pairing for one taster could be less enjoyable to another. While there are many books, magazines and websites with detailed guidelines on how to pair food and wine, most food and wine experts believe that the most basic element of food and wine pairing is understanding the balance between the "weight" of the food and the weight (or body) of the wine. Heavy, robust wines like Cabernet Sauvignon can overwhelm light delicate dish like a quiche while light bodied wines like Pinot grigio would be similarly overwhelmed by a hearty stew. Beyond weight, flavors and textures can either be contrasted or complemented. From there a food and wine pairing can also take into consideration the sugar, acid, alcohol and tannins of the wine and how they can be accentuated or minimized when paired with certain types of food.

History

For most of history, wine has been a regular accompaniment to meals. Often the culinary and winemaking traditions of a region would evolve together, creating a natural pairing between the local wine and the local cuisine.

Wine has had a long history of being served as an accompaniment to food. The early history of wine has it origins as another dietary staple and a beverage that was often more sanitary than the local water supply. There is little evidence that much serious thought was given to pairing particular dishes to particular wines with most likely whatever wine was available being used. However, as culinary traditions in a region developed, so too did local winemaking tradition. Many pairings that are considered "classics" today emerged from the centuries old relationship between a region's cuisine and their wines. In Europe, lamb was a staple meat of the diet for many areas that today are leading wine regions. The red wines of regions such as , Greece, Rioja, Ribera del Duero, Rhone and Provence are considered classic pairings with the lamb dishes found in the local cuisines of those regions. In Italy, the intimate connection between food and wine is deeply embedded in the culture and is exemplified by the country's wine. Historically, Italians rarely dined without wine and a region's wine was crafted to be "food friendly", often with bright acidity. While some Italian wines may seem tannic, lean or tart by themselves they often will show a very different profile when paired with boldly flavored Italian foods.

There have been some historical anecdotes that have related to food and wine pairing before modern times. One anecdote often attributed to British wine merchants is "*Buy on an apple and sell on cheese*" meaning that if a wine taste good when paired with a raw, uncooked apple it must be truly good and pairing any wine with cheese will make it more palatable to the average consumer and easier to sell. The principles behind this anecdote lies in the food pairing properties of both fruit and cheeses. Fruits that are high in sugar and acidity (such as the malic acid in green apples) can make wines taste metallic and thin bodied. In contrast, hard cheeses such as cheddar can soften the tannins in wines and make them taste fuller and fruitier. Another historical anecdote, still repeated today, is "*White wine with fish; Red wine with meat*". The root of this adage rests on the principle of matching the body (weight) of the wine with the weight of the food. Meat was generally heavier and "red" in colour so it was assumed that a red wine (which was usually heavier than white wine) paired better. Similarly fish was generally light and "white" in colour so it was often paired with white wine. This adage has become out dated somewhat due to the variety of wine styles prevalent in modern winemaking where there are now many "heavy" white wines such as "New World" oaky Chardonnay that can have more body than lighter reds such as Pinot noir or Italian Merlots.

Modern History

In restaurants, sommeliers are trained to recommend wine pairings to go with the restaurant's fare.

In recent years, the popularity and interest in food and wine pairings have increased and taken on new connotations. Industries have sprung up with print publications and media dedicated to expounding on the principles and ideals of pairing the perfect wine with the perfect dish. In the restaurant industry, there is often a dedicated individual or staff of sommeliers who are trained to recommend wine pairings with the restaurant's fare. The origins of this recent phenomenon can be traced to the United States in the 1980s when Neo-Prohibitionists prompted the wine industry to reexamine the context of wine-drinking as a component of dining rather than as just an alcoholic beverage meant for consumption and intoxication. Winemakers started to emphasize the kind of food dishes that their wines would go well with, some even printing pairing suggestions on back wine labels. Food magazines began to suggest particular wines with recipes and restaurants would offer multi-course dinners matched with a specific wine for each course.

Today there are multiple sources for detailed guidelines and tips on food and wine pairing. But many wine drinkers select wine pairings based on instinct, the mood of the meal or simply a desire to drink a particular wine at the moment they desire to eat a particular meal. The subjective nature of taste makes it possible to drink any kind of wine with any kind of food and have an enjoyable experience. Wine expert Mark Oldman has noted "*Food and Wine pairing can be like sex and pizza: even when it's bad, it can still be pretty good*" and gives the example of wedding cake with a dry sparkling wine. A very dry wine with a very sweet food is, according to Oldman, "*the equivalent of nails on a chalkboard*" and is not a "good pairing" according to most guidelines but the atmosphere of the occasion and the subjective nature of taste can trump any rule or guideline. Today, many wine experts and advocates in the realm of food and wine pairing try to focus on the more objective physical aspects of food that have an effect on the palate, altering (or enhancing) the perception of various aspects of the wine.

Matching Weight

In food and wine pairings, the most basic element considered is "weight"-the balance between the weight of the food (a heavy, red sauce pasta versus a more delicate salad) and the weight or "body" of the wine (a heavy Cabernet Sauvignon versus a more delicate Pinot grigio). In wine tasting, body is determined primarily by the alcohol level of the wine and can be influence by the perceptions of tannins (from the grape skins or oak) and extract (the dissolved solids in the wine derived from winemaking processes like extended maceration and sur lie aging). An oaked Chardonnay from a warm wine region, such as Australia will be "heavier" in body than a stainless steel fermented Chardonnay from a cooler wine region such as Chablis. Pairing heavy wines with light dishes or vice versa, can result in one partner overwhelming the other. The "weight" of a food can also be described in terms of the intensity of its flavors-such as delicate and more subtle flavors versus dishes that have more robust and hearty flavors. A key to pairing upon this principle is to identify the dominant flavor of the dish. Sauces can be the dominant flavor instead of the meat or main component. While poached fish is usually light bodied and better served with a light white, if the fish is served with a heavy cream sauce it could be better balanced with a fuller bodied white wine or light red.

Weights of Wine

Gruner Veltliner is often made as a light bodied wine. Below is a rough guideline of the various weights of wines. Winemaker and regional style as well as oak treatment can cause a wine to be lighter or heavier in body. For example, Pinot noir can vary from being very light to more medium bodied. Another example is the influence of regional climates. Warmer climate wine regions tend to produce wines with higher alcohol levels and thus more fuller bodied wines so that a Sauvignon blanc from California may have a heavier weight than a Sauvignon blanc from the Loire.

Lighter Whites

Pinot gris, Pinot blanc, Riesling, Sauvignon blanc, Chablis, Champagne and sparkling wines, Gruner Veltliner, Vinho Verde

Medium to Heavy Whites

Oaked Sauvignon blanc, Alsatian wines, Albarino, White Bordeaux (Semillon), White Burgundy, Rhone whites (Viognier, Roussanne, Marsanne) and New World Chardonnay.

Lighter Reds

Beaujolais, Dolcetto, some Pinot noir

Medium Reds

Chianti, Barbera, Chinon, Rioja, Cabernet franc, Merlot, Malbec, Zinfandel, some Pinot noir

Heavier Reds

Syrah, Brunello di Montalcino, Cabernet Sauvignon, Port, Barbaresco and Barolo

Focus of the Pairing

While a perfect balance where both food and wine are equally enhanced is theoretically possible, typically a pairing will have a more enhancing influence on one or the other. Master Sommelier Evan Goldstein notes that food and wine pairing is like two people having a conversation "*one must listen while the other speaks or the result is a muddle*". This means either the food or the wine will be the dominant focus of the pairing, with the other serving as a complement to enhance the enjoyment of the first. In regards to weight and intensity, if the focus of the pairing is the wine then a more ideal balance will be a food that is slightly lighter in weight to where it will not compete for attention with the wine but not too light to where it is completely overwhelm. If the focus of the pairing is to highlight a dish then the same thought would apply in pairing a wine.

Complement and Contrast

The racy acidity of sparkling wine and the rich texture of foie gras is an example of a contrasting pairing.

After considering weight, pairing the flavors and texture can be dealt with using one of two main strategies—complement or contrast.

The first strategy tries to bring wine together with dishes that complement each other such as an earthy, Burgundian Pinot noir with an earthy, mushroom dish.

The second strategy operates under the truism that "opposites attract" and brings together food and wine that have contrasting traits such as a crisp, acidic Sauvignon blanc and a fish with a creamy, lemon sauce. The crisp, acidity of the wine serves as a contrast that can cut through the creaminess of the sauce and give a different, refreshing sensation for the palate as oppose to what a complementary pairing, such as a creamy, buttery Chardonnay would bring. For most of history, the "complementary strategy" was the prevailing thought on food and wine pairing. In the 1980s, as more people started to discover and experiment with pairings, the idea of using contrast started to gain more favour. It follows the same idea that the "salty/sweet" pairing does in cooking (such as salty peanut butter with sweet jelly).

The same food may be complemented *or* contrasted: a hard, nutty cheese such as Hirtenkase should have "a nutty, slightly sweet wine with it," *or* a full bodied red wine.

Physical Properties of Wine

While it is often said that "taste is subjective", there are quantifiable taste characteristics (like bitter, sweet, salty or sour) that can be perceived and measured as low, moderate or high—such as measuring the sweetness of honey or the saltiness of oysters. Flavors, such as butterscotch, char and strawberry, are more personal and can't be quantifiable. Flavors are either perceived to be present or not. The perception of flavors is linked to our sense of smell, while tastes come from the sensory glands of the taste buds. Though individual sensitivity to the different taste "senses" can vary, wine experts will often recommend pairings based on these more objective measurements rather than the more subjective concept of "flavors". In wine there are three basic tastes-bitter, sweet and sour. These three taste can each be identified with a primary component of the wine-tannins (bitter), residual sugar (sweet) and acidity (sour). A fourth component, alcohol, is identified in wine tasting with a perception of "heat" or hotness in the back of the mouth and is the primary factor influencing the body of the wine. The residual heat of the alcohol can be considered in food pairing with some ingredients minimizing the heat of the wine while some will accentuate it.

Acidity

The acidity of salad dressing and tomatoes can cancel some of the tartness in a Beaujolais wine, allowing the fruit to be more noticeable.

Acidity is a dominant player in any food and wine pairing due to the pronounced and complex ways that it can heighten the perception of flavors. In wine tasting, acidity is perceived by a mouth watering response by the salivary glands. This mouth watering can also serve to stimulate the appetite. In wine there are three main acids that have their own associated flavors-malic (green apples), lactic (milky) and tartaric (bitter). In dishes that are fatty, oily, rich or salty, acidity in wine can "cut" (or standout and contrast) through the heaviness and be a refreshing change of pace on the palate. In cooking, acidity is often used in similar fashions such as a lemon wedges with a briny seafood dish such as oysters. The acidity of the lemon juices can make the oysters seem less briny. A wine that is less tart than the dish it is served with will taste thin and weak. A wine that comes across as "too tart" on its own maybe soften when paired when an acidic and tart dish. The complementing "tartness" of the food and wine cancels each other out and allows the other components (fruit of the wine, other flavors of the food) to be more noticeable.

Sweetness

The sweetness of wines is determined by the amount of residual sugar left in the wine after the fermentation process. Wines can be bone dry (with the sugars fully fermented into alcohol), off-dry (with a hint of sweetness), semi-dry (medium-sweet) and dessert level sweetness (such as the high sugar content in Sauternes and Tokays). Sweet wines often need to be sweeter than the dish they are served with. Vintage brut champagne paired with sweet, wedding cake can make the wine taste tart and weak while the cake will have off flavors. In food pairings, sweetness balances spice and heat. It can serve as a contrast to the heat and alleviate some of the burning sensation caused by peppers and spicy Asian cuisine. It can accentuate the mild sweetness in some foods and can also contrast with salt such as the European custom of pairing salty Stilton cheese with a sweet Port. Sweetness in a wine can balance tartness in food, especially if the food has some sweetness (such as dishes with sweet and sour sauces).

Bitterness

The proteins and fats in cheese can soften the perception of tannins in wine making a wine seem less bitter and more fruity.

The bitterness associated with wine is usually derived from a wine's tannins. Tannins add a gritty texture and chalky, astringent taste. It can enhance the perception of "body" or weight in the wine. Tannins are normally derived from the skins and stems of the grapes themselves (leeched out during the maceration process) or from contact with oak during barrel aging. Tannins react to proteins. When paired with dishes that are high in proteins and fats (such as red meat and hard cheeses), the tannins will bind to the proteins and come across as softer. In the absence of protein from the food, such as some vegetarian dishes, the tannins will react with the proteins on the tongue and sides of the mouth—accentuating the bitterness and having a drying effect on the palate. Various cooking methods, such as grilling and blackening can add a bitter "char" component to the dish that will allow it to play well with a tannic wine. While fish oils can make tannic wines taste metallic or off. Bitter tannic wines like Barolo and Cabernet Sauvignon can overwhelm a lot of foods but can be soften by fatty foods with a lot of proteins such as hard cheeses or meats. The dry tannins also serve as a cleansing agent on the palate by binding to the grease and oils left over in the mouth. Spicy and sweet foods can accentuate the dry, bitterness of tannins and make the wine seem to have off flavors.

Alcohol

Alcohol is the primary factor in dictating a wine's weight and body. Typically the higher the alcohol level, the more weight the wine has. An increase in alcohol content will increase the perception of density and texture. In food and wine pairing, salt and spicy heat will accentuate the alcohol and the perception of "heat" or hotness in the mouth. Conversely, the alcohol can also magnify the heat of spicy food making a highly alcoholic wine paired with a very spicy dish one that will generate a lot of heat for the taster.

Other Pairing Principles

Beyond the basic guidelines listed above, food pairings can dive even further into matching several layers of texture and flavors. The term "bridge ingredients" refer to ingredients and flavors that have certain affinities to wine pairing (such as slow cooked onions to creamy wines, etc). It can also refer to using particular herbs and spices perceived in the wine (such as rosemary in some Cabernet Sauvignon) and adding them to the dish as an ingredient. Their presence in a dish may increase the likelihood that the certain wines will pair well.

ODOR

An odor or odour is a volatilized chemical compound, generally at a very low concentration, that humans or other animals perceive by the sense of olfaction. Odors are also called smells, which can refer to both pleasant and unpleasant odors. The terms fragrance, scent, and aroma are used primarily by the food and cosmetic industry to describe a pleasant odor, and are sometimes used to refer to perfumes. In contrast, malodorous, stench, reek, and stink are used specifically to describe unpleasant odors.

Terminology

In the United Kingdom, "odour" refers to scents in general. In the US, "odor" has a more negative connotation; "scent" or "aroma" are used for pleasant smells.

Basics

The widest range of odors consists of organic compounds, although some inorganic substances, such as hydrogen sulfide and ammonia, are also odorants. The perception of an odor effect is a two-step process. First, there is the physiological part; the sense of the stimulus by receptors in the nose. After that, the psychological part follows. The stimuli are processed by the region of the human brain which is responsible for smelling. Because of this, an objective and analytical measure of odor is impossible. While odor feelings are very personal perceptions, individual reactions are related to gender, age, state of health, and private affectations. Common odors that people are used to, such as their own body odor, are less noticeable to individuals than external or uncommon odors.

For most people, the process of smelling gives little information concerning the ingredients of a substance. It only offers information related to the emotional impact. Experienced people, however, such as flavorists and perfumers, can pick out individual chemicals in complex mixes through smell alone.

Odor Analysis

In Germany, the concentrations of odorants have since the 1870's been defined by the "Olfaktometrie", which helps to analyze the human sense of smell using the following parameters: odor substance concentration, intensity of odor, and hedonic assessment.

To establish the odor concentration, an olfactometer is used which employs a panel of human noses as sensors. In the olfactometry testing procedure, a diluted odorous mixture and an odor-free gas (as a reference) are presented separately from sniffing ports to a group of panelists, which are housed in an odor neutral room. They are asked to compare the gases emitted from each sniffing port, after which the panelists are asked to report the presence of odor together with a confidence level such as guessing, inkling, or certainty of their assessment. The gas-diluting ratio is then decreased by a factor of two (*i.e.* chemical concentration is increased by a factor of two). The panelists are asked to repeat their judgment. This continues for a number of dilution levels. The responses of the panelists over a range of dilution settings are used to calculate the concentration of the odor in terms of European Odor Units. The main panel calibration gas used is Butan-1-ol., which at a certain diluting gives 1 ouE/m^3.

General Survey

The analytic methods could be subdivided into the physical, the gas chromatographical, and the chemosensory method.

When measuring odor, there is a difference between the emission and immission measurements. During the emission measurement, the odor concentration in the air is so high that the so called "Olfaktometer" is needed to thin the assay. Because of this, all measurement methods based on thinning assays are called "olfaktometrical methods." On the contrary, an "Olfaktormeter" is rarely used during the immission measurement. The same measuring principals are used, but the judgment of the air assay happens without thinning the assay.

Measurement

Different aspects of odor can be measured through a number of quantitative methods, such as assessing concentration or apparent intensity.

Odor Concentration

The measurement of odor concentration is the most widespread method to quantify odors. It is standardized in CEN EN 13725:2003. The method is based on dilution of an odor sample to the odor threshold (the point at which the odor is only perceptible to 50 per cent of the test panel). The numerical value of the odor concentration is equal to the dilution factor that is necessary to reach the odor threshold. Its unit is the *European Odor Unit*, OU_E. Therefore, the odor concentration at the odor threshold is 1 OU_E by definition.

To establish the odor concentration, an olfactometer is used which employs a panel of test persons. A diluted odorous mixture and an odor-free gas (as a reference) are presented from sniffing ports to a group of panelists. In comparing the gases emitted from each port, the panelists are asked to report the presence of odor. The gas-diluting ratio is then decreased by a factor of 1.4 or two (*i.e.* the concentration is increased accordingly). The panelists are asked to repeat their judgment. This continues for a number of dilution levels. The responses of the panelists over a range of dilution settings are used to calculate the concentration of the odor in terms of European Odor Units (OU_E/m).

The test persons must fulfill certain requirements, for example regarding their sensitivity of odor perception. The main panel calibration gas to verify this requirement used is n-Butanol.

Odor Intensity

Odor intensity can be divided into the following categories according to intensity:

- 0—no odor
- 1—very weak (odor threshold)
- 2—weak
- 3—distinct
- 4—strong
- 5—very strong
- 6—intolerable

This method is most often applied by having a dilution series tested by a panel of independent observers who have been trained to differentiate intensity.

Hedonic Assessment

Hedonic assessment is the process of scaling odors on a scale ranging from extremely unpleasant via neutral up to extremely pleasant. There is no correlation between this method and the method of measuring the odor intensity. However, the hedonic perception of some odors may change from pleasant to unpleasant with increasing concentration and intensity.

Odor Type

This is a verbal characterization of the sensed odor by the test person, such as disgusting, caustic, ruffling, etc. There are no more applications needed than a test person to run this method. The evaluation of the odor type could be an emission or an immission method. It has a great impact on evaluating the source of the odor emission.

Emission Measurement

The following details have to be differentiated while the emission is measured:

First there is the odor time slice (Result = Part of "odor hours per year" per area). Then there is the olfactory flag scope (Result = Current scope at actual meteorology situation). And last but not least there is the harassment exaltation by questionings (Result = differentiated acquisition harassments).

Sampling Technique

There are two main odor sampling techniques, the direct odor sampling and the indirect odor sampling technique.

Direct Odor Sampling

Air will be sampled at the source and fed straight into the olfactometer for assessment by an odor panel. The following problems can be associated with this technique: Odor panel members need to be seated in an odor neutral environment, thus they need to be housed in a separate area. This is difficult to achieve when assessing odor released from, for example factories, where the odor can be emitted from a stack on the end of a production line. This means that the odor sample collected needs to be transported from the stack to the unit where the odor panel sits. This can sometimes be on the other side of the factory plant. The sample then must therefore pass through a very long sample line to the olfactometer. This can have influences on the sample quality, can have potential air blockages due to water condensation or other operational procedures. Therefore most odor annoyance assessment companies use the indirect air sampling method.

Indirect Odor Sampling

Indirect odor sampling is done with the use of odor (air) sampling bags, which are made from an odor neutral material *e.g.* Teflon. The odor sample bags are connected to an air sampling line which is then, for example, hooked up to a stack. The air stream is then sampled and stored in the odor sample bag and can then be analyzed in a suitable environment (*e.g.* in an odor laboratory).

The indirect method is used to sample a wide variety of odor sources. From stacks on the end of a factory line, water surfaces or ambient air surroundings.

Each odor source has its own set of problems when sampled; these problems need to be overcome in order to collect a representative sample of the odor source. The following problems can be encountered:

Vacuum

Vacuum can be overcome by placing the odor sample bag in vacuum container which can be placed under vacuum. If the vacuum is higher than the vacuum at the source, the odor sample will collect in the bag.

High Temperatures and Moisture Content

High temperatures and high moisture contents inside the odor source leads to complications when sampled. When the sample leaves the source, it will cool down and produce condensate in the sample line and or odor sample bag. This can lead to growth of bacteria or when drying out release more odor, thus alter the odor concentration of the sample. The same is true when sampling in high moisture conditions. A way round the problem is to use a stack dilution probe through which an inert gas (for example dry nitrogen) can be fed that dries the sample stream. This prevents the moisture condensing in the sample line and or the odor sample bag.

High Lethal Gas Concentrations

Sometimes odor sources emit a high concentration of gases that are lethal to man. These samples must be diluted to a safe level, before being presented to the odor panel. This pre-dilution can be done in a stack-dilution probe, by the addition of an inert gas or on a dilution device for example an extra olfactometer.

Odor Concentrations

More often than not, odor sampled at the source is higher then the ambient odor concentration. In a few cases the odor concentration can be so high that panelists will make a positive identification even if the olfactometer is diluting the odor sample in its upper dilution range. The sample must then be pre-diluted to make a sensible reading, this pre-dilution can again be done with a stack-dilution probe, by the addition of an inert gas or on a dilution device for example an extra olfactometer.

Large Odor Emitting Surface (Land or Water)

When a large surface is emitting odor, for example a sewage treatment plant, a fixed dimension "hood" can be used. In one end of the hood, clean air is blown in at a known rate, and on the other end, a sample is collected via the indirect method. If a large land surface is emitting odor, for example a bio filter (a big concrete basin filled with wood chip through which the factories waste air is pumped), a section can be cornered off with plastic (*e.g.* Teflon) (of which the dimensions are known). The air from the factory will inflate the plastic (lift it up) and an odor sample can be taken from under the plastic via the direct air method.

All involved sampling parts have to be made out of olfactory neutral materials. Principally every sampling has to meet the logically requirements, has to be defined, standardized, meaningful and reproducible. This is needed to make different measurements comparable. Odorant concentrations scaled in either GG/m^3 or ouE/m^3 aren't convincing while comparing different emissions of different plants. Because of this instead of comparing different concentrations, different emission mass currents of the emitted freight are compared.

Legislative Provisions Associated with Odors

When legislation for environmental protection in Germany first began, it raised the question of the evaluation of different odors. Since that time, the following laws have been made:

1. "refinery guideline" (early 1970s)
2. federal emission protection law (1974)
3. technical guideline to keep the air fresh
4. olfactory emission guideline (early 1980s until 1998)

Controls at the point of the emission, like plural vitrification against aircraft noise, drop out. Terms of transmission could be marginally changed by establishing ramparts, plantings and so on, but the objective efficiency of those controls is likely minimal. But the subjective efficiency of a plantings is remarkable.

The choice of the location is the most important control, that means keeping an adequate distance to the nearest receptor and paying attention to the meteorology conditions, such as the prevailing wind direction. Reduction of the emission, by way of dilution of a small emission concentration with large air flow volumes, could be an effective and economic alternative, instead of reducing the emission with different controls.

Encapsulating of olfactory relevant asset areas is the best known method to reduce the emission, but it is not the most suitable one. Different matters need to be considered by encapsulation. Within an enclosure a damp and oppressive atmosphere can arise, so that the inner materials of the capsule produce a high degree of mechanical stress. Not to let the explosion hazard slide.

For encapsulation to be viable, there must be some way to exhaust the spent air. When emission is avoided through capsuling, odorants remain inside the medium and tend to leak at the next suitable spot. In any case, capsuling is never really gas-proof, and at some spots substances may leak out at considerably higher concentrations. There are three different ways exhausted air may be treated:

- chemical treatment

- physical treatment
- biological treatment

Adsorption as Separating Process

Adsorption is a thermo separation process, which is characterized by the removal of molecules out of a fluid phase at a solid surface. Molecules of a gas-or fluid mixture are taken up by a solid with a porous interface surface. The solid matter is called the adsorbant, the adsorbed fluid is called the adsorbate. There are two types of adsorption, physisorption and chemisorption. The type of force driving the adsorption process is different between the two.

Physisorption

A special type of adsorption is physisorption. The difference between physisorption and chemisorption is that the adsorbed molecule is tied up with the substrate by physical forces, defined here as forces which do not cause chemical bonds. Such interactions are mostly unfocused in contrast to chemical bonds. "Van-Der-Waals"—forces are a special type of such physical forces. These forces are characterized by electrostatic interactions between induced, fluctuating dipoles. To be more specific you have to call those forces "London's Dispersal forces." A so called dipole moment occurs because of fluctuations in the distribution of electrons around individual atoms. The temporary mean value of this force is however zero. Even though it's only a mere transient dipole moment, this moment can cause a nonparallel dipole moment in an adjacent molecule. Operating forces of this nature are in inverse proportion to the sixth power of the distance between those molecules. These forces occur in almost every chemical system, but are relatively weak.

Physisorption is an exothermic and reversible reaction. Obviously stronger strengths accrue through the interaction between solid dipoles at polar surfaces or reflexive loadings, appearing in electric conductive surfaces. Such interactions could be defined as a chemisorption because of their strength.

Chemisorption

In many reactions, physisorption is a pre-cursor to chemisorption. Compared to physisorption, chemisorption is not reversible and requires a larger activation energy. Usually the bond energy is about 800 kJ/mol. For physisorption the bond energy is only about 80 kJ/mol. A monomolecular layer could be maximally adsorbed. Strong bonds between the adsorbative molecules and the substrate could lead to the point that their intermolecular bonds partly or completely detach. In such a case you have to call this a dissociation. Those molecules are in a highly reactive state. This is the basis of heterogeneous catalysis. The substrate is then called catalytic converter. The differences between Chemisorption and Physisorption extends beyond an increased activation energy. An important criteria for chemisorption is the chemical mutation of the absorbent. Thereby it is possible that you have to deal with a chemisorption in a few combinations with a relatively low bond energy, for example 80 kJ/mol, as a physisorption could be another combination with a bond energy even by 100 kJ/mol. The interaction with different adsorbative molecules is very different. The surface could be taken by substances, which point out a very high bond energy with the substrate, and as a consequence of this the wanted reaction is impossible. Because of that feature those substances are called catalytic converter venom. Heat is released during that process too.

Loading of the Adsorben

During the adsorption of a molecule, energy—the heat of adsorption—is released. This energy is the difference of the enthalpy of the adsorben in the fluid or gaseous phase and the its corresponding enthalpy on the surface of the adsorbant. With an increase of the loading on the surface of the adsorbant the bond energy decreases in the area of the monomolecular covering. For higher loading this value approaches zero. This implies that there is a limit for the loading of an adsorbant. The procedure of turning back that process is called desorption. Adsorption as a separating process is a challenging process, in the case of finding the eligible adsorbents, which could link as multilateral as possible.

Types of Odors

Some odors such as perfumes and flowers are sought after, elite varieties commanding high prices. Whole industries have developed products to remove unpleasant odors. The perception of odors is also very much dependent upon circumstance and culture. The odor of cooking processes may be pleasurable while cooking but not necessarily after the meal.

The odor molecules send messages to the limbic system, the area of the brain that governs emotional responses. Some believe that these messages have the power to alter moods, evoke distant memories, raise their spirits, and boost self-confidence. This belief has led to the concept of "aromatherapy" wherein fragrances are claimed to cure a wide range of psychological and physical problems. Aromatherapy claims fragrances can positively affect sleep, stress, alertness, social interaction, and general feelings of well-being. However, the evidence for the effectiveness of aromatherapy consists mostly of anecdotes and lacks controlled scientific studies to back up its claims.

With some fragrances, such as those found in perfume, scented shampoo, scented deodorant, or similar products, people can be allergic to the ingredients. The reaction, as with other chemical allergies, can be anywhere from a slight headache to anaphylactic shock, which can result in death.

Unpleasant odors can arise from specific industrial processes, adversely affecting workers and even residents downwind of the industry. The most common sources of industrial odor arise from sewage treatment plants, refineries, specific animal rendering plants and industries processing chemicals (such as sulfur) which have odorous characteristics. Sometimes industrial odor sources are the subject of community controversy and scientific analysis.

The Study of Odors

The study of odors is a growing field but is a complex and difficult one. The human olfactory system can detect many thousands of scents based on only very minute airborne concentrations of a chemical. The sense of smell of many animals is even better. Some fragrant flowers give off *odor plumes* that move downwind and are detectable by bees more than a kilometer away.

The study of odors can also get complicated because of the complex chemistry taking place at the moment of a smell sensation. For example iron metal objects are perceived to have an odor when touched although iron vapor pressure is negligible. According to a 2006 study this smell is the result of aldehydes (for example nonanal) and ketones (example: 1-octen-3-one) released from the human skin on contact with ferrous ions that are formed in the sweat-mediated corrosion of iron. The same chemicals are also associated with the smell of blood as ferrous iron in blood on skin produces the same reaction.

Pheromones

Pheromones are odors that are used for communication. A female moth may release a pheromone that can entice a male moth that is several kilometers away. Honeybee queens constantly release pheromones that regulate the activity of the hive. Workers can release such smells to call other bees into an appropriate cavity when a swarm moves in or to "sound" an alarm when the hive is threatened.

In mammals some pathway of Pheromones identification lay in vomeronasal organ and some—in odor receptors.

Advanced Technology

There are hopes that advanced technology could do everything from test perfumes to help detect cancer or explosives by detecting specific scents, but as of yet artificial noses are still problematic. The complex nature of the human nose, its ability to detect even the most subtle of scents, is at the present moment difficult to replicate.

Most artificial or electronic nose instruments work by combining output from an array of non-specific chemical sensors to produce a finger print of whatever volatile chemicals it is exposed to. Most electronic noses need to be "trained" to recognize whatever chemicals are of interest for the application in question before it can be used. The training involves exposure to chemicals with the response being recorded and statistically analyzed, often using multivariate analysis and neural network techniques, to "learn" the chemicals. Many current electronic nose instruments suffer from problems with reproducibility with varying ambient temperature and humidity.

WINE TASTING DESCRIPTORS

The use of wine tasting descriptors allow the taster an opportunity to put into words the aromas and flavors that they experience and can be used in assessing the overall quality of wine. Many wine writers, like Karen MacNeil in her book *The Wine Bible,* note that the difference between casual drinkers and serious wine tasters is the focus and systematic approach to tasting wine with an objective description of what they are sensing. The primary source of a person's ability to taste wine is derived from their olfactory senses. A taster's own personal experiences play a significant role in conceptualizing what they are tasting and attaching a description to that perception. The individual nature of tasting means that descriptors may be perceived differently among various tasters.

The following is a list of wine tasting descriptors and a common meaning of the terms.

- *Accessible:* A wine that is easy to drink without an overwhelming sense of tannin, acidity or extract.
- *Aftertaste:* A term for the taste left on the palate after wine has been swallowed. "Finish" is a synonym.
- *Aggressive:* A wine with harsh and pronounced flavors. The opposite of a wine described as "smooth" or "soft"
- *Alcoholic:* A wine that has an out of balanced presence of too much alcohol.
- *Astringent:* An overly tannic white wine.
- *Acidic:* A wine with a noticeable sense of acidity.
- *Aroma:* The smell of a wine. The term is generally applied to younger wines, while the term *Bouquet* is reserved for more aged wines.
- *Austere:* A wine that is dominated by harsh acidity or tannin and is lacking the fruit needed to balance those components.

- *Autolytic:* Aroma of "yeasty" or acacia-like floweriness commonly associated with wines that have been aged *sur lie*.
- *Baked:* A wine with a high alcohol content that gives the perception of stewed or baked fruit flavors. May indicate a wine from grapes that were exposed to the heat of the sun after harvesting.
- *Balanced:* A wine that incorporates all its main components—tannins, acid, sweetness, and alcohol—in a manner where no one single component stands out.
- *Big:* A wine with intense flavor, or high in alcohol.
- *Biscuity:* A wine descriptor often associated with Pinot noir dominated-Champagne. It is sense of yeasty or bread dough aroma and flavors.
- *Bite:* A firm and distinctive perception of tannins or acidity. This can be a positive or negative attribute depending on whether the overall perception of the wine is balanced.
- *Bitter:* An unpleasant perception of tannins.
- *Blowzy:* An exaggerated fruity aroma. Commonly associated with lower quality fruity wines.
- *Body:* The sense of alcohol in the wine and the sense of feeling in the mouth.
- *Bouquet:* The layers of smells and aromas perceived in a wine.
- *Bright:* When describing the visual appearance of the wine, it refers to high clarity, very low levels of suspended solids. When describing fruit flavors, it refers to noticeable acidity and vivid intensity.
- *Buttery:* A wine that has gone through malolactic fermentation and has a rich, creamy mouthfeel with flavors reminiscent of butter.
- *Cassis:* The French term for the flavors associated with black currant. In wine tasting, the use of cassis over black currant typically denotes a more concentrated, richer flavor.
- *Cedarwood:* A collective term used to describe the woodsy aroma of a wine that has been treated with oak.
- *Charming:* A subjective term used to describe a wine with

a range pleasing properties but nothing that stands out in an obvious fashion.

- *Chewy:* The sense of tannins that is not overwhelming.
- *Cheesy:* An aroma element characteristic of aged Champagne that develops after an extended period of aging. It is associated with the aroma of aged, nutty cheeses such as gouda and is caused by a small amount of butyric acid that is created during fermentation and later develops into an ester known as ethyl butyrate.
- *Chocolaty:* A term most often used of rich red wines such as Cabernet Sauvignon and Pinot noir that describes the flavors and mouthfeel associated with chocolate—typically dark.
- *Cigar-box:* An term used to describe the tobacco aromas derived from oak influence.
- *Citrous:* A wine with the aromas and flavor from the citrus family of fruits.
- *Classic:* A subjective term used to denote a wine of exceptional quality that display the typicity of its varietal(s), displays layers of complexity and is very well balanced.
- *Clean:* A wine that is not demonstrating any obvious faults or unwanted aromas and flavors.
- *Clear:* A wine with no visible particular matter.
- *Closed:* A wine that is not very aromatic.
- *Cloves:* An aroma associated with oak treatment that gives the perception of cloves. It is caused by the creation of eugenic acid by the toasting of the oak barrels.
- *Cloying:* A wine with a sticky or sickly sweet character that is not balanced with acidity.
- *Coarse':* A term for a wine with a rough texture or mouthfeel. Usually applies the perception of tannins.
- *Coconut:* Aroma perception of coconut derived from treatment in American oak.
- *Compact:* Opposite of "open knit". A wine with a dense perception fruit that is balanced by the weight of tannins and acidity.
- *Complete:* Similar to the description of a "balanced wine" but more encompassing a wine that all the main

components-acidity, alcohol, fruit and tannins-in long along with a pleasing mouthfeel and long finish.

- *Complex:* A wine that gives a perception of being multi-layered in terms of flavors and aromas.
- *Concentrated:* Intense flavors.
- *Connected:* A sense of the wine's ability to relay its place of origin or *terroir*
- *Concoction:* Usually a derogatory term used to refer to a wine that seems to have many different components that are "thrown together" rather than integrating into one cohesive profile.
- *Cooked:* A term similar to "bake" where the fruit flavors of the wine seemed like they have been cooked, baked or stewed. It may also indicate that grape concentrate was adding to the must during fermentation.
- *Corked:* A tasting term for a wine that has cork taint
- *Creamy:* A term used to describe the perception of a warm, creamy mouthfeel. In sparkling wines, the sense of creaminess arises from a combination of the finesse of the mousse and the results of malolactic fermentation. The perception of creaminess is generally picked up at the back of the throat and through the finish of the wine.
- *Crisp:* A pleasing sense of acidity in the wine.
- *Crust:* Sediment, generally potassium bitartrate, that adheres to the inside of a wine bottle
- *Definition:* A wine that not only is well balanced but also gives a clear expression of its grape variety or place of origin.
- *Delicate:* A term used similar to charm but more often relates to the more subtle notes of a wine.
- *Depth:* A term used to denote a wine with several layers of flavor. An aspect of complexity.
- *Dirty:* A wine with off flavors and aromas that most likely resulted from poor hygiene during the fermentation or bottling process
- *Dry:* A wine that is lacking the perception of sweetness.
- *Dried up:* A wine that has lost some of its freshness or fruitiness due to extended aging.

- *Earthy:* This can mean a wine with aromas and flavor reminiscent of earth—such as forest floor or mushrooms. It can also refer to the drying impression felt on the palate caused by high levels of geosmin that occur naturally in grapes.
- *Easy:* A term that can be synonymous with "approachable" but more commonly refers to a wine that is simple and straightforward without much complexity but still enjoyable to drink.
- *Edgy:* A wine with a noticeably level of acidity that heightens the flavors on the palate. Maybe synonymous with "nervy"
- *Elegant:* A term to describe a wine that possess finesse with subtle flavors that are in balance.
- *Expansive:* A wine that is considered "big" but still accessible.
- *Expressive:* A wine with clearly projected aromas and flavors.
- *Fallen over:* A wine that, at a relatively young age, has already went past its peak (or optimal) drinking period and is rapidly declining in quality is said to have "fallen over".
- *Farmyard:* A generally more positive term than "Barnyard" used to describe the earthy and vegetal undertones that some Chardonnay and Pinot noir develop after maturing in the bottle. While for some wine drinkers this may not be very appealing, for others this maybe a sign that the wine has entered its peak drinking period.
- *Fat:* A wine that is full in body and has a sense of viscosity. A wine with too much fat that is not balanced by acidity is said to be "flabby" or "blowzy"
- *Feminine:* Describes a wine that emphasize more it delicate flavors, silky textures and subtle aromas rather strength, weight and intensity of fruit.
- *Finish:* The sense and perception of the wine after swallowing.'
- *Finesse:* A very subjective term used to describe a wine of high quality that is well balanced.
- *Firm:* A stronger sense of tannins.

- *Flabby:* A lacking sense of acidity.
- *Flat:* In relation to sparkling wines, flat refers to a wine that has lost its effervescence. In all other wines the term is used interchangeably with "flabby" to denote a wine that is lacking acidity, particularly on the finish.
- *Fleshy:* A wine with a noticeable perception of fruit and extract.
- *Foxy:* A tasting term for the musty odor and flavor of wines made from *Vitis labrusca* grapes native to North America, usually a negative term.
- *Fresh:* A positive perception of acidity.
- *Fruit:* The perception of the grape characteristics and sense of body that is unique to the varietal.
- *Full:* A term usually used in context of wine with heavy weight or body due to its alcohol content. It can also refer to a wine that is full in flavor and extract
- *Grapey:* A wine with the aromas and flavors reminiscent of grape flavoring—such as those associated with grape jelly. The Muscat family of grapes often produce wines that are described as "grapey".
- *Grassy:* A term used to describe an herbaceous or vegetal element of a wine—ranging from freshly mown lawn grass to lemon grass flavors.
- *Green:* Overly acidic wine. Typically used to describe a wine made from unripe fruit.
- *Gutsy:* A wine with noticeable body, extract and fruit.
- *Hard:* Overly tannic wine.
- *Harsh:* Similar to "coarse" but usually used in a more derogatory fashion to denote a wine that has unbalanced tannins and acidity.
- *Heavy:* A wine that is very alcoholic with too much sense of body.
- *Herbaceous:* The herbal, vegetal aromas and flavors that maybe derived from varietal characteristics or decisions made in the winemaking process-such as harvesting under-ripened grapes or using aggressive extraction techniques for a red wine fermented in stainless steel.
- *Hollow:* A wine lacking the sense of fruit.

- *Hot:* Overly alcoholic wine.
- *Inky:* A term that may refer to a wine's dark colouring and opacity.
- *Jammy:* A wine that is rich in fruit but maybe lacking in tannins.
- *Lean:* The sense of acidity in the wine that lacks a perception of fruit.
- *Legs:* The tracks of liquid that cling to the sides of a glass after the contents have been swirled. Often said to be related to the alcohol or glycerol content of a wine. Also called *tears*.
- *Lemony:* A term referring to the tangy acidity of a wine with fruit flavors reminiscent of lemons.
- *Lightstruck:* A tasting term for a wine that has had long exposure to Ultraviolet light causing "*wet cardboard*" type aroma and flavor.
- *Linalool:* The characteristic flowery-peach aroma associated with Muscat and Riesling wines. It derived from the chemical compound linalool.
- *Liquorice:* A term used to describe the concentrated flavor from rich sweet wines such as those of Monbazillac AOC which are produced by botrytized grapes.
- *Liveliness:* A term used to describe a wine with slight carbonation and fresh, bright acidity.
- *Luscious:* Similar to "voluptuous" but more commonly associated with sweet wines that have a rich, concentrated mouthfeel.
- *Mature:* A wine that has aged to its peak point of quality.
- *Mean:* A wine without sufficient fruit to balance the tannins and/or acidity of the wine, making it unbalance and unpleasant to drink.
- *Meaty:* A wine with a rich, full body (and often pronounce tannins and extract) that gives the drinker the impression of being able to "chew" it.
- *Mellow:* A wine with a soft texture that is nearing the peak of its maturity.
- *Midpalate:* A tasting term for the feel and taste of a wine when held in the mouth.

- *Minerality:* A sense of mineral-ness in the wine, flavors of slate, shist, silex, etc
- *Nervy:* A wine with a noticeable amount of acidity but is still balanced with the rest of the wine's components
- *Nose:* A tasting term for the aroma or bouquet of a wine.
- *Oaky:* A wine with a noticeable perception of the effects of oak. This can include the sense of vanilla, sweet spices like nutmeg, a creamy body and a smoky or toasted flavor.
- *Oily:* A generally full bodied wine with a viscous mouthfeel. If the wine is lacking acidity, this term maybe used in conjunction with flabby.
- *Opulent:* A rich tasting wine with a pleasing texture mouthfeel that is well balanced.
- *Oxidized:* A generally negative term describing a wine that has experienced too much exposure to oxidation. A wine that has been oxidized, is considered faulty and may exhibit sherry-like odors.
- *Oxidative:* Unlike "oxidized", this is generally a more positive term describing a wine that has experienced constrained exposure to oxidation over the course of its aging process. The aromas and flavors that develop as a wine oxidatively matures can range from nuttiness, biscuity and butteriness to more spicy notes.
- *Palate:* A tasting term for the feel and taste of a wine in the mouth.
- *Peak:* The point where a wine is at its most ideal drinking conditions for an individual taster. This is a very subjective determination as for some tasters a wine will be at its peak when the fruit is still fresh and young while for some tasters the peak will arrive when a wine has matured in flavor.
- *Peppery:* A wine with the aromas and flavors reminiscent of the fruit from the Piper family of plants such as black peppercorn associated with Syrah and Grenache based wine or the aroma of crushed white pepper associated with Gruner Veltliner.
- *Perfume:* A generally positively used to describe an aspect of a wine's aroma or bouquet.

- *Plummy:* A wine with the juicy, fresh fruit flavors of plum
- *Polished:* A wine that is very smooth to drink, with no roughness in texture and mouthfeel. It is also well balanced.
- *Powerful:* A wine with a high level of alcohol that is not excessively alcoholic.
- *Prickly:* A wine with some slight residual carbonic gas, though not necessarily to the point of the wine being considered a sparkling wine. Some very young white wines (such as Vinho Verde) and dry rosé may be described as "prickly"
- *Racy:* A wine with noticeable acidity that is well balanced with the other components of the wine.
- *Reticent:* A wine that is not exhibiting much aroma or bouquet characteristics perhaps due to its youth. It can be described as the sense that a wine is "holding back".
- *Rich:* A sense of sweetness in the wine that is not excessively sweet.
- *Robust:* A term with similar connotations as "aggressive" except that "robust" is more commonly applied to older, mature wines while "aggressive" tends to describe younger wines.
- *Round:* A wine that has a good sense of body that is not overly tannic.
- *Sassy:* A wine with bold, brash and audacious flavors.
- *Sharp:* A term normally used to describe the acidity of a wine though it can refer to the degree of bitterness derived from a wine's tannin.
- *Sherrylike:* A term used to describe a non-Sherry wine that exhibits oxidized aromas that may have been caused by excessive amounts of acetaldehyde.
- *Short:* A wine with well develop aromas and mouthfeel but has a finish that is little to non-existent due to the fruit quickly disappearing after swallowing.
- *Smokey:* A wine exhibiting the aromas and flavors of the various types of smoke, such as tobacco smoke, roasting fire smoke and a toasty smoke derived from oak influences.
- *Smooth:* A wine with a pleasing texture. Typically refers to a wine with soft tannins.

- *Soft:* A wine that is not overly tannic.
- *Spicy:* A wine with aromas and flavors reminiscent of various spices such as black pepper and cinnamon. While this can be a characteristic of the grape varietal, many spicy notes are imparted from oak influences.
- *Stalky:* A woody, green herbaceous note in a wine.
- *Structure:* A term used to describe the solid components of a wine-acidity, sugar, density of fruit flavors and phenolic compounds such as tannins in relation to the overall balance and body of the wine.
- *Supple:* A wine that is not overly tannic.
- *Sweet:* A wine with a noticeable sense of sugar levels.
- *Tannic:* A wine with aggressive tannins.
- *Tart:* A wine with high levels of acidity.
- *Texture:* A tasting term for the mouthfeel of wine on the palate.
- *Thin:* A wine that is lacking body or fruit
- *Tight:* A wine with a significant presence of tannins that is restraining the other qualities of the wine, such as fruit and extract, from being more noticeable. A "tight wine" is expected to age well as the tannins soften to reveal these other qualities.
- *Toasty:* A sense of the charred or smoky taste from an oaked wine.
- *Transparency:* The ability of a wine to clearly portray all unique aspects of its flavor—fruit, floral, and mineral notes. The opposite would be a wine where flavors are diffused and thoroughly integrated.
- *Typicity:* A wine tasting term used to describe how much a wine expresses the typical characteristics of the varietal.
- *Unoaked:* Also known as *unwooded*, refers to wines that have been matured without contact with wood/oak such as in aging barrels.
- *Undertone:* The more subtle nuances, aromas and flavors of wine.
- *Upfront:* A wine with very perceivable characteristics and quality that do not require much thought or effort to discover.

- *Vanilla:* An oak induced characteristic aroma reminiscent of vanilla.
- *Vegetal:* A wine with aromas and flavor reminiscent of vegetation as oppose to fruit or floral notes.
- *Vivid:* A wine with very expressive ripe, fruit flavors.
- *Voluptuous:* A wine with a full body and rich texture.
- *Warm:* A wine with noticeable but balanced alcohol as opposed to a wine with excessive alcohol that maybe described as "hot". It can also refer to a creamy texture derived from oak treatment.
- *Watery:* A wine that is excessively "thin" in body and fruit.
- *Yeasty:* Often uses synonymously with "biscuity" and can describe a wine with aromas and flavor reminiscent of bread dough or biscuits.
- *Young:* Wine that is not matured and usually bottled and sold within a year of its vintage.
- *Zesty:* A wine with noticeable acidity and usually citrus notes.
- *Zippy:* A wine with noticeable acidity that is balanced with enough fruit structure so as to not taste overly acidic.

11

Food University, Research Institutes and Centres: Select Case Studies

WAGENINGEN UNIVERSITY AND RESEARCH CENTRE

Wageningen University and Research Centre (also known as Wageningen UR; abbreviation: WUR) is a research and higher education concern which consistis of Wageningen University, the Van Hall-Larenstein School of Higher Professional Education, and the former agricultural research institutes (Dienst Landbouwkundig Onderzoek) from the Dutch Ministry of Agriculture. With its combination of knowledge and experience in higher eduction and research Wageningen UR aims to train specialists (BSc, MSc and PhD) in life sciences and through its research contribute actively to solving scientific, social and commercial problems in the field of life sciences and natural resources. Wageningen UR approaches national and international research topics from the perspectives of various disciplines, with an integrated approach and in close collaboration with governments, companies, stakeholder organisations, citizens, and other knowledge institutions. The research should pay attention to the balance between three priorities in society: *economics, culture and nature.*

It is based in the Dutch city of Wageningen.

Wageningen University

Wageningen University was established in 1918 and was the successor of the Agricultural School founded in 1876.

Wageningen University provides education and generates knowledge in the field of life sciences and natural resources. Wageningen UR aims to make a real contribution to our quality of life. Quality of life means to the university both an adequate supply of safe and healthy food and drink, on the one hand, and the chance to live, work and play in a balanced ecosystem with a large variety of plants and animals.

The university has about 6,000 students from 98 countries. Its core business are life and agricultural sciences.

ECTS-Label

Wageningen University was the first Dutch University or school that was allowed to use the ECTS label. This label is awarded by the European Commission and guarantees the quality of the study programme. An important consideration is that the University consequently applies the European Credit Transfer System; this promotes the mobility of students within Europe and prevents study delay. The label is a prestigious acknowledgment of the international character of the University. Out of 56 European applications in 2005, only 3 ECTS labels were awarded.

BSc Programmes

Wageningen University offers 18 BSc programmes (2005-2006). The language of instruction is partly Dutch, partly English. For some BSc programmes the language of instruction is English. The programmes start each year in September, they last three years and consist of 180 ECTS credits. The programmes are in the field of economy and society, health, life sciences and technology, nature and environment, animals and plants.

Economy and Society

- Business and consumer studies
- Economics and policy
- International development studies

Health

- Health and society
- Nutrition and health

Technology

- Agro technology
- Biotechnology
- Food technology
- Molecular life sciences

Nature and Environment

- Soil, water and air
- Forest and nature conservation
- International land and water conservation
- Landscape architecture and spatial planning
- Environmental studies

Animals and Plants

- Biology
- Biological Production Sciences
- Animal sciences
- Plant sciences

MSc Programmes

Wageningen University offers a wide range of MSc programmes (2005-2006). The language of instruction is English. The programmes start each year in September, they last two years and consist of 120 ECTS credits. Most programmes offer various specialisations and possibilities for majors.

- Agricultural and Bioresource Engineering
- Animal Sciences and Aquaculture
- Aquaculture and Fisheries
- Bioinformatics
- Biology
- Biotechnology
- Communication Science
- Earth System Science
- Environmental Sciences
- European Masters Degree in Food Studies
- Food Quality Management

- Food Safety
- Food Technology
- Forest and Nature Conservation
- Geo-information Science
- Hydrology and Water Quality
- International Development Studies
- International Land and Water Management
- Landscape Architecture and Planning
- Leisure, Tourism and Environment
- Management of Agro-ecological Knowledge and Social Change
- Management, Economics and Consumer Studies
- Meteorology and Air Quality
- Molecular Life Sciences
- Nutrition and Health
- Organic Agriculture
- Plant Biotechnology
- Plant Sciences
- Soil Science
- Urban Environmental Management

PhD Programme

The PhD programme is a four-year programme which consists of a research component and a smaller education component. To apply for a PhD position, the applicant must contact one of the Graduate Schools of Wageningen University. In order to guarantee adequate supervision, the research subject must fit in the research programme of a Graduate School. The four-year PhD programme consists of a research component (conducting research under supervision and writing a thesis) and a smaller education component (up to 15 per cent of the total PhD time) Upon completion of PhD programme, the PhD student is expected to be:

- Able to function as an independent scientist.
- Able to integrate his/her own work in the theoretical framework of his/her discipline(s) in a broader area of research and able to communicate this in a scientific or general setting.

- Competent in identifying priority areas of research and in formulating questions and experimental hypotheses pertinent to this research.

Research Institutes

The following research institutes are part of Wageningen UR:

- Agricultural Economics Research Institute (LEI)
- Agrotechnology and Food Sciences Group (AFSG)
- Alterra—Research Institute for the Green World (ALTERRA)
- International Institute for Land Reclamation and Improvement (ILRI)
- Animal Sciences Group (ASG)
- Applied Plant Research (PPO)
- Central Institute for Animal Disease Control (CIDC-Lelystad)
- International Agricultural Centre (IAC)
- Plant Research International (PRI)
- RIKILT-Institute of Food Safety

Van Hall-Larenstein School of Higher Professional Education

The Van Hall-Larenstein School was formed out of a merger of the Van Hall Instituut and Larenstein, School of Professional Education.

Van Hall Larenstein offers 14 bachelor's degree programmes and 6 professional master's degree programmes to a total of 4,400 students of 20 nationalities. The study programmes are mainly in Dutch.

Student Activities and Associations

M.S.V. Alchimica is a study association for students of Molecular Life Sciences. For over 35 years it has been organising many different activities for its members.

CODON is a study association for students of Biotechnology

and Bioinformatics established by the first students of *Bioprocesstechnology* on 16 September 1991. At that time the association carried the name "BiPS" which was later changed to CODON. The association's main language is English.

Nicolas Appert (study association) is the study association for the students of Food technology and Food Safety of both the University and the Van Hall Larenstein College. It was founded on September 6, 1962, and was named after the French food scientist Nicolas Appert.

Heeren XVII is the study association for the students of Agrotechnology and Agricultural and Bioresource Engineering. It was founded on April 1, 1965.

Genius Loci is the study association for the students of Landscape architecture and spatial planning. The name reflects the atmosphere of a certain place. It was founded on February 15, 1990 after the merger of two older study programmes.

Licere is the study association for the students of the MSc Leisure, Tourism and Environment. It was founded in 2006 and its name means 'Leisure time' in Latin.

Mercurius is the study association for the students of the BSc's Business and Consumer Studies and Economics and Policy and for the students of the MSc Management, Economics and Consumer Studies.

De Veetelers is the study association of students Animal Sciences (both BSc and MSc). The name means literally Animal Breeders and was founded in 1962.

Notable Alumni

- Reinier van den Berg
- Jan Just Bos
- Gerrit Braks
- Staf Depla
- Jeroen Dijsselbloem
- Willem Hilbrand van Dobben
- Louise Fresco
- Volkert van der Graaf
- Wilbert Hetterscheid

- Gerrit Hiemstra
- Martijn Katan
- Ruud Koopmans
- Ferdinand Antoon Langguth Oliviera
- Joan Leemhuis-Stout
- Helga van Leur
- Jan Odink
- Theo Quené
- Rudy Rabbinge
- Ramsewak Shankar
- Bernard Slicher van Bath
- Harry Snelders
- Katja Staartjes, first Dutch woman on Mount Everest
- Onno van de Stolpe, CEO of Galapagos Genomics.
- Rob Urgert
- Jan Valckenier Suringar
- Cees Veerman
- Joris Voorhoeve
- Anne Vondeling
- Marijke Vos
- Henk Vredeling
- Simon Vroemen
- Karel Vuursteen, CEO of Heineken
- Clive West
- Onno Wijnands
- Mamoudou H. DICKO

INSTITUTE OF FOOD TECHNOLOGISTS

The Institute of Food Technologists or IFT is an international, non-profit professional organization for the advancement of food science and technology. It is the largest of food science organizations in the world, encompassing 22,000 members worldwide as of 2006. It is referred to as "THE Society of Food Science and Technology." Its current president is Sheri Schellhaass of General Mills.

As food technology grew from the individual family farm to the factory level, including the slaughterhouse for meat and poultry processing, the cannery for canned foods, and bakeries for bread as examples, the need to have personnel trained for the food industries did also. Literature such as Upton Sinclair's *The Jungle* in 1906 about slaughterhouse operations would be a factor in the establishment of the Food and Drug Administration (FDA) later that year. The United States Department of Agriculture was also interested in food technology and research was also being done at agricultural colleges in the United States, including the Massachusetts Institute of Technology (MIT), the University of Illinois at Urbana-Champaign, the University of Wisconsin-Madison, and the University of California, Berkeley . By 1935, two MIT professors, Samuel C. Prescott and Bernard E. Proctor decided that it was time to hold an international conference regarding this . A detailed proposal was presented to MIT President Karl Taylor Compton in 1936 was presented with $1500 of financial aid from MIT for a meeting to be held from June 30 to July 2, 1937 with Compton asking how many people would be in attendance at this meeting. Prescott replied with "fifty or sixty people." 500 people actually attended the event.

This meeting proved so successful that in early 1938 that a second conference would be held in 1939. Initially led by George J. Hucker of the New York State Agricultural Experiment Station (part of Cornell University) in Geneva, New York, a small group meeting was held on August 5, 1938 on forming an organization with an expanded group meeting in New York City on January 16, 1939 to further discuss this. The second conference was held at MIT June 29 to July 1, 1939 with Proctor as conference chair. 600 people attended this event. At the final session, the chairman of the session Fred C. Blanck of the United States Department of Agriculture, proposed that an organization be established as the Institute of Food Technologists. This was approved unanimously. Its first officers were Prescott as President, Roy C. Newton of Swift and Company in Chicago, Illinois as Vice President, and Hucker as Secretary-Treasurer . By 1949, IFT had 3,000 members.

IFT Growth

Regional sections were established in IFT as early as 1940 in northern California (San Francisco, Bakersfield, Sacramento) . The first IFT Award, the Nicholas Appert Award was established in 1942 by IFT's

Chicago section with additional awards being established since then. For the first ten years, the IFT's officers were established to be President, Vice-President, Secretary, and Treasurer. In 1949, the IFT moved into offices in Chicago and created a permanent position of Executive Secretary to run daily organizational operations. Retired U.S. Army Colonel Charles S. Lawrence was named the first Executive Secretary, a position he would hold until 1961 when he was replaced by Calvert L. Willey. During Willey's term as Executive Director (Executive Secretary 1961-1966), IFT would grow from 6,000 members in 1961 to 23,000 members in 1987 . Additionally, IFT Divisions were first established in 1971 with the Refrigerated and frozen foods division . The IFT Student Division was established in 1975, and was reorganized in 1984 to be the IFT Student Association with the chairperson serving as a member of the IFT Executive Committee.

IFT Today

Today there are a total of twenty-eight divisions and fifty-four sections within IFT. Members from these sections and divisions make up the governing body of the Institute with each section having councilors determined by sectional membership, which 1 councilor per 200 members of that section while each division has one councilor. Additionally the council has the President, President-Elect, Immediate Past President, Treasurer, and as an *ex-officio* member with the Executive Vice President (Executive Director 1966-1999). The Executive Committee until July 2007 consisted of the main officers, six representatives elected by the IFT council, six representatives elected by the IFT membership, the Executive Vice President, and the chair of the student association. Since July 2007, it consists of a Board of Directors made up of the President, President-Elect, Past-President, Treasurer, and Executive Vice President along with twelve Board of Directors that have four of these elected every three years. This change was approved by the IFT members in a change to the constitution in April 2007. IFT also communicates with the news media, using seventy university-based scientists to discuss the scientific perspective on food issues. IFT is also active in the international level by its membership in the International Union of Food Science and Technology, headquartered in Oakville, Ontario, Canada. Education has always been an issue of IFT, going as far back as 1941, with desires to have uniform education standards in food technology. Education standards for undergraduate students were approved by IFT in 1966 for food science and technology. These standards were revised and updated in 1977, 1992, and 2001. Future issues in food technology include microsensors and biosensors in food, antimicrobials in food packaging to extend the shelf-life of food, and nanotechnology creation for food microencapsulation In March 2007, about 200 members participated in a strategic leadership forum in Chicago that dealt with changes within IFT leadership, including membership, future success of the profession, the value of IFT membership, and communications.

IFT Annual Meeting and Food Expo

The largest gathering for IFT is the annual IFT Annual Meeting and Food Expo. In 2007, the event was held at McCormick Place in Chicago, IL. Approximately 24,000 people from 75 countries attended and over 1000 exhibitors presented. In 2008, the Expo will be held in New Orleans for the first time since 2005.

IFT Concerns

One constant that IFT has concern with is food safety, mainly with pathogens in food and how to counter these harmful bacteria. The global trade of food is also concerned with food safety and food security, specifically with ingredient availability and consumer tastes worldwide. . Since 1996, issues such as obesity, food allergens, and bioterrorism have become main issues that IFT are dealing with. A recent issue IFT has been addressing is the 2006 North American E. coli outbreak dealing with the outbreak of *Escherichia coli O157:H7* (*E. coli O157:H7*) in spinach (September and October 2006), green onions, and lettuce..

IFT Awards

All awards except the Loncin prize have this reference listed below.

- Nicholas Appert Award (1942)—IFT's highest honour. Lifetime contributions to food technology.
- Babcock-Hart Award (1948)—Significant contributions in food technology that resulted in public health improvements through some aspect of nutrition.
- Bor S. Luh International Award (1956—International Award from 1956 to 2004)—Individual member or institution that had outstanding efforts in one of the following in food technology: 1) international exchange of ideas, 2) better international understanding, and/or 3) practical successful technology transfer to an economically depressed area in a developed or developing nation.
- Food Technology Industrial Achievement Award (1959)—Developing an outstanding food process or product that represents a significant advance in food technology or food production.
- Samuel Cate Prescott Award (1964)—Outstanding ability of research in food technology. This is for researchers who are under 36 years of age or who are a maximum ten years after earning their highest academic degree whichever is later.
- William V. Cruess Award (1970)—Excellence in teaching food science and technology. It is the only award that students can vote.
- Carl R. Fellers Award (1984)—For IFT members who have brought honour and recognition to food science through achievements in areas other than education, research, development, and technology transfer.
- Calvert L. Willey Award (1989)—For meritorious and imaginative service in IFT.
- Stephen S. Chang Award for Lipid or Flavor Science (1993)—For contribution in lipid and flavor science.
- IFT Industrial Scientist Award (1994)—For technical contributions that advances the food industry.

- Marcel Loncin Research Prize (1994) —For basic research involving chemistry, physics, and food engineering applied to food research and food quality.
- IFT Research and Development Award (1997)—For significant research and development contributions to better understanding within food science, food technology, and nutrition.
- Elizabeth Fleming Stier Award (1997)—Humanitarian and unselfish dedication that results in significant contributions to the well-being in the food industry, academia, general public, or students.
- Bernard L. Oser Award (2000)—For contributing to the scientific knowledge of food ingredient safety or in leadership in establishing food ingredient safety evaluation or regulation.
- Myron Solberg Award (2004)—For providing leadership in establishing, successfully developing, and continuing a cooperative organization involving academia, government, and industry.

IFT Divisions

These are divisions of interest by the IFT Members

- Aquatic Food Products—seafood
- Biotechnology
- Carbohydrate
- Citrus products
- Dairy Foods—dairy products
- Education
- Extension and outreach
- Fermented Food and Beverage—fermented foods (cheese, meat, and alcohol)
- Food chemistry
- Food engineering
- Food Law and regulations—legal aspects in food processing and technology.
- Food microbiology

- Food packaging
- Foodservice—restaurants.
- Fruits and Vegetables
- IFT Student Association
- International
- Marketing and Management
- Muscle Foods—meat, poultry, and seafood
- Nonthemal Processing—food processing through non-heating methods.
- Nutraceuticals and Functional Foods—Combining nutrition and medicine for healthy living
- Nutrition
- Product development—Developing foods from the laboratory to the general public.
- Quality Assurance—food quality
- Refrigerated and Frozen Foods—foods that are processed through refrigeration and freezing.
- Religious and Ethnic Foods—Kosher, Halal, and non-traditional foods.
- Sensory evaluation
- Toxicology and Safety Evaluation—food allergy mainly.

IFT Sections

These are usually cities, states, and regions. If a region is mentioned, a city in that region is mentioned which include areas surrounded by the city.

- AK-SAR-BEN—Nebraska
- Alamo—San Antonio, Texas
- Bluegrass—Kentucky
- Bonneville—Utah
- British—Great Britain
- British Columbia—Canadian province
- Cactus—Arizona
- Central New Jersey—subsection of New York Section
- Central Valley—subsection of Northern California section
- Chicago

- Dogwood—North Carolina
- Florida
- Great Lakes—Michigan
- Great Plains—South Dakota, subsection of Minnesota section
- Hawaii
- Indiana
- Intermountain—Idaho
- Iowa
- Japan
- Kansas City—Kansas and western Missouri
- Keystone—Harrisburg, Pennsylvania
- Lake Erie—Cleveland, Ohio
- Long Island—Brooklyn and Queens, New York
- Longhorn—Dallas, Texas
- Louisiana Gulf Coast—New Orleans, Louisiana
- Magnolia—Mississippi
- Maryland—includes Delaware and West Virginia
- Mexico
- Mid-South—Memphis, Tennessee and eastern Arkansas
- Minnesota
- Monterey Bay—subsection of Northern California section
- New York—New York City boroughs of Manhattan, Staten Island, and Bronx, and parts of northern New Jersey.
- Northeast—Maine, Massachusetts, New Hampshire, Rhode Island, and Vermont
- Northern California—San Francisco, Sacramento, Bakersfield
- Northwest Ohio—Toledo, Ohio
- Nutmeg—Connecticut
- Ohio Valley—Cincinnati and Columbus, Ohio
- Oklahoma
- Oregon
- Ozark—Western Arkansas and southwest Missouri
- Philadelphia—Includes eastern Pennsylvania
- Pittsburgh—Includes western Pennsylvania

- Puget Sound—Seattle, Washington and Alaska
- Rocky Mountain—Colorado and University of Wyoming
- San Joaquin Valley—subsection of Northern California section
- Southeastern—Alabama, Georgia, and South Carolina (Dixie: 1950-2007)
- South Florida—Miami, Florida
- Southern California—Los Angeles and San Diego
- St. Louis—includes southern Illinois and eastern Missouri
- Volunteer—All of Tennessee except Memphis
- Washington, D.C.—includes Virginia
- Wisconsin

IFT Publications

The Institute also has many publications that are both in print and online that are shown below:

- *Food Technology* (1947)—IFT's monthly magazine available to all members. Besides the monthly columns dealing with food processing, food packaging, ingredients, laboratory, nutraceuticals, and consumer trends, it also focues on in-depth issues such as biotechnology, food safety, and nutrition.
- *Journal of Food Science* or *JFS* (Founded in 1936 as *Food Research*)—IFT's premier scientific journal dealing with all peer-reviewed aspects of food science (food chemistry, food microbiology, food engineering, food processing, food packaging, nutrition, and sensory evaluation. It annually publishes over 500 papers and 3000 pages with authors from 90 countries and readers from 50 countries internationally. The Library of Congress in 2004 put *JFS* as one of only 300 journals as a high priority ... (ISSN).

Both *Food Technology* and the *Journal of Food Science* can be accessed in print or online. Other publications are shown below:

- *The Journal of Food Science Education* or *JSFE* (2002—Online only)—This journal deals with innovations, techniques, and methods to improve education in food science and technology.

- *Comprehensive Reviews in Food Science and Food Safety* (2002—Online only)—This journal, published online on a quarterly basis, deals with a broad review of a narrowly defined topic relating to food science and technology, including physiology, economics, history, nutrition, microbiology, engineering, processing, and genetics. All aspects of these reviews are studied, including strengths, weaknesses, and research differences in order to present insightful investigation, including interpretation, summary, and conclusion.
- *IFT Weekly Newsletter*—A weekly e-mail newsletter sent out every Wednesday about issues within food industry, including food companies, food research, food regulations, and IFT itself.
- *Nutraceutical Newsletter*—An e-mail newsletter similar to the *IFT Weekly Newsletter* though it deals specifically with nutraceuticals.
- IFT Books—The Institute has four book publishers that offer membership discounts, including IFT Press (a partnership with Blackwell Publishing), Grey House Publishing, Marcel Dekker, and John Wiley and Sons, Inc., on food science and technology.
- *Annual Meeting Abstracts*—These are abstracts from the *IFT Annual Meeting and Food Expo* that can be accessed online or purchased in print .
- Three annual buyers guides are also published: *Food Technology Buyer's Guide*, *Nutraceutical Buyer's Guide*, and *Annual Guide to Food Industry Services*. These are usually published as a supplement to *Food Technology* magazine.

INTERNATIONAL UNION OF FOOD SCIENCE AND TECHNOLOGY

The International Union of Food Science and Technology (IUFoST) (pronounced "eye-you-fost") is an international non-governmental organization established in 1962 to devote the advancement of food science and technology. Its has as its members national food science and technology institutes and societies. Initially called the International Congress of Food Science and Technology, it would become known as IUFoST after the 1970 meeting in Washington, DC. It is a full member of the International Council for Science (ICSU).

IUFoST's Roles

- Every two years, it hosts a World Congress of Food Science and Technology. The next world congress is in Shanghai, China in 2008. The most recent conference at Nantes, France during September 17-21, 2006 had over 1000 papers presented with delegates from 62 countries.
- It is cosponsor of regional and national conferences, short courses, and workshops. This has included over 80 symposia with subsequent publications afterwards.
- Liaision with the United Nations' Food and Agricultural Organization and World Health Organization on international food science and technology issues.

Publications

- *Newsline,* published three times a year
- *Lebensmittel-Wissenschaft and Technologie* (lwt—food science and technology in (German)), an official journal of IUFoST, published by Elsevier in Amsterdam.
- *International Review of Food Science and Technology,* an annual review published by Sovereign Publications.
- *World of Food Science*, online journal jointly published by IUFoST and the Institute of Food Technologists (IFT)

IUFoST Fellows

- International Academy of Food Science and Technology, a distinguished list of food scientists and technologists world wide. This includes the current and past presidents of IUFoST, the current and past secretaries-general of

IUFoST; and other scientists world wide. As of 2006, there are 131 fellows from 36 countries.

IUFoST World Congresses

Year	*City*	*Country*
1962	London	United Kingdom
1966	Warsaw	Poland
1970	Washington, DC	United States
1974	Madrid	Spain
1978	Kyoto	Japan
1983	Dublin	Ireland
1987	Singapore	Singapore
1991	Toronto	Canada
1995	Budapest	Hungary
1999	Sydney	Australia
2001	Seoul	South Korea
2003	Chicago, Illinois	United States
2006	Nantes	France
2008	Shanghai	China—scheduled for October 19-24
2010	Cape Town	South Africa

Governing Council

The current president of IUFoST is David R. Lineback (University of Maryland, College Park) of the United States. Its president-elect is Geoffrey Campbell-Platt of the United Kingdom while its immediate past president is Alan Mortimer of Australia.

Membership

Its members include the Institute of Food Technologists in the United States and the European Federation of Food Science and Technology.

Headquarters

IUFoST is headquartered in Oakville, Ontario, Canada, a suburb of Toronto.

EUROPEAN FEDERATION OF FOOD SCIENCE AND TECHNOLOGY

The European Federation of Food Science and Technology (EFFoST) is a European-based non-governmental organization devoted to the advancement of food science and technology. It consists of eighty different societies in 21 different European countries. They are a regional group of the International Union of Food Science and Technology.

EFFoST's Roles

1. Increase closer contacts among the academic, government, and industrial areas in food.
2. Enhance rapid technology transfer to increase economic competitiveness in Europe.
3. Promote continuing education within food science and technology.
4. Standardize food law and its enforcement throughout Europe.
5. Maintain collaborative relationships within the European food industry on Knowledge sharing.

Publications

- *Innovative Food Science and Emerging Technologies* (*ISFET*), the official scientific journal of EFFoST.
- *Trends in Food Science and Technology* (*TIFS*), a mini peer-review scientific journal that is more involved in product development than basic research.
- *Food Processing Intelligence* (*FPI*), the official EFFoST book published twice a year on the status of the food industry in Europe.
- Position papers on various issues in European food science and technology.
- General information on foods is provided on the website Food-Info, which is run as a project of EFFoST in co-operation with Wageningen University, the European

Hygienic Engineering and Design Group (EHEDG) and the student organisation FISEC
- Also in co-operation with Food-Info a new site on (international) jobs in food and nutrition has been launched.

Executive Committee

The Executive Committee consists of a President, Past Presdident, President-Elect, four other elected officials, and twelve Members-At-Large.

Headquarters

EFFoST is headquartered at Wageningen University in Wageningen, Netherlands.

INSTITUTE OF FOOD RESEARCH

The Institute of Food Research (IFR) is one of the worlds leading agricultural and food science research institutions. It prides itself upon its wealth of research, regularly contributing to both Government Policies and Industry bodies as well as public relations.

Foundation

Based in Norwich, IFR was originally founded in 1968; however its research can be traced back almost 100 years to the Low Temperature Research Station in . The Institute was based on four sites across the UK (Shinfield, Langford, Long Ashton and Colney in Norwich) until 1999, when it was decided that the research centres should all consolidate on the Norwich Research Park . IFR receives great support from its parent research council the Biotechnology and Biological Sciences Research Council (BBSRC).

IFR has a significant presence in the UK, Europe and worldwide. Its vast research base extends the focus of extremely topical subjects, such as food security, diet and health including obesity and healthy aging.

Current Focus

IFR is the UKs only integrated basic science provider focused on food and is a world leading contributor to harnessing food for health and controlling food-related disease. The outcomes of its work feed into national and international strategies and deliver advice and solutions for UK Government, public sector bodies, regulatory authorities, industry and consumers.

IFR's science is focussed around four themes, investigating the interplay between the host (including the Gastrointestinal Tract mucosa, the immune system, and the complex microflora), and ingested food (including its composition, physical structure, the presence of pathogenic micro-organisms, and allergenicity). This central goal integrates the research activities across IFR.

The central aim of IFR's research into the Integrated Biology of the GI Tract is to understand how the homeostasis between the GI tract microflora and the gut epithelium is maintained and how perturbation of this system can have consequence for health. This exciting approach, including the critical food dimension, differentiates our GI tract programme from work undertaken in a more medical setting.

IFR's research into Plant Natural Products and Health aims to elucidate the fundamental mechanisms in humans by which diets that contain certain natural products help to maintain health and prevent the onset of chronic disease.

IFR consider their food chemistry and biophysics approach to be unique in the UK, and IFR's research into Food Structure and Health brings together expertise in food biopolymers, colloid and interface science with those of protein biochemistry, molecular modelling and gut epithelial biology, and has the overall aim of understanding how food structures behave in the gut and affect the breakdown and availability of macro-and micro-nutrients in food.

In studying Foodborne Bacterial Pathogens IFR aim to discover new aspects of the biology of bacterial pathogens with the long-term goal of reducing the burden of foodborne disease in humans. Combining state-of-the-art molecular microbiology with mathematical biology reveals patterns in pathogen responses that are currently hidden by massive levels of complexity.

IFR Extra

IFR is currently making its expertise available to small and medium-sized companies in the food industry. The Institute realises that these companies often demand a rapid response. IFR Extra has been set up to ensure this happens.

IFR undertakes short-term applied research, trouble-shooting and specialist analysis, making its expertise available and affordable to all companies. The Institute also provides access to food experts on a consultancy basis. IFR Extra is available to bigger companies they have worked with previously and it is now able to offer an extended range of services.

The Institute has its own laboratories and staff, it can also access a full range of experts and resources and can put together multidisciplinary teams quickly. These teams can be enhanced by the expertise and capabilities of carefully selected partners.

The Norwich Vision

IFR is a major contributor to 'The Norwich Vision.' There are three major areas of scientific excellence at Norwich—the pillars:

- Environmental science
- Plant and microbial genetics
- Food, diet and health

These inter-relate to address major 21st Century 'Grand Challenges' of global food security, (bio) energy, healthy ageing and living with environmental change. The pillars are highly distinctive in the UK and comprise a unique geographical juxtaposition world-wide.

Three 'bridges' of creativity and innovation link the pillars, delivering major new agenda-setting interdisciplinary research and training in key areas:

- The Earth and Life Systems Alliance (ELSA)
- Food and Health Alliance (the partnership between IFR and the University)

- Plant and microbial natural products

CENTRAL FOOD TECHNOLOGICAL RESEARCH INSTITUTE

India's Central Food Technological Research Institute (CFTRI) is based in Mysore and was opened on 21 October 1950.

CFTRI Mysore is one of the 40 national research laboratories set up under the aegis of the Council of Scientific and Industrial Research (CSIR) India.

The current Director of the Institute is Dr. V. Prakash.

Departments

- Biochemistry and Nutrition
- Fermentation Technologies and Bio-engineering
- Flour Milling, Baking and Confectionery Technologies
- Food Engineering
- Food Microbiology
- Food Packaging Technology
- Food Protectants and Infestation Control
- Food Safety and Analytical Quality Control Laboratories
- Food Science and Technology Information Services (FOSTIS)
- Fruit and Vegetable Technology
- Grain Science and Technology
- Human Resource Development
- Lipid Science and Traditional Foods
- Meat, Fish and Poultry Technology
- Plant Cell Biotechnology
- Plantation Products, Spices and Flavour Technology
- Protein Chemistry and Technology
- Sensory Science

After 50 years of tryst with India's food destiny, CFTRI today stands out among the largest and most diversified technology laboratories in the world. From an institution that started with just 3 persons, it has now evolved into a solid pool of talent and knowledge, with over 300 scientists, technologists and engineers, and over 400 technicians, skilled workers and support staff on its payroll. Its multi-disciplinary spread (across 16 R&D departments) covers almost every field of scientific investigation connected with foods and their relationship to humans, including the cutting edge area of food biotechnology.

The institute today features a catalogue of R&D achievements (with over 300 products/processes/equipment designs and close to 160 commercialised technologies); a whole matrix of solutions for the industry and society; an impressive portfolio of research productivity (with high-science content publications and patents); a worldwide web of nationally and internationally leading agencies; an extensive and highly responsive constituency of users from both the public and private sectors; and an impressive overall performance profile. But it is not in the tradition of the institute to rest on its laurels, and it continues to seek ever greater challenges in ever greater fulfilment of its vision.

Vision

The institute has come a long way not only chronologically, but even in terms of its perception of its own role as well as its perspectives, in tune with the radical changes taking place on India's economic, social and technological fronts. The food crisis that inspired the birth of the institute is far behind us now, and so are the fetters to India's global economic growth.

Today India is the second largest food producer in the world and can make a difference to itself and to the world with its own advancements in food technology which automatically brings CFTRI to the fore. In every category of our food crop—grains, pulses, oilseeds, spices, plantation produce, fruits and vegetables—and even in meat, fish and poultry, there is now a remarkable uptrend in production, and CFTRI is obviously playing a valuable role in developing and transferring technologies to derive the optimal processed value from this favourable situation for the benefit of farmers, processors and consumers.

Distinctive Strengths

Five decades of intensive and extensive experience in providing specialised knowledge services to industry and society, marked by a series of technology breakthroughs and seminal basic research Multi-disciplinary knowledge-base spanning 16 R&D departments Solutions in all specialities and super specialities of science and technology, known to influence food and manufacture of food products Total solutions under one roof for interdisciplinary synergy value and user convenience State-of-the-art and comprehensive resource base, including top-notch human resources at the cutting edge of food knowledge, an ultra-sophisticated pilot plant, a nationally referred analytical quality control laboratory and a high-tech information resource centre. Outstanding credentials in developing sub-regional, regional and international programmes for training in wide ranging areas of food and agri-products processing: Internationally referred centre of excellence for postgraduate studies and research in food-centric disciplines Impressive productivity in terms of processes and products as well as patents and papers High technology and high science with a human face Location in a developing country that enables development and transfer of technology with the right perception of the unique needs for technology and self-reliance in other developing countries Solid record in consultative and contract research Competence and confidence to venture into unexplored and unexposed lines of scientific investigation

Major R&D Programmes

Development of appropriate and relevant technologies, and upgradation of traditional technologies to increase the efficiency of food industries and reduce/eliminate post-harvest losses of perishables and durables Development of food products and processes for efficient conservation, protection and processing of food and other agro-economic resources Development of technology packages for value added convenience food products Development of export-oriented valued added products from horticultural and plantation produce Basic research related to food additives/preservatives, micronutrients, food toxicity and safety, food microbiology, enzymatic and molecular biology, bioactive substances and food packaging Development of food products/processes from untapped food sources Exploration of new ways of processing nutritious foods Establishment and management of food industry units Development of energy-efficient and cost-effective equipment for modernisation of primary processing industries Design and fabrication of prototype food processing machinery Development of protocols for post-harvest handling and transportation of fruits and vegetables for export and inland markets Development of new methods of packaging food materials and processed food products for extended shelf-life.

Major Support Programmes Planning and implementation of an imaginative information dissemination programme in post-harvest technology Assistance to agro-based food industries and new entrepreneurs through consultancy, contract research and other technical services covering installation, production, quality control and day-to-day management Development of human resources through varieties of training programmes for carrying out R&D and scientific management in the area of food processing and preservation

Major Upcoming R&D Projects Rationalisation of energy input in food processing systems Reduction of post-harvest losses in raw materials Upgradation of technologies for food quality, safety and productivity, with special reference to traditional foods Speciality foods for focus groups Technology applications with long term and export perspectives

Bibliography

Acosta-Baladon, A.N., 1996. Agricultural uses of occult precipitation. Agrometeorological

Adams, M.R., 1980, *Small Scale Production of Vinegar from Bananas*, TDRI Post Harvest Technology Publication, G132, Natural Resources Institute, Chatham, Kent, UK.

Adams, M.R., 1982, *Kick start for Village Vinegar in Papua New Guinea*, AT Journal, 9, (2), IT Publications, London, UK.

Agricultural and food scientists. U.S. Department of Labor. Bureau of Labor Statistics. Occupational Outlook Handbook.

Agricultural Research, Livelihoods, and Poverty: Studies of Economic and Social Impacts in Six Countries Edited by Michelle Adato and Ruth Meinzen-Dick (2007),Johns Hopkins University Press Food Policy Report (Brief)

Alexandratos, N. 1995.. The outlook for world food and agriculture to year 2010. *In* Nurul Islam ed. *Population and food in the early twenty-first century*. Washington, DC, International Food Policy Research Institute.

Alexandratos, N. 1996. China's projected cereals deficits in a world context. *Agricultural Economics,* 15: 1-16.

Alston, J.M., Norton, G.W. & Pardey, P.G. 1995. *Science under scarcity: principles and practices for agricultural research evaluation and priority setting*. Ithaca, USA, Cornell University Press.

Anderson, B. 1994. *Sustainable agricultural development and the environment: toward an optimal solution*. Canadian International Development Agency, October.

Anderson, J.R. & Purcell, D.L. 1996. International policies for agricultural research investments. In *Conference proceedings for global agricultural science policy for the twenty-first century*. 16-28

August 1996, Melbourne, Australia.

Antle, J. & Capalbo, S. 1988. An introduction to recent developments in production theory and productivity measurement. In S. Capalbo and J. Antle, eds. *Agricultural productivity measurement and explanation.* Washington, DC, Resources for the Future.

Antle, J. M. 1983. Infrastructure and aggregate agricultural productivity: international evidence. *Economic Development and Cultural Change,* 31(3):609-619.

Arnade, C.A. 1997. *Agriculture growth sources: a look at 77 countries,* Staff Paper No. 9709. Washington, DC, US Department of Agriculture, Economic Research Service

Asenso-Okyere, W.K., Benneh, G. & Tims, W., eds. 1997. *Sustainable food security in West Africa.* Dordrecht, Kluwer Academic Publishers.

Auraujo, B., Chambas, C.G. & Foirry, J.P. 1997. Conséquences de l'ajustement des finances publiques sur l'agriculture marocaine et tunisienne. Unpublished FAO study, March.

Badiane, O. & Delgado, C. 1995. *A 2020 vision for food, agriculture, and the environment in Sub-Saharan Africa.* Food, Agriculture, and the Environment Discussion Paper 4. Washington, DC, International Food Policy Research Institute.

Basch, E.M. and Ulbricht, C.E. 2005. Natural standard herb & supplement handbook[:]The clinical bottom line.St. Louis: Elsevier Mosby. 963p.

Battaglia, R., Pfannhauser, W., and Murkovic, M. (eds.) 2001. *Who's who in food chemistry. Europe.* 2nd ed. rev. expanded. Berlin; NY: Springer, 241 p.

Bazzaz, F., and W. Sombroek (Eds.), 1996a. Global climate change and agricultural production.

Bazzaz, F.A., and W.G. Sombroek, 1996b. Global climatic change and agricultural production: an assessment of current knowledge and critical gaps. In: F. Bazzaz and W. Sombroek (Eds.), 1996a, 199-235.

Binswanger, H. & Pingali, P. 1988. Technological priorities for farming in Sub-Saharan Africa. *World Bank Research Observer,* 3: 81-98.

Block, S.A. 1995. The recovery of agricultural productivity in Sub-Saharan Africa. *Food Policy,* 20(5):385-405.

Boserup, E. 1965. *The conditions of agricultural growth: the economics of agrarian change under population pressure.* London, George Allen & Unwin.

Brown, L. 1995. *Who will feed China? Wake-up call for a small planet* New York, USA, W.W. Norton.

Bruce, J.W. & Migot-Adholla, S.E. 1994. *Searching for land tenure security in Africa.* Dubuque, Iowa, USA, Kendall-Hunt, for the World Bank.

Byerlee, D. & Heisey, P.W. 1996. Past and potential impacts of maize research in sub-Saharan Africa: a critical assessment. *Food Policy,* 21(3):255-277.

Chavas, J.P. & Cox, T. 1992. A nonparametric analysis of the influence of research on agricultural productivity. *American Journal of Agricultural Economics,* 74: 583-591.

Cleaver, K.M. & Schreiber, G.A. 1994. *Reversing the spiral: the population, agriculture, and environment nexus in Sub-Saharan Africa.* Washington, DC, The World Bank,

Coelli, T. & Battese, G. 1996. Identification of factors which influence the technical efficiency of Indian farmers. *Australian Journal of Agricultural Economics,* 40 (August): 103-128.

Cousin et al (2005). "Foodborne Mycotoxins: Chemistry, Biology, Ecology, and Toxicology", *Foodborne Pathogens: Microbiology and Molecular Biology.* Caister Academic Press.

Craig, B.J., Pardey, P.G. & Roseboom, J. 1997. International productivity patterns: accounting for input quality, infrastructure and research. *American Journal of Agricultural Economics,* 79(November):1064-1076.

Creti, J. Fenicia, P. Aureli. 1990. Occurrence of Clostridium botulinum in the soil of the vicinity of Rome. Curr. Microbiol. 20:317.

Crosby, Alfred W.: *The Columbian Exchange : Biological and Cultural Consequences of 1492.* Praeger Publishers, 2003 (30th Anniversary Edition).

Crosson, P. & Anderson, J.R. 1995. *Achieving a sustainable agricultural system in Sub-Saharan Africa.* World Bank, Building

Blocks for Africa 2025 Paper No. 2. March.

Davidson, Alan. *The Oxford Companion to Food*. 2nd ed. UK: Oxford University Press, 2006.

Davis, Donald R., and Hugh D. Riordan (2004) Changes in USDA Food Composition Data for 43 Garden Crops, 1950 to 1999. Journal of the American College of Nutrition, Vol. 23, No. 6, 669-682.

Davis, J.A. 1981. Comparison of procedures to estimating returns to research using production function. *Australian Journal of Agricultural Economics,* 25: 60-71.

Emanuel, W.R., H.H. Shugart and M.P. Stevenson. 1985. Climatic change and the broad scale distribution of terrestrial ecosystem complexes. Clim. Change 7:29:43

Environmental Affairs, 1986. Report of the Director-General: Environmental Affairs for the period 1 April 1984 to 31 March 1985. Environmental Affairs Authority, Republic of South Africa, Pretoria, 100 pp.

Evenson, R.E. & McKinsey, Jr., J.W. 1991. Research, extension, infrastructure, and productivity change in Indian agriculture. *In* R.E. Evenson and C.E. Pray, eds. *Research and productivity in Asian agriculture*, p. 158-184. Ithaca, USA, Cornell University Press.

Fan, S. & Agcaoili-Sombilla, M. 1997. Why projections of China's future food supply and demand differ. *Australian Journal of Agricultural and Resource Economics,* 4:169-190.

Fan, S. 1996. Research investment, input quality, and the economic returns to Chinese agricultural research. Paper presented to the post-conference workshop on agricultural productivity and R&D policy in China, *Conference proceedings for global agricultural science policy for the twenty-first century*. 16-28 August 1996, Melbourne, Australia.

FAO. 1993. *Agriculture: towards 2010.* Rome.

FAO. 1996a. *World food summit plan of action*. Rome, FAO.

FAO. 1996b. Investment in agriculture: evolution and prospects. *World food summit technical background documents*, 10. Rome, FAO.

FAO. 1997a. *Sustainable food security: people, institutions, knowledge and environment.* Rome, FAO, Sustainable Development Department.

FAO. 1997b. *AGROSTAT* Database. Rome, FAO, Statistics Division.

FAO. 1999. FAOSTAT Database. Http://apps.fao.org. Accessed 23 March 1999.

Fellows, P.J., 1993, *Food Processing Technology*, Woodhead Publishing, Cambridge, UK

Fennema, O.R., Ed. (1985). *Food Chemistry - Second Edition, Revised and Expanded.* New York: Marcel Dekker, Inc.

Ferrando, R., 1981, *Traditional and Non-Traditional Foods*, FAO Publications, Via delle Terme di Caracalla, Rome, Italy.

Find a Food. USDA Search by food name or using a coding scheme. Provides detailed list of vitamins and nutrients.

Fischer, G., and H. van Velthuizen, 1996. Climate change and global agricultural potential project: A case study of Kenya. IIASA Working Paper WP-96-71, Laxenburg, Austria. 96 pp.

Fischer, G., K. Frohberg, M.L. Parry and C. Rozenzweig, 1994. Climate change and world food supply, demand and trade. Who benefits, who loses? Global Envir. Change, 4(1):7-23.

Fischer, G., K. Frohberg, M.L. Parry and C. Rozenzweig, 1996. The potential effects of Climate Change on World Food Production and Security. In: F. Bazzaz and W. Sombroek (Eds.), 1996a, 199-235.

Frisvold, G. & Ingram, K. 1995. Sources of agricultural productivity growth and stagnation in Sub-Saharan Africa. *Agricultural Economics,* 13:51-61.

Fulginiti, L.E. & Perrin, R.K. 1993. Prices and productivity in agriculture. *The Review of Economics and Statistics,* 75(August, no. 3):471-482.

Fulginiti, L.E. & Perrin, R.K. 1997. LDC agriculture: nonparametric Malmquist productivity indexes. *Journal of Development Economics,* 53:373-390.

Griliches, Z. 1963. The sources of measured productivity growth: United States agriculture 1940-1960. *Journal of Political Economics,* 71: 331-346.

Guernsey, L. (1993). Many colleges clear their tables of steak, substitute fruit and pasta. Chronicle of Higher Education, 39(26), A30.

Guerrant, N.B., Dutcher, R.A. (1940) *Journal of Nutrition* 20:589.

Hall, L. M. McCroskey, B. J. Pincomb, C. L. Hatheway. 1985. Isolation of an organism resembling Clostridium baratii which produces a type F botulinal toxin from an infant with botulism. J. Clin. Microbiol. 21: 654–655.

Hayami, Y. & Ruttan, V.W. 1985. *Agricultural development: an international perspective*. Baltimore, USA. The Johns Hopkins University Press.

Hazell, P. & Fan, S. 1998. Balancing regional development priorities to achieve sustainable and equitable agricultural growth. Washington, DC, International Food Policy Research Institute. Prepared for the AAEA International Conference on Agricultural Intensification, Economic Development, and the Environment, July 31-August 1, Salt Lake City.

Heisey, P.W. & Mwangi, W. 1996. *Fertilizer use and maize production in Sub-Saharan Africa*. CIMMYT Economics Working Paper 96-01. Mexico, D.F., International Maize and Wheat Improvement Center (CIMMYT).

Helms, S., R. Mendelsohn and J. Neuman, 1996. The impact of climate change on agriculture, Editorial essay. Climatic Change, 33:1-6.

Hillison J. (1996). The Origins of Agriscience: Or Where Did All That Scientific Agriculture Come From?. *Journal of Agricultural Education*.

Hohn, Thomas M.. "Trichothecene-resistant transgenic plants <internet>". Retrieved on 13 August 2007.

Home Food Preservation. Food Science Department, Pennsylvania State University. A searchable database of full-text resources including extension, USDA, and industry publication.

Hu, F. & Antle, J.M. 1993. Agricultural policy and productivity: international evidence. *Review of Agricultural Economics*, 15(September, no. 3):495-505.

Huang, J., Rosegrant, M. & Rozelle, S. 1996. Public investment, technological change, and reform: a comprehensive accounting of Chinese agricultural growth. Paper presented to the post-conference workshop on agricultural productivity and R&D policy in China, *Conference proceedings for global agricultural science policy for the twenty-first century*. 16-28 August 1996, Melbourne, Australia.

Huffman, W. & Evenson, R.E. 1993. *Science for agriculture: a long-term perspective*. Ames, USA, Iowa State University Press.

Jahnke, H. E., Kirschke, D. & Lagemann, J. 1987. *The impact of agricultural research in tropical Africa: a study of the collabouration between international and national research systems*. CGIAR Study Paper Number 21. Washington, DC, The World Bank.

Jamison, D. & Lau, L. 1982. *Farmer education and farm efficiency*. Washington, DC, World Bank.

Kalirajan, K.P., Obwona, M.B. &. Zhao, S. 1996. A decomposition of total factor productivity growth: the case of Chinese agricultural growth before and after reform. *American Journal of Agricultural Economics*, 78:331-338.

Kawagoe, T., Hayami, Y. & Ruttan, V.W. 1985. The intercountry agricultural production function and productivity differences among countries. *Journal of Development Economics*, 19(Sept-Oct):113-132.

Lachaal, L. 1994. Subsidies, endogenous technical efficiency and the measurement of production growth. *Journal of Agriculture and Applied Economics*, 26 (July): 299-310.

Larson, B.A. & Frisvold, G.B. 1996. Fertilizers to support agricultural development in Sub-Saharan Africa: what is needed and why. *Food Policy*, 21:509-525.

Lin, J.Y. 1992. Rural reforms and agricultural growth in China. *The American Economic Review*, 82 (March):34-51.

Lin, J.Y. 1998. *How did China feed itself in the past? How will China feed itself in the future?* Mexico, DF, CIMMYT Economic Program, Second Distinguished Economist Lecture, January.

Mankiw, N.G., Romer, D. & Weil, D.N. 1992. A contribution to the empirics of economic growth. *Quarterly Journal of Economics*, 107 (May): 407-437.

Maredia, M. & Byerlee, D. 1996. Efficiency of wheat improvement research: a comparative analysis of national and international research systems in developing countries. *In*

Conference proceedings for global agricultural science policy for the twenty-first century. 16-28 August 1996, Melbourne, Australia.

McMillan, J., Whalley, J. & Zhu, L. 1989. The impact of China's economic reforms on agricultural productivity growth. *Journal of Political Economy,* 97 (August):781-807.

Messer, E., Cohen, M.J. & D'Costa, J. 1998. Food from peace: breaking the links between conflict and hunger. 2020 Brief 50. Washington, DC, The International Food Policy Research Institute.

Mitchell, D.O. & Ingco, M.D. 1995. Global and regional food demand and supply prospects. In N. Islam, ed. *Population and Food in the Early Twenty-First Century*. Washington, DC, International Food Policy Research Institute.

Mullen, J. & Cox, T. 1995. The returns from research in Australian broadacre agriculture. *Australian Journal of Agricultural Economics,* 39:105-28.

Mullen, J. & Strappazzon, L. 1996. The relationship between investment and productivity growth: Australian broadacre agriculture. *In Conference proceedings for global agricultural science policy for the twenty-first century*. 16-28 August 1996, Melbourne, Australia.

Nelson, R. 1964. Aggregate production functions and medium-range growth projections. *American Economic Review*, 54 (September): 575-606.

Nelson, R. 1981. Research on productivity growth and productivity differences: dead ends and new departures. *Journal of Economic Literature,* 19 (September): 1029-1064.

Oldeman, L.R. 1992. Global extent of soil degradation. In *Biannual Report 1991-1992,* p.19-36. Wageningen, Netherlands, International Soil Reference and Information Center.

Pardey, P., Roseboom, J. & Anderson, J. 1989. *Agricultural research policy: international quantitative perspectives*. Cambridge, UK: Cambridge University Press.

Pardey, P.G. & Alston, J.M. 1995. *Revamping agricultural R&D.* Washington, DC, International Food Policy Research Institute (IFPRI).

Peterson, W. 1987. *International land quality indexes.* Department of Agricultural and Applied Economics Staff Paper P87-10, University of Minnesota.

Pingali, P.L. & Heisey, P.W. 1996. Cereal crop productivity in developing countries: past trends and future prospects. *Conference Proceedings, Global Agricultural Science Policy for the Twenty-First Century*, Melbourne, Australia, 26-28 August.

Pinstrup-Andersen, P. & Pandya-Lorch, R. 1998. Food security and sustainable use of natural resources: a 2020 vision. *Ecological Economics*.

Pinstrup-Andersen, P. 1994. *World food trends and future food security*. Washington, DC, International Food Policy Research Institute.

Pinstrup-Andersen, P., Lundberg, M. & Garrett, L. 1995. *Foreign assistance to agriculture: a win-win proposition*. Washington, DC, International Food Policy Research Institute.

Pray, C.E. & Evenson, R.E. 1991. Research effectiveness and the support base for national and international agricultural research and extension programs. In R.E. Evenson and C.E. Pray, eds. *Research and Productivity in Asian Agriculture*. Ithaca, USA Cornell University Press.

Rosegrant, M.W. & Ringler, C. 1997. World food market into the 21st century: environmental and resource constraints and policies. *The Australian Journal of Agricultural and Resource Economics,* 41 (September): 401-428.

Salmon, D.C. 1991. Rice productivity and the returns to rice research in Indonesia. In R.E. Evenson and C.E. Pray, eds. *Research and Productivity in Asian Agriculture*, p. 133-157. Ithaca, USA, Cornell University Press.

Schultz, T.W. 1964. *The economic value of education*. New York, Columbia University Press.

Shane, M., Roe, T. & Gopinath, M. 1998. *US agricultural growth and productivity: an economywide perspective.* Market and Trade Economics Division, Economic Research Service, US Department of Agriculture, Washington, DC. Agricultural Economic Report No. 758.

Shapouri, S. & Rosen, S., eds. 1998. *Food security assessment*. International Agriculture and Trade Report No. GFA-10. Economic Research Service, US Department of Agriculture, Washington, DC.

Thirtle, C. 1996. Productivity and returns to levy-funded R&D in UK sugar: an alternative approach to rate of return calculations. In *Conference proceedings for global agricultural science policy for the twenty-first century*. 16-28 August 1996, Melbourne, Australia.

Thirtle, C., Hadley, D. & Townsend, R. 1995. Policy induced innovation in Sub-Saharan African agriculture: a multilateral Malmquist productivity index approach. *Development Policy Review*, 13(4): 323-342.

Thrupp, L.A. 1997. *Linking biodiversity and agriculture: challenges and opportunities for sustainable food security*. Washington, DC, World Resources Institute.

Trueblood, M.A. & Coggins, J. 1997. Nonparametric estimates of intercountry agricultural efficiency and productivity. Economic Research Service, US Department of Agriculture, Washington, DC. Mimeo.

Trueblood, M.A. 1991. *Agricultural production functions estimated from aggregate intercountry observations: a selected survey*. Agriculture and Trade Analysis Division, Economic Research Service, US Department of Agriculture, Washington, DC. Staff Report No. AGES 9132.

Wen, G.J. 1993. Total factor productivity change in China's farm sector: 1952-1989. *Economic Development and Cultural Change*, 42:1-41.

Wiens, T.B. 1983. Price adjustment, the responsibility system, and agricultural productivity. *American Economic Review*, 73 (May):319-324.

Wiggens, S. 1998. African farming seen from village studies: changes from the 1970s to the 1990s. Paper for the ASA-Supported Conference on Africa and globalisation: towards the millennium, University of Central Lancashire at Preston, 24-26 April 1998.

Wong, L. 1989. Agricultural productivity in China and India: a comparative analysis. *Canadian Journal of Agricultural Economics*, 37: 77-93.

World Bank. 1998. *World development indicators 1998*. Washington, DC, The World Bank.

Index